AF595314

Calculus of Variations, Optimal Control, and Mathematical Biology: A Themed Issue Dedicated to Professor Delfim F. M. Torres on the Occasion of His 50th Birthday

Calculus of Variations, Optimal Control, and Mathematical Biology: A Themed Issue Dedicated to Professor Delfim F. M. Torres on the Occasion of His 50th Birthday

Editors

Natália Martins
Ricardo Almeida
Cristiana João Soares da Silva
Moulay Rchid Sidi Ammi

MDPI • Basel • Beijing • Wuhan • Barcelona • Belgrade • Manchester • Tokyo • Cluj • Tianjin

Editors

Natália Martins
University of Aveiro
Aveiro, Portugal

Ricardo Almeida
University of Aveiro
Aveiro, Portugal

Cristiana João Soares da Silva
Instituto Universitário de
Lisboa (ISCTE-IUL)
Lisbon, Portugal

Moulay Rchid Sidi Ammi
Moulay Ismail University
Errachidia, Morocco

Editorial Office
MDPI
St. Alban-Anlage 66
4052 Basel, Switzerland

This is a reprint of articles from the Special Issue published online in the open access journal *Axioms* (ISSN 2075-1680) (available at: https://www.mdpi.com/journal/axioms/special_issues/delfim).

For citation purposes, cite each article independently as indicated on the article page online and as indicated below:

LastName, A.A.; LastName, B.B.; LastName, C.C. Article Title. *Journal Name* **Year**, *Volume Number*, Page Range.

ISBN 978-3-0365-6856-0 (Hbk)
ISBN 978-3-0365-6857-7 (PDF)

Contents

About the Editors

Natália Martins

Natália Martins, PhD in Mathematics (2006, University of Aveiro, Portugal), is an Associate Professor in the Department of Mathematics at the University of Aveiro (UA), Portugal. She has worked in the Department of Mathematics of UA since 1992 and was Vice-Director of the department from 2008 to 2019. She was also a member of the Pedagogical Council of UA from 2009 to 2016, where she was elected Vice-President, a position she held for more than 5 years between 2011 and 2016. Since 2022, she has been the scientific coordinator of the group "System and Control group" of the research unit CIDMA (Center for Research and Development in Mathematics and Applications), which is a research team hosted at the Department of Mathematics of the University of Aveiro. Her research interests include Time Scale Calculus, Fractional Calculus, Calculus of Variations, and Optimal Control Theory. She is the author of more than 55 papers in international journals, conference papers, and book chapters. Natália Martins belongs to the Editorial Board of *Fractal and Fractional*, *Axioms*, and *International Journal of Dynamical Systems and Differential Equations*.

Ricardo Almeida

Ricardo Almeida received his bachelor's and master's degrees in Mathematics from the University of Porto, Portugal, and his Ph.D. degree in Mathematics from the University of Aveiro, Portugal. He is currently an Assistant Professor at the University of Aveiro. His research interests include fractional calculus, calculus of variations, and optimal control theory. He has published 5 books and more than 90 papers in international journals. He is the Associate Editor in the journal *Fractal and Fractional* and is an Editor for *AIMS Mathematics* and *Axioms*. He is the adjunct coordinator of the Center for Research and Development in Mathematics and Applications (CIDMA) since 2019.

Cristiana João Soares da Silva

Cristiana João Soares da Silva received her Ph.D. degree in Mathematics from the University of Aveiro, in Portugal, and from the University of Orléans, in France, in 2010. From 2011 to 2018, she was a post-doctoral researcher at the Center for Research and Development in Mathematics and Applications (CIDMA) of the University of Aveiro and, between 2019 and 2022, she had a Researcher position at CIDMA. Since September 2022, she is Assistant Professor at Iscte—Instituto Universitário de Lisboa, in Lisbon, Portugal. Her research interests include optimal control theory, optimization methods, and mathematical modeling of infectious diseases.

Moulay Rchid Sidi Ammi

Moulay Rchid Sidi Ammi is a Full Professor of Mathematics at the University Moulay Ismail, Morocco (UMI), and is responsible for the R&D Unit AMNEA and Laboratory MAIS. His main research areas are EDPs; calculus of variations; optimal control; optimization; fractional derivatives and integrals; dynamic equations on time scales; and mathematical biology. He has written outstanding scientific and pedagogical publications in international journals ranked in the 1st quartile by ISI Web of Science and Scopus. He has strong experience in graduate and post-graduate student supervision and teaching in mathematics. Moreover, he has been team leader and member in several national and international R&D projects.

Preface to "Calculus of Variations, Optimal Control, and Mathematical Biology: A Themed Issue Dedicated to Professor Delfim F. M. Torres on the Occasion of His 50th Birthday"

This publication is a Special Issue of the journal *Axioms* entitled "Calculus of Variations, Optimal Control and Mathematical Biology: A Themed Issue Dedicated to Professor Delfim F. M. Torres on the Occasion of His 50th birthday". This Special Issue is dedicated to Professor Delfim F. M. Torres on his 50th birthday, as a recognition of his significant contributions to Mathematics, in particular regarding the calculus of variations, optimal control, and mathematical biology. Professor Torres is a distinguished University Professor, a highly cited researcher in Mathematics (in the top 1% for Mathematics in the Web of Science in 2015, 2016, 2017, and 2019), and a lifetime member of the American Mathematical Society. He is one of the founders of fractional variational analysis and fractional optimal control, and has tremendously contributed to the theory of variational analysis, with applications to many other fields such as optimization, optimal control, time-scale analysis, and mathematical epidemiology and biology. Professor Torres is the recipient of several international awards, including the 325 Years of Fractional Calculus Award, which is a testament to the high regard in which his achievements in the area of fractional calculus and its applications are held. He has been included in the World's Top 2% Scientists by Stanford University in 2020, 2021, and 2022, both as a career-long highly cited researcher and as a single calendar year highly cited researcher. He is considered by Thomson Reuters as one of the World's Most Influential Scientific Minds, has won a Publons Peer Review Award as a world's top peer reviewer, and was recognized in the top 1% of reviewers. He has also won a Sentinel of Science Award. Many of his research works have been considered top papers in their area, served as research highlights, and been awarded prizes. Besides being a great scholar, he is also a great teacher who has already supervised twenty-four Ph.D. students from all over the world. This Special Issue covers many of Professor Torres' research interests, which include several areas of pure and applied mathematical sciences, such as approximations and expansions, biology, and other natural sciences, calculus of variations, and optimal control, optimization, difference and functional equations, fluid mechanics, functional analysis, game theory, economics, social and behavioral sciences, general measure and integration, the mechanics of deformable solids, number theory, numerical analysis, operations research, mathematical programming, ordinary differential equations, partial differential equations, quantum theory, real functions, systems and control theory, fractional calculus and its applications, integral equations and transforms, higher transcendental functions and their applications, q-series and q-polynomials, inventory modeling and optimization, dynamic equations on time scales, and mathematical modeling.

This book comprises an editorial and fifteen original research papers that have been carefully reviewed.

Natália Martins, Ricardo Almeida, Cristiana João Soares da Silva, and Moulay Rchid Sidi Ammi
Editors

Editorial

Editorial for the Special Issue of Axioms "Calculus of Variations, Optimal Control and Mathematical Biology: A Themed Issue Dedicated to Professor Delfim F. M. Torres on the Occasion of His 50th Birthday"

Natália Martins [1,*,†], Ricardo Almeida [1,†], Cristiana J. Silva [1,2,†] and Moulay Rchid Sidi Ammi [3,†]

1 Center for Research and Development in Mathematics and Applications, Department of Mathematics, University of Aveiro, 3810-193 Aveiro, Portugal
2 Department of Mathematics, Instituto Universitário de Lisboa (ISCTE-IUL), Av. das Forças Armadas, 1649-026 Lisbon, Portugal
3 MAIS Laboratory, AMNEA Group, FST Errachidia, Moulay Ismaïl University of Meknès, P.O. Box 509 Boutalamine, Errachidia 52000, Morocco
* Correspondence: natalia@ua.pt
† These authors contributed equally to this work.

Citation: Martins, N.; Almeida, R.; Silva, C.J.; Sidi Ammi, M.R. Editorial for the Special Issue of Axioms "Calculus of Variations, Optimal Control and Mathematical Biology: A Themed Issue Dedicated to Professor Delfim F. M. Torres on the Occasion of His 50th Birthday". *Axioms* **2023**, *12*, 110. https://doi.org/10.3390/axioms12020110

Received: 13 January 2023
Accepted: 17 January 2023
Published: 20 January 2023

1. Introduction

This publication is an editorial for the Special Issue of Axioms "Calculus of Variations, Optimal Control and Mathematical Biology: A Themed Issue Dedicated to Professor Delfim F. M. Torres on the Occasion of His 50th birthday". This Special Issue is dedicated to Professor Delfim F. M. Torres on the occasion of his 50th birthday, as recognition of his significant contributions to Mathematics, in particular in the calculus of variations, optimal control, and mathematical biology. Professor Torres is a distinguished University Professor, a highly cited researcher in Mathematics (in the top 1% for Mathematics in the Web of Science in the years 2015, 2016, 2017 and 2019), and a lifetime member of the American Mathematical Society. He is one of the founders of fractional variational analysis and fractional optimal control, and has made tremendous contributions to the theory of variational analysis, with applications to many other fields such as optimization, optimal control, time-scale analysis, and mathematical epidemiology and biology. Professor Torres is the recipient of several international awards, including the 325 Years of Fractional Calculus Award, which is testament to the high regard in which his achievements in the area of fractional calculus and its applications are held. He has been included in the World's Top 2% Scientists by Stanford University in the years 2020, 2021 and 2022, both as a career-long highly cited researcher and as a single calendar year highly cited researcher. He is considered by Thomson Reuters as one of the World's Most Influential Scientific Minds, has won a Publons Peer Review Award as a world's top peer reviewer, and was recognized in the top 1% of reviewers. He has also won a Sentinel of Science Award. Many of his research works have been considered top papers in their area, served as research highlights and been awarded prizes. Besides being a great scholar, he is also a great teacher who has already supervised twenty-four Ph.D. students from all over the world. This Special Issue covers many of Professor Torres' research interests, which include several areas of pure and applied mathematical sciences, such as approximations and expansions, biology and other natural sciences, calculus of variations and optimal control, optimization, difference and functional equations, fluid mechanics, functional analysis, game theory, economics, social and behavioral sciences, general measure and integration, the mechanics of deformable solids, number theory, numerical analysis, operations research, mathematical programming, ordinary differential equations, partial differential equations, quantum theory, real functions, systems and control theory, fractional calculus and its applications, integral equations and transforms, higher transcendental functions and their applications, q-series and q-polynomials, inventory modeling and optimization, dynamic equations on time scales, and mathematical modeling.

This Special Issue comprises fifteen original research papers that have been carefully reviewed. In the next section, we briefly describe the main contributions of these fifteen papers. We finalize this editorial with a brief biography of Delfim F. M. Torres.

2. Contributions

In this section, we provide an overview of the scientific contributions of the papers that constitute this Special Issue, gathering them into five mathematical research areas: Calculus of variations, Optimal control, Mathematical biology, Fractional calculus, and Differential geometry. We remark that some of the papers can be integrated in more than one of these research topics.

2.1. Calculus of Variations

In the paper *On a Non–Newtonian Calculus of Variations*, Delfim F. M. Torres presents, for the first time in the literature, a non-Newtonian calculus of variations that involves the minimization of a function defined by a non-Newtonian integral with a Lagrangian depending on the non-Newtonian derivative. The main result of this paper is a first-order necessary optimality condition of a Euler–Lagrange type that each solution of a non-Newtonian variational problem, with admissible functions taking positive values only, must satisfy. The non-Newtonian calculus of variations introduced in this paper provides a natural framework for dealing with multiplicative functions that arise in economics, physics and biology.

In the article *Weighted Generalized Fractional Integration by Parts and the Euler–Lagrange Equation*, Zine et al. introduce the right-weighted generalized fractional derivative in the Riemann–Liouville sense, and also introduce its associated integral operator, proving their main properties, and, in particular, their integration by parts formula. With the new general integration by parts formula, the authors obtain an appropriate weighted Euler–Lagrange equation for dynamic optimization, extending those existing in the literature. The paper finishes with an illustration of the obtained theoretical results in the quantum mechanics setting.

2.2. Optimal Control

The article *Maximum Principle and Second-Order Optimality Conditions in Control Problems with Mixed Constraints*, by Arutyunov et al., concerns the optimality conditions for a smooth optimal control problem with an endpoint and mixed constraints. Under the normality assumption, second-order necessary optimality conditions based on the Robinson stability theorem are derived. The main novelty is that only a local regularity with respect to the mixed constraints is required, instead of the conventional stronger global regularity hypothesis. This affects the maximum condition. Therefore, the normal set of Lagrange multipliers in question satisfies the maximum principle, albeit along with the modified maximum condition in which the maximum is taken over a reduced feasible set. In the second part of this work, the case of abnormal minimizers, that is, when the full rank of controllability matrix condition is not valid, is addressed. The same type of reduced maximum condition is obtained.

In the paper *Minimum Energy Problem in the Sense of Caputo for Fractional Neutral Evolution Systems in Banach Spaces*, by Ech-chaffani et al., a class of fractional neutral evolution equations on Banach spaces involving Caputo derivatives is investigated. Conditions for the controllability of the fractional-order system and conditions for existence of a solution to an optimal control problem of minimum energy are established. The main results are proved with the help of fixed-point and semigroup theories.

2.3. Mathematical Biology

In the paper *Global Stability Condition for the Disease-Free Equilibrium Point of Fractional Epidemiological Models*, Almeida et al. present a new method to study the global asymptotic stability of dynamical systems described by fractional differential equations. The usual approach involves the determination of an appropriate Lyapunov function, very laborious computations and the application of LaSalle's invariance principle. The method proposed by the authors uses only basic results from matrix theory and some well-known results from fractional-order differential equations. To illustrate the applicability of the theoretical results, the authors exemplify the procedure with three epidemiological models: a fractional-order SEIR model with classical incidence function, a fractional-order SIRS model with a general incidence function, and a fractional-order model for HIV/AIDS.

In the research *Hybrid Method for Simulation of a Fractional COVID-19 Model with Real Case Application*, by Din et al., a mathematical analysis for the novel coronavirus responsible for COVID-19 is provided. The fractional-order analysis is carried out using a non-singular kernel type operator known as the Atangana–Baleanu–Caputo (ABC) derivative. The model, adopting available information about the disease from Pakistan in the period from 9 April to 2 June 2020, is parametrized. The authors obtain the required solution with the help of a hybrid method, which is a combination of the decomposition method and the Laplace transform. Furthermore, a sensitivity analysis is carried out to evaluate the parameters that are more sensitive to the basic reproduction number of the model. The obtained results are compared with real data of Pakistan, and numerical plots are presented at various fractional orders.

In the paper *Fractional Dynamics of a Measles Epidemic Model*, Abboubakar et al. propose a fractional mathematical model for the transmission dynamics of measles, considering a Caputo fractional derivative. The main goal of this work is to compare the dynamics of a measles epidemic model with integer and Caputo fractional derivatives. The epidemic model considers vaccination and hospitalization of infected individuals. A local and global stability analysis of the equilibrium points is proved, and the theoretical results are illustrated through numerical simulations using the Adams-type predictor–corrector iterative scheme. The numerical simulations demonstrate that the fractional model shows different quantitative behavior than the model with integer derivatives.

The paper *Pattern Formation Induced by Fuzzy Fractional-Order Model of COVID-19*, by Alnahdi et al., proposes a mathematical model for COVID-19 using a fuzzy fractional evolution equation, stated in Caputo's sense for order $\alpha \in (1, 2)$. The model considers six compartments: susceptible, exposed, totally infected, asymptotically infected, totally recovered individuals and reservoir. The existence and uniqueness of the solution of the model is proved using Schauder's fixed point theorem. Moreover, the dynamic behavior of the model is studied by combining the fuzzy Laplace approach with the Adomian decomposition transform.

In the paper *Modeling the Impact of the Imperfect Vaccination of the COVID-19 with Optimal Containment Strategy*, Benahmadi et al. propose a mathematical model applied to COVID-19, aiming to investigate the impact of vaccination in the control of the disease's spread. The compartmental model is given by a system of nine ordinary differential equations. After computing the basic reproduction number, a local and global stability analysis of the disease's free equilibrium is presented. An optimal control problem is proposed wherein the aim is to find the optimal control strategy under imperfect vaccination. Three controls are introduced in the model, representing awareness of taking the vaccine, movement restrictions for susceptible and vaccinated individuals, and improvement of the vaccine's efficacy, respectively. In the numerical simulations, the model is calibrated to real data from a vaccination campaign in Morocco between 1 February 2021 and 25 March 2021. The numerical solutions show that to reduce the impact of imperfect vaccination, a longer awareness campaign is needed to engage the population in vaccination. On the other hand, restrictions on population mobility should not be long lasting. Moreover, to ensure the full protection of the health population, vaccination efficacy must be increased by 30% in the first 50 days.

In the paper *Determining COVID-19 Dynamics Using Physics Informed Neural Networks*, by Malinzi et al., the physics informed neural networks (PINNs) framework is applied to COVID-19 epidemics, using a compartmental SIRD (susceptible–infected–recovered–death) mathematical model. The main goal of this work is to find patterns of the transmission dynamics of the disease, which involves predicting the infection, recovery and death rates and therefore, predicting the actively infected, totally recovered, susceptible and deceased individuals at any given time. First, the PINNs framework's application to the SIRD model is validated through numerical simulations using *Mathematica*, and after, the SIRD model is tested and validated using a real COVID-19 dataset.

In the paper by Francesca Acotto and Ezio Venturino, entitled *A Note on an Epidemic Model with Cautionary Response in the Presence of Asymptomatic Individuals*, the authors studied an epidemiological model, taking into consideration some demographic features, and also the case in which the illness appears in two forms, asymptomatic and symptomatic. Because of the presence of asymptomatic individuals, fear drives a reduction in social contact, lowering the overall transmission rate. With some numerical simulations, it is shown that with an increase in information regarding the propagation of the disease, the number of infected individuals decreases.

2.4. Fractional Calculus

In *Riemann–Liouville Fractional Sobolev and Bounded Variation Spaces* by Leaci and Tomarelli, a study of fractional derivatives on Sobolev spaces is carried out. The fractional operators are given by the mean value between the left and right fractional operators. Some problems, such as embeddings or compactness properties, the Abel equation, and semigroup properties, are considered. These methods can be applied into fractional variational models for image analysis.

In the article *On Periodic Fractional (p,q)-Integral Boundary Value Problems for Sequential Fractional (p,q)-Integrodifference Equations* by Soontharanon and Sitthiwirattham, questions on the existence and uniqueness of solution for a fractional (p,q)-integrodifference equation with periodic fractional (p,q)-integral boundary condition are studied. The proofs are based on Banach and Schauder's fixed point theorems. Furthermore, some properties of the fractional (p,q)-integral are obtained. The paper ends with some illustrative examples.

The paper *Existence Results for a Multipoint Fractional Boundary Value Problem in the Fractional Derivative Banach Space* by Boucenna et al., deals with a class of nonlinear implicit fractional differential equations subject to nonlocal boundary conditions, expressed in terms of nonlinear integro-differential equations. Using the Krasnoselsky fixed-point theorem, via the Kolmogorov–Riesz criteria, the existence of solutions in a specific fractional derivative Banach space is established. Then, two numerical examples are given to illustrate the theoretical results.

2.5. Differential Geometry

The paper *Local Structure of Convex Surfaces near Regular and Conical Points*, by Plakhov, deals with the limiting behavior of a part of a surface when it is cut by a plane. More precisely, given a point on a convex surface, and a plane of support Π to the surface at this point, the author considers the plane parallel with Π cutting a part of the surface. Two cases are studied: when the point is regular, and when it is singular conical. This work follows on from a 1995 conjecture by Buttazzo, Ferone, and Kawohl, "Minimum problems over sets of concave functions and related questions", published in *Math. Nachr.*, already solved by the author under some assumptions, and continued in this current work.

3. Short Biography of Delfim F. M. Torres

Delfim Fernando Marado Torres was born 16 August 1971 in Nampula, Mozambique. Since March 2015, he has been a full Professor of Mathematics at the University of Aveiro

(UA), Director of the Research Unit CIDMA, the largest Portuguese research center in Mathematics, and Coordinator of its Systems and Control Group. He obtained a PhD in Mathematics from UA in 2002, and Habilitation in Mathematics from UA in 2011. His main research areas are calculus of variations and optimal control, optimization, fractional derivatives and integrals, dynamic equations on time scales and mathematical biology. Torres has written outstanding scientific and pedagogical publications. In particular, he has co-authored two books with Imperial College Press [1,2], three books with Springer [3–5], and edited several research books, such as [6–8]. He has strong experience in graduate and postgraduate student supervision and teaching in mathematics. Moreover, he has been team leader and member in several national and international R&D projects, including EU projects and networks. Since 2013, he has been the Director of the Doctoral Programme Consortium in Mathematics and Applications (MAP-PDMA) of the Universities of Minho, Aveiro, and Porto. Prof. Torres has been married since 2003, and has one daughter and two sons.

Author Contributions: Conceptualization, N.M., R.A., C.J.S. and M.R.S.A.; methodology, N.M., R.A., C.J.S. and M.R.S.A.; writing—original draft preparation, N.M., R.A., C.J.S. and M.R.S.A.; writing—review and editing, N.M., R.A., C.J.S. and M.R.S.A. All authors have read and agreed to the published version of the manuscript.

Funding: This work was partially funded by Portuguese funds through the CIDMA—Center for Research and Development in Mathematics and Applications, and the Portuguese Foundation for Science and Technology (FCT—Fundação para a Ciência e a Tecnologia), reference UIDB/04106/2020.

Data Availability Statement: Not applicable.

Acknowledgments: The authors deeply thank Axioms (ISSN 2075-1680) and Luna Shen, Managing Editor of Axioms, for all the support given to the Special Issue "Calculus of Variations, Optimal Control and Mathematical Biology: A Themed Issue Dedicated to Professor Delfim F. M. Torres on the Occasion of His 50th birthday". The authors of this editorial are also deeply grateful to all colleagues and friends of Delfim F. M. Torres for submitting high quality articles to this Special Issue.

Conflicts of Interest: The authors declare no conflict of interest.

References

1. Malinowska, A.B.; Torres, D.F.M. *Introduction to the Fractional Calculus of Variations*; Imperial College Press: London, UK, 2012.
2. Almeida, R.; Pooseh, S.; Torres, D.F.M. *Computational Methods in the Fractional Calculus of Variations*; Imperial College Press: London, UK, 2015.
3. Almeida, R.; Tavares, D.; Torres, D.F.M. *The Variable-Order Fractional Calculus of Variations*; SpringerBriefs in Applied Sciences and Technology; Springer: Cham, Switzerland, 2019.
4. Malinowska, A.B.; Torres, D.F.M. *Quantum Variational Calculus*; SpringerBriefs in Electrical and Computer Engineering; Springer: Cham, Switzerland, 2014.
5. Malinowska, A.B.; Odzijewicz, T.; Torres, D.F.M. *Advanced Methods in the Fractional Calculus of Variations*; SpringerBriefs in Applied Sciences and Technology; Springer: Cham, Switzerland, 2015.
6. Agarwal, P.; Nieto, J.J.; Ruzhansky, M.; Torres, D.F.M. *Analysis of Infectious Disease Problems (COVID-19) and Their Global Impact*; Infosys Science Foundation Series in Mathematical Sciences; Springer: Singapore, 2021.
7. Agarwal, P.; Nieto, J.J.; Torres, D.F.M. *Mathematical Analysis of Infectious Diseases*; Academic Press: London, UK, 2022.
8. Tchemisova, T.V.; Torres, D.F.M.; Plakhov, A. *Dynamic Control and Optimization*; Springer Nature Switzerland AG: Cham, Switzerland, 2022.

 MDPI

Article

A Note on an Epidemic Model with Cautionary Response in the Presence of Asymptomatic Individuals

Francesca Acotto † and **Ezio Venturino** *,†,‡

Dipartimento di Matematica "Giuseppe Peano", Università di Torino, Via Carlo Alberto 10, 10123 Torino, Italy
* Correspondence: ezio.venturino@unito.it
† These authors contributed equally to this work.
‡ Member of the INdAM research group GNCS.

Abstract: We analyse a simple disease transmission model accounting for demographic features and an illness appearing in two forms, asymptomatic and symptomatic. Its main feature is the epidemic-induced fear of the population, for which contacts are reduced, responding to increasing symptomatic numbers. We find that in the presence of asymptomatic individuals, if the progression rate to symptomatic is high, protection measures may prevent the whole population becoming infected. The results also elucidate the importance of assessing transmission rates as quickly as possible.

Keywords: population dynamics; vertical transmission; epidemic fear

MSC: 95D30; 95D25

1. Introduction

Mathematical models for the spread of a communicable disease date back almost a century, from the original work of Kermack and McKendric. From the early classical models with no demographics, where the population at risk was fixed in size, the models evolved to encompass size-varying populations [1,2]. A thorough review of infection models, mainly for diseases affecting humans, is provided by [3]. Other reviews with more recent developments in the field appear in [4,5]. In addition, stochastic models for these situations can be developed [6]. These types of models have also been adapted for various situations [7], also including, possibly, the spread of epidemics among interacting populations, from which originated the so-called ecoepidemic models, see the fairly recent review [8].

The first model that incorporated the human behavior response to an epidemic spread is [9], an SIR-type system [10]. It models the fact that when the epidemic is spreading, people react by reducing contacts, in order to not be infected, e.g., by using protective means or distancing [11]. The use of vaccines, when available, would be another option. However, more recently, other issues have arisen, such as the anti-vaccination attitude of parts of populations. Some studies investigating this phenomenon have been undertaken [12,13].

The still ongoing COVID-19 epidemic [14] has highlighted another feature of these transmissible diseases. Namely, there is a relevant role played by asymptomatics. In fact, in the earlier phases of this pandemic, its spread was mainly due to contacts among susceptibles and asymptomatic infected individuals. A similar situation is exemplified by the Spanish flu of the XX century [15], or the most recent SARS. In the latter viral shedding outbreaks only for advanced stages of the disease cause respiratory symptoms occur, but for SARS-CoV-2, the infected can also spread the disease in the early stages, when they are asymptomatic [16]. There are also other diseases that do not show symptoms promptly. For instance, a measles-infected person is contagious in the very first days after getting the disease, the average latent period is 14 days, while the symptoms appear later, with an average infectious period of one week [3]. In the case of pneumonic plague, experiments

Citation: Acotto, F.; Venturino, E. A Note on an Epidemic Model with Cautionary Response in the Presence of Asymptomatic Individuals. *Axioms* **2023**, *12*, 62. https://doi.org/10.3390/axioms12010062

Academic Editor: Valery Y. Glizer

Received: 16 November 2022
Revised: 22 December 2022
Accepted: 30 December 2022
Published: 6 January 2023

with mice indicate that the initial 36 h of infection show fast bacterial replication in the lungs, but no host immune responses or obvious disease symptoms appear, ref. [17]. In addition, primary septicemic plague, the second most diffused, starts with no palpable lymph nodes but with bacteremia [18]. For COVID-19, many models are now available, see, e.g., [19], where a highly detailed SPEIQHRD model is presented and validated using data from various different countries; Ref. [20], which introduces an SEIRD model and is identified for different USA states; or [21], which empirically considers the chaotic and cyclic behavior of the epidemics, verified by actual data.

However, we stress that this paper is not at all concerned with this specific epidemic. Rather, we have mentioned COVID-19 as well as measles just as paradigms for diseases where asymptomatic people appear. In summary, the salient feature of the epidemics that we consider here consists of the presence of asymptomatic people who are able to spread it. The present paper is also a theoretical investigation; therefore, the use of real data is not our concern here. The goal of the paper focuses on people's possible response to the spread of a generic epidemic. In this sense, it, rather, represents a model for possible human behaviors. Therefore, empirical data on specific contagious diseases are not needed as they are not used. Nor are other possible numerical schemes for the forecasting of the disease incidence of relevance for our purposes.

Along these lines, we would like to investigate a model in which people respond to higher numbers of the infected, which are recognized as disease carriers, i.e., symptomatic individuals. However, the disease is essentially transmitted by the asymptomatic people, because it is assumed that the symptomatic ones, once discovered, are isolated so as not to render them vehicles of propagation. We propose a model for this effect. Additionally, it also incorporates demographic features of the population. Some other specific properties of the system introduced here are the following. Essentially, it is a variation of the classical SI infection model, in which asymptomatic individuals are also accounted for, by splitting the infected (the class I of the SI model) into asymptomatics and symptomatics. The latter, however, are assumed to be isolated; therefore, it can also be interpreted as an SIR model, in which the R class denotes the set of individuals removed from circulation and, therefore, are not able to spread the disease any longer, rather than those recovered from the disease. Two variants are proposed: without and with vertical transmission. The models are fully analysed, determining all their possible equilibria, feasibility and stability. Their explicit coordinate expressions are determined, except for coexistence, for which we provide sufficient conditions for its feasibility. Their transcritical bifurcations are investigated both analytically and numerically, assessing the critical parameter values for which they occur. Implications for people's behavior are discussed in the final sections.

The most important result appears to be the fact that in the presence of a high progression rate from asymptomatic to symptomatic, the SAI model proposed here is able to preserve some susceptibles from the contagion, in spite of being an SI model for which everyone becomes infected.

The main findings of this investigation show that in the presence of asymptomatic individuals, people's voluntary means to reduce their own possible contagion must be undertaken more strictly, by suitably lowering the overall transmission term, than in the case where the epidemic's symptoms are immediately manifested. Our simulations also allow the quantification, if sufficient information on the spread of the disease is available, of the number of susceptibles that remain unaffected by the epidemic. They also show the importance of the prompt broadcasting of this information by the authorities.

2. Materials and Methods

As mentioned in the Introduction, we consider here a human population which reproduces and experiences an epidemic. Thus, it is divided into susceptibles S and infected individuals and, in turn, subdivided among those that do not show symptoms, but still can spread the disease, A, and the symptomatic ones I, who are isolated as recognized virus carriers. The latter, thus, do not contribute to the diffusion of the pathogenic agent among the population. We also assume that the susceptibles can react to the presence of the disease, by reducing their contacts. However, this behavior is influenced by the number of the people recognized as diseased, i.e., the Is; the higher this number, the lower the contact rate must be. The model contains only three compartments, because in the end its behavior will be compared with another classical model concerned with people's cautionary response, where asymptomatic individuals are absent, as elaborated at length in Section 5.

The demographic features incorporate reproduction, natural mortality and competition for resources. All these involve, in principle, the three subsets into which the total population is partitioned.

Two possibilities arise, considering reproduction: namely, that the disease is or is not passed onto the offsprings. In the first model, we do not consider vertical transmission, so asymptomatic individuals reproduce, but their offsprings are healthy and, therefore, are accounted for among the new recruits of susceptibles. Alternatively, vertical transmission hypothesizes that newborns from asymptomatic people appear in the asymptomatic class as well. The two models are constructed and analysed in the following subsections.

2.1. The SAI Model without Vertical Transmission

Model Equations

Using the notation introduced in the above preamble, the system reads:

$$\begin{aligned} \frac{dS}{dt} &= r(S+A) - mS - c_{SS}S^2 - c_{SA}SA - c_{SI}SI - \frac{\alpha SA}{1+\beta I}, \\ \frac{dA}{dt} &= -mA - c_{AA}A^2 - c_{AS}AS - c_{AI}AI + \frac{\alpha SA}{1+\beta I} - \pi A, \\ \frac{dI}{dt} &= -(m+\mu)I - c_{II}I^2 - c_{IS}IS - c_{IA}IA + \pi A. \end{aligned} \tag{1}$$

The first equation for susceptibles contains reproduction at rate r, which is related to both the susceptible and asymptomatic classes, first term. Here, we implicitly assume that the disease does not affect the asymptomatic reproduction rate. Susceptibles are subject to natural mortality m, second term, as well as intraspecific competition, the next three terms, due to other susceptible individuals, at rate c_{SS}, or to asymptomatic ones, at rate c_{SA}, or finally by infected, at rate c_{SI}. The last term is the epidemiological one, which accounts for the disease spread. In view of our assumptions, it has two main features. The disease transmission is modeled via a mass action term in the numerator, accounting for the "successful" contacts between the susceptibles and the unrecognized disease carriers, i.e., the asymptomatic ones. The denominator, instead, accounts for the preventive measures, so that susceptibles tend to reduce their intermingling with other people when more and more diseased individuals are identified, i.e., it must decrease with an increasing number of symptomatic individuals I. The transmission parameter α measures how many contacts there are in the time unit, as well as how many of them yield a new case of the infection. Instead, β can be considered the weight and relevance that susceptibles give to the information about the disease spread.

The second equation models the asymptomatic dynamics. The first term contains the natural mortality, assuming that at this stage the disease does not cause deaths. The next three terms denote the intracompartment competition and the corresponding ones with the other two population classes. The susceptible individuals that become infected in the process described in the first equation appear here as new recruits, while the last term denotes transition toward a more serious form of the disease, and, thereby, showing symptoms.

This very same term, the last one in the third equation, appears in the infected dynamics, as the only input in this class. Here, in addition to the natural mortality, the first term also contains the disease-related one, μ. The following three terms again represent the competition between individuals of this class as well as with the other two compartments.

Tables 1 and 2 summarize the meaning of the parameters, all assumed to be non-negative.

Table 1. Interpretation and dimensions of demographic parameters.

Parameters	Interpretation	Dimensions
$c_{XY}, X, Y \in \{S, A, I\}$	competition pressure of Y on X	$\frac{1}{[t]}$
m	natural mortality rate	$\frac{1}{[t]}$
r	natural birth rate	$\frac{1}{[t]}$

Table 2. Interpretation and dimensions of epidemiological parameters.

Parameters	Interpretation	Dimensions
α	transmission rate	$\frac{1}{[t]}$
β	inhibitory effect coefficient	-
μ	disease-related mortality rate	$\frac{1}{[t]}$
π	progression rate from asymptomatics to symptomatics	$\frac{1}{[t]}$

2.2. Boundedness

Note, first of all, that if $r < m$, by summing the three equations of (1) and dropping most of the negative terms, we obtain

$$\frac{dT}{dt} = \frac{d(S + A + I)}{dt} \leq (r - m)(S + A) - (m + \mu)I,$$

which entails that the population $T = S + A + I$ will eventually vanish, implying also that each one of its subclasses does as well, as they are necessarily non-negative. Hence, $S \to 0^+$, $A \to 0^+$ and $I \to 0^+$. Unless otherwise stated, from now on, we assume

$$r > m. \tag{2}$$

On the other hand, even in case where (2) holds, the system trajectories are bounded. Indeed, considering again the total population T, summing the equations in (1), but retaining some of the quadratic terms for an arbitrary $\eta > 0$, we obtain

$$\frac{dT}{dt} + \eta T \leq \Pi_S(S) + \Pi_A(A) + \Pi_I(I),$$

where the functions on the right hand side are concave parabolae:

$$\Pi_S(S) = [r + \eta - c_{SS}S]S, \quad \Pi_A(A) = [r + \eta - c_{AA}A]A, \quad \Pi_I(I) = [\eta - c_{II}I]I.$$

By replacing the latter with their maxima, namely, evaluating each one of them, respectively, at the abscissae

$$S_m = \frac{r + \eta}{2c_{SS}}, \quad A_m = \frac{r + \eta}{2c_{AA}}, \quad I_m = \frac{\eta}{2c_{II}},$$

so that

$$\Pi_S(S) \leq \Pi_S(S_m) = \frac{(r + \eta)^2}{4c_{SS}}, \quad \Pi_A(A) \leq \Pi_A(A_m) = \frac{(r + \eta)^2}{4c_{AA}}, \quad \Pi_I(I) \leq \Pi_I(I_m) = \frac{\eta^2}{4c_{II}},$$

we obtain the bound for the differential inequality

$$\frac{dT}{dt} + \eta T \leq M, \quad M = \frac{(r+\eta)^2}{4c_{SS}} + \frac{(r+\eta)^2}{4c_{AA}} + \frac{\eta^2}{4c_{II}}.$$

It then follows that

$$T(t) \leq \max\left\{T(0), \frac{M}{\eta}\right\},$$

which is the desired result; since the total population is bounded, each subpopulation must be bounded as well because it cannot have negative values.

2.3. Model Equilibria

The model allows only three possible equilibria. Two are easily found: the population collapse $E_0 = (0,0,0)$ and the disease-free point,

$$E_S = \left(\frac{r-m}{c_{SS}}, 0, 0\right),$$

which are always admissible in view of the assumption (2).

2.3.1. Endemic Coexistence

The final allowed equilibrium point is coexistence of the three population classes, with the disease becoming endemic. To assess this point, we need to study the three equilibrium equations of (1). They give three surfaces in the S, A, I phase space. To understand their shape, we intersect them with parallel planes to the coordinate planes.

The first surface $\Sigma^{(1)}$

The first surface $\Sigma^{(1)}$,

$$\Sigma^{(1)}: \quad r(S+A) - mS - c_{SS}S^2 - c_{SA}SA - c_{SI}SI - \frac{\alpha SA}{1+\beta I} = 0, \tag{3}$$

arises from the corresponding equilibrium equation of (1). On the plane $S = 0$, this surface intersects the first quadrant of the A-I plane only on the I axis.

On the plane $A = 0$, in addition to $S = 0$, i.e. the I axis, the intersection is the straight line

$$c_{SS}S + c_{SI}I = r - m\,. \tag{4}$$

There are two intersection points with the coordinate axes:

$$I_0 = \frac{r-m}{c_{SI}}, \quad S_0 = \frac{r-m}{c_{SS}},$$

both are strictly positive in view of (2).

On the generic plane $I = h > 0$, the intersection is the conic

$$r(S+A) - mS - c_{SS}S^2 - c_{SA}SA - c_{SI}Sh - \tilde{\alpha}SA = 0\,, \quad \tilde{\alpha} = \frac{\alpha}{1+\beta h}. \tag{5}$$

To study this, it should be observed that the matrix of this quadratic form is

$$\mathbf{M}_{\mathbf{h}}^{\Sigma^{(1)}} = \begin{bmatrix} -c_{SS} & -\frac{1}{2}(c_{SA}+\tilde{\alpha}) & \frac{1}{2}(r-m-c_{SI}h) \\ -\frac{1}{2}(c_{SA}+\tilde{\alpha}) & 0 & \frac{r}{2} \\ \frac{1}{2}(r-m-c_{SI}h) & \frac{r}{2} & 0 \end{bmatrix},$$

whose determinant is

$$\Delta_h^{\Sigma^{(1)}} = \frac{r}{4}(rc_{SS} - (c_{SA} + \widetilde{\alpha})(r - m - c_{SI}h)).$$

The conic is nondegenerate whenever $\Delta_h \neq 0$, i.e., if and only if

$$r \neq \frac{(m + c_{SI}h)(c_{SA} + \widetilde{\alpha})}{c_{SA} + \widetilde{\alpha} - c_{SS}}, \tag{6}$$

a condition that we now assume. The principal minor of order two is always negative

$$\delta_h^{\Sigma^{(1)}} = \begin{vmatrix} -c_{SS} & -\frac{1}{2}(c_{SA} + \widetilde{\alpha}) \\ -\frac{1}{2}(c_{SA} + \widetilde{\alpha}) & 0 \end{vmatrix} = -\frac{1}{4}(c_{SA} + \widetilde{\alpha})^2 < 0$$

so that the conic is a hyperbola. It intersects the A axis only at the origin, while the intersection with the S axis is given by the point

$$S_0^h = \frac{r - m - c_{SI}h}{c_{SS}}$$

which is positive if and only if $h < I_0$. Thus, in such a case, the two intersections with the coordinate axes are $(0,0)$ and $\left(S_0^h, 0\right)$. On the other hand, the origin is the only intersection when $h \geq I_0$.

The conic (5) can be explicitly written as

$$A_h(S) = \frac{(r - m - c_{SS}S - c_{SI}h)S}{(c_{SA} + \widetilde{\alpha})S - r}, \tag{7}$$

so that its vertical asymptote is

$$S = S_\infty^h = \frac{r}{c_{SA} + \widetilde{\alpha}} = \frac{r(1 + \beta h)}{c_{SA}(1 + \beta h) + \alpha}.$$

Since S_∞^h is strictly positive, for our only case of interest $h > 0$, there are two possible situations for the conic (5)

- If $h \geq I_0$, it follows that $S_0^h \leq 0$; then, $A_h(S) > 0$ if and only if $0 < S < S_\infty^h$. The conic is positive and increasing only from the origin to the vertical asymptote.
- If $h < I_0$, the conic crosses the point $\left(S_0^h, 0\right)$. In such case, we have the following three possibilities.
  - If $S_0^h < S_\infty^h$, then $A_h(S) > 0$ if and only if $S_0^h < S < S_\infty^h$. The conic is a hyperbola which is positive and increasing in the $(I = h)$ S-A plane only from the zero to the asymptote.
  - If $S_\infty^h < S_0^h$, then $A_h(S) > 0$ if and only if $S_\infty^h < S < S_0^h$. The conic is a hyperbola which is positive and decreasing only from the asymptote to $\left(S_0^h, 0\right)$.
  - Finally, if $S_0^h = S_\infty^h$, then the conic is degenerate because (6) fails to hold, having, in this case,
$$r = \frac{(m + c_{SI}h)(c_{SA} + \widetilde{\alpha})}{c_{SA} + \widetilde{\alpha} - c_{SS}}.$$

Further, in the positive cone of the S–I plane, as h increases S_0^h decreases linearly, moving along the line segment (4), starting from S_0 and vanishing when $h = I_0$, while S_∞^h increases, starting from S_∞ and tending asymptotically to rc_{SA}^{-1}.

Thus, in the plane S–I, the trajectories of S_0^h and S_∞^h never intersect if $S_0 < S_\infty$, while if $S_0 \geq S_\infty$, they intersect on the line segment (4) at a point on the $I = h$ plane with

$$h = \frac{-(c_{SI}(c_{SA} + \alpha) - \beta c_{SA}(r - m) + \beta c_{SS} r) + \sqrt{D}}{2\beta c_{SI} c_{SA}},$$

with

$$D = (c_{SI}(c_{SA} + \alpha) - \beta c_{SA}(r - m) + \beta c_{SS} r)^2 + 4\beta c_{SI} c_{SA}((r - m)(c_{SA} + \alpha) - c_{SS} r).$$

The second surface $\Theta^{(1)}$

The second surface $\Theta^{(1)}$, from the corresponding equilibrium equation of (1)

$$\Theta^{(1)}: \quad -m - c_{AA}A - c_{AS}S - c_{AI}I + \frac{\alpha S}{1 + \beta I} - \pi = 0, \tag{8}$$

meets the plane $S = 0$ on the segment with negative intersections with the coordinate axes

$$c_{AA}A + c_{AI}I = -m - \pi, \quad \widehat{I}_0 = -\frac{m + \pi}{c_{AI}} < 0, \quad \widehat{A}_0 = -\frac{m + \pi}{c_{AA}} < 0, \tag{9}$$

so that no feasible portion exists.

On the plane $A = 0$, the intersection is

$$m + m\beta I + c_{AS}S + c_{AS}\beta SI + c_{AI}I + c_{AI}\beta I^2 - \alpha S + \pi + \pi\beta I = 0. \tag{10}$$

The matrix of this conic is

$$\mathbf{M}^{\Theta^{(1)}} = \begin{bmatrix} 0 & \frac{1}{2}c_{AS}\beta & \frac{1}{2}(c_{AS} - \alpha) \\ \frac{1}{2}c_{AS}\beta & c_{AI}\beta & \frac{1}{2}(\beta(m + \pi) + c_{AI}) \\ \frac{1}{2}(c_{AS} - \alpha) & \frac{1}{2}(\beta(m + \pi) + c_{AI}) & m + \pi \end{bmatrix},$$

with determinant

$$\Delta^{\Theta^{(1)}} = \frac{1}{4}\alpha\beta(c_{AS}c_{AI} - c_{AI}\alpha - \beta c_{AS}(m + \pi)).$$

It is nondegenerate for $\Delta \neq 0$, i.e., if and only if

$$\alpha \neq c_{AS}\left(1 - \frac{\beta c_{AS}(m + \pi)}{c_{AI}}\right). \tag{11}$$

If (11) holds, since

$$\delta^{\Theta^{(1)}} = \begin{vmatrix} 0 & \frac{1}{2}c_{AS}\beta \\ \frac{1}{2}c_{AS}\beta & c_{AI}\beta \end{vmatrix} = -\frac{1}{4}c_{AS}^2\beta^2 < 0$$

it is a hyperbola. Its intersections with the coordinate axes are

$$\widehat{S}_0 = \frac{m + \pi}{\alpha - c_{AS}}, \tag{12}$$

which is positive only if

$$\alpha > c_{AS}, \tag{13}$$

and the roots of the quadratic

$$c_{AI}\beta I^2 + ((m + \pi)\beta + c_{AI})I + m + \pi = 0,$$

which, following Descartes' rule, are both negative. Writing the hyperbola explicitly as

$$S(I) = -\frac{c_{AI}\beta I^2 + [(m+\pi)\beta + c_{AI}]I + m + \pi}{c_{AS}(1+\beta I) - \alpha}, \tag{14}$$

we find its vertical asymptote

$$\widehat{I}_\infty = \frac{\alpha - c_{AS}}{\beta c_{AS}} > 0$$

in view of (13). In such case, the hyperbola is positive if and only if $0 \leq I < \widehat{I}_\infty$. Then, the hyperbola increases from $\left(0, \widehat{S}_0\right)$ to the asymptote $I = \widehat{I}_\infty$. If (13) fails to hold, no portion of the conic lies in the first quadrant.

On the plane $I = h$, recalling (5), the surface $\Theta^{(1)}$ gives the straight line

$$c_{AA}A + (c_{AS} - \widetilde{\alpha})S = -m - \pi - c_{AI}h, \tag{15}$$

with the following intersections with the coordinate axes

$$\widehat{A}_0^h = -\frac{m + \pi + c_{AI}h}{c_{AA}}, \quad \widehat{S}_0^h = \frac{m + \pi + c_{AI}h}{\widetilde{\alpha} - c_{AS}}.$$

The former is always negative, the latter is positive if and only if $\widetilde{\alpha} > c_{AS}$, which is equivalent to

$$h < \widehat{I}_\infty. \tag{16}$$

Consequently, the surface $\Theta^{(1)}$ does no intersect the plane $I = h$ if $h \geq \widehat{I}_\infty$, while it crosses the half-line portion in the positive cone of the line joining the point $\left(\widehat{S}_0^h, 0\right)$, with $\left(0, \widehat{A}_0^h\right)$, if $h < \widehat{I}_\infty$.

Further, as h increases, $\widehat{A}_0^h$ decreases linearly to $-\infty$, while $\widehat{S}_0^h$ grows along the hyperbola (10), in the first quadrant of the S-I plane, starting from $\widehat{S}_0$, recall (12), and tending asymptotically to $\widehat{I}_\infty$.

The third surface Γ

Here, we have

$$\Gamma: \quad -(m+\mu)I - c_{II}I^2 - c_{IS}IS - c_{IA}IA + \pi A = 0, \tag{17}$$

On the plane $S = \ell$, we obtain the conic

$$c_{II}I^2 + c_{IA}IA + (m + \mu + c_{IS}\ell)I - \pi A = 0, \tag{18}$$

whose matrix is

$$\mathbf{M}_\ell^\Gamma = \begin{bmatrix} c_{II} & \frac{c_{IA}}{2} & \frac{1}{2}(m+\mu+c_{IS}\ell) \\ \frac{c_{IA}}{2} & 0 & -\frac{\pi}{2} \\ \frac{1}{2}(m+\mu+c_{IS}\ell) & -\frac{\pi}{2} & 0 \end{bmatrix},$$

with

$$\Delta_\ell^\Gamma = -\frac{\pi}{4}(\pi c_{II} + c_{IA}(m + \mu + c_{IS}\ell)) < 0.$$

Since Δ_ℓ^Γ is always negative, the conic is always nondegenerate. Further, the principal minor of order two is always negative, indicating that the conic is a hyperbola:

$$\delta_\ell^\Gamma = \begin{vmatrix} c_{II} & \frac{c_{IA}}{2} \\ \frac{c_{IA}}{2} & 0 \end{vmatrix} = -\frac{1}{4}c_{IA}^2.$$

This hyperbola intersects the axes only at the origin in the feasible range. We can write it explicitly as

$$A_\ell(I) = \frac{(m+\mu+c_{II}I+c_{IS}\ell)I}{\pi - c_{IA}I}. \tag{19}$$

Its vertical asymptote is independent of ℓ,

$$\widetilde{I}_\infty = \frac{\pi}{c_{IA}} > 0.$$

Hence, in the positive cone of the I-A plane, $A_\ell(I) > 0$ if and only if $0 < I < \widetilde{I}_\infty$. Thus, the conic (18) raises up from the origin to the asymptote, for every value of ℓ.

It is also easily seen that on $A = 0$, the surface Γ intersects the first quadrant of the S–I plane only on the S horizontal axis.

The possible intersections of $\Sigma^{(1)}$, $\Theta^{(1)}$ and Γ

Thus, on the $S = 0$ coordinate plane, $\Sigma^{(1)}$ and $\Theta^{(1)}$ meet at a point $Q_{I=0}$ if the condition

$$\widehat{S}_0 < S_0 \tag{20}$$

is satisfied. In such case, on the plane $A = 0$, the segment joining S_0 and I_0 meets at the point $Q_{A=0}$, the hyperbola generated by the intersection of the surface $\Theta^{(1)}$ with $A = 0$. The intersection of $\Sigma^{(1)}$ and $\Theta^{(1)}$ always exists, provided the condition (20) holds, because $\Sigma^{(1)}$ raises up to the vertical asymptote, while $\Theta^{(1)}$ has a positive slope on the $I = h$ planes. The curve, $\rho = \Sigma^{(1)} \cap \Theta^{(1)}$, thus joins the points $Q_{I=0}$ and $Q_{A=0}$. However, the former has a positive value of A, on the coordinate plane $I = 0$, the latter lies on the plane $A = 0$, and, thus, they lie in the upper half space in which Γ partitions the phase space S–A–I, the latter in the lower one. Hence, the line ρ intersects Γ and this intersection point represents the endemic coexistence equilibrium.

2.4. Equilibria Stability

Local Stability

The Jacobian matrix associated with the system (1) is

$$\mathbf{J} = \begin{bmatrix} J_{1,1} & r - c_{SA}S - \dfrac{\alpha S}{1+\beta I} & -c_{SI}S + \dfrac{\alpha\beta SA}{(1+\beta I)^2} \\ -c_{AS}A + \dfrac{\alpha A}{1+\beta I} & J_{2,2} & -c_{AI}A - \dfrac{\alpha\beta SA}{(1+\beta I)^2} \\ -c_{IS}I & -c_{IA}I + \pi & J_{3,3} \end{bmatrix},$$

where

$$J_{1,1} = r - m - 2c_{SS}S - c_{SA}A - c_{SI}I - \frac{\alpha A}{1+\beta I},$$

$$J_{2,2} = -m - 2c_{AA}A - c_{AS}S - c_{AI}I + \frac{\alpha S}{1+\beta I} - \pi$$

and

$$J_{3,3} = -m - \mu - 2c_{II}I - c_{IS}S - c_{IA}A.$$

For point E_0, we immediately obtain the eigenvalues $r - m$, $-m - \pi < 0$ and $-m - \mu < 0$, so that the ecosystem collapses if

$$r < m \tag{21}$$

in line with the earlier considerations.

In addition, at the disease-free point E_S, the eigenvalues can all be determined analytically:

$$-r + m\,, \quad -m - \frac{(r-m)(c_{AS} - \alpha)}{c_{SS}} - \pi, \quad -m - \mu - \frac{c_{IS}}{c_{SS}}(r - m)\,.$$

In view of (2), the first and third eigenvalues are negative, and the feasibility conditions reduce to

$$\pi + m > \frac{(r-m)(\alpha - c_{AS})}{c_{SS}}. \tag{22}$$

For the coexistence endemic equilibrium E_{SAI}, we rely on numerical simulations. Observe that, using the equilibrium equations, the diagonal terms of the Jacobian simplify, becoming

$$J_{11} = -\frac{rA}{S} - c_{SS}S, \quad J_{22} = -c_{AA}A, \quad J_{11} = -\frac{\pi A}{I} - c_{II}I.$$

Thus, the trace is negative. The remaining two Routh–Hurwitz conditions are too complicated to shed any further light on and are not analysed further.

Table 3 summarizes the information gathered on the three equilibrium points and their local stability.

Table 3. Summary of equilibria and local stability for model (1).

Equilibria	Existence Conditions	Stability
$E_0 = (0,0,0)$	-	stable if $r < m$
$E_S = \left(\frac{r-m}{c_{SS}}, 0, 0\right)$	$r > m$	stable if (22)
$E_{SAI} = (S^*, A^*, I^*)$	sufficient: (20)	numerical simulations

2.5. *SAI Model with Vertical Transmission*

The model is:

$$\begin{aligned} \frac{dS}{dt} &= rS - mS - c_{SS}S^2 - c_{SA}SA - c_{SI}SI - \frac{\alpha SA}{1+\beta I}, \\ \frac{dA}{dt} &= rA - mA - c_{AA}A^2 - c_{AS}AS - c_{AI}AI + \frac{\alpha SA}{1+\beta I} - \pi A, \\ \frac{dI}{dt} &= -(m+\mu)I - c_{II}I^2 - c_{IS}IS - c_{IA}IA + \pi A. \end{aligned} \tag{23}$$

The variables and parameters retain the same meanings as (1), which is recalled in Tables 1 and 2. Thus, the description of (1) holds in this case as well. The only change with respect to model (1) is the fact that reproduction of asymptomatic individuals leads to new offsprings in the very same class.

2.6. *Preliminary Analysis*

In view of the fact that (1) and (23) differ only in one term, the considerations that lead to the system disappearance if (2) is not satisfied, as well as the boundedness of the trajectories, can be repeated using the same steps and show that these results hold here as well. Thus, in order to ensure that the model solutions do not vanish, the condition (2) must be imposed here as well.

Further, the equilibria E_0 and E_S are the same as the corresponding ones of (1), with the same feasibility condition for the latter, namely, (2).

2.6.1. Equilibria

In addition to endemic coexistence, we find also the point $E_{AI}^{\pm}$, where the susceptibles vanish. To investigate the latter, the second equilibrium equation yields

$$A = \frac{1}{c_{AA}}(r - m - c_{AI}I - \pi). \tag{24}$$

Substituting (24) into the third equilibrium equation, we find the following quadratic equation I:

$$c_2 I^2 + c_1 I + c_0 = 0, \quad c_2 = c_{II} - \frac{c_{IA}c_{AI}}{c_{AA}},$$
$$c_1 = m + \mu + \frac{c_{IA}r - c_{IA}m - c_{IA}\pi + c_{AI}\pi}{c_{AA}}, \quad c_0 = -\frac{\pi}{c_{AA}}(r - m - \pi).$$

Its roots are

$$I_{\pm} = \frac{c_{AA}}{2(c_{II}c_{AA} - c_{IA}c_{AI})}\left(-m - \mu - \frac{c_{IA}r - c_{IA}m - c_{IA}\pi + c_{AI}\pi}{c_{AA}} \pm \sqrt{\Delta}\right), \tag{25}$$

with

$$\Delta = \left(m + \mu + \frac{c_{IA}r - c_{IA}m - c_{IA}\pi + c_{AI}\pi}{c_{AA}}\right)^2 + \frac{4\pi}{c_{AA}}\left(c_{II} - \frac{c_{IA}c_{AI}}{c_{AA}}\right)(r - m - \pi). \tag{26}$$

Thus, there are two possible equilibria

$$E_{AI}^{\pm} = \left(0, \frac{r - m - c_{AI}I_{\pm} - \pi}{c_{AA}}, I_{\pm}\right)$$

with feasibility conditions

$$r > m + c_{AI}I_{\pm} + \pi \tag{27}$$

and

$$\frac{1}{c_{II}c_{AA} - c_{IA}c_{AI}}\left(-m - \mu - \frac{c_{IA}r - c_{IA}m - c_{IA}\pi + c_{AI}\pi}{c_{AA}} \pm \sqrt{\Delta}\right) > 0. \tag{28}$$

2.6.2. The Endemic Coexistence Equilibrium

Again, we study this point through the intersection of suitable surfaces, arising from the equilibrium equations of (23). Note that since the last equation in (23) is unchanged with respect to the same one in (1), the resulting surface is once again represented by the function Γ, already investigated at the end of Section 2.3.1.

The first surface $\Sigma^{(2)}$

From the first equilibrium equation of (23), we have

$$\Sigma^{(2)} : r - m - c_{SS}S - c_{SA}A - c_{SI}I - \frac{\alpha A}{1 + \beta I} = 0. \tag{29}$$

On the plane $A = 0$, we obtain

$$c_{SS}S + c_{SI}I = r - m, \tag{30}$$

which meets the coordinate axes at the points with nonvanishing abscissae given by

$$S_0 = \frac{r - m}{c_{SS}}, \quad I_0 = \frac{r - m}{c_{SI}}. \tag{31}$$

Thus, the intersection with the coordinate plane $A = 0$ is the line segment joining the points $(0, I_0)$ and $(S_0, 0)$.

Similarly, on the plane $I = 0$, we find the point A_0 and the line

$$A_0 = \frac{r-m}{c_{SA}+\alpha}, \quad c_{SS}S + (c_{SA}+\alpha)A = r - m\,, \tag{32}$$

whose feasible portion joins the points $(0, A_0)$ and $(S_0, 0)$.

On the plane $S = \ell$, we obtain the conic

$$r - m - c_{SS}\ell + (\beta(r-m-c_{SS}\ell) - c_{SI})I - (c_{SA}+\alpha)A - \beta c_{SA}AI - \beta c_{SI}I^2 = 0\,, \tag{33}$$

whose coefficient matrix is

$$\mathbf{M}_\ell^{\Sigma(2)} = \begin{bmatrix} -c_{SI}\beta & -\frac{1}{2}c_{SA}\beta & \frac{1}{2}(\beta(r-m-c_{SS}\ell)-c_{SI}) \\ -\frac{1}{2}c_{SA}\beta & 0 & -\frac{1}{2}(c_{SA}+\alpha) \\ \frac{1}{2}(\beta(r-m-c_{SS}\ell)-c_{SI}) & -\frac{1}{2}(c_{SA}+\alpha) & r-m-c_{SS}\ell \end{bmatrix},$$

with determinant

$$\Delta_\ell^{\Sigma(2)} = \frac{1}{4}\alpha\beta(\alpha c_{SI} + \beta c_{SA}(r-m-c_{SS}\ell) + \alpha c_{SI}c_{SA})\,.$$

Now, if $\ell \le S_0$, we have that $\Delta_\ell^{\Sigma(2)}$ is always positive and the conic is nondegenerate. Since

$$\delta_\ell^{\Sigma(2)} = \begin{vmatrix} -c_{SI}\beta & -\frac{1}{2}c_{SA}\beta \\ -\frac{1}{2}c_{SA}\beta & 0 \end{vmatrix} = -\frac{1}{4}c_{SA}^2\beta^2 < 0$$

the conic is a hyperbola. The intersections on this $S = \ell$ plane with the coordinate axes are the points

$$A_0^\ell = \frac{r-m-c_{SS}\ell}{c_{SA}+\alpha}$$

positive for

$$r > m + c_{SS}\ell \tag{34}$$

and the roots of the quadratic

$$\beta c_{SI}I^2 + (c_{SI} - (r-m-c_{SS}\ell)\beta)I - (r-m-c_{SS}\ell) = 0\,,$$

namely,

$$I_0^{\ell\,\pm} = \frac{(r-m-c_{SS}\ell)\beta - c_{SI} \pm \sqrt{\Delta_I^\ell}}{2\beta c_{SI}},$$

with

$$\Delta_I^\ell = (\beta(r-m-c_{SS}\ell) + c_{SI})^2\,.$$

Recalling (31), the roots explicitly are

$$I_0^{\ell\,+} = \frac{r-m-c_{SS}\ell}{c_{SI}}, \quad I_0^{\ell\,-} = -\frac{1}{\beta} = I_0^- < 0.$$

Note that if (34) is not satisfied, no portion of the conic lies in the feasible cone. In addition, both A_0^ℓ and $I_0^{\ell\,+}$ are positive if and only if $\ell < S_0$. As ℓ increases, both $I_0^{\ell\,+}$ and A_0^ℓ decrease linearly, respectively, along the segments (30) and (32), starting, respectively, from I_0 and A_0, and coalescing into the origin when $\ell = S_0$.

Writing the conic in explicit form

$$A_\ell(I) = -\frac{\beta c_{SI} I^2 + (c_{SI} - (r - m - c_{SS}\ell)\beta)I - (r - m - c_{SS}\ell)}{c_{SA}(1 + \beta I) + \alpha}. \tag{35}$$

the hyperbola is seen to have the vertical asymptote at $I = I_\infty = -(c_{SA} + \alpha)(c_{SA}\beta)^{-1} < 0$, always being negative and independent of ℓ. Thus, if $\ell < S_0$, the hyperbola is positive, with a feasible branch on the plane $S = \ell$ joining the points

$$\left(I_0^{\ell+}, 0\right), \quad \left(0, A_0^\ell\right)$$

on the coordinate axes. This branch is concave because

$$I_0^- = -\frac{1}{\beta} > -\frac{1}{\beta}\left(1 + \frac{\alpha}{c_{SA}}\right) = I_\infty < 0.$$

The second surface $\Theta^{(2)}$

This surface has the expression

$$\Theta^{(2)} : r - m - \pi - c_{AA}A - c_{AS}S - c_{AI}I + \frac{\alpha S}{1 + \beta I} = 0. \tag{36}$$

On the plane $S = 0$, we obtain the straight line

$$c_{AA}A + c_{AI}I = r - m - \pi. \tag{37}$$

with intersections with the axes at the points

$$\widehat{I}_0 = \frac{r - m - \pi}{c_{AI}}, \quad \widehat{A}_0 = \frac{r - m - \pi}{c_{AA}},$$

giving the segment joining $(0, \widehat{I}_0)$ and $(\widehat{A}_0, 0)$. These are both positive if

$$r > m + \pi. \tag{38}$$

If this condition does not hold, there is no feasible intersection.

Recalling (5), the intersection with the plane $I = h$ gives

$$c_{AA}A + (c_{AS} - \widetilde{\alpha})S = r - m - \pi - c_{AI}h. \tag{39}$$

Assuming (38), the intersections with the coordinate axes are

$$\widehat{A}_h = \frac{r - m - \pi - c_{AI}h}{c_{AA}}, \quad \widehat{S}_h = \frac{r - m - \pi - c_{AI}h}{c_{AS} - \widetilde{\alpha}}.$$

Note that on $I = 0$, these intersections become $\widehat{A}_h = \widehat{A}_0$, found above, and $\widehat{S}_0 = (r - m - \pi)(c_{AS} - \alpha)^{-1}$. The positivity of the latter is given by

$$c_{AS} > \alpha. \tag{40}$$

On the plane $A = 0$ we once again obtain a conic section

$$\beta c_{AI}I^2 + \beta c_{AS}SI + (c_{AI} - \beta(r - m - \pi))I + (c_{AS} - \alpha)S - (r - m - \pi) = 0, \tag{41}$$

whose matrix is

$$\mathbf{M}^{\Theta^{(2)}} = \begin{bmatrix} \beta c_{AI} & \frac{1}{2}\beta c_{AS} & \frac{1}{2}(c_{AI} - \beta(r-m-\pi)) \\ \frac{1}{2}\beta c_{AS} & 0 & \frac{1}{2}(c_{AS} - \alpha) \\ \frac{1}{2}(c_{AI} - \beta(r-m-\pi)) & \frac{1}{2}(c_{AS} - \alpha) & -(r-m-\pi) \end{bmatrix},$$

with determinant

$$\Delta^{\Theta^{(2)}} = \frac{1}{4}\alpha\beta(c_{AS}\beta(r-m-\pi) + c_{AS}c_{AI} - c_{AI}\alpha).$$

It is nondegenerate if and only if

$$\alpha \neq c_{AS}\left(1 + \frac{\beta(r-m-\pi)}{c_{AI}}\right). \tag{42}$$

Since

$$\delta^{\Theta^{(2)}} = \begin{vmatrix} \beta c_{AI} & \frac{1}{2}\beta c_{AS} \\ \frac{1}{2}\beta c_{AS} & 0 \end{vmatrix} = -\frac{1}{4}c_{AS}^2\beta^2 < 0$$

the conic section is again a hyperbola. Its intersections with the axes are the points

$$\widehat{S}_0 = \frac{r-m-\pi}{c_{AS}-\alpha}, \quad \widehat{I}_0^{\pm} = \frac{\beta(r-m-\pi) - c_{AI} \pm \sqrt{\Delta_I}}{2\beta c_{AI}},$$
$$\Delta_I = (c_{AI} - \beta(r-m-\pi))^2 + 4\beta c_{AI}(r-m-\pi) = (\beta(r-m-\pi) + c_{AI})^2,$$

the latter being the roots of the quadratic

$$\beta c_{AI} I^2 + (c_{AI} - \beta(r-m-\pi))I - (r-m-\pi) = 0.$$

Note that these roots are in fact

$$\widehat{I}_0^{+} = \widehat{I}_0, \quad \widehat{I}_0^{-} = -\frac{1}{\beta}.$$

Writing this conic explicitly as

$$S(I) = -\frac{\beta c_{AI} I^2 + (c_{AI} - \beta(r-m-\pi))I - (r-m-\pi)}{c_{AS}(1+\beta I) - \alpha}, \tag{43}$$

we observe that it has a vertical asymptote at

$$\widehat{I}_\infty = \frac{\alpha - c_{AS}}{c_{AS}\beta},$$

which is positive if (40) does not hold.

The possible intersections of $\Sigma^{(2)}$, $\Theta^{(2)}$ and Γ

We now briefly also discuss for this case the sufficient conditions for an intersection. Now, an intersection between $\Sigma^{(2)}$ and $\Theta^{(2)}$ can be guaranteed if the corresponding intersections with the coordinate axes are suitably arranged, imposing some kind of interlacing properties between the corresponding coordinates.

More specifically, assuming all intersections with the coordinate axes for both $\Sigma^{(2)}$ and $\Theta^{(2)}$ to be positive, imposing

$$S_0 > \widehat{S}_0, \quad A_0 < \widehat{A}_0, \quad I_0 < \widehat{I}_0, \tag{44}$$

the two surfaces meet on the S–A coordinate plane at the point $W_- = (S_-, A_-, 0)$,

$$S_- = \frac{1}{\delta_-}[(r-m)c_{AA} - (r-m-\pi)(c_{SA}+\alpha)],$$
$$A_- = \frac{1}{\delta_-}[c_{SS}(r-m-\pi) - (c_{AS}-\alpha)(r-m)],$$
$$\delta_- = c_{SS}c_{AA} - (c_{AS}+\alpha)(c_{SA}+\alpha),$$

and also at a point on the S–I coordinate plane, say $E_+ = (S_+, 0, I_+)$

$$S_+ = \frac{1}{c_{SS}}[(r-m)c_{SI}I_+],$$

where I_+ a root of the quadratic

$$\kappa_2 I^2 + \kappa_1 I + \kappa_0 = 0, \quad \kappa_0 = \frac{1}{c_{SS}}(r-m)(c_{AS}-\alpha) - (r-m-\pi),$$
$$\kappa_1 = \frac{1}{c_{SS}}(r-m)(c_{AS}-\alpha) + c_{AI} - \beta(r-m-\pi), \quad \kappa_2 = \beta\left(c_{AI} - \frac{c_{AS}c_{SI}}{c_{SS}}\right),$$

and clearly also on the line ρ_1 joining these two points. In addition, imposing instead the reverse inequalities

$$S_0 < \widehat{S}_0, \quad A_0 > \widehat{A}_0, \quad I_0 > \widehat{I}_0, \tag{45}$$

$\Sigma^{(2)}$ and $\Theta^{(2)}$ intersect each other again on the line ρ_1. Both these sets of conditions (44) and (45) represent sufficient intersection conditions, and more cases could arise and also lead to other intersection lines; however, we do not examine them further.

In addition, the intersection line ρ_1 meets Γ because $W_- = (S_-, A_-, 0)$ lies in the upper semispace generated by Γ, while $E_+ = (S_+, 0, I_+)$ in the lower one. Hence, since the intersection exists, it provides the population values of the endemic equilibrium for (23).

Table 4 summarizes these findings.

2.6.3. Local Stability

The Jacobian here has slight differences with respect to the one of (1), namely, it is

$$\widehat{\mathbf{J}} = \begin{bmatrix} \widehat{J}_{1,1} & -c_{SA}S - \dfrac{\alpha S}{1+\beta I} & -c_{SI}S + \dfrac{\alpha\beta SA}{(1+\beta I)^2} \\ -c_{AS}A + \dfrac{\alpha A}{1+\beta I} & \widehat{J}_{2,2} & -c_{AI}A - \dfrac{\alpha\beta SA}{(1+\beta I)^2} \\ -c_{IS}I & -c_{IA}I + \pi & \widehat{J}_{3,3} \end{bmatrix},$$

with

$$\widehat{J}_{1,1} = r - m - 2c_{SS}S - c_{SA}A - c_{SI}I - \frac{\alpha A}{1+\beta I},$$
$$\widehat{J}_{2,2} = r - m - 2c_{AA}A - c_{AS}S - c_{AI}I + \frac{\alpha S}{1+\beta I} - \pi$$

and

$$\widehat{J}_{3,3} = -m - \mu - 2c_{II}I - c_{IS}S - c_{IA}A.$$

For equilibrium E_0 the stability condition is unchanged with respect to model (1), namely (21). For E_S, instead, there is a change in the second eigenvalue, for which, instead of (22), the stability changes in

$$\pi > \left(1 - \frac{c_{AS}-\alpha}{c_{SS}}\right)(r-m). \tag{46}$$

For the equilibrium $E_{AI}^{\pm}$, one eigenvalue factorizes to give the first stability condition

$$r < m + c_{SA}A_{\pm} + c_{SI}I_{\pm} + \frac{\alpha A_{\pm}}{1 + \beta I_{\pm}}. \tag{47}$$

Applying the Routh–Hurwitz conditions to the remaining minor, we obtain from the condition on the trace

$$r < 2m + \pi + \mu + 2c_{AA}A_{\pm} + 2c_{II}I_{\pm} + c_{AI}I_{\pm} + c_{IA}A_{\pm}, \tag{48}$$

while the one on the determinant provides the last stability condition

$$c_{AI}A_{\pm}(-c_{IA}I_{\pm} + \pi) > (r - m - 2c_{AA}A_{\pm} - c_{AI}I_{\pm} - \pi)(m + \mu + 2c_{II}I_{\pm} + c_{IA}A_{\pm}). \tag{49}$$

For the coexistence equilibrium, using the equilibrium equations we can simplify the diagonal terms of the Jacobian, which become

$$\widehat{J}_{11} = -c_{SS}S, \quad \widehat{J}_{22} = -c_{AA}A, \quad \widehat{J}_{33} = -\frac{\pi A}{I} - c_{II}I,$$

immediately showing that the trace is negative. Stability hinges then on the remaining two Routh–Hurwitz conditions, which are very much involved and are neither going to be stated, nor investigated.

Table 5 summarizes these findings.

Table 4. Summary of equilibria and their feasibility for model (23).

Equilibria	Existence Conditions
$E_0 = (0,0,0)$	-
$E_S = \left(\frac{r-m}{c_{SS}}, 0, 0\right)$	$r > m$
$E_{AI}^{\pm} = \left(0, \frac{r - m - c_{AI}I_{\pm} - \pi}{c_{AA}}, I_{\pm}\right)$	(27), (28)
$E_{SAI} = (S_*, A_*, I_*)$	Sufficient: (38), (40) and either one of (44) or (45)

Table 5. Summary of equilibria and their local stability for model (23).

Equilibria	Stability Conditions
E_0	$r < m$
E_S	(46)
$E_{AI}^{\pm}$	(47), (48), (49)
E_{SAI}	numerical simulations

3. Results

We proposed two models for the phenomenon of contact reduction in the case of an epidemic's spread. The novelty here is represented by the fact that the fear inducing individuals' intermingling reduction is based on the number of symptomatic cases, not just on the "infected", which also includes the asymptomatic individuals.

The numerical experiments were all performed using our own codes written in Matlab, using the intrinsic function ode45 (or ode15s) for the differential equations integration. In the simulations, we always took the following hypothetical initial conditions:

$$S(0) = 10{,}000, \quad A(0) = 1, \quad I(0) = 0. \tag{50}$$

In addition, the very same hypothetical competition rates are used, namely

$$c_{SS} = 0.0004, \quad c_{AA} = 0.0005, \quad c_{II} = 0.0006, \quad c_{SA} = 0.0008, \quad c_{SI} = 0.0003, \\ c_{AS} = 0.0001, \quad c_{AI} = 0.0002, \quad c_{IS} = 0.0009, \quad c_{IA} = 0.0007. \tag{51}$$

Two models are presented, differing in the way disease transmission occurs between parents and offsprings. In the former, vertical transmission is not allowed, a fact that, instead, is present in the second model.

We present the simulation results from these models in Figures 1–7.

In Figure 1, the disease-free point is attained for model (1) and the hypothetical parameters

$$\alpha = 0.005, \quad \beta = 0.006, \quad \mu = 0.06, \quad \pi = 7, \quad m = 0.2, \quad r = 0.7. \tag{52}$$

For model (1), in Figure 2, coexistence is obtained using the hypothetical parameter values

$$\alpha = 0.5, \quad \beta = 0.6, \quad \mu = 0.3, \quad \pi = 1, \quad m = 0.4, \quad r = 3. \tag{53}$$

In Figure 3, the hypothetical parameters are

$$\alpha = 0.5, \quad \beta = 6, \quad \mu = 0.5, \quad \pi = 1, \quad m = 0.4, \quad r = 3. \tag{54}$$

The hypothetical parameters of Figure 4 are

$$\alpha = 0.5, \quad \beta = 0.6, \quad \mu = 0.3, \quad \pi = 4, \quad m = 2, \quad r = 3. \tag{55}$$

Note that in the former case, the asymptomatics attain the highest value, followed by the susceptibles. In the second case, instead, the disease seems to affect less of the population; the most populated compartment is the one of the susceptibles, followed by the asymptomatic individuals. Figure 4 shows an instance in which the largest population is represented by the symptomatic individuals and the second largest are the asymptomatics.

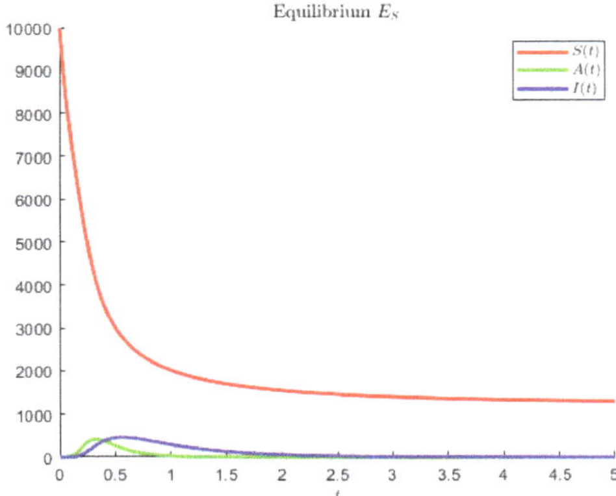

Figure 1. Disease-free point for model (1) obtained with parameter values (52), (51) and initial conditions (50).

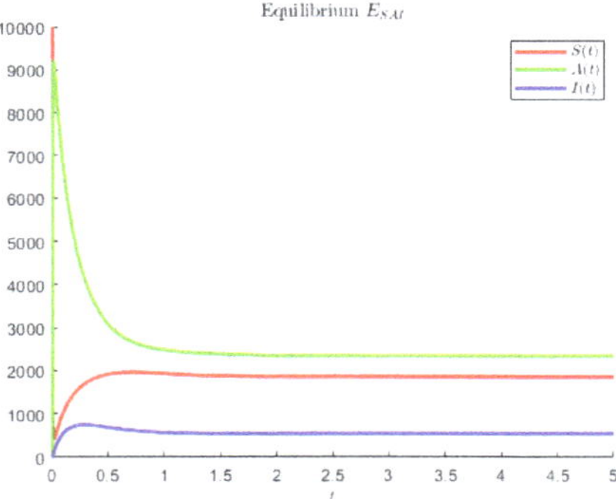

Figure 2. Coexistence equilibrium for model (1) obtained with parameter values (53), (51) and initial conditions (50).

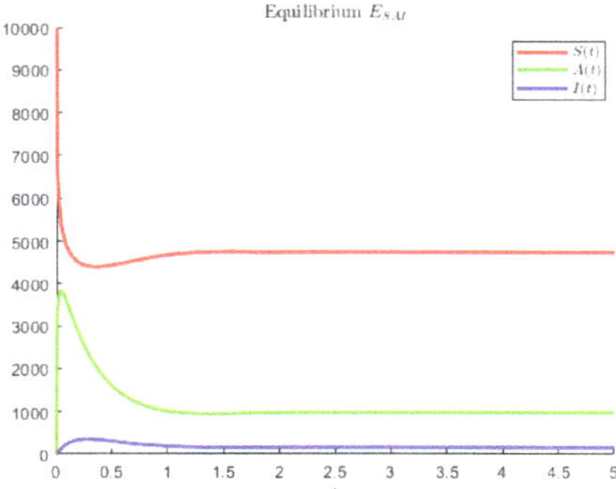

Figure 3. Coexistence equilibrium for model (1) obtained with parameter values (54), (51) and initial conditions (50).

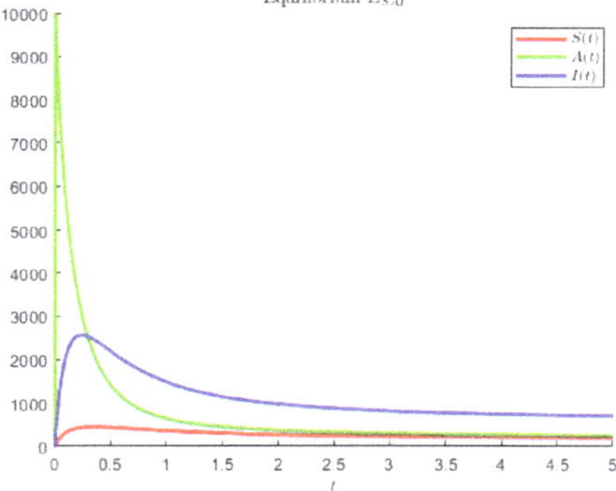

Figure 4. Coexistence equilibrium for model (23) obtained with parameter values (55), (51) and initial conditions (50).

For the model (23) with vertical transmission, we also show three instances of the endemic equilibrium. The hypothetical parameters of Figure 5 are (51) and

$$\alpha = 0.5, \quad \beta = 6, \quad \mu = 0.5, \quad \pi = 2, \quad m = 0.4, \quad r = 3, \tag{56}$$

while those hypothetical of Figure 6, in addition to (51), instead being

$$\alpha = 0.5, \quad \beta = 0.6, \quad \mu = 0.5, \quad \pi = 2.5, \quad m = 0.4, \quad r = 3 \tag{57}$$

and those hypothetical parameters for Figure 7, in which the asymptomatics represent the most numerous class, are

$$\alpha = 1.6, \quad \beta = 0.6, \quad \mu = 0.3, \quad \pi = 4, \quad m = 0.4, \quad r = 3 \tag{58}$$

The disease-free point is also attained with the following hypothetical parameters

$$\alpha = 0.5, \quad \beta = 0.6, \quad \mu = 0.3, \quad \pi = 4, \quad m = 0.4, \quad r = 3.$$

The resulting figure is extremely close to Figure 1 and, therefore, it is not shown.

In case of Figure 5, the most populated compartment is the susceptibles, followed by the asymptomatic individuals. In Figure 6, the asymptomatic prevail, followed by the symptomatic individuals, so that in this case the disease has been contracted by a larger proportion of the population. Figure 7 shows, instead, the case in which the asymptomatics represent the largest class.

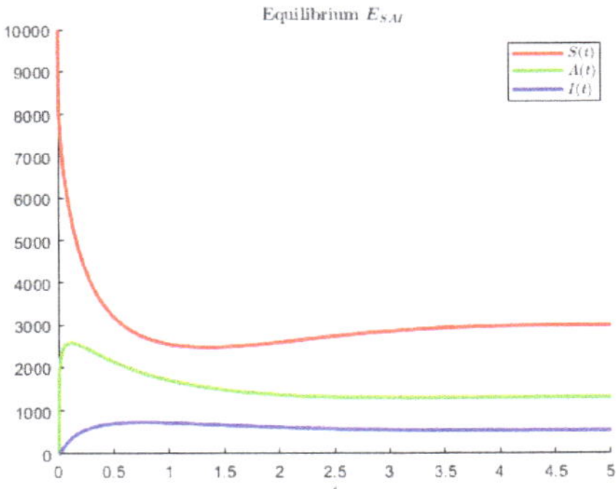

Figure 5. Coexistence equilibrium for model (23) obtained with parameter values (56), (51) and initial conditions (50).

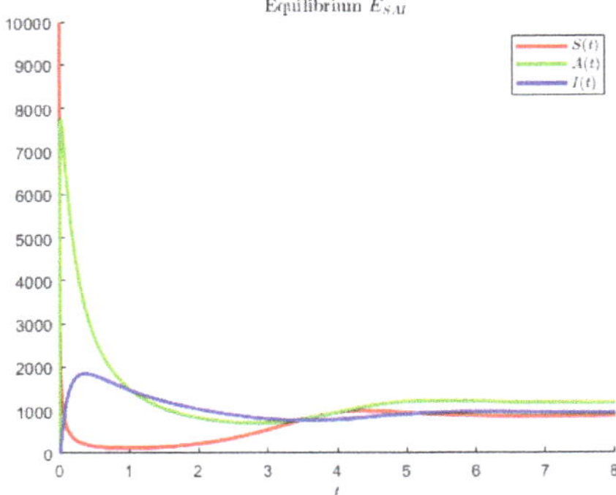

Figure 6. Coexistence equilibrium for model (23) obtained with parameter values (57), (51) and initial conditions (50).

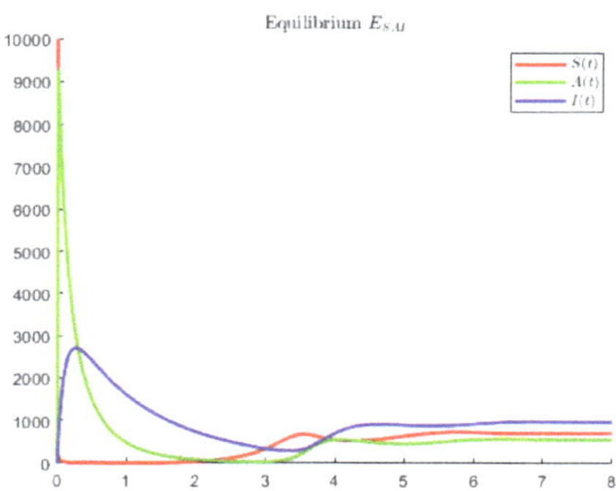

Figure 7. Coexistence equilibrium for model (23) obtained with parameter values (58), (51) and initial conditions (50).

4. Global Behavior of the Systems

On the basis of the previous results, comparing the feasibility and stability conditions of the various equilibria, we can conjecture the existence of bifurcations relating them. We show analytically their presence, but we can go even further, completely assessing the models behavior.

The first step in this direction is to recall that both models' trajectories are confined to a compact set, as discussed in Sections 2.2 and 2.6. In the second place, using Sotomayor's theorem [22], we show that a chain of transcritical bifurcations ties the various systems' equilibria.

Model (1) implies that no two such equilibria may appear simultaneously to be stable and feasible, and, therefore, when locally asymptotically stable, they are also globally asymptotically stable. Therefore, these systems move from "disappearance" to the susceptible-only, i.e., disease-free, point, and, finally, to the endemic case.

In order to apply Sotomayor's theorem, some preliminary calculations are needed. We slightly change our notation, denoting now by $f(S,A,I) = (f_1(S,A,I), f_2(S,A,I), f_3(S,A,I))^T$ both systems' (1) and (23) right-hand sides, and by Df their Jacobian.

4.1. Application of Sotomayor's Theorem for Model (1)

We now determine the second partial derivatives of f with respect to the variables S, A and I, i.e., the elements of $D^2 f$:

$$\frac{\partial^2 f_1}{\partial S^2} = -2c_{SS}, \quad \frac{\partial^2 f_1}{\partial I^2} = -\frac{2\alpha\beta^2 SA}{(1+\beta I)^3}, \quad \frac{\partial^2 f_1}{\partial S\partial A} = \frac{\partial^2 f_1}{\partial A\partial S} = -c_{SA} - \frac{\alpha}{1+\beta I},$$
$$\frac{\partial^2 f_1}{\partial S\partial I} = \frac{\partial^2 f_1}{\partial I\partial S} = -c_{SI} + \frac{\alpha\beta A}{(1+\beta I)^2}, \quad \frac{\partial^2 f_1}{\partial A\partial I} = \frac{\partial^2 f_1}{\partial I\partial A} = \frac{\alpha\beta S}{(1+\beta I)^2}, \quad \frac{\partial^2 f_2}{\partial A^2} = -2c_{AA},$$
$$\frac{\partial^2 f_2}{\partial I^2} = \frac{2\alpha\beta^2 SA}{(1+\beta I)^3}, \quad \frac{\partial^2 f_2}{\partial S\partial A} = \frac{\partial^2 f_2}{\partial A\partial S} = -c_{AS} + \frac{\alpha}{1+\beta I}, \quad \frac{\partial^2 f_2}{\partial S\partial I} = \frac{\partial^2 f_2}{\partial I\partial S} = -\frac{\alpha\beta A}{(1+\beta I)^2},$$
$$\frac{\partial^2 f_2}{\partial A\partial I} = \frac{\partial^2 f_2}{\partial I\partial A} = -c_{AI} - \frac{\alpha\beta S}{(1+\beta I)^2}, \quad \frac{\partial^2 f_3}{\partial I^2} = -2c_{II}, \quad \frac{\partial^2 f_3}{\partial S\partial I} = \frac{\partial^2 f_3}{\partial I\partial S} = -c_{IS},$$
$$\frac{\partial^2 f_3}{\partial A\partial I} = \frac{\partial^2 f_3}{\partial I\partial A} = -c_{IA}, \quad \frac{\partial^2 f_1}{\partial A^2} = \frac{\partial^2 f_2}{\partial S^2} = \frac{\partial^2 f_3}{\partial S^2} = \frac{\partial^2 f_3}{\partial A^2} = \frac{\partial^2 f_3}{\partial S\partial A} = \frac{\partial^2 f_3}{\partial A\partial S} = 0.$$

Furthermore, by differentiating the components of f with respect to r, we find

$$f_r = \left[\frac{\partial f_1}{\partial r}, \frac{\partial f_2}{\partial r}, \frac{\partial f_3}{\partial r}\right]^T = [S+A, 0, 0]^T,$$

whose Jacobian, differentiating with respect to the variables S, A and I, is

$$Df_r = \begin{bmatrix} 1 & 1 & 0 \\ 0 & 0 & 0 \\ 0 & 0 & 0 \end{bmatrix}. \tag{59}$$

4.1.1. Transcritical Bifurcation $E_0 \to E_S$

Consider the equilibrium point E_0 and choose r as the bifurcation parameter. Evaluating the Jacobian at the equilibrium E_0, we obtain

$$J = Df(E_0, r) = \begin{bmatrix} r-m & m & 0 \\ 0 & -m-\pi & 0 \\ 0 & \pi & -m-\mu \end{bmatrix},$$

having the eigenvalues $\lambda_1 = r - m$, $\lambda_2 = -m - \pi$ and $\lambda_3 = -m - \mu$; thus, two eigenvalues have a negative real part and the first one vanishes by taking as the critical bifurcation value

$$r_0 = m. \tag{60}$$

Its right v and left w eigenvectors corresponding to the zero eigenvalue are, therefore

$$v = [1, 0, 0]^T, \quad w = \left[\frac{m+\pi}{m}, 1, 0\right]^T.$$

The only nonvanishing terms in $D^2 f(E_0, m)$ are

$$\begin{gathered} \frac{\partial^2 f_1}{\partial S^2} = -2c_{SS}, \quad \frac{\partial^2 f_1}{\partial S \partial A} = \frac{\partial^2 f_1}{\partial A \partial S} = -c_{SA} - \alpha, \quad \frac{\partial^2 f_1}{\partial S \partial I} = \frac{\partial^2 f_1}{\partial I \partial S} = -c_{SI}, \\ \frac{\partial^2 f_2}{\partial A^2} = -2c_{AA}, \quad \frac{\partial^2 f_2}{\partial S \partial A} = \frac{\partial^2 f_2}{\partial A \partial S} = -c_{AS} + \alpha, \quad \frac{\partial^2 f_2}{\partial A \partial I} = \frac{\partial^2 f_2}{\partial I \partial A} = -c_{AI}, \\ \frac{\partial^2 f_3}{\partial I^2} = -2c_{II}, \quad \frac{\partial^2 f_3}{\partial S \partial I} = \frac{\partial^2 f_3}{\partial I \partial S} = -c_{IS}, \quad \frac{\partial^2 f_3}{\partial A \partial I} = \frac{\partial^2 f_3}{\partial I \partial A} = -c_{IA}. \end{gathered} \tag{61}$$

Furthermore, recalling (59), we have

$$f_r(E_0, m) = [0, 0, 0]^T, \quad Df_r(E_0, m) = Df_r.$$

Finally, the components of $D^2 f(E_0, m)(v, v)$ are

$$\begin{aligned} D^2 f_1(E_0, m)(v, v) &= -2c_{SS} v_1^2 - 2(c_{SA} + \alpha) v_1 v_2 - 2c_{SI} v_1 v_3, \\ D^2 f_2(E_0, m)(v, v) &= -2c_{AA} v_2^2 - 2(c_{AS} - \alpha) v_1 v_2 - 2c_{AI} v_2 v_3, \\ D^2 f_3(E_0, m)(v, v) &= -2c_{II} v_3^2 - 2c_{IS} v_1 v_3 - 2c_{IA} v_2 v_3. \end{aligned}$$

Thus, the three conditions required by Sotomayor's Theorem for a transcritical bifurcation are met; indeed

$$\begin{aligned} w^T f_r(E_0, m) = 0, \quad w^T [Df_r(E_0, m) v] &= w_1 v_1 = \frac{m+\pi}{m} \neq 0, \\ w^T \left[D^2 f(E_0, m)(v, v)\right] &= -2c_{SS} w_1 v_1^2 = -\frac{2c_{SS}(m+\pi)}{m} \neq 0. \end{aligned}$$

4.1.2. Transcritical Bifurcation $E_S \to E_{SAI}$

Now consider the E_S and choose r as the bifurcation parameter.

The Jacobian evaluated at E_S is

$$J = Df(E_S, r) = \begin{bmatrix} -\dfrac{c_{SS}(m+\pi)}{\alpha - c_{AS}} & J_{1,2} & -\dfrac{c_{IS}(m+\pi)}{\alpha - c_{AS}} \\ 0 & J_{2,2} & 0 \\ 0 & \pi & -\dfrac{(\alpha - c_{AS})(m+\mu) + c_{IS}(m+\pi)}{\alpha - c_{AS}} \end{bmatrix},$$

with

$$J_{1,2} = \frac{(c_{SS} - c_{AS} - c_{SA})m + (c_{SS} - c_{SA} - \alpha)\pi}{\alpha - c_{AS}}, \quad J_{2,2} = -m - \pi + (\alpha - c_{AS})\frac{r - m}{c_{SS}}.$$

Its eigenvalues are

$$\lambda_1 = -\frac{c_{SS}(m+\pi)}{\alpha - c_{AS}}, \quad \lambda_2 = J_{22}, \quad \lambda_3 = -\frac{(\alpha - c_{AS})(m+\mu) + c_{IS}(m+\pi)}{\alpha - c_{AS}}.$$

Now, λ_2 vanishes by choosing the critical value

$$r^{\dagger} = m + \frac{c_{SS}(m+\pi)}{\alpha - c_{AS}}, \tag{62}$$

while the remaining ones are negative, if $\alpha > c_{AS}$; or the first one is positive, if $\alpha < c_{AS}$. The left eigenvector is $w = [0,1,0]^T$, while the right one is

$$v = \begin{bmatrix} \frac{1}{c_{SS}(m+\pi)}\left(\frac{((c_{SS}-c_{AS}-c_{SA})m+(c_{SS}-c_{SA}-\alpha)\pi)((\alpha-c_{AS})(m+\mu)+c_{IS}(m+\pi))}{(\alpha-c_{AS})\pi} - c_{SI}(m+\pi)\right) \\ \dfrac{(\alpha - c_{AS})(m+\mu) + c_{IS}(m+\pi)}{(\alpha - c_{AS})\pi} \\ 1 \end{bmatrix}.$$

Recalling (59), we further have

$$f_r\left(E_S, r^{\dagger}\right) = \left[\frac{m+\pi}{\alpha - c_{AS}}, 0, 0\right]^T, \quad Df_r\left(E_S, r^{\dagger}\right) = Df_r.$$

However, in spite of having $w^T f_r(E_S, r^{\dagger}) = 0$, Sotomayor's Theorem is inconclusive because the last condition is not satisfied, namely $w^T[Df_r(E_S, r^{\dagger})v] = 0$.

Remark 1. *This calculation is very interesting, because in spite of the fact that the analysis is undecided concerning the existence of the transcritical bifurcation, the simulations below will show that it does indeed take place. Thus, it is an example of the fact that the conditions in Sotomayor's Theorem are sufficient but not necessary.*

4.2. Application of Sotomayor's Theorem for Model (23)

The proof follows pretty much the one of model (1). We outline only the basic changes. Once again, we use the same notation for f and Df, which, here, denote the right hand side and the Jacobian of (23). The only changes in the Jacobian are the elements

$$\frac{\partial f_1}{\partial A} = -c_{SA}S - \frac{\alpha S}{1+\beta I}, \quad \frac{\partial f_2}{\partial A} = r - m - 2c_{AA}A - c_{AS}S - c_{AI}I + \frac{\alpha S}{1+\beta I} - \pi.$$

It further turns out that $D^2 f$ is the same as the one of model (1). Here, instead, we find

$$f_r = \left[\frac{\partial f_1}{\partial r}, \frac{\partial f_2}{\partial r}, \frac{\partial f_3}{\partial r}\right]^T = [S, A, 0]^T,$$

whose Jacobian—differentiating with respect to the variables S, A and I—is now

$$Df_r = \begin{bmatrix} 1 & 0 & 0 \\ 0 & 1 & 0 \\ 0 & 0 & 0 \end{bmatrix}.$$

4.2.1. Transcritical Bifurcation $E_0 \to E_S$

We consider the equilibrium point E_0 and we choose r as the bifurcation parameter. By evaluating the Jacobian, we obtain

$$J = Df(E_0, r) = \begin{bmatrix} r - m & 0 & 0 \\ 0 & -\pi & 0 \\ 0 & \pi & -m-\mu \end{bmatrix},$$

having the eigenvalues $\lambda_1 = r - m$, $\lambda_2 = -\pi$ and $\lambda_3 = -m - \mu$, two of them being negative. The first one vanishes by taking, again, $r_0 = m$ as in (60), the same the critical bifurcation threshold as for model (1). The left and right eigenvectors are, respectively

$$v = [1,0,0]^T, \quad w = [1,0,0]^T.$$

The elements of $D^2 f(E_0, r_0)$ are the same as the first model, because $D^2 f$ is the same; consequently, also, are the components of

$$D^2 f(E_0, r_0)(v,v) = \left[D^2 f_1(E_0, r_0)(v,v), D^2 f_2(E_0, r_0)(v,v), D^2 f_3(E_0, r_0)(v,v)\right]^T.$$

Furthermore, recalling (59), we have

$$f_r(E_0, r_0) = [0,0,0]^T, \quad Df_r(E_0, r_0) = Df_r.$$

Thus, the three conditions required by Sotomayor's Theorem are met; indeed

$$w^T f_r(E_0, m) = 0, \quad w^T[Df_r(E_0, m)v] = w_1 v_1 = 1 \neq 0,$$
$$w^T\left[D^2 f(E_0, m)(v,v)\right] = -2c_{SS} w_1 v_1^2 = -2c_{SS} \neq 0.$$

Hence, at $r = r_0 = m$, there is a transcritical bifurcation for which E_0 becomes E_S.

4.2.2. Transcritical Bifurcation $E_S \to E_{SAI}$

We consider the equilibrium point E_S and we choose r as the bifurcation parameter. From the evaluation of the Jacobian, for

$$c_{AS} \neq c_{SS} + \alpha \tag{63}$$

we obtain

$$J = Df(E_S, r) = \begin{bmatrix} -\dfrac{c_{SS}\pi}{c_{SS} - c_{AS} + \alpha} & -\dfrac{(c_{AS} + \alpha)\pi}{c_{SS} - c_{AS} + \alpha} & -\dfrac{c_{IS}\pi}{c_{SS} - c_{AS} + \alpha} \\ 0 & J_{22} & 0 \\ 0 & \pi & -m - \mu - \dfrac{c_{IS}\pi}{c_{SS} - c_{AS} + \alpha} \end{bmatrix},$$

having the eigenvalues

$$\lambda_1 = -\frac{c_{SS}\pi}{c_{SS} - c_{AS} + \alpha}, \quad \lambda_2 = J_{22}, \quad \lambda_3 = -m - \mu - \frac{c_{IS}\pi}{c_{SS} - c_{AS} + \alpha}.$$

The second eigenvalue vanishes by taking as the critical bifurcation threshold

$$r^{\ddagger} = m + \frac{c_{SS}\pi}{c_{SS} - c_{AS} + \alpha} > m, \tag{64}$$

the latter inequality following by requiring $\lambda_1 < 0$. Note that $r^{\dagger}$ for model (1) and $r^{\ddagger}$ for model (23) differ; compare (62) and (64). The left and right eigenvectors are, respectively

$$v = \begin{bmatrix} \frac{c_{SI}}{c_{SS}} - \frac{c_{AS}+\alpha}{c_{SS}}\left(\frac{m+\mu}{\pi} + \frac{c_{IS}}{c_{SS}-c_{AS}+\alpha}\right) \\ \frac{m+\mu}{\pi} + \frac{c_{IS}}{c_{SS}-c_{AS}+\alpha} \\ 1 \end{bmatrix}, \quad w = [0,1,0]^T.$$

The nonvanishing elements of $D^2 f(E_S, r^{\ddagger})$ are exactly those of (61) with the exception of

$$\frac{\partial^2 f_1}{\partial A \partial I} = \frac{\partial^2 f_1}{\partial I \partial A} = \frac{\alpha\beta\pi}{c_{SS}-c_{AS}+\alpha}, \quad \frac{\partial^2 f_2}{\partial A \partial I} = \frac{\partial^2 f_2}{\partial I \partial A} = -c_{AI} - \frac{\alpha\beta\pi}{c_{SS}-c_{AS}+\alpha}, \quad \frac{\partial^2 f_3}{\partial I^2} = -2c_{II}.$$

Further, recalling once again (59), we have

$$f_r\left(E_S, r^{\ddagger}\right) = \left[\frac{\pi}{c_{SS}-c_{AS}+\alpha}, 0, 0\right]^T, \quad Df_r\left(E_S, r^{\ddagger}\right) = Df_r.$$

Finally, the components of

$$D^2 f\left(E_S, r^{\ddagger}\right)(v,v) = \left[D^2 f_1\left(E_S, r^{\ddagger}\right)(v,v), D^2 f_2\left(E_S, r^{\ddagger}\right)(v,v), D^2 f_3\left(E_S, r^{\ddagger}\right)(v,v)\right]^T$$

are

$$\begin{aligned} D^2 f_1\left(E_S, r^{\ddagger}\right)(v,v) &= -2c_{SS}v_1^2 - 2(c_{SA}+\alpha)v_1v_2 - 2c_{SI}v_1v_3 + \frac{2\alpha\beta\pi}{c_{SS}-c_{AS}+\alpha}v_2v_3, \\ D^2 f_2\left(E_S, r^{\ddagger}\right)(v,v) &= -2c_{AA}v_2^2 - 2(c_{AS}-\alpha)v_1v_2 - 2\left(c_{IA} + \frac{\alpha\beta\pi}{c_{SS}-c_{AS}+\alpha}\right)v_2v_3, \\ D^2 f_3\left(E_S, r^{\ddagger}\right)(v,v) &= -2c_{II}v_3^2 - 2c_{IS}v_1v_3 - 2c_{IA}v_2v_3. \end{aligned}$$

The first condition required to use Sotomayor's Theorem is $w^T f_r(E_S, r^{\ddagger}) = 0$. We would also need the nonvanishing of the following quantities:

$$w^T\left[Df_r\left(E_S, r^{\ddagger}\right)v\right] = v_2 = \frac{m+\mu}{\pi} + \frac{c_{IS}}{c_{SS}-c_{AS}+\alpha}, \tag{65}$$

$$\begin{aligned} w^T\left[D^2 f\left(E_S, r^{\ddagger}\right)(v,v)\right] &= D^2 f_2\left(E_S, r^{\ddagger}\right)(v,v) = -2c_{AA}\left(\frac{m+\mu}{\pi} + \frac{c_{IS}}{c_{SS}-c_{AS}+\alpha}\right)^2 \\ &-2(c_{AS}-\alpha)\left(\frac{c_{SI}}{c_{SS}} - \frac{c_{AS}+\alpha}{c_{SS}}\left(\frac{m+\mu}{\pi} + \frac{c_{IS}}{c_{SS}-c_{AS}+\alpha}\right)\right)\left(\frac{m+\mu}{\pi} + \frac{c_{IS}}{c_{SS}-c_{AS}+\alpha}\right) \\ &-2\left(c_{IA} + \frac{\alpha\beta\pi}{c_{SS}-c_{AS}+\alpha}\right)\left(\frac{m+\mu}{\pi} + \frac{c_{IS}}{c_{SS}-c_{AS}+\alpha}\right). \end{aligned} \tag{66}$$

The first one is satisfied and shows that either a transcritical or a pitchfork bifurcation is possible. The remaining ones are needed for a transcritical bifurcation. Thus, if these quantities are different from 0 for $r = r^{\ddagger}$, there is a transcritical bifurcation from E_S to E_{SAI}.

In case (66) vanishes, we further investigate the situation. We can observe that (65) is zero if and only if

$$\pi = -\frac{(m+\mu)(c_{SS}-c_{AS}+\alpha)}{c_{IS}}, \tag{67}$$

while (66) is zero if and only if either (67) holds or

$$c_{AA} = \frac{\pi(c_{SS} - c_{AS} + \alpha)}{(m+\mu)(c_{SS} - c_{AS} + \alpha) + \pi c_{IS}} \times \left\{ (\alpha - c_{AS}) \left[\frac{c_{SI}}{c_{SS}} - \frac{c_{AS} + \alpha}{c_{SS}} \left(\frac{m+\mu}{\pi} + \frac{c_{IS}}{c_{SS} - c_{AS} + \alpha} \right) \right] - c_{AI} - \frac{\alpha\beta\pi}{c_{SS} - c_{AS} + \alpha} \right\}, \quad (68)$$

and, in this second case, we must assume

$$\pi \neq -\frac{(m+\mu)(c_{SS} - c_{AS} + \alpha)}{c_{IS}}. \quad (69)$$

Now, because $w = [0,1,0]^T$, the third derivatives simplify, namely

$$w^T[D^3 f(E_S, r^\ddagger)(v,v,v)] = D^3 f_2(E_S, r^\ddagger)(v,v,v).$$

The only nonvanishing third partial derivatives of f_2 are

$$\frac{\partial^3 f_2}{\partial I^3} = -\frac{6\alpha\beta^3 SA}{(1+\beta I)^4},$$

$$\frac{\partial^3 f_2}{\partial I^2 \partial S} = \frac{\partial^3 f_2}{\partial S \partial I^2} = \frac{\partial^3 f_2}{\partial I \partial S \partial I} = \frac{2\alpha\beta^2 A}{(1+\beta I)^3}, \quad \frac{\partial^3 f_2}{\partial I^2 \partial A} = \frac{\partial^3 f_2}{\partial A \partial I^2} = \frac{\partial^3 f_2}{\partial I \partial A \partial I} = \frac{2\alpha\beta^2 S}{(1+\beta I)^3}$$

and

$$\frac{\partial^3 f_2}{\partial S \partial A \partial I} = \frac{\partial^3 f_2}{\partial S \partial I \partial A} = \frac{\partial^3 f_2}{\partial A \partial S \partial I} = \frac{\partial^3 f_2}{\partial A \partial I \partial S} = \frac{\partial^3 f_2}{\partial I \partial S \partial A} = \frac{\partial^3 f_2}{\partial I \partial A \partial S} = -\frac{\alpha\beta}{(1+\beta I)^2}.$$

Of these, upon evaluation at $(E_S, r^\ddagger)$, the only nonvanishing ones are those in the last two groups, namely

$$\frac{\partial^3 f_2(E_S, r^\ddagger)}{\partial I^2 \partial A} = \frac{\partial^3 f_2(E_S, r^\ddagger)}{\partial A \partial I^2} = \frac{\partial^3 f_2(E_S, r^\ddagger)}{\partial I \partial A \partial I} = \frac{2\alpha\beta^2\pi}{c_{SS} - c_{AS} + \alpha}$$

and

$$\frac{\partial^3 f_2(E_S, r^\ddagger)}{\partial S \partial A \partial I} = \frac{\partial^3 f_2(E_S, r^\ddagger)}{\partial S \partial I \partial A} = \frac{\partial^3 f_2(E_S, r^\ddagger)}{\partial A \partial S \partial I} = \frac{\partial^3 f_2(E_S, r^\ddagger)}{\partial A \partial I \partial S} = \frac{\partial^3 f_2(E_S, r^\ddagger)}{\partial I \partial S \partial A} = \frac{\partial^3 f_2(E_S, r^\ddagger)}{\partial I \partial A \partial S} = -\alpha\beta.$$

Consequently

$$w^T[D^3 f(E_S, r^\ddagger)(v,v,v)] = D^3 f_2(E_S, r^\ddagger)(v,v,v) = \sum_{j_1, j_2, j_3 = 1}^{3} \frac{\partial^3 f_2(E_S, r^\ddagger)}{\partial x_{j_1} \partial x_{j_2} \partial x_{j_3}} v_{j_1} v_{j_2} v_{j_3}$$
$$= 3\frac{\partial^3 f_2(E_S, r^\ddagger)}{\partial A \partial I^2} v_2 v_3^2 + 6\frac{\partial^3 f_2(E_S, r^\ddagger)}{\partial S \partial A \partial I} v_1 v_2 v_3 = 6\alpha\beta v_2 v_3 \left(\frac{\beta\pi}{c_{SS} - c_{AS} + \alpha} v_3 - v_1 \right).$$

Substituting into this expression the components of v we explicitly have

$$w^T[D^3 f(E_S, r^\ddagger)(v,v,v)] = 6\alpha\beta \left(\frac{m+\mu}{\pi} + \frac{c_{IS}}{c_{SS} - c_{AS} + \alpha} \right) \times \left[\frac{\beta\pi}{c_{SS} - c_{AS} + \alpha} - \frac{c_{SI}}{c_{SS}} + \frac{c_{AS} + \alpha}{c_{SS}} \left(\frac{m+\mu}{\pi} + \frac{c_{IS}}{c_{SS} - c_{AS} + \alpha} \right) \right]. \quad (70)$$

Still assuming (63),we can observe that (70) is zero if and only if (67) holds or, alternatively

$$\beta = \frac{c_{SI}c_{SS}\pi - (c_{AS} - \alpha)c_{SI}\pi - (m + \mu)(c_{AS} + \alpha)(c_{SS} - c_{AS} + \alpha) - c_{IS}c_{SS}\pi}{c_{SS}\pi^2}. \quad (71)$$

In summary, for the case (63), since Sotomayor's Theorem gives only sufficient conditions, we cannot conclude anything about the bifurcation from E_S for $r = r^{\ddagger}$ being a saddle-node. Instead, a sufficient condition for a transcritical bifurcation to exist is that the conditions (67) and (68) are both not satisfied; alternatively, a pitchfork bifurcation occurs if (67) and (71) are not satisfied but (68) is verified.

We, finally, also investigate the case for which (63) does not hold, i.e., $c_{SS} - c_{AS} + \alpha = 0$. The Jacobian evaluated at E_S gives $\widehat{J}_{22} = -\pi$. Taking now π as the bifurcation parameter, with threshold value $\pi_0 = 0$, we find the right and left eigenvectors $v = [v_1, v_2, 0]^T$ $w = [0, 1, 0]^T$, with

$$v_1 = -\frac{c_{SS}}{c_{SA} + \alpha} < 0, \quad v_2 = 1 > 0.$$

In addition, calculating the derivative with respect to the bifurcation parameter, $f_\pi = [0, -A, A]^T$ so that $f_\pi(E_S, \pi) = [0, 0, 0]^T$ and $w^T f_\pi(E_S, \pi) = 0$. Upon evaluation of the Jacobian Df_π, it also follows that $w^T Df_\pi(E_S, \pi)v = [0, -1, 0]^T v = -v_2 > 0$. Since $w_1 = w_3 = 0$, it is enough to calculate just the partial derivatives of the second component of f:

$$w^T D^2 f(v, v) = \frac{\partial^2 f_2}{\partial S^2} v_1^2 + 2\frac{\partial^2 f_2}{\partial A \partial S} v_1 v_2 + \frac{\partial^2 f_2}{\partial A^2} v_2^2 = -2c_{AS} v_1 v_2 - 2c_{AA} A v_2^2$$

so that

$$w^T D^2 f(v, v)|_{(E_S, \pi)} = -2c_{AS} v_1 v_2 > 0$$

and, therefore, there is also no transcritical bifurcation in this case.

Remark 2. *Note that if we try to apply the same technique to the situation* $\alpha - c_{AS} = 0$ *for (1), we obtain* $J_{22} = -m - \pi$, *so that in this case the threshold value for the bifurcation parameter* π *would be negative,* $\pi_0^{(1)} = -m < 0$, *and, therefore, not biologically feasible.*

4.3. Numerical Simulations for the Bifurcations

In addition, we also show numerically the two transcritical bifurcation diagrams, indicating in particular that the coexistence equilibrium in both models originates from the disease-free equilibrium, which, in turn, arises from the origin when the population reproduction rate overcomes its mortality rate, as it also appears from the theoretical analysis. Note that Figures 8 and 9 qualitatively appear to be the same, but their vertical axes differ quite a bit.

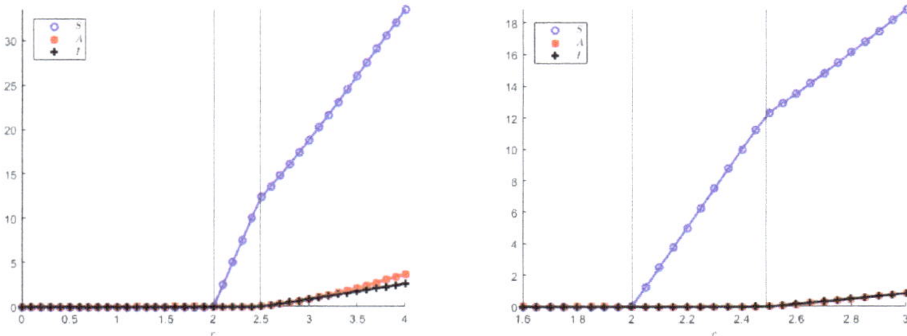

Figure 8. (**Left**): transcritical bifurcations for model (1) obtained with parameter values (72), (55) and initial conditions (50). (**Right**): zoom of the left image showing two bifurcations as r changes; the first from E_0 to E_S when $r = r_0 = 2$ and the second from E_S to E_{SAI} when $r = r^{\dagger} = 2.4898$.

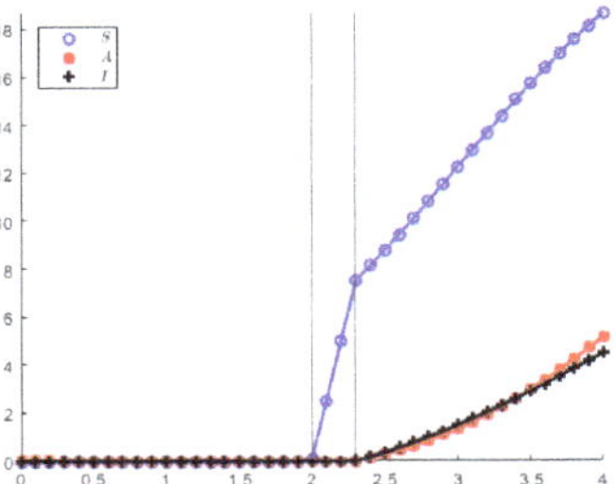

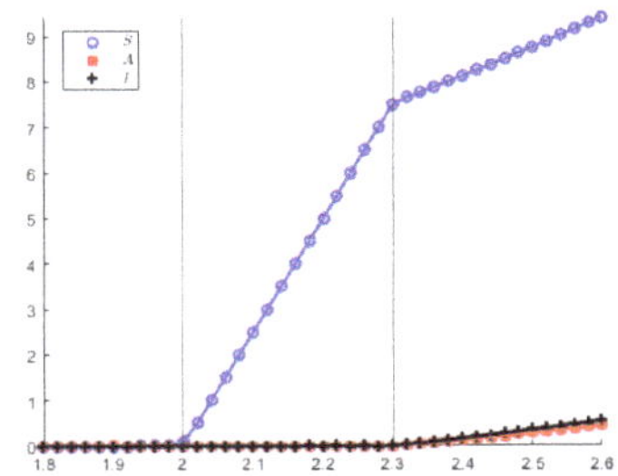

Figure 9. (**Left**): transcritical bifurcations for model (23) obtained with parameter values (72), (55) and initial conditions (50). (**Right**): zoom of the left image showing two bifurcations as r changes; the first one from E_0 to E_S when $r = r_0 = 2$ and the second from E_S to E_{SAI} when $r = r^{\ddagger} = 2.3019$.

The transcritical bifurcation parameters for both model (1) and (23), still using (55) are:

$$c_{SS} = 0.04, \quad c_{AA} = 0.05, \quad c_{II} = 0.06, \quad c_{SA} = 0.08, \quad c_{SI} = 0.03, \\ c_{AS} = 0.01, \quad c_{AI} = 0.02, \quad c_{IS} = 0.09, \quad c_{IA} = 0.07. \tag{72}$$

5. Discussion

This investigation has been prompted by the recent COVID-19 pandemic, in which, at least at its outbreak, the role of asymptomatic individuals was apparently fundamental. We consider the general population response to such an event. The now classical Capasso–Serio model [9] was the first to encompass this feature. Although more generally formulated, it also contains three compartments: susceptibles, infected and removed.

The specific form for the model [9] that we adopt in our simulations for comparison purposes is the following:

$$\begin{aligned} \frac{dS}{dt} &= -\frac{\alpha SI}{1+\beta I^2}, \\ \frac{dI}{dt} &= \frac{\alpha SI}{1+\beta I^2} - \gamma I, \\ \frac{dR}{dt} &= \gamma I. \end{aligned} \tag{73}$$

As already mentioned in the Introduction, the SAI models (1) and (23) proposed here from the epidemiological viewpoint are an extension of the classical SI model or as an SIR model in which R stands for removed rather than recovered. Since, in the SI case, removed do not appear, in a sense (1) and (23) share its properties. Introducing the asymptomatics A in (1) and (23), the infected are split among A and symptomatics I. System (1) or (23) can be compared with the Capasso–Serio model by means of the following matches:

$$S : (1) \to S : (73), \quad \text{(the susceptible classes)}$$
$$A : (1) \to I : (73), \quad \text{(the classes that can transmit the disease)}$$
$$I : (1) \to R : (73), \quad \text{(removed from circulation and unable to transmit the disease).}$$

In (1) and (23), I are recognized as disease carriers. In order to possibly not get the disease, the susceptibles take the size of symptomatic I as an index by which to measure the reduction in their contacts with all other individuals. In (73), instead, both I and R are recognized as disease carriers, but R are isolated and cannot produce new cases of the disease. Here, the contact reductions are based on the size of the infected class I. Both I for (1) and (23) as well as R for (73), represent sinks for the dynamical system. In the comparison between (1) (or (23)) and (73), only the people that can spread the disease are relevant, respectively, A and I. However, the important point is that the people can only react to the infected they see, i.e., I in both models. Note, indeed, that in spite of the above matching among compartments, we cannot completely identify the asymptomatics A of our case with the infected in [9], nor our symptomatic individuals I with the removed of [9], because the "fear" response function in our model depends inversely on symptomatics, while in [9], it does on the infected class. In [9], this dependence has to be quadratic to push the disease transmission to zero, meaning a large contact reduction for a large number of infected. There would (almost) be symmetrization if, in [9], the fear would be induced by the removed. However, it is clearly assumed in [9] that in such a case the infected are recognizable as disease carriers, in contrast to the asymptomatics of the model (1) (or (23)), and, therefore, are seen by the susceptibles as potential contagion sources.

We now compare the two models, ours and [9], in several situations. Firstly, the full cases, respectively, SAI and SIR. In the following figures, we show comparisons between these compartments in [9] and our models. Note that to make the comparison fair in our models, we set all demographic parameters of type c_{EB}, with $E, B \in \{S, I, A\}$, to zero, because in [9] no demographics is present. In this case, the models (1) and (23) coincide, so from now on we can just refer to one of them. The remaining hypothetical reference parameter values are the following

$$\alpha = 0.5, \quad \beta = 1, \quad \mu = 0, \quad m = 0, \quad r = 0. \tag{74}$$

In both Figures 10 and 11, while in the classical model (73) ultimately almost the whole population is affected, a good number of susceptibles are preserved in the SAI model (1). For the (73) model, the higher the removal rate γ, the faster the disease affects the population, Figure 10. Instead, for the SAI model, the higher the progression rate π, the higher the number of susceptibles that are preserved from the disease. A similar effect is noted if the disease contact rate α increases, Figure 11.

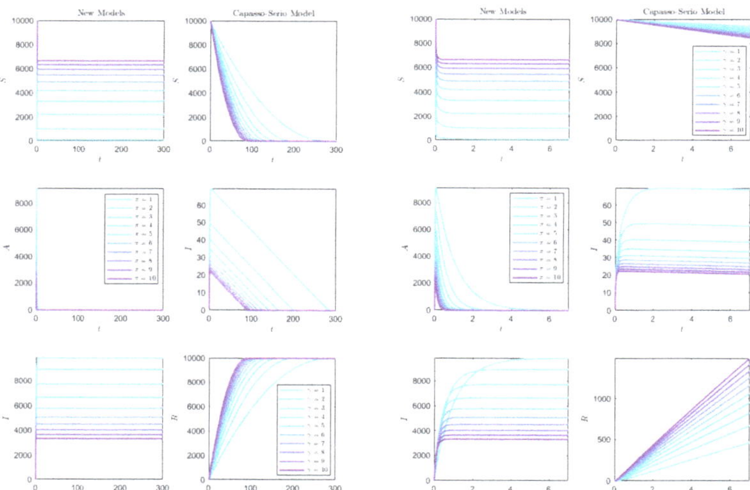

Figure 10. Here, the disease transmission rate is fixed and lower than the progression to the symptomatic/removed classes, with $\alpha = 0.5 < \pi = \gamma \in [1, \ldots, 10]$. Comparison between (1) (or equivalently (23)) on the left column, and the Capasso–Serio model (73) on the right column, in terms of the progression from asymptomatic to symptomatic π for (1) (as well as (23)) and of the removal rate γ for (73), with parameter values (74) and initial conditions (50). Left frame: the simulations to show the settling of the systems. Right frame: the blow up of the initial instants to better show the transients.

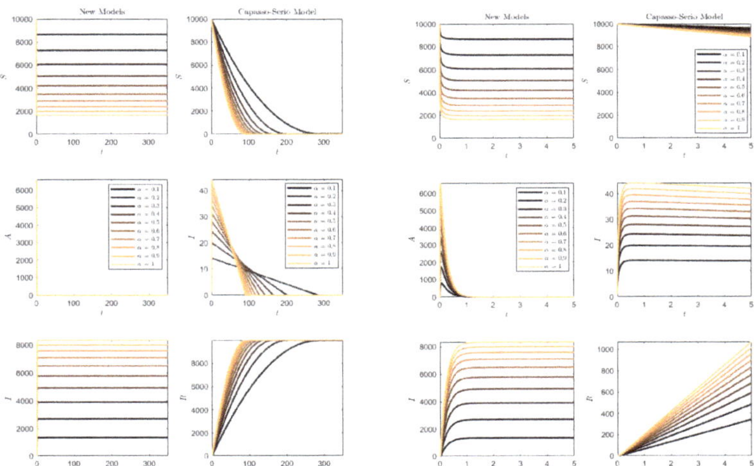

Figure 11. Here, $\pi = \gamma = 5$. Comparison between (1) (or, equivalently, (23)) on the left column, and the Capasso–Serio model (73) on the right column, in terms of the disease transmission rate $\alpha \in [0.1, \ldots, 1.0]$ so that, again, it is below the progression to symptomatic or removal rates, $\alpha < \pi = \gamma$ with parameter values (74) and initial conditions (50). Left frame: the simulations to show the settling of the systems. Right frame: the blow up of the initial instants to better show the transients.

Figure 12 contains the simulation of the opposite conditions: higher transmission rate than progression/removal rates. In this situation, the susceptibles of the SAI model (1) are quickly depleted and the whole population quickly becomes symptomatic. For (73), this occurs much more slowly, and the slower the smaller the transmission rate is.

We next compare the two types of models as different versions of the SI model, with just asymptomatics in (1) and recognizable infected for (73). We, thus, take $\pi = \gamma = 0$ to prevent progression respectively to symptomatics or removed. Figure 13 contains the results of the simulations in this case. In the case of model (73), the R compartment is initially empty and, in this case, clearly remains empty throughout the simulation. For the SAI model, there is no progression to symptomatics and, also, this compartment is empty. It is interesting to note that the lower values of α once again have a delaying effect on the epidemics' propagation for (73), the more marked the lower the value of the rate α. However, in the SAI model, everyone is very quickly affected and the susceptibles are quickly depleted. The important remark is that the behavior remains in agreement with the ones found in the former, Figure 12, since $\pi = 0 < \alpha$.

We also compared the two models (1) and (73), interpreting the former as an extension of the SI model, allowing in it two classes of individuals affected by the disease: asymptomatic and symptomatic. This is performed by allowing progression from A to I, while there is no removal rate in the classical model. Hence, $\pi \neq 0$ and $\gamma = 0$. Figure 14 shows the results. It is clearly seen that for $\alpha < \pi$, the SAI model preserves again part of the susceptibles, while this does not occur for (73), and for $\alpha > \pi$ in both models the whole population is affected.

We finally consider a less relevant simulation, for the reverse case $\pi = 0$ and $\gamma = 0.05$. This is not a fair comparison as on one hand we have an SA model (the SI model with all asymptomatics) versus a full SIR model. In the former case, everybody becomes asymptomatic and in the latter, almost everyone also contracts the disease, within a timespan that is longer the lower the contact rate. The results are not shown, as they coincide with Figure 13.

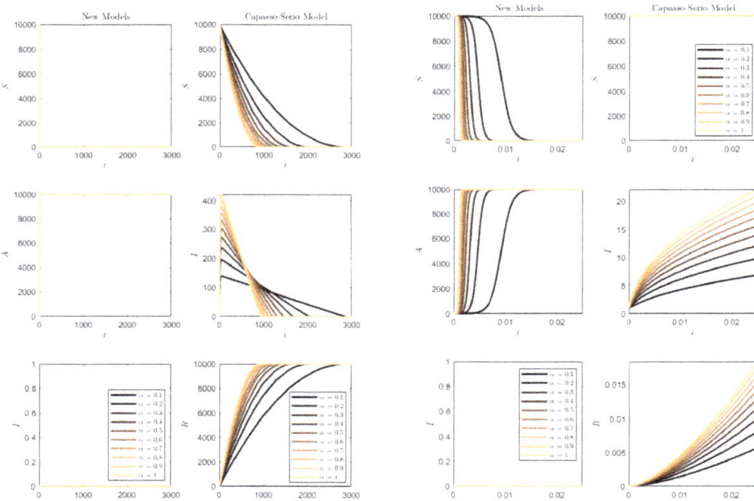

Figure 12. Here, we illustrate the case $0 = \pi < \alpha \in [0.1 \ldots, 1.0]$, $\gamma = 0.05$. In this case, in the SAI model, the whole population is quickly affected, while for (73), the disease propagates at a lower speed with a lower transmission rate. In the long run, here as well, the susceptible class is depleted. Left frame: the simulations to show the settling of the systems. Right frame: the time interval is shorter to better show the transients in the SAI model. Other parameter values are (74) and initial conditions (50).

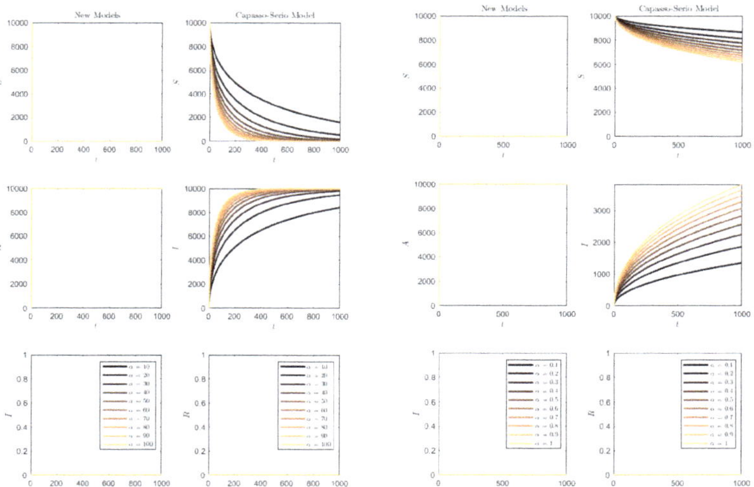

Figure 13. Here, $\pi = \gamma = 0$, so that both systems become SI models. Comparison between (1) (or, equivalently, (23)) on the left column, and the Capasso–Serio model (73), on the right column. Left frame: $\alpha \in [10, \ldots, 100]$. Right frame: $\alpha \in [0.1, \ldots, 1.0]$. The other parameter values are given in (74) and initial conditions (50).

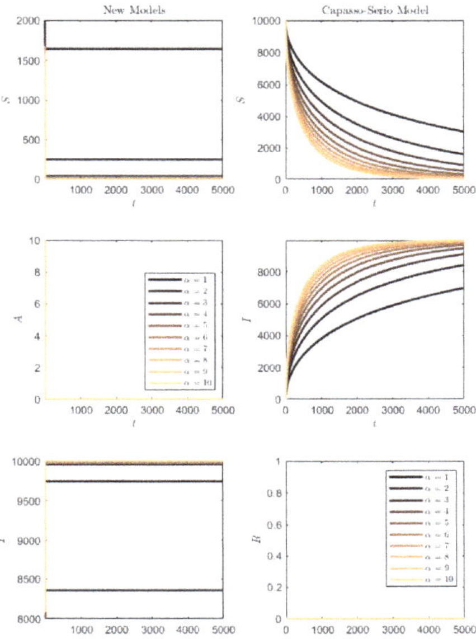

Figure 14. Here, $\pi = 5, \gamma = 0$ and $\alpha \in [1.0, 10]$. It is clearly seen that for $\alpha < 5$ in the SAI model the susceptibles are preserved, while for transmission rates higher than this threshold, $\alpha > \pi = 5$, all individuals eventually become symptomatic. The same occurs in the (73) model independently of α, at a lower pace. Other parameter values are (74) and initial conditions (50).

The simulations for model (1), or equivalently (23), show a much higher impact of the disease. For increasing disease contact rate, the susceptibles are very much reduced, and symptomatic people rise to high values, Figure 11; instead, the asymptomatic ones quickly vanish. Higher progression rates from asymptomatic to symptomatic are beneficial, because they increase the number of susceptibles and sensibly reduce the symptomatic individuals.

6. Conclusions

In this paper, we analysed a simple disease transmission system with some demographic features. The illness is assumed to develop at first in an asymptomatic form. The model accounts for epidemic-induced fear in the population, for which measures are taken to reduce contacts. The main novelty is represented by the fact that susceptibles respond not to a large number of asymptomatic infected, but just to the size of the symptomatic individuals compartment.

The demographic part of the model accounts for inter- and intraspecific pressures among the various compartments, but this is not symmetric. Specifically, such pressure could be reduced in the class of symptomatic infected I, because they are known to be sick and are, therefore, supposedly being cared for. Alternatively, it could be higher, meaning that being debilitated by the disease, they feel more the competition of other compartments. They instead exert a pressure on the other classes; this could be interpreted, e.g., as a burden, namely, the costs for the society to hospitalize them. This is very approximate and could be modified and expanded, if needed.

However, our main focus lies on the epidemiology. The susceptibles become infected by the asymptomatic in a mild form, migrating, indeed, into the asymptomatic class. This could be criticized and improved, but it is essential in order to compare the results with the classical model (73). Considering other infection mechanisms that may lead directly from susceptibles to symptomatic individuals, this would indeed significantly change the model and make the comparison less fair. In this way, both in (1) as well as in (23), because in the absence of demographics they coincide, the transition between compartments is the same used in (73). Thus, the simulations' results can be compared in an adequate way. There is only a minor mathematical change of a technical nature, in that the response function depends inversely on $1 + \beta I^2$ in (73), and this is necessary to push it to zero for large values of I, while for the SAI case it is enough for it to be inversely proportional to $1 + \beta I$, in view of the fact that in the numerator no I appears.

The main finding in the simulations is that the SAI model introduced here, in spite of being an SI-type model with the infected individuals split between two classes, asymptomatic and symptomatic, may prevent some susceptibles from contracting the disease, in the proper situation, in contrast to the classical SI model. Specifically, this occurs if the progression rate from asymptomatic to symptomatic is above the contact rate. The progression rate plays, thus, a fundamental role. The explanation for this situation lies in the fact that for a high progression rate, the asymptomatic class is fast depleted so that the symptomatic reaches high numbers quickly. As a consequence, the transmission rate in the SAI model quickly approaches zero, because the denominator grows and the numerator is reduced. Specifically, for the single susceptible the infected washout rate should exceed their recruitment rate i.e.,

$$\frac{\alpha I}{1 + \beta I} < \pi.$$

Thus the minimum weight that the individuals should give on the information about the symptomatics, can be assessed by finding the critical threshold $\beta^{\dagger}$

$$\beta^{\dagger} = \frac{\alpha}{\pi} - \frac{1}{I}$$

where I is the number of observed symptomatics. In this way for $\beta > \beta^{\dagger}$ the ratio of transmission to progression rates falls below one, and a number of susceptibles are preserved from getting the disease, the higher the farther β is from the critical threshold. The transmission rate of model (73), instead, contains the infected I also in the numerator. Thus, although the denominator grows quickly because it contains the term $1 + \beta I^2$, the transmission rate reaches zero at a slower pace than the SAI model does. This phenomenon represents an instance of the well-known fact that diseases that severely affect individuals preserve instead the whole community, while they significantly impact the population if they are mild at the individual level.

This analysis shows that the determination of whether a disease is asymptomatic and the assessment of the progression and transmission rates proves fundamental for its possible containment. In addition, in the presence of asymptomatic diseased individuals, the individual protection measures should have a higher impact, measured by a larger weight coefficient β, when just a few symptomatics appear in the population. The simulations of Figures 10 and 11 also help in quantifying the disease impact on the population, if a reliable measure of the contact rate α, as well as of the transition rate to symptomatic π, exist.

Overall, this investigation reveals the importance of properly assessing these rates as soon as possible. It also stresses that accounting for asymptomatics in the individual response as well as in the epidemic control is of utmost importance.

Author Contributions: Both authors have contributed to the paper in each and every one of its parts. All authors have read and agreed to the published version of the manuscript.

Funding: This work was partially supported by the local research project "Metodi numerici per l'approssimazione e le scienze della vita" of the Dipartimento di Matematica "Giuseppe Peano", Università di Torino.

Institutional Review Board Statement: Not applicable.

Informed Consent Statement: Not applicable.

Data Availability Statement: Not applicable.

Acknowledgments: The authors dedicate the paper to, and congratulate, Delfim F. M. Torres on his 50th birthday. The authors thank the referees for their valuable comments.

Conflicts of Interest: The author declares no conflict of interest.

References

1. Gao, Q.L.; Hethcote, H.W. Disease transmission models with density dependent demographics. *J. Math. Biol.* **1992**, *30*, 717–731. [CrossRef] [PubMed]
2. Mena-Lorca, J.; Hethcote, H.W. Dynamic models of infectious diseases as regulator of population sizes. *J. Math. Biol.* **1992**, *30*, 693–716. [CrossRef] [PubMed]
3. Hethcote, H.W. The mathematics of infectious diseases. *SIAM Rev.* **2000**, *42*, 599–653. [CrossRef]
4. Agarwal, P.; Nieto, J.J.; Torres, D.F.M. *Mathematical Analysis of Infectious Diseases*; Elsevier: Amsterdam, The Netherlands, 2022.
5. Trejos, D.Y.; Valverde, J.C.; Venturino, E. Dynamic of infectious diseases: A review of the main biological aspects and their mathematical translation. *Appl. Math. Nonlinear Sci.* **2021**, *7*, 1–26. [CrossRef]
6. Zine, H.; Danane, J.; Torres, D.F.M. Stochastic SICA epidemic model with jump Lévy processes. *Math. Anal. Infect. Dis.* **2022**, 61–72. [CrossRef]
7. Marziano, V.; Poletti, P.; Guzzetta, G.; Ajelli, M.; Manfredi, P.; Merler, S. The impact of demographic changes on the epidemiology of herpes zoster: Spain as a case study. *Proc. R. Soc. Biol. Sci.* **2015**, *282*, 20142509. [CrossRef] [PubMed]
8. Venturino, E. Ecoepidemiology: A more comprehensive view of population interactions. *Math. Model. Nat. Phenom.* **2016**, *11*, 49–90. [CrossRef]
9. Capasso, V.; Serio, G. A Generalization of the Kermack-Mckendrick Deterministic Epidemic Model. *Math. Biosci.* **1978**, *42*, 43–61. [CrossRef]
10. Bohner, M.; Streipert, S.; Torres, D.F.M. Exact solution to a dynamic SIR model. *Nonlinear Anal. Hybrid Syst.* **2022**, *32*, 228–238. [CrossRef]
11. d'Onofrio, A.; Manfredi, P. Behavioral SIR models with incidence-based social-distancing. *Chaos Solitons Fractals* **2022**, *159*, 112072. [CrossRef]
12. d'Onofrio, A.; Manfredi, P.; Poletti, P. The Interplay of Public Intervention and Private Choices in Determining the Outcome of Vaccination Programmes. *PLoS ONE* **2012**, *7*, e45653. [CrossRef] [PubMed]
13. Manfredi, P.; d'Onofrio, A. *Modeling the Interplay between Human Behavior and the Spread of Infectious Diseases*; Springer: Berlin/Heidelberg, Germany, 2013.
14. Silva, C.J.; Cantin, G.; Cruz, C.; Fonseca-Pinto, R.; Passadouro, R.; Dos Santos, E.S.; Torres, D.F. Complex network model for COVID-19: Human behavior, pseudo-periodic solutions and multiple epidemic waves. *J. Math. Anal. Appl.* **2022**, *514*, 125171. [CrossRef] [PubMed]
15. Reid, A.H.; Taubenberger, J.K.; Fanning, T.G. The 1918 Spanish influenza: Integrating history and biology. *Microbes Infect.* **2001**, *3*, 81–87. [CrossRef]
16. Matta, S.; Arora, V.K.; Chopra, K.K. Lessons to be learnt from 100 year old 1918 influenza pandemic viz a viz 2019 corona pandemic with an eye on NTEP. *Indian J. Tuberc.* **2020**, *67*, S132–S138. [CrossRef]
17. Pechous, R.D.; Sivaraman, V.; Stasulli, N.M.; Goldman, W.E. Pneumonic Plague: The Darker Side of *Yersinia pestis*. *Trends Microbiol.* **2016**, *24*, 190–197. [CrossRef]
18. Ansari, I.; Grier, G.; Byers, M. Deliberate release: Plague—A review. *J. Biosaf. Biosecur.* **2020**, *2*, 10–22. [CrossRef] [PubMed]
19. Friji, H.; Hamadi, R.; Ghazzai, H.; Besbes, H.; Massoud, Y. A Generalized Mechanistic Model for Assessing and Forecasting the Spread of the COVID-19 Pandemic. *IEEE Access* **2021**, *9*, 13266–13285. [CrossRef] [PubMed]
20. Menda, K.; Laird, L.; Kochenderfer, M.J.; Caceres, R.S. Explaining COVID-19 outbreaks with reactive SEIRD models. *Sci. Rep.* **2021**, *11*, 17905. [CrossRef] [PubMed]
21. Mangiarotti, S.; Peyre, M.; Zhang, Y.; Huc, M.; Roger, F.; Kerr, Y. Chaos theory applied to the outbreak of COVID-19: An ancillary approach to decision making in pandemic context. *Epidemiol. Infect.* **2020**, *148*, E95. [CrossRef] [PubMed]
22. Perko, L. *Differential Equations and Dynamical Systems*; Springer: New York, NY, USA, 2001.

MDPI

Article

Minimum Energy Problem in the Sense of Caputo for Fractional Neutral Evolution Systems in Banach Spaces

Zoubida Ech-chaffani [1,†], Ahmed Aberqi [2,†], Touria Karite [3,*,†] and Delfim F. M. Torres [4,†]

1 LAMA Laboratory, Faculty of Sciences Dhar El Mahraz, Sidi Mohamed Ben Abdellah University, Avenue My Abdallah Km 5 Route d'Imouzzer, Fes BP 1796, Morocco; echzoubida@gmail.com
2 LAMA Laboratory, Department of Electrical Engineering & Computer Science, National School of Applied Sciences, Sidi Mohamed Ben Abdellah University, Avenue My Abdallah Km 5 Route d'Imouzzer, Fes BP 1796, Morocco; ahmed.aberqi@usmba.ac.ma
3 Laboratory of Engineering, Systems and Applications, Department of Electrical Engineering & Computer Science, National School of Applied Sciences, Sidi Mohamed Ben Abdellah University, Avenue My Abdallah Km 5 Route d'Imouzzer, Fes BP 1796, Morocco
4 Center for Research and Development in Mathematics and Applications (CIDMA), Department of Mathematics, University of Aveiro, 3810-193 Aveiro, Portugal; delfim@ua.pt
* Correspondence: touria.karite@usmba.ac.ma
† These authors contributed equally to this work.

Abstract: We investigate a class of fractional neutral evolution equations on Banach spaces involving Caputo derivatives. Main results establish conditions for the controllability of the fractional-order system and conditions for existence of a solution to an optimal control problem of minimum energy. The results are proved with the help of fixed-point and semigroup theories.

Keywords: fractional control systems; neutral evolution systems; controllability; optimal control; minimum energy; Banach fixed point theorem

MSC: 26A33; 34K40; 49J30; 93B05

Citation: Ech-chaffani, Z.; Aberqi, A.; Karite, T.; Torres, D.F.M. Minimum Energy Problem in the Sense of Caputo for Fractional Neutral Evolution Systems in Banach Spaces. *Axioms* **2022**, *11*, 379. https://doi.org/10.3390/axioms11080379

Academic Editor: Valery Y. Glizer

Received: 6 July 2022
Accepted: 29 July 2022
Published: 31 July 2022

Publisher's Note: MDPI stays neutral with regard to jurisdictional claims in published maps and institutional affiliations.

1. Introduction

A neutral system is a system where time-delays play an important role. Precisely, such delays appear in both state variables and their derivatives. A delay in the derivative is called "neutral", which makes the system more complex than a classical one where the delays only occur in the state. Neutral delays do not only occur in physical systems, but they also appear in control systems, where they are sometimes added to improve the performance. For instance, a wide range of neutral-type control systems are expressed by

$$\frac{d}{dt}[y(t) - Ky_t] = Ly_t + Bu(t), \quad t \geq 0, \quad y_0(\cdot) = f_0(\cdot), \tag{1}$$

where $y_t : [-1,0] \to \mathbb{C}^n$ is defined by $y_t(s) = y(t+s)$; for $f \in H^1([-1,0],\mathbb{C}^n)$, the difference operator K is given by $Kf = A_{-1}f(-1)$ with A_{-1} a constant $n \times n$ matrix. The delay operator L is defined by

$$Lf = \int_{-1}^{0} [A_2(\theta)f'(\theta) + A_3(\theta)f(\theta)]d\theta$$

with A_2 and A_3 $n \times n$ matrices whose elements belong to $L_2(-1,0)$; B is a constant $n \times r$ matrix; and the control u is an L_2-function [1].

Nowadays, many researchers have investigated neutral differential equations in Banach spaces [2–4]. This interest is explained by the fact that neutral-argument differential equations have interesting applications in real-life problems: they appear, e.g., while modeling networks containing lossless transmission lines or in super-computers. Moreover, second-order neutral equations play an important role in automatic control and in aeromechanical systems, where inertia plays a central role [5–7].

Controllability plays an inherent crucial role in finite and infinite-dimensional systems, being one of the primary concepts in control theory, along with observability and stability. This concept has also attracted many authors; see, for instance, [8–10].

In the last two decades, several researchers have been interested in exploring the concept of controllability for fractional systems [11–13]. This is natural because fractional differential equations are considered a valuable tool in modeling various real-world dynamic systems, including physics, biology, socio-economy, chemistry and engineering [14–16].

It turns out that system (1) can also be studied in the fractional sense, e.g., being expressed by

$$\begin{cases} {}^C D_t^q [y(t) - K y_t] = L y(t) + B u(t), \quad t \in [0, T], \\ y_0(\cdot) = f_0(\cdot), \end{cases}$$

where ${}^C D_t^q$ denotes the Caputo fractional derivative of order q. The existence of solutions to fractional differential equations for neutral systems involving Caputo or other fractional operators, like Riemann–Liouville fractional derivatives, has been paid much attention [17–19]. Recently, some achievements regarding the existence and uniqueness of mild solutions to fractional stochastic neutral differential systems in a finite dimensional space have been made [20]. Other works are consecrated to demonstrate existence of a mild solution for neutral fractional inclusions of the Sobolev type [21].

In [22], Sakthivel et al. examined the exact controllability of fractional differential neutral systems by establishing sufficient conditions via a fixed-point analysis approach. Later on, Sakthivel et al. investigated the weak controllability of fractional dynamical systems of order $1 < q < 2$ using sectorial operators and Krasnoselskii's fixed-point theorem [23]. Using the same techniques as the previous authors, Qin et al. have studied the controllability and optimal control of fractional dynamical systems of order $1 < q < 2$ in Banach spaces [24]. Yan and Jia used stochastic analysis theory and fixed-point theorems with the strongly continuous α-order cosine family to study an optimal control problem for a class of stochastic fractional equations of order $\alpha \in (1, 2]$ in Hilbert spaces [25]. In 2021, Zhou and He obtained, via the contraction principle and Shauder's fixed-point theorem, a set of sufficient conditions for the exact controllability of a class of fractional systems [26]. More recently, Xi et al. studied the approximate controllability of fractional neutral hyperbolic systems using Sadovskii's fixed point theorem while constructing a Cauchy sequence and a control function [27]. Dineshkumar et al. addressed the problem of approximate controllability for neutral stochastic fractional systems in the sense of Hilfer, treating the problem using Schauder's fixed-point theorem and extending the obtained results to the case of nonlocal conditions [28]. In [29], Ma et al. analyzed the weak controllability of a fractional neutral differential inclusion of the Hilfer type in Hilbert spaces using Bohnenblust–Karlin's fixed point theorem. The concept of complete controllability is studied in [30] by Wen and Xi, where they establish sufficient conditions to assure this type of controllability.

Here, we let $(X, |\cdot|)$ be a Banach space, and we denote the Banach space of continuous functions by $\mathcal{C}(0, T; X)$ with the norm $|x| = \sup_{t \in J} |x(t)|$. Our main goal is to explore the concepts of controllability and optimal control for the following general evolution fractional system:

$$\begin{cases} {}^C D_t^\nu [x(t) - h(t, x_t)] = \mathcal{A} x(t) + \mathcal{B} u(t), \quad t \in (0, T], \\ x(0) = x_0 \in D(\mathcal{A}), \end{cases} \tag{2}$$

where ${}^C D_t^\nu$ denotes the fractional derivative of order $\nu \in (0, 1)$ in the sense of Caputo, $h : [0, T] \times \mathcal{C}(0, T; X) \to X$ is a given continuous function, and the dynamic of the system $\mathcal{A} : D(\mathcal{A}) \subseteq X \to X$ is a linear, closed operator with dense domain $D(\mathcal{A})$ generating a compact and uniformly bounded C_0 semigroup $\{\mathcal{T}(t)\}_{t\geq 0}$ on X. The control function $u(\cdot)$ is given in $L^2(0,T;U)$, with U a reflexive Banach space, and the control operator $\mathcal{B} \in \mathcal{L}(U,X)$ is a linear continuous bounded operator, i.e., there exists a constant $M_1 > 0$ such that

$$|\mathcal{B}| \leq M_1. \tag{3}$$

Our main aim is to be able to obtain a set of sufficient conditions assuring the controllability of system (2) and, afterwards, to consider an associated optimal control problem and prove existence of a solution.

The rest of this paper is organized as follows. In Section 2, the definitions of Caputo fractional derivative and mild solutions for system (2) are recalled. Our main result on the controllability of (2) is proved in Section 3. In Section 4, we prove the existence of a control giving minimum energy on a closed convex set of admissible controls. Section 5 is consecrated to the analysis of a concrete example, illustrating the applicability of our main results. We end with Section 6, which contains conclusions and points out some possible future directions of research.

2. Background

In this section, basic definitions, notations, and lemmas are introduced to be used throughout the paper. In particular, we recall the main properties of fractional calculus [31,32] and useful properties of semigroup theory [33].

Throughout the paper, let $\mathcal{A}$ be the infinitesimal generator of a compact and uniformly bounded C_0 semi-group $\{\mathcal{T}(t)\}_{t\geq 0}$. Let $0 \in \varrho(\mathcal{A})$, where $\varrho(\mathcal{A})$ denotes the resolvent of $\mathcal{A}$. Then, for $0 \leq \mu \leq 1$, the fractional power $\mathcal{A}^\mu$ is defined as a closed linear operator on its domain $D(\mathcal{A}^\mu)$. For a compact semi-group $\{\mathcal{T}(t)\}_{t\geq 0}$, the following properties are useful in this paper:

(i) There exists $M_T \geq 1$ such that

$$M_T = \sup_{t\geq 0} |\mathcal{T}(t)|; \tag{4}$$

(ii) For any $\mu \in (0,1]$, there exists $\mathbb{L}_\mu > 0$ such that

$$|\mathcal{A}^\mu \mathcal{T}(t)| \leq \frac{\mathbb{L}_\mu}{t^\mu}, \quad 0 \leq t \leq T. \tag{5}$$

Now we recall the notion of a Caputo fractional derivative.

Definition 1 (See [32]). *The left-sided Caputo fractional derivative of order $\nu > 0$ of a function $z \in L^1([0,T])$ is*

$${}_0^C D_t^\nu z(t) = \frac{1}{\Gamma(\kappa-\nu)} \int_0^t (t-s)^{\kappa-\nu-1} \frac{d^\kappa}{ds^\kappa} z(s) ds, \tag{6}$$

where $t \geq 0$, $\kappa - 1 < \nu < \kappa$, $\kappa \in \mathbb{N}$, and $\Gamma(\cdot)$ is the gamma function.

Using the probability density function and its Laplace transform [34] (see also [35,36]), we recall the definition of a mild solution for system (2).

Definition 2 (See [34]). *Let $u \in U$ for $t \in]0,T]$. A function $x \in \mathcal{C}(0,T;X)$ is said to be a mild solution of system (2) if*

$$\begin{aligned} x(t,u) = {} & S_\nu(t)[x_0 - h(0,x_0)] + h(t,x_t) + \int_0^t (t-s)^{\nu-1} \mathcal{A} K_\nu(t-s) h(s,x_s) ds \\ & + \int_0^t (t-s)^{\nu-1} K_\nu(t-s) \mathcal{B} u(s) ds, \end{aligned} \tag{7}$$

where $S_\nu(\cdot)$ and $K_\nu(\cdot)$ are the characteristic solution operators defined by

$$S_\nu(t) = \int_0^\infty \phi_\nu(\Theta)\mathcal{T}(t^\nu\Theta)\,\mathrm{d}\Theta \quad \textit{and} \quad K_\nu(t) = \nu\int_0^\infty \Theta\phi_\nu(\Theta)\mathcal{T}(t^\nu\Theta)\,\mathrm{d}\Theta$$

with

$$\phi_\nu(\Theta) = \frac{1}{\nu}\Theta^{-1-\frac{1}{\nu}}\psi_\nu\left(\Theta^{-\frac{1}{\nu}}\right)$$

and

$$\psi_\nu(\Theta) = \frac{1}{\pi}\sum_{n=1}^{\infty}(-1)^{n-1}\Theta^{-\nu n-1}\frac{\Gamma(n\nu+1)}{n!}\sin(n\pi\nu), \quad \Theta\in(0,\infty),$$

the probability density. In addition, we have

$$\int_0^\infty \psi_\nu(\Theta)d\Theta = 1 \textit{ and } \int_0^\infty \Theta^\Lambda\phi_\nu(\Theta)d\Theta = \frac{\Gamma(1+\Lambda)}{\Gamma(1+\nu\Lambda)}, \quad \Lambda\in[0,1].$$

Remark 1. *The solution $x(t,u)$ of (2) is considered in the weak sense, and, when there are no ambiguities, it is denoted by $x_u(t)$. We denote by $x_u(T)$ the mild solution of system (2) at the final time T.*

The following properties of $S_\nu(\cdot)$ and $K_\nu(\cdot)$ will be used throughout the paper.

Lemma 1 (See [34]).

1. *For any $t\geq 0$, the operators $S_\nu(t)$ and $K_\nu(t)$ are linear and bounded, i.e.,*

$$|S_\nu(t)y| \leq M_T|y| \quad \textit{and} \quad |K_\nu(t)y| \leq \frac{\nu M_T}{\Gamma(1+\nu)}|y|$$

for any $y\in X$ where $M_T = \sup_{t\geq 0}|\mathcal{T}(t)|$.

2. *For $t>0$, if $\mathcal{T}(t)$ is compact, then $S_\nu(t)$ and $K_\nu(t)$ are both compact operators.*

Lemma 2 (See [34]). *For any $x\in X$, $\varsigma\in(0,1)$ and $\mu\in(0,1]$ we have*

(i) $\mathcal{A}K_\nu(t)x = \mathcal{A}^{1-\varsigma}K_\nu(t)\mathcal{A}^\varsigma x,\ 0\leq t\leq a;$

(ii) $|\mathcal{A}^\mu K_\nu(t)| \leq \dfrac{\nu\mathbb{L}_\mu}{t^{\nu\mu}}\dfrac{\Gamma(2-\mu)}{\Gamma(1+\nu(1-\mu))},\ 0<t\leq a.$

3. Controllability

Following [37], let us define the meaning of controllability for our system (2).

Definition 3. *System (2) is said to be controllable in X on $[0,T]$ if for any given initial state $x_0\in X$ and any desired final state $x_d\in X$, there exists a control $u(\cdot)\in L^2(0,T;U)$ such that the mild solution $x\in\mathcal{C}(0,T;X)$ of system (2) satisfies $x_u(T)=x_d$.*

To prove controllability, we make use of the following assumptions (A_1) and (A_2):

(A_1) $\mathcal{T}(t)$ is compact for every $t>0$;

(A_2) The function $h:[0,T]\times\mathcal{C}(0,T;X)\to X$ is continuous, and there exists a constant $\varsigma\in]0,T[$ and $H,H_1>0$ such that $h\in D(\mathcal{A}^\varsigma)$, and for any $z,y\in\mathcal{C}(0,T;X)$, $t\in[0,T]$, the function $\mathcal{A}^\varsigma h(\cdot,z)$ is strongly measurable and $\mathcal{A}^\varsigma h(t,\cdot)$ satisfies the Lipschitz condition

$$|\mathcal{A}^\varsigma h(t,z) - \mathcal{A}^\varsigma h(t,y)| \leq H\|z-y\| \tag{8}$$

and

$$|\mathcal{A}^\varsigma h(t,z)| \leq H_1(\|z\|+1). \tag{9}$$

Let $H_\nu : L^2(0,T;U) \to X$ be the linear operator defined by

$$H_\nu u = \int_0^T (T-s)^{\nu-1} K_\nu(T-s)\mathcal{B}u(s)\mathrm{d}s.$$

By construction, this operator is invertible. Indeed, because H_ν takes values in the cokernel $L^2(0,T;U)/kerH_\nu$, then it is injective. It is also surjective because $L^2(0,T;U)/kerH_\nu \simeq ImH_\nu$ (see [38,39]). The inverse operator H_ν^{-1} takes values in $L^2(0,T;U)/\ker H_\nu$. Thus, there exists a positive constant $M_2 \geq 0$ such that

$$\left|H_\nu^{-1}\right|_{\mathcal{L}\left(X, L^2(0,T;U)/\ker H_\nu\right)} \leq M_2. \tag{10}$$

Let $r \geq 0$. Note that $B_r = \{x \in \mathcal{C}(0,T;X) : \|x\| \leq r\}$ is a bounded closed and convex subset in $\mathcal{C}(0,T;X)$.

Theorem 1. *If (A_1) and (A_2) are fulfilled, then the evolution system (2) is controllable in $[0,T]$ provided*

$$\left[|\mathcal{A}^{-\varsigma}| + \frac{\mathbb{L}_{1-\varsigma}\Gamma(1+\varsigma)}{\varsigma\Gamma(1+\nu\varsigma)}T^{\nu\varsigma} + \frac{MM_TM_1}{\Gamma(1+\nu)}T^\nu\left(|\mathcal{A}^{-\varsigma}| + \frac{\mathbb{L}_{1-\varsigma}\Gamma(1+\varsigma)}{\varsigma\Gamma(1+\nu\varsigma)}T^{\nu\varsigma}\right)\right]H < 1. \tag{11}$$

Proof. For any function x, we define the control

$$\begin{aligned} u_x(t) = H_\nu^{-1}\Big[& x_d - S_\nu(t)[x_0 - h(0,x_0)] - h(T,x_T) \\ & - \int_0^T (T-s)^{\nu-1}\mathcal{A}K_\nu(T-s)h(s,x_s)\mathrm{d}s\Big](t). \end{aligned} \tag{12}$$

We shall prove that $G : \mathcal{C}(0,T;X) \to \mathcal{C}(0,T;X)$, defined by

$$\begin{aligned} (Gx(t)) &= S_\nu(t)[x_0 - h(0,x_0)] + h(t,x_t) + \int_0^t (t-s)^{\nu-1}\mathcal{A}K_\nu(t-s)h(s,x_s)\mathrm{d}s \\ &+ \int_0^t (t-s)^{\nu-1}K_\nu(t-s)\mathcal{B}u_x(s)\mathrm{d}s, \quad t \in [0,T], \end{aligned} \tag{13}$$

has a fixed point x for the control u_x steering system (2) from x_0 to x_d in time T. From (3), (10), Lemma 1 and (i) of Lemma 2, we have

$$\begin{aligned} |\mathcal{B}u_x(t)| \leq MM_1\Big(& |x_d| + M_T\Big[|x_0| + |h(0,x_0)|\Big] + |h(T,x_T)| \\ & + \int_0^T (T-s)^{\nu-1}\Big|\mathcal{A}^{1-\varsigma}K_\nu(T-s)\mathcal{A}^\varsigma h(s,x_s)\Big|ds\Big). \end{aligned}$$

In view of (9) and (ii) of Lemma 2, it follows that

$$\begin{aligned} |\mathcal{B}u_x(t)| &\leq MM_1\Bigg(|x_d| + M_T\Big[|x| + \Big(r+1\Big)H_1|\mathcal{A}^{-\varsigma}|\Big] + \Big(r+1\Big)H_1|\mathcal{A}^{-\varsigma}| \\ &\quad + \frac{\nu\mathbb{L}_{1-\varsigma}\Gamma(1+\varsigma)}{\Gamma(1+\nu\varsigma)}H_1\Big(r+1\Big)\int_0^T (T-s)^{\nu\varsigma-1}ds\Bigg) \\ &\leq MM_1\Bigg(|x_d| + M_T\Big[|x| + \Big(r+1\Big)H_1|\mathcal{A}^{-\varsigma}|\Big] + \Big(r+1\Big)H_1|\mathcal{A}^{-\varsigma}| \\ &\quad + \frac{\mathbb{L}_{1-\varsigma}}{\Gamma(1+\varsigma)}H_1\Big(r+1\Big)T^{\nu\varsigma}\Bigg). \end{aligned}$$

Let

$$\mathcal{Y} = MM_1\Bigg(|x_d| + M_T\Big[|x| + \Big(r+1\Big)H_1|\mathcal{A}^{-\varsigma}|\Big] + \Big(r+1\Big)H_1|\mathcal{A}^{-\varsigma}| + \frac{\mathbb{L}_{1-\varsigma}}{\Gamma(1+\varsigma)}H_1\Big(r+1\Big)T^{\nu\varsigma}\Bigg).$$

It follows that

$$|\mathcal{B}u_x(t)| \leq \mathcal{Y}. \tag{14}$$

In order to show that G has a unique fixed point on B_r, we will proceed in two steps.
Step I: $Gx \in B_r$ whenever $x \in B_r$. For any fixed $x \in B_r$ and $0 \leq t \leq T$, we have

$$\begin{aligned}|(Gx(t))| \leq& |S_\nu(t)[x_0 - h(0,x_0)]| + |h(t,x_t)| + \int_0^t \Big|(t-s)^{\nu-1}\mathcal{A}K_\nu(t-s)h(s,x_s)\Big|\,ds \\ &+ \int_0^t (t-s)^{\nu-1}|K_\nu(t-s)\mathcal{B}u_x(s)|ds.\end{aligned}$$

From Lemma 1, (9), and (i) of Lemma 2, it results that

$$\begin{aligned}|(Gx(t))| \leq& M_T\Big[r + \Big(r+1\Big)H_1|\mathcal{A}^{-\varsigma}|\Big] + \Big(r+1\Big)H_1|\mathcal{A}^{-\varsigma}| \\ &+ \int_0^t (t-s)^{\nu-1}\Big|\mathcal{A}^{1-\varsigma}K_\nu(t-s)\mathcal{A}^{\varsigma}h(s,x_s)\Big|\,ds \\ &+ \frac{\nu M_T}{\Gamma(1+\nu)}\int_0^t (t-s)^{\nu-1}|\mathcal{B}u_x|ds.\end{aligned}$$

Now, by using (ii) of Lemma 2, we get

$$\begin{aligned}|(Gx(t))| \leq& M_T[r + H|\mathcal{A}^{-\varsigma}|\Big(r+1\Big)] + H|\mathcal{A}^{-\varsigma}|\Big(r+1\Big)\Big| \\ &+ \frac{\nu\mathbb{L}_{1-\varsigma}\Gamma(1+\varsigma)}{\Gamma(1+\nu\varsigma)}H\Big(r+1\Big)\int_0^t (t-s)^{\nu\varsigma-1}ds \\ &+ \frac{\nu M_T}{\Gamma(1+\nu)}\int_0^t (t-s)^{\nu-1}\Big|\mathcal{B}u_x(s)\Big|ds.\end{aligned}$$

According to (14), one has

$$\begin{aligned}|(Gx(t))| \leq& M_T\Bigg[r + H|\mathcal{A}^{-\varsigma}|\Big(r+1\Big)\Bigg] + H|\mathcal{A}^{-\varsigma}|\Big(r+1\Big)| \\ &+ \frac{\nu\mathbb{L}_{1-\varsigma}\Gamma(1+\varsigma)}{\varsigma\Gamma(1+\nu\varsigma)}H\Big(r+1\Big)T^{\nu\varsigma} + \frac{M_T}{\Gamma(1+\nu)}\mathcal{Y}T^{\nu}.\end{aligned}$$

By choosing

$$\begin{aligned}r =& M_T\Bigg[r + \Big(r+1\Big)H_1|\mathcal{A}^{-\varsigma}|\Bigg] + \Big(r+1\Big)H_1|\mathcal{A}^{-\varsigma}| \\ &+ \frac{\nu\mathbb{L}_{1-\varsigma}\Gamma(1+\varsigma)}{\varsigma\Gamma(1+\nu\varsigma)}H_1\Big(r+1\Big)T^{\nu\varsigma} + \frac{\nu M_T}{\Gamma(1+\nu)}\mathcal{Y}T^{\nu},\end{aligned}$$

we get that $Gx \in B_r$ whenever $x \in B_r$.
Step II: G is a contraction on B_r. For any $v, w \in B_r$ and $0 \leq t \leq T$, in accordance with (12), we obtain

$$
\begin{aligned}
|(Gv)(t)-(Gw)(t)| \leq & \Big|h(t,v_t)-h(t,w_t)\Big| \\
& + \int_0^t (t-s)^{\nu-1}\Big|\mathcal{A}r_\nu(t-s)\Big(h\big(s,v(s)\big)-h\big(s,w(s)\big)\Big)\Big|\,ds \\
& + \int_0^t (t-s)^{\nu-1}\Big|r_\nu(t-s)\mathcal{B}H_\nu^{-1}\Big[h(T,v_T)-h(T,w_T)+\int_0^T (T-\tau)^{\nu-1} \\
& \times \mathcal{A}K_\nu(T-\tau)\Big(h\big(\tau,v(\tau)\big)-h\big(\tau,w(\tau)\big)\Big)d\tau\Big](s)\Big|ds.
\end{aligned}
$$

Considering Lemma 2 and (A_2), we get

$$
\begin{aligned}
|(Gv)(t)-(Gw)(t)| \leq & H|\mathcal{A}^{-\varsigma}|v-w| + \frac{\nu\mathbb{L}_{1-\varsigma}\Gamma(1+\varsigma)}{\Gamma(1+\nu\varsigma)}H|v-w|\int_0^t (t-s)^{\nu\varsigma-1}ds \\
& + \frac{\nu M M_T M_1}{\Gamma(1+\nu)}\int_0^t (t-s)^{\nu-1}\Big[\Big|h(T,v_T)-h(T,w_T)\Big| \\
& + \int_0^t (T-\tau)^{\nu-1}\Big|\mathcal{A}^{1-\varsigma}K_\nu(t-\tau)\mathcal{A}^{\varsigma}\Big[h\big(\tau,v(\tau)\big)-h\big(\tau,w(\tau)\big)\Big]\Big|d\tau\Big]ds.
\end{aligned}
$$

From (8), we obtain that

$$
\begin{aligned}
|(Gv)(t)-(Gw)(t)| \leq & H|\mathcal{A}^{-\varsigma}|v-w| + \frac{\mathbb{L}_{1-\varsigma}\Gamma(1+\varsigma)}{\varsigma\Gamma(1+\nu\varsigma)}H|v-w|T^{\nu\varsigma} \\
& + \frac{\nu M M_T M_1}{\Gamma(1+\nu)}\int_0^T (t-s)^{\nu-1}\Big[H|\mathcal{A}^{-\varsigma}|v-w| \\
& + \frac{\mathbb{L}_{1-\varsigma}\Gamma(1+\varsigma)}{\varsigma\Gamma(1+\nu\varsigma)}H|v-w|T^{\nu\varsigma}\Big]ds \\
\leq & H|\mathcal{A}^{-\varsigma}|v-w| + \frac{\mathbb{L}_{1-\varsigma}\Gamma(1+\varsigma)}{\varsigma\Gamma(1+\nu\varsigma)}H|v-w|T^{\nu\varsigma} \\
& + \frac{M M_T M_1}{\Gamma(1+\nu)}T^{\nu}\Big[|\mathcal{A}^{-\varsigma}| + \frac{\mathbb{L}_{1-\varsigma}\Gamma(1+\varsigma)}{\varsigma\Gamma(1+\nu\varsigma)}T^{\nu\varsigma}\Big]H|v-w| \\
= & \Big[|\mathcal{A}^{-\varsigma}| + \frac{\mathbb{L}_{1-\varsigma}\Gamma(1+\varsigma)}{\varsigma\Gamma(1+\nu\varsigma)}T^{\nu\varsigma} + \frac{M M_T M_1}{\Gamma(1+\nu)}T^{\nu} \\
& \Big(|\mathcal{A}^{-\varsigma}| + \frac{\mathbb{L}_{1-\varsigma}\Gamma(1+\varsigma)}{\varsigma\Gamma(1+\nu\varsigma)}T^{\nu\varsigma}\Big)\Big]H|v-w|.
\end{aligned}
$$

From Theorem 1, we have

$$
\Big[|\mathcal{A}^{-\varsigma}| + \frac{\mathbb{L}_{1-\varsigma}\Gamma(1+\varsigma)}{\varsigma\Gamma(1+\nu\varsigma)}T^{\nu\varsigma} + \frac{M M_T M_1}{\Gamma(1+\nu)}T^{\nu}\Big(|\mathcal{A}^{-\varsigma}| + \frac{\mathbb{L}_{1-\varsigma}\Gamma(1+\varsigma)}{\varsigma\Gamma(1+\nu\varsigma)}T^{\nu\varsigma}\Big)\Big]H < 1;
$$

it follows that

$$
|(Gv)(t)-(Gw)(t)| < |v-w|,
$$

that is, G is a contraction on B_r. We conclude from the Banach fixed-point theorem that G has a unique fixed point x in $\mathcal{C}(0,T;X)$. Then, by injecting u_x in (7), we have

$$
\begin{aligned}
x_{u_x}(T) &= S_\nu(T)[x_0 - h(0,x_0)] + h(T,x_T) + \int_0^T (T-s)^{\nu-1}\mathcal{A}K_\nu(T-s)h(s,x_s)\mathrm{d}s \\
&\quad + \int_0^T (T-s)^{\nu-1}K_\nu(T-s)\mathcal{B}u_x(s)\mathrm{d}s, \\
&= S_\nu(T)[x_0 - h(0,x_0)] + h(T,x_T) + \int_0^T (T-s)^{\nu-1}\mathcal{A}K_\nu(T-s)h(s,x_s)\mathrm{d}s \\
&\quad + H_\nu H_\nu^{-1}\Big[x_d - S_\nu(T)[x_0 - h(0,x_0)] - h(T,x_T) \\
&\quad - \int_0^T (T-s)^{\nu-1}\mathcal{A}K_\nu(T-s)h(s,x_s)\mathrm{d}s\Big] \\
&= x_d
\end{aligned}
$$

and system (2) is exactly controllable, which completes the proof. □

We have shown, under assumptions (A_1) and (A_2), and with the help of Schauder's fixed-point theorem, that the neutral system (2) is controllable when condition (11) holds. It would be interesting to clarify if the obtained control is unique in the sense that any control that allows reaching the state x_d is such that the associated state x is a fixed point of the operator G. This uniqueness question is relevant but remains open.

4. Optimal Control

Now, we consider the problem of steering system (2) from the state x_0 to a target state x_d in time T with minimum energy. We prove the existence of solution to such an optimal control problem when the set of admissible controls is closed and convex.

Let $\mathcal{U}_{ad}$ be the nonempty set of admissible controls defined by

$$
\mathcal{U}_{ad} = \Big\{u \in L^2(0,T;U) : x_u(T) = x_d\Big\}.
$$

We shall prove that $\mathcal{U}_{ad}$ is closed. For that, let us consider a sequence u_n in $\mathcal{U}_{ad}$ such that $u_n \to u$ strongly in $L^2(0,T;U)$, so

$$
\begin{aligned}
x_{u_n}(T) &= S_\nu(T)[x_0 - h(0,x_0)] + h(T,x_T) + \int_0^T (T-s)^{\nu-1}\mathcal{A}K_\nu(T-s)h(s,x_s)\mathrm{d}s \\
&\quad + \int_0^T (T-s)^{\nu-1}K_\nu(T-s)\mathcal{B}u_n(s)\mathrm{d}s.
\end{aligned}
$$

Put

$$
\mathcal{Q}u = \int_0^T (T-s)^{\nu-1}\mathcal{A}K_\nu(T-s)h(s,x_s)\mathrm{d}s + \int_0^T (T-s)^{\nu-1}K_\nu(T-s)\mathcal{B}u_n(s)\mathrm{d}s.
$$

Since $\mathcal{Q}u$ is continuous, then $\mathcal{Q}u_n \to \mathcal{Q}u$ strongly in X. We also have that $h : [0,T] \times \mathcal{C}(0,T;X) \to X$ is continuous; then $x_{u_n}(T) \to x_u(T)$ in X, but $x_{u_n}(T) \in \{x_d\}$, which is closed. Therefore, $x_u(T) \in \{x_d\}$, which means that $u \in \mathcal{U}_{ad}$. Hence, $\mathcal{U}_{ad}$ is closed.

For a desired state x_d, our optimal control problem consists of finding within $\mathcal{U}_{ad}$ a control minimizing the functional

$$
J(u) = \frac{\varsigma}{2}\int_0^T |x_u(t) - x_d|_X^2 dt + \frac{\varepsilon}{2}\int_0^T |u(t)|_U^2 dt,
$$

where $x_u(\cdot)$ is the mild solution of system (2) associated with u. The parameters ε and ς are non-negative constants. Precisely, our optimal control problem is:

$$
\begin{cases} \inf\limits_{u \in \mathcal{U}_{ad}} J(u), \\ \text{s.t. (2).} \end{cases} \tag{15}
$$

The following result gives a necessary condition for the existence of an optimal control to our minimum energy problem.

Theorem 2. *Let $\mathcal{U}_{ad}$ be closed and convex. If $1 - H|\mathcal{A}^{-\varsigma}| > 0$, then there exists a $u^\star \in \mathcal{U}_{ad}$ solution to the optimal control problem* (15).

Proof. Let $|u_p|^2 \leq \frac{2}{\varepsilon}J(u_p)$ with $(u_p)_{p\in\mathbb{N}}$ bounded. Then there exists a subsequence, still denoted $(u_p)_{p\in\mathbb{N}}$, that converges weakly to a limit $u^\star$. If $\mathcal{U}_{ad}$ is closed and convex, then $\mathcal{U}_{ad}$ is closed for the weak topology, which implies that $u^\star \in \mathcal{U}_{ad}$. Let x_p be the unique solution of system (2) associated with u_p, and let $x^\star$ be the unique solution of system (2) associated with $u^\star$. Then,

$$\begin{aligned}
|x_p(t) - x^*(t)| \leq{}& |h(t, x_p(t)) - h(t, x^\star(t))| \\
&+ \left|\int_0^t (t-s)^{\nu-1}\mathcal{A}K_\nu(t-s)[h(s, x_p(s)) - h(s, x^\star(s))]\mathrm{d}s\right| \\
&+ \left|\int_0^t (t-s)^{\nu-1}K_\nu(t-s)\mathcal{B}[u_p(s) - u^\star(s)]\mathrm{d}s\right| \\
\leq{}& H|\mathcal{A}^{-\varsigma}||x_p(t) - x^\star(t)| \\
&+ \int_0^t (t-s)^{\nu-1}\left|\mathcal{A}^{1-\varsigma}K_\nu(t-s)[\mathcal{A}^\varsigma h(s, x_p(s)) - \mathcal{A}^\varsigma h(s, x^\star(s))]\right|\mathrm{d}s \\
&+ \left|\int_0^t (t-s)^{\nu-1}K_\nu(t-s)\mathcal{B}[u_p(s) - u^\star(s)]\mathrm{d}s\right|, \quad t \in [0, T].
\end{aligned} \tag{16}$$

This leads us to

$$\begin{aligned}
(1 - H|\mathcal{A}^{-\varsigma}|)|x_p(t) - x^*(t)| \leq{}& \frac{\nu\Gamma(1+\varsigma)}{\Gamma(1+\nu\varsigma)}\mathbb{L}_{1-\varsigma}\int_0^t (t-s)^{\nu\varsigma-1}H|x_p(t) - x^\star(t)|\mathrm{d}s \\
&+ \left|\int_0^t (t-s)^{\nu-1}K_\nu(t-s)\mathcal{B}[u_p(s) - u^\star(s)]\mathrm{d}s\right|,
\end{aligned} \tag{17}$$

$t \in [0, T]$. Set $\mathcal{K}' = \dfrac{1}{1 - H|\mathcal{A}^{-\varsigma}|}$. Then,

$$\begin{aligned}
|x_p(t) - x^*(t)| \leq{}& \mathcal{K}'\frac{\nu\Gamma(1+\varsigma)}{\Gamma(1+\nu\varsigma)}\mathbb{L}_{1-\varsigma}\int_0^t (t-s)^{\nu\varsigma-1}H|x_p(t) - x^*(t)|\mathrm{d}s \\
&+ \mathcal{K}'\left|\int_0^t (t-s)^{\nu-1}K_\nu(t-s)\mathcal{B}[u_p(s) - u^\star(s)]\mathrm{d}s\right|, \quad t \in [0, T].
\end{aligned} \tag{18}$$

Using the Gronwall lemma, we obtain that

$$\begin{aligned}
|x_p(t) - x^*(t)| \leq{}& \mathcal{K}'\left|\int_0^t (t-s)^{\nu-1}K_\nu(t-s)\mathcal{B}[u_p(s) - u^\star(s)]\mathrm{d}s\right| \\
&\exp\left(\mathcal{K}'\frac{\nu\Gamma(1+\varsigma)}{\Gamma(1+\nu\varsigma)}\mathbb{L}_{1-\varsigma}H\int_0^t (t-s)^{\nu\varsigma-1}\mathrm{d}s\right) \\
\leq{}& \mathcal{K}'\left|\int_0^t (t-s)^{\nu-1}K_\nu(t-s)\mathcal{B}[u_p(s) - u^\star(s)]\mathrm{d}s\right| \\
&\exp\left(\mathcal{K}'\frac{\Gamma(1+\varsigma)}{\varsigma\Gamma(1+\nu\varsigma)}\mathbb{L}_{1-\varsigma}HT^{\nu\varsigma}\right).
\end{aligned} \tag{19}$$

Now, by the weak convergence, $u_p \rightharpoonup u^*$ in $L^2(0,T,U)$, and from Lemma 1, we obtain that

$$\left|\int_0^t (t-s)^{\nu-1} K_\nu(t-s)\mathcal{B}[u_p(s)-u^\star(s)]\mathrm{d}s\right| \\ \leq \frac{\nu M_T M_1}{\Gamma(1+\nu)}\int_0^t (t-s)^{\nu-1}\left|u_p(s)-u^\star(s)\right|_{L^2(0,T,U)}\mathrm{d}s, \quad (20)$$

from which $x_p \to x^\star$ strongly in $L^2(0,T;X)$. Hence,

$$\lim_{n\to\infty}\int_0^T |x_p(t)-x_d|_X^2 dt = \int_0^T |x(t)-x_d|_X^2 dt.$$

Using the lower semi-continuity of norms, the weak convergence of $(u_p)_n$ gives

$$|u^\star| \leq \lim_{n\to\infty}\inf|u_p|.$$

Therefore, $J(u^\star) \leq \lim_{n\to\infty}\inf J(u_p)$, leading to $J(u^\star) = \inf_{u\in\mathcal{U}_{ad}} J(u_p)$, which establishes the optimality of $u^\star$. □

We have just proved the existence of an optimal control for a closed convex set of admissible controls. In Section 5, our main results are illustrated with the help of an example.

5. An Application

In this section we illustrate the results given by our Theorems 1 and 2.

Let $X = L^2((0,1);\mathbb{R})$ and consider the fractional differential system

$$\begin{cases} {}^C D_t^{1/2}\Big(y(t,z)-h(t,y_t)\Big) = \Delta y(t,z)+\mathcal{B}u(t,z), & t\in[0,1], \\ y(t,0)=y(t,1)=0, & t\in[0,1], \end{cases} \quad (21)$$

where the order ν of the fractional derivative is equal to $\frac{1}{2}$, and the function $h:[0,1]\times\mathcal{C}\to X$ is given by

$$h(t,y_t)(x) = \int_0^1 \mathcal{F}(x,z)u_t(v,z)dz, \quad (22)$$

where $\mathcal{F}$ is assumed to satisfy the following conditions:

(a) The function $\mathcal{F}(x,z)$, $x,z\in[0,1]$, is measurable and

$$\int_0^1\int_0^1 \mathcal{F}^2(x,z)dz < \infty;$$

(b) The function $\partial x\mathcal{F}(x,z)$ is measurable, $\mathcal{F}(0,z)=\mathcal{F}(1,z)=0$, and

$$\left(\int_0^1\int_0^1 \big(\partial x\mathcal{F}(x,z)\big)^2 dzdx\right)^{1/2} < \infty.$$

Let $\mathcal{A}: D(\mathcal{A})\subseteq X\to X$ be defined by $\mathcal{A}x = -x''$ with the domain

$$D(\mathcal{A}) = \left\{x(\cdot)\in X : x, x' \text{ absolutely continuous}, x''\in X, x(0)=x(1)=0\right\}.$$

We begin by proving that the assumption (A_1) holds. Indeed, operator $\mathcal{A}$ is self-adjoint, with a compact resolvent, and generating an analytic compact semi-group $\mathcal{T}(t)$. Furthermore, the eigenvalues of $\mathcal{A}$ are $\Lambda_p = p^2\pi^2$, $p\in\mathbb{N}$, with corresponding normalized eigenvectors $e_p(z) = \sqrt{\frac{2}{\pi}}\sin(p\pi z)$, $\{e_i\}_{i=1}^\infty$ forming an orthonormal basis of X. Then,

$$\mathcal{A}x = -\sum_{p=1}^{p=\infty} \Lambda_p(x,e_p)e_p, \quad x \in D(\mathcal{A}),$$

and

$$\mathcal{T}(t)x(s) = \sum_{i=1}^{i=\infty} \exp(\Lambda_i t)(x,e_i)e_i(s), \quad x \in X.$$

Note that $\mathcal{T}(\cdot)$ is a uniformly stable semi-group and $\|\mathcal{T}(t)\|_{L^2[0,1]} \leq \exp(-t)$. The following properties hold:

(i) $\mathcal{A}^{-\frac{1}{2}}x = \sum_{p=1}^{\infty} \frac{1}{p}(x,e_p)e_p;$

(ii) The operator $\mathcal{A}^{\frac{1}{2}}$ is given by

$$\mathcal{A}^{\frac{1}{2}}x = \sum_{p=1}^{\infty} p(x,e_p)e_p$$

and $D(\mathcal{A}^{\frac{1}{2}}) = \left\{ x(\cdot) \in X, \sum_{p=1}^{\infty} p(x,e_p)e_p \in X \right\}.$

Clearly, (4), (5), and (A_1) are satisfied.

Under our assumptions (a) and (b) on $\mathcal{F}$, (8) and (9) are also satisfied, and assumption (A_2) also holds.

Let U be a reflexive Banach space. We consider the control operator $\mathcal{B} : U \to X$ defined by

$$\mathcal{B}u = \sum_{p=1}^{p=\infty} \Lambda_p(\bar{u},e_p)e_p,$$

where

$$\bar{u} = \begin{cases} u_p, & p = 1,2,\dots N, \\ 0, & p = N+1, N+2,\dots \end{cases}$$

We see that $\mathcal{B}$ is a bounded continuous operator with $M_1 = N\Lambda_N$. For $N \in \mathbb{N}$ and $H_{1/2} : L^2([0,1],U) \to X$ given by

$$H_{1/2}u = \int_0^1 (1-s)^{1/2} P_{1/2}(1-s)\mathcal{B}u(s)ds,$$

we have

$$\begin{aligned}
H_{1/2}u &= \int_0^1 (1-s)^{1/2} \frac{1}{2} \int_0^{\infty} \Theta\phi_{1/2}(\Theta) T((1-s)^{1/2}\Theta)\mathcal{B}u(s)d\Theta\, ds \\
&= \int_0^1 (1-s)^{1/2} \frac{1}{2} \int_0^{\infty} \Theta\phi_{1/2}(\Theta) \sum_{i=1}^{i=\infty} \exp(\Lambda_i(1-s)^{1/2}\Theta)(\mathcal{B}u,e_i)e_i(s)d\Theta\, ds \\
&= \int_0^1 (1-s)^{1/2} \sum_{i=1}^{\infty} \int_0^{\infty} \frac{1}{2}\Theta\phi_{1/2}(\Theta) \sum_{j=0}^{\infty} \frac{\Lambda_i(1-s)^{1/2}\Theta)^j}{j!}(u,e_i)e_i(s)d\Theta\, ds \\
&= \int_0^1 (1-s)^{1/2} \sum_{i=1}^{\infty}\sum_{j=0}^{\infty} \frac{(\Lambda_i(1-s)^{1/2})^j}{\Gamma(1/2+\frac{1}{2}j)}(u,e_i)e_i(s)ds \\
&= \sum_{i=1}^{\infty}\sum_{j=0}^{\infty} \int_0^1 \frac{\Lambda_i^j}{\Gamma(\frac{1}{2}+\frac{1}{2}j)}(1-s)^{\frac{1+j}{2}}(u,e_i)e_i(s) \\
&= \sum_{i=1}^{\infty}\sum_{j=0}^{\infty} \frac{2\Lambda_i^j}{\Gamma(\frac{1}{2}+\frac{1}{2}j)(3+j)}(u,e_i)e_i(s).
\end{aligned}$$

Applying Theorem 1, we deduce that the fractional differential system (21) is controllable. Moreover, for function h defined as in (22) with the Lipshitz constant $H < \frac{1}{\left|\mathcal{A}^{-\frac{1}{2}}\right|}$, we conclude from Theorem 2 that there exists a control steering the system, in one unit of time, from a given initial state to a given terminal state with minimum energy.

6. Conclusions

Using the Banach fixed-point theorem, we have obtained a set of sufficient conditions for the controllability of a class of fractional neutral evolution equations involving the Caputo fractional derivative of order $\alpha \in]0,1[$ (cf. Theorem 1). The result is proved in two major steps: (i) in the first step, we proved that the operator G defined by (13) is an element of the bounded closed and convex subset B_r, (ii) while in the second, we proved that G is a contraction on the same subset B_r. Moreover, we formulated a minimum energy optimal control problem and proved conditions assuring the existence of a solution for the optimal control problem $\inf_{u \in \mathcal{U}_{ad}} J(u)$ subject to (2) (cf. Theorem 2). An example was given illustrating the two main results.

Our work can be extended in several directions: (i) to a case of enlarged controllability using different fractional derivatives; (ii) by developing methods to determine the control predicted by our existence theorem, e.g., by using RHUM and penalization approaches [10,40,41]; (iii) or by giving applications of neutral systems to epidemiological problems [42,43]. Many other questions remain open, as is the case of regional controllability and regional discrete controllability for problems of the type considered here. A strong motivation behind the investigation of neutral evolution systems, such as (2) considered here, comes from physics, since they describe well various physical phenomena as fractional diffusion equations. However, neutral systems are difficult to study, since such control systems contain time-delays not only in the state but also in the velocity variables, which make them intrinsically more complicated. The limitations of the method we proposed here is that we are not able to provide conditions under which the optimal control is unique. Additionally, we do not have an explicit form for it.

Author Contributions: Conceptualization, A.A.; methodology, Z.E.-c., A.A., T.K. and D.F.M.T.; validation, Z.E.-c., A.A., T.K. and D.F.M.T.; formal analysis, Z.E.-c., A.A., T.K. and D.F.M.T.; investigation, Z.E.-c., A.A., T.K. and D.F.M.T.; writing—original draft preparation, Z.E.-c., A.A., T.K. and D.F.M.T.; writing—review and editing, Z.E.-c., A.A., T.K. and D.F.M.T. All authors have read and agreed to the published version of the manuscript.

Funding: T.K. and D.F.M.T. were partially funded by FCT, project UIDB/04106/2020 (CIDMA).

Institutional Review Board Statement: Not applicable.

Informed Consent Statement: Not applicable.

Data Availability Statement: Not applicable.

Acknowledgments: This research is part of Ech-chaffani's Ph.D., which is being carried out at Sidi Mohamed Ben Abdellah, Fez, under the scientific supervision of Aberqi. It was essentially finished during a one-month visit of Karite to the Department of Mathematics of University of Aveiro, Portugal, April and May 2022. The authors are very grateful to two anonymous referees for many suggestions and invaluable comments.

Conflicts of Interest: The authors declare no conflict of interest. The funders had no role in the design of the study; in the collection, analyses, or interpretation of data; in the writing of the manuscript; or in the decision to publish the results.

References

1. Rabah, R.; Sklyar, G.M.; Barkhayev, P.Y. On the exact controllability and observability of neutral type systems. *Commun. Math. Anal.* **2014**, *17*, 279–294. [CrossRef]
2. Chandrasekaran, S.; Karunanithi, S. Existence results for neutral functional integrodifferential equations with infinite delay in Banach spaces. *J. Appl. Math. Inform.* **2015**, *33*, 45–60. [CrossRef]
3. Huang, H.; Fu, X. Optimal control problems for a neutral integro-differential system with infinite delay. *Evol. Equ. Control Theory* **2022**, *11*, 177–197. [CrossRef]

4. Harisa, S.A.; Ravichandran, C.; Sooppy Nisar, K.; Faried, N.; Morsy, A. New exploration of operators of fractional neutral integro-differential equations in Banach spaces through the application of the topological degree concept. *AIMS Math.* **2022**, *7*, 15741–15758. [CrossRef]
5. Baker, C.T.H.; Bocharov, G.A.; Rihan, F.A. Neutral delay differential equations in the modelling of cell growth. *J. Egypt. Math. Soc.* **2008**, *16*, 133–160.
6. Baculikova, B. Oscillatory behavior of the second order functional differential equations. *Appl. Math. Lett.* **2017**, *72*, 35–41. [CrossRef]
7. Bazighifan, O.; Elabbasy, E.M.; Moaaz, O. Oscillation of higher-order differential equations with distributed delay. *J. Inequal. Appl.* **2019**, *55*, 1–9. [CrossRef]
8. Lions, J.L. *Optimal Control of Systems Governed by Partial Differential Equations*; Springer: New York, NY, USA, 1971.
9. Li, X.; Yong, J. *Optimal Control Theory for Infinite Dimensional Systems, Systems Control: Foundations Applications*; Birkhäuser Boston, Inc.: Boston, MA, USA, 1995. [CrossRef]
10. Karite, T.; Boutoulout, A. Global and regional constrained controllability for distributed parabolic linear systems: RHUM Approach. *Numer. Algebr. Control Optim.* **2021**, *11*, 555–566. [CrossRef]
11. Ge, F.; Chen, Y.Q.; Kou, C. *Regional Analysis of Time-Fractional Diffusion Processes*; Springer: Cham, Switzerland, 2018. [CrossRef]
12. Guechi, S.; Debbouche, A.; Torres, D.F.M. Approximate controllability of impulsive non-local non-linear fractional dynamical systems and optimal control. *Miskolc Math. Notes* **2018**, *19*, 255–271. [CrossRef]
13. Almeida, R.; Tavares, D.; Torres, D.F.M. *The Variable-Order Fractional Calculus of Variations*; Springer: Cham, Switzerland, 2019. [CrossRef]
14. Hilfer, R. *Applications of Fractional Calculus in Physics*; World Scientific Publishing: Singapore, 2000. [CrossRef]
15. Magin, R.L. *Fractional Calculus in Bioengineering*; Begell House Inc.: New York, NY, USA, 2006.
16. Tabanfar, Z.; Ghassemi, F.; Bahramian, A.; Nouri, A.; Ghaffari Shad, E.; Jafari, S. Fractional-order systems in biological applications: Estimating causal relations in a system with inner connectivity using fractional moments. In *Emerging Methodologies and Applications in Modelling*; Radwan, A.G., Khanday, F.A., Said, L.A., Eds.; *Fractional-Order Design*; Academic Press: Cambridge, MA, USA, 2022; pp. 275–299.
17. Mophou, G.M.; N'Guerekata, G.M. Existence of mild solutions of some semilinear neutral fractional functional evolution equations with infinite delay. *Appl. Math. Comput.* **2010**, *216*, 61–69. [CrossRef]
18. Shu, X.B.; Lai, Y.; Chen, Y. The existence of mild solutions for impulsive fractional partial differential equations. *Nonlinear Anal. TMA* **2011**, *74*, 2003–2011. [CrossRef]
19. Abada, N.; Benchohra, M.; Hammouche, H. Existence and controllability results for nondensely defined impulsive semilinear functional differential inclusions. *J. Differ. Equ.* **2009**, *246*, 3834–3863. [CrossRef]
20. Ahmadova, A.; Mahmudov, N.I. Existence and uniqueness results for a class of fractional stochastic neutral differential equations. *Chaos Solitons Fractals* **2020**, *139*, 110253. [CrossRef]
21. Kavitha, K.; Vijayakumar, V.; Shukla, A.; Sooppy Nisar, K.; Udhayakumar, R. Results on approximate controllability of Sobolev-type fractional neutral differential inclusions of Clarke subdifferential type. *Chaos Solitons Fractals* **2021**, *151*, 111264. [CrossRef]
22. Sakthivel, R.; Mahmudov, N.I.; Nieto, J.J. Controllability for a class of fractional-order neutral evolution control systems. *Appl. Math. Comput.* **2012**, *218*, 10334–10340. [CrossRef]
23. Sakthivel, R.; Ganesh, R.; Ren, Y.; Anthoni, S.M. Approximate controllability of nonlinear fractional dynamical systems. *Commun. Nonlinear Sci. Numer. Simul.* **2013**, *18*, 3498–3508. [CrossRef]
24. Qin, H.; Zuo, X.; Liu, J.; Liu, L. Approximate controllability and optimal controls of fractional dynamical systems of order $1 < q < 2$ in Banach spaces. *Adv. Differ. Equ.* **2015**, *73*, 1–17. [CrossRef]
25. Yan, Z.; Jia, X. Optimal controls for fractional stochastic functional differential equations of order $\alpha \in (1,2]$. *Bull. Malays. Math. Sci. Soc.* **2018**, *41*, 1581–1606. [CrossRef]
26. Zhou, Y.; He, J.W. New results on controllability of fractional evolution systems with order $\alpha \in (1,2)$. *Evol. Equ. Control Theory* **2021**, *10*, 491–509. [CrossRef]
27. Xi, X.-X.; Hou, M.; Zhou, X.-F.; Wen, Y. Approximate controllability of fractional neutral evolution systems of hyperbolic type. *Evol. Equ. Control Theory* **2022**, *11*, 1037–1069. [CrossRef]
28. Dineshkumar, C.; Udhayakumar, R.; Vijayakumar, V.; Kottakkaran Sooppy, N. A discussion on the approximate controllability of Hilfer fractional neutral stochastic integro-differential systems. *Chaos Solitons Fractals* **2021**, *142*, 110472. [CrossRef]
29. Ma, Y.-K.; Kavitha, K.; Albalawi, W.; Shukla, A.; Sooppy Nisar, K.; Vijayakumar, V. An analysis on the approximate controllability of Hilfer fractional neutral differential systems in Hilbert spaces. *Alex. Eng. J.* **2022**, *61*, 7291–7302. [CrossRef]
30. Wen, Y.; Xi, X.-X. Complete controllability of nonlinear fractional neutral functional differential equations. *Adv. Contin. Discrete Models* **2022**, *33*, 1–11. [CrossRef]
31. Podlubny, I. *Fractional Differential Equations, Mathematics in Science and Engineering*; Academic Press, Inc.: San Diego, CA, USA, 1999.
32. Kilbas, A.A.; Srivastava, H.M.; Trujillo, J.J. *Theory and Applications of Fractional Differential Equations*; Elsevier Science B.V.: Amsterdam, The Netherlands, 2006.
33. Pazy, A. *Semi-Groups of Linear Operators and Applications to Partial Differential Equations*; Springer: New York, NY, USA, 1983. [CrossRef]

34. Zhou, Y.; Jiao, F. Existence of mild solutions for fractional neutral evolution equations. *Comput. Math. Appl.* **2010**, *59*, 1063–1077. [CrossRef]
35. Kumar, S.; Sukavanam, N. Approximate controllability of fractional order semilinear systems with bounded delay. *J. Differ. Equ.* **2012**, *252*, 6163–6174. [CrossRef]
36. El-Borai, M.M. Some probability densities and fundamental solutions of fractional evolution equations. *Chaos Solitons Fractals* **2002**, *14*, 433–440. [CrossRef]
37. Ge, F.; Zhou, H.; Kou, C. Approximate controllability of semilinear evolution equations of fractional order with nonlocal and impulsive conditions via an approximating technique. *Appl. Math. Comput.* **2016**, *275*, 107–120. [CrossRef]
38. Carmichael, N.; Pritcbard, A.J.; Quinn, M.D. State andparameter estimation problems for nonlinear systems. *Appl. Math. Optim.* **1982**, *9*, 133–161. [CrossRef]
39. Quinn, M.D.; Carmichael, N. An approach to nonlinear control problems using fixed-point methods, degree theory and pseudo-inverses. *Numer. Funct. Anal. Optim.* **1985**, *7*, 197–219. [CrossRef]
40. Karite, T.; Boutoulout, A.; Torres, D.F.M. Enlarged controllability and optimal control of sub-diffusion processes with Caputo fractional derivatives. *Progr. Fract. Differ. Appl.* **2020**, *6*, 81–93. [CrossRef]
41. Karite, T.; Boutoulout, A.; Torres, D.F.M. Enlarged controllability of Riemann-Liouville fractional differential equations. *J. Comput. Nonlinear Dynam.* **2018**, *13*, 090907. [CrossRef]
42. Khan, A.; Ullah, H.; Zahri, M.; Humphries, U.W.; Karite, T.; Yusuf, A.; Ullah, H.; Fiza, M. Stationary distribution and extinction of stochastic coronavirus (COVID-19) epidemic model. *Fractals* **2022**, *30*, 2240050. [CrossRef]
43. Zarin, R.; Khan, A.; Inc, M.; Humphries, U.W.; Karite, T. Dynamics of five grade leishmania epidemic model using fractional operator with Mittag-Leffler kernel. *Chaos Solitons Fractals* **2021**, *147*, 110985. [CrossRef]

MDPI

Article

Fractional Dynamics of a Measles Epidemic Model

Hamadjam Abboubakar [1,2,*], Rubin Fandio [2], Brandon Satsa Sofack [2] and Henri Paul Ekobena Fouda [2]

1 Depatment of Computer Engineering, University Institute of Technology of Ngaoundéré, University of Ngaoundéré, Ngaoundéré P.O. Box 455, Cameroon
2 Department of Physics, Laboratory of Biophysics, Faculty of Science, The University of Yaoundé 1, Yaoundé P.O. Box 812, Cameroon; rubinkitou0@gmail.com (R.F.); satsabrandon@gmail.com (B.S.S.); hekobena@gmail.com (H.P.E.F.)
* Correspondence: h.abboubakar@gmail.com or hamadjam.abboubakar@uni-konstanz.de; Tel.: +237-694-52-3111

Abstract: In this work, we replaced the integer derivative with Caputo derivative to model the transmission dynamics of measles in an epidemic situation. We began by recalling some results on the local and global stability of the measles-free equilibrium point as well as the local stability of the endemic equilibrium point. We computed the basic reproduction number of the fractional model and found that is it equal to the one in the integer model when the fractional order $\nu = 1$. We then performed a sensitivity analysis using the global method. Indeed, we computed the partial rank correlation coefficient (PRCC) between each model parameter and the basic reproduction number R_0 as well as each variable state. We then demonstrated that the fractional model admits a unique solution and that it is globally stable using the Ulam–Hyers stability criterion. Simulations using the Adams-type predictor–corrector iterative scheme were conducted to validate our theoretical results and to see the impact of the variation of the fractional order on the quantitative disease dynamics.

Keywords: measles; mathematical model; global sensitivity analysis; partial rank correlation coefficient (PRCC); fractional derivative; Caputo derivative; Ulam–Hyers stability

MSC: 92D30; 26A33

Citation: Abboubakar, H.; Fandio, R.; Sofack, B.S.; Ekobena Fouda, H.P. Fractional Dynamics of a Measles Epidemic Model. *Axioms* **2022**, *11*, 363. https://doi.org/10.3390/axioms11080363

Academic Editor: Natália Martins

Received: 27 June 2022
Accepted: 21 July 2022
Published: 26 July 2022

1. Introduction

Measles, also called rubeola or morbilli, is an infectious illness caused by the *Morbillivirus* of the family of *Paramyxoviridae* [1,2]. It principally affects children below five years of age and has high mortality [2,3]. Despite the availability of a vaccine against the measles virus, this illness remained a health problem that concerns the World Health Organization (WHO). Indeed, in 2017, about 110,000 people died from measles, particularly children below the age of 6 [2,4]. Table 1 depicts the 10 countries that have mainly been affected by a global measles outbreak, while Figure 1 shows the global repartition of measles.

Several works based on mathematical modeling have been proposed to study the transmission dynamics of several diseases. These works are mainly based on the SIR-type compartmental modeling [5–12]. The interest in using mathematical modeling to study the measles transmission dynamics and to find control measures that permit preventing or stopping measles outbreaks is increasing [13]. Many authors have proposed and analyzed different mathematical models of measles based on compartmental modeling [2,14–22]. Among these works, only [2] took into account the effects of hospitalization of the infected individuals. Indeed, most of them consider the traditional SEIR-compartmental models.

Table 1. Top 10 countries with global measles outbreaks [23].

Rank	Country	Number of Cases	Rank	Country	Number of Cases
1	Nigeria	17,794	6	Democratic Republic of the Congo	1907
2	India	5874	7	Afghanistan	1621
3	Somalia	4772	8	Liberia	1495
4	Ethiopia	3403	9	Cameroon	1373
5	Pakistan	2677	10	Ivory Coast	1152

Figure 1. World repartition of measles in 2019 [24].

Fractional calculus has been a useful tool to model, predict, and forecast epidemic outbreaks for the last 20 years. Indeed, as fractional calculus was predicted by Leibniz to be a paradox, it has become a central interest for many researchers in various fields, such as engineering sciences [25], mathematical epidemiology [26,27], physics [28], and economics [29]. The most commonly known fractional operators are Caputo derivatives and their variants [30–38], the Caputo–Fabrizio derivative [39], Atangana–Baleanu derivative [40,41], and piece-wise derivative [42]. The kernels of some of these mentioned operators have different characteristics. For example, the Caputo operator is defined with the power law-type kernel (nonlocal but singular), and that of Caputo–Fabrizio has an exponentially decaying (nonsingular) kernel, while the Atangana–Baleanu operator in the Caputo sense has a Mittag–Leffler-type kernel [43]. In a recent work [44], Atangana formulated and studied a compartmental model that could be used to depict the survival of fractional calculus. The fact that the Caputo operator has a memory effect and the Caputo derivative of a constant function is equal to zero [45] means that this derivative is the most used.

Concerning the transmission dynamics of measles, few authors have used fractional derivatives [22,46–49]. In [46]. Farman et al. employed a fraction Caputo operator on a SEIR epidemic model to control measles for infected populations. Ogunmiloro et al. [47] studied a mathematical model describing the transmission dynamics of measles with a double vaccination dose, treatment, and two groups of measles-infected and measles-induced encephalitis-infected humans with relapse under the fractional Atangana–Baleanu–Caputo (ABC) operator. Qureshi, in [22], proposed a new epidemiological system for the measles epidemic using the Caputo fractional derivative with a memory effect.

The objective of this work was to compare, from a quantitative point of view, the dynamics of an epidemic model of measles with integer derivatives and fractional derivatives (in the sense of Caputo). To achieve our goal, we extended the model by Olumuyiwa et al. [2], which consists of a six-compartmental model integrating vaccinated and hospitalized individuals, by replacing the integer derivative with the Caputo derivative. The theoretical analysis of the fractional model was performed by a classical method and consists of the algebraic determination of the basic reproduction number R_0, which depends on the fractional order ν, the proof of the local and global stabilities of the disease-free equilibrium, as well as the local stability of the endemic equilibrium. To determine the model parameters that have a great influence on the measles epidemic in Nigeria, we performed a global sensitivity analysis of the model by computing the partial rank correlation coefficients between the basic reproduction number (as well as state variables) and the model parameters. We proved the existence and uniqueness of the solutions of the fractional model as well as its global stability using the Ulam–Hyers method. We then constructed a numerical scheme based on the Adams-type predictor–corrector iterative scheme [50,51], and finally, performed a numerical simulation to see the impact of the variation of the fractional order on the disease dynamics.

The paper is presented as follows: Section 2 is devoted to the model formulation and basic results. Section 3 is devoted to the global sensitivity analysis. In Section 4, we recall some definitions and useful results concerning fractional calculus. We also formulated the fractional measles model with the Caputo derivative and performed asymptotic stability of equilibrium points. Then, we provide the proof of existence, the uniqueness of the solution, and the global stability of the fractional model. The numerical scheme is also presented in this section. Section 5 is devoted to the numerical simulations. A conclusion rounds up the paper.

2. Model Formulation and Basic Results

In [2], Olumuyiwa et al. proposed and studied the following $\mathcal{SVEIHR}$ compartmental model with the integer derivative

$$\begin{cases} \frac{d}{dt}\mathcal{S}(t) &= \Lambda + \alpha\mathcal{V} - \beta\mathcal{S}\mathcal{I} - \overbrace{(c+d)}^{k_1}\mathcal{S}, \\ \frac{d}{dt}\mathcal{V}(t) &= c\mathcal{S} - \overbrace{(d+\alpha)}^{k_2}\mathcal{V}, \\ \frac{d}{dt}\mathcal{E}(t) &= \beta\mathcal{S}\mathcal{I} - \overbrace{(d+\gamma)}^{k_3}\mathcal{E}, \\ \frac{d}{dt}\mathcal{I}(t) &= \gamma\mathcal{E} - \overbrace{(d+a+\phi)}^{k_4}\mathcal{I}, \\ \frac{d}{dt}\mathcal{H}(t) &= \phi\mathcal{I} - \overbrace{(\sigma+a+d)}^{k_5}\mathcal{H}, \\ \frac{d}{dt}\mathcal{R}(t) &= \sigma\mathcal{H} - d\mathcal{R}. \end{cases} \tag{1}$$

to model the transmission dynamics of measles in Nigeria. In Equation (1), $\mathcal{S}$ denotes the susceptible population, $\mathcal{V}$ is the vaccinated population, $\mathcal{E}$ is the total number of latent persons (infected but not infectious), $\mathcal{I}$ is the total number of infected persons, $\mathcal{H}$ is the total number of hospitalized persons, and $\mathcal{R}$ is the total number of recovered persons. The description of the model parameters and their values are consigned in Table 2.

Table 2. Biological description of model parameters and their numerical values [2].

Parameter	Description	Values (per Week)
Λ	Recruitment rate	68,027
α	Rate of loss of vaccine immunity	0.003286
β	Transmission Rate	10^{-9}
c	Vaccination rate	10^{-6}
d	Natural death rate	0.000309
γ	Rate of progression from $\mathcal{E}$ to $\mathcal{I}$	0.5
a	Disease-induced rate	0.033720
ϕ	Rate of progression from $\mathcal{I}$ to $\mathcal{H}$	0.036246
σ	Rate of progression from $\mathcal{H}$ to $\mathcal{R}$	0.062366

The following subset of $\mathbb{R}^6$

$$\Sigma = \left\{ (\mathcal{S}, \mathcal{V}, \mathcal{E}, \mathcal{I}, \mathcal{H}, \mathcal{R})^t \in \mathbb{R}^6_+ : \mathcal{N} := \mathcal{S} + \mathcal{V} + \mathcal{E} + \mathcal{I} + \mathcal{H} + \mathcal{R} \leq \frac{\Lambda}{d} \right\}$$

is positively invariant for system Equation (2), which defines a dynamical system. Since the state variable $\mathcal{R}$ only appears in the last equation of Equation (1), it is sufficient to study the following reduced system

$$\begin{cases} \dfrac{d}{dt}\mathcal{S}(t) &= \Lambda + \alpha\mathcal{V} - \beta\mathcal{S}\mathcal{I} - \overbrace{(c+d)}^{k_1}\mathcal{S}, \\ \dfrac{d}{dt}\mathcal{V}(t) &= c\mathcal{S} - \overbrace{(d+\alpha)}^{k_2}\mathcal{V}, \\ \dfrac{d}{dt}\mathcal{E}(t) &= \beta\mathcal{S}\mathcal{I} - \overbrace{(d+\gamma)}^{k_3}\mathcal{E}, \\ \dfrac{d}{dt}\mathcal{I}(t) &= \gamma\mathcal{E} - \overbrace{(d+a+\phi)}^{k_4}\mathcal{I}, \\ \dfrac{d}{dt}\mathcal{H}(t) &= \phi\mathcal{I} - \overbrace{(\sigma+a+d)}^{k_5}\mathcal{H}. \end{cases} \tag{2}$$

The model Equation (2) admits two nonnegative equilibrium points: the disease-free equilibrium $\mathbb{E}_0 = (\mathcal{S}_0, \mathcal{V}_0, 0, 0, 0, 0)^t = \left(\frac{(d+\alpha)\Lambda}{(d+\alpha+c)d}, \frac{c\Lambda}{(d+\alpha+c)d}, 0, 0, 0, 0 \right)^t$ and a unique endemic equilibrium $\mathbb{E}_1 = (\mathcal{S}_1, \mathcal{V}_1, \mathcal{E}_1, \mathcal{I}_1, \mathcal{H}_1, \mathcal{R}_1)^t$, where

$$\begin{cases} \mathcal{I}_1 &= \dfrac{\Lambda\gamma}{R_0 k_3 k_4}(R_0 - 1), \\ \mathcal{S}_1 &= \dfrac{k_2\Lambda}{c\alpha + k_2(k_1 + \beta\mathcal{I}_1)}, \\ \mathcal{V}_1 &= \dfrac{c}{k_2}\mathcal{S}_1, \\ \mathcal{E}_1 &= \dfrac{\beta}{k_3}\mathcal{S}_1\mathcal{I}_1, \\ \mathcal{H}_1 &= \dfrac{\phi}{k_5}\mathcal{I}_1, \end{cases} \tag{3}$$

with R_0, which denotes the basic reproduction number expressed as follows:

$$R_0 = \frac{\beta\gamma}{k_3 k_4} \frac{(d+\alpha)\Lambda}{(d+\alpha+c)d}. \tag{4}$$

From Equation (3), it follows that:

Proposition 1. *The model Equation (2) admits a unique endemic equilibrium point* $\mathbb{E}_1 = (\mathcal{S}_1, \mathcal{V}_1, \mathcal{E}_1, \mathcal{I}_1, \mathcal{H}_1, \mathcal{R}_1)^t$ *if and only if* $R_0 > 1$.

Theorem 1 ([2])**.** *(i)* *The disease-free equilibrium point* $\mathbb{E}_0$ *is locally and globally asymptotically stable if and only if* $R_0 \leq 1$;

(ii) *The unique endemic equilibrium point* $\mathbb{E}_1$ *is locally stable whenever* $R_0 > 1$.

Remark 1. *As suggested in [44], the epidemic spread can also be evaluated by computing the so-called threshold "strength number". Indeed, the strength number permits knowing, in an epidemic period, if there is the possibility for a renewal process [44]. In the case of epidemic measles, model Equation (9), this threshold is equal to zero, which implies that the spread does not have a renewal process.*

3. Uncertainty and Global Sensitivity Analysis

In [2], the authors performed a local sensitivity analysis by computing the sensitivity indices of R_0 against model parameters. The disadvantage of this kind of sensitivity analysis (SA) is that the sensitivity index is calculated by varying only one parameter while the remaining parameters are fixed. Considering the combined variability from all input parameters simultaneously, we performed a global sensitivity analysis to examine the model's response to parameter variation in the parameter space. To this aim, we computed the partial rank correlation coefficient (PRCC) between R_0 (as well as each state variable of the model) and each model parameter. PRCC is the best and most reliable sensitivity analysis method that provides monotonicity between parameters and the model output when we want to measure the nonlinear (but monotonic) relationship between two variables [52,53]. The Latin hypercube sampling (LHS) was used as a sampling technique [54] with the number of runs equal to 5000. Each model parameter was supposed to be random with uniform distribution and their mean values are listed in Table 2. The most influential parameter is the one with the PRCC less than -0.5 or greater than $+0.5$ [53]. The results of the SA are depicted in Figures 2–4.

From Figure 2, it is clear that the parameters β, a, and ϕ have the highest influence on R_0. This suggests that individual protection combined with efficient treatment may potentially be the most effective strategy to reduce the basic reproduction number.

From Figure 3, the parameters with the highest influence on $\mathcal{S}$ are β, a, and ϕ; the ones with the highest influence on $\mathcal{V}$ are β, α, and c, while the ones with the highest influence on $\mathcal{R}$ are a, ϕ, and σ.

From Figure 4, the parameters with the highest influence on $\mathcal{E}$ are Λ and γ; the ones with the highest influence on $\mathcal{I}$ are Λ, a, and ϕ, while the ones with the highest influence on $\mathcal{H}$ are a, ϕ, and σ. In general, mass vaccination combined with personal protection and care as soon as the first symptoms appear would make it possible to effectively fight measles.

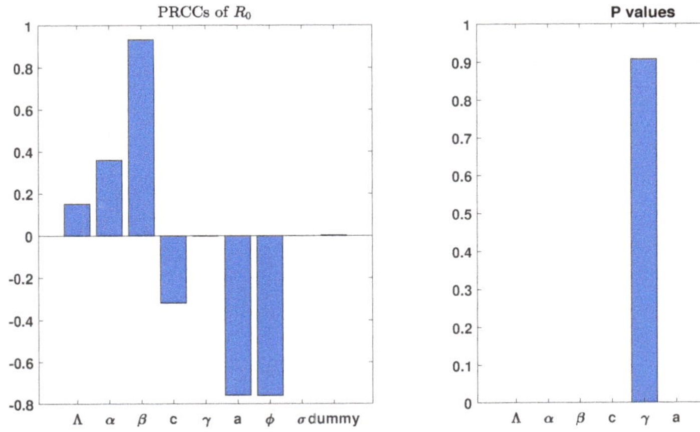

Figure 2. Partial rank correlation coefficients between the basic reproduction number R_0 and model parameters.

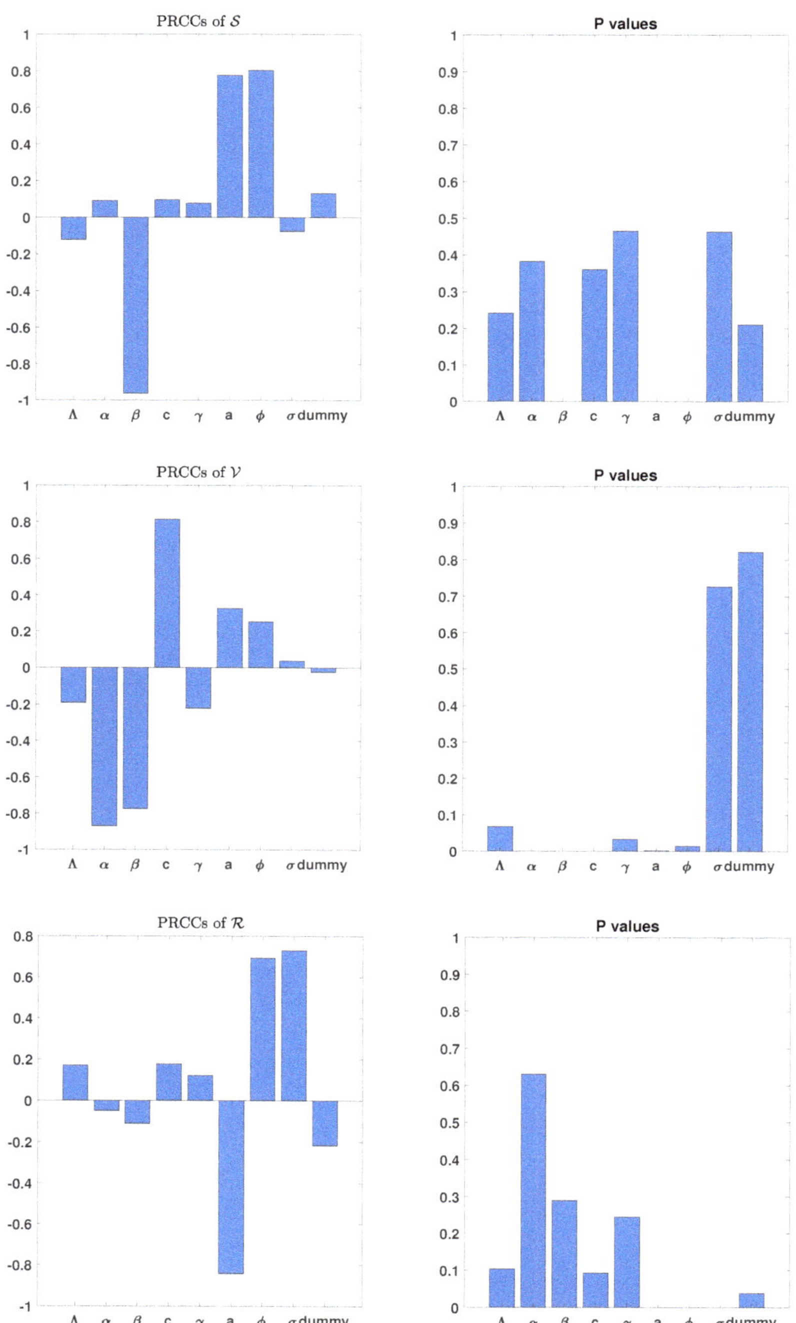

Figure 3. Partial rank correlation coefficients between uninfected state variables of the model and model parameters.

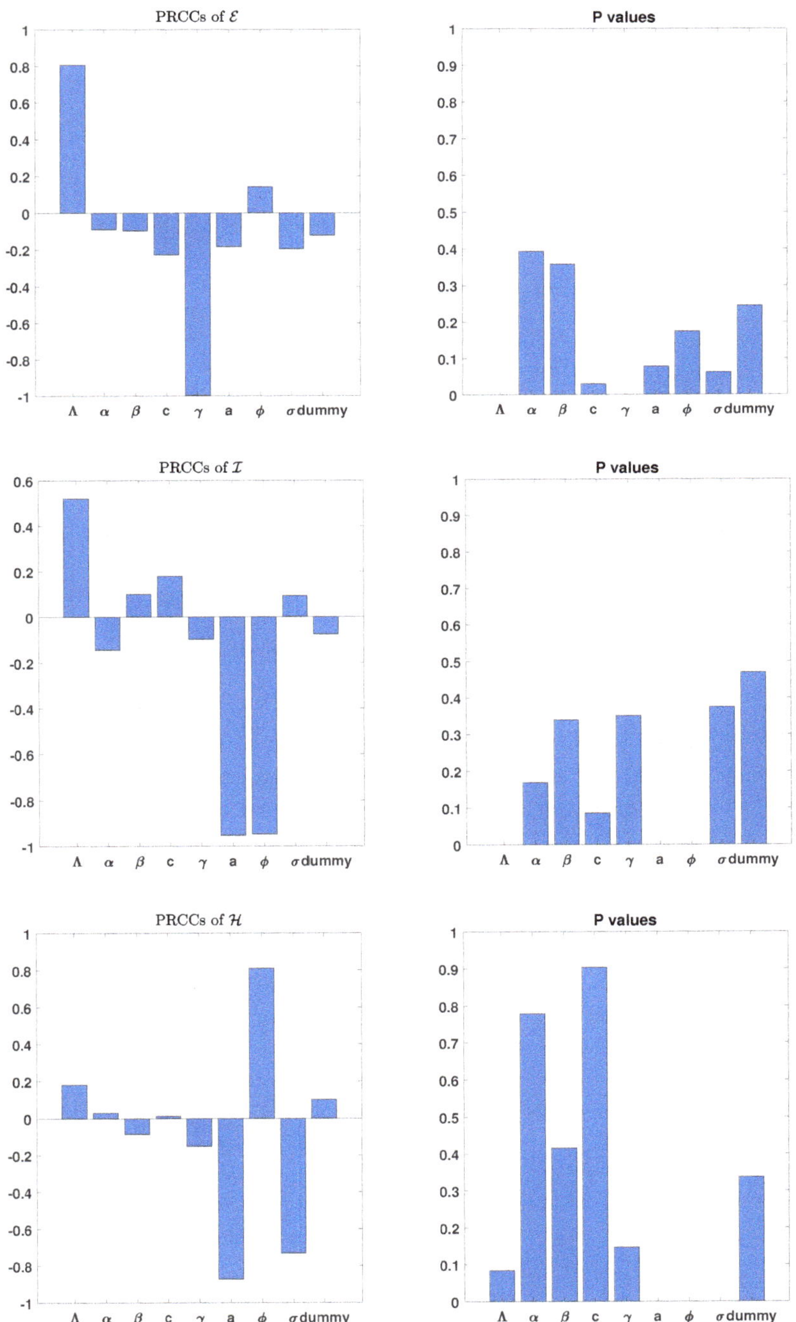

Figure 4. Partial rank correlation coefficients between infected state variables of the model and model parameters.

4. The Fractional Model and Its Analysis

4.1. Primarily Definition and Results of Fractional Calculus

Now, we present some main properties of the fractional-order differential equations.

Definition 1 ([55,56]). *The Caputo fractional derivative of a function $f \in C([a,b],\mathbb{R})$ can be written as follows:*

$$ {}_0^C\mathbb{D}_t^{\nu}\psi(t) = \chi^{\mu-\nu} f^{(\mu)}(t), \quad \beta > 0, \tag{5} $$

where ${}_0^C\mathbb{D}^{\nu}$ is the Caputo operator, $0 < \nu \leq 1$ is the fractional order, μ is the smallest integer not equal to ν, and χ^{ϕ} is the Riemann–Liouville integral operator of order ϕ, defined as follows

$$ \chi^{\phi}\Upsilon(t) = \frac{1}{\Gamma(\phi)} \int_0^t (t-\tau)^{\phi-1}\Upsilon(\tau)\mu\tau, \quad \phi > 0. \tag{6} $$

Theorem 2 ([55,57]). *Consider the m-dimensional system*

$$ \begin{cases} \frac{\theta^{\nu} z}{\theta t^{\nu}} & = & \omega z, \\ z(0) & = & z_0, \end{cases} \tag{7} $$

where ω is a squared constant matrix of order $m \times m$, and $0 < \nu < 1$. Let us denote by γ_i, $i = 1, 2, \ldots, m$ the eigenvalues of ω.

- *$|arg(\gamma_i)| > \frac{\nu\pi}{2}$, $i = 1, 2, \ldots, m$ iff $z = 0$ is asymptotically stable;*
- *$z = 0$ is stable if $|arg(\gamma_i)| \geq \frac{\nu\pi}{2}$, $i = 1, 2, \ldots, m$, and eigenvalues with $|arg(\gamma_i)| = \frac{\nu\pi}{2}$ have the same algebraic and geometric multiplicities.*

Theorem 3 ([55,57]). *Let us consider the following system:*

$$ \begin{cases} \frac{\mu^{\nu} z}{\mu t^{\nu}} & = & \Psi(z), \\ z(0) & = & z_0 \text{ with } 0 < \nu < 1 \text{ and } z \in \mathbb{R}^{\nu}. \end{cases} \tag{8} $$

Solving equation $\Psi(z) = 0$ permits obtaining the equilibrium points of system Equation (8). If all of the eigenvalues γ_i of $J = \frac{\partial \psi(z^*)}{\partial z}$ satisfy $|arg(\gamma_i)| > \frac{\nu\pi}{2}$, then the equilibrium point z^* is locally asymptotically stable (LAS).

4.2. The Fractional Model with Caputo Operator

By replacing integer derivative with Caputo fractional derivative in system (1), we obtain the following system:

$$ \begin{cases} {}_0^C\mathbb{D}^{\nu}\mathcal{S}(t) & = & \Lambda^{\nu} + \alpha^{\nu}\mathcal{V} - \beta^{\nu}\mathcal{S}\mathcal{I} - (c^{\nu} + d^{\nu})\mathcal{S}, \\ {}_0^C\mathbb{D}^{\nu}\mathcal{V}(t) & = & c^{\nu}\mathcal{S} - (d^{\nu} + \alpha^{\nu})\mathcal{V}, \\ {}_0^C\mathbb{D}^{\nu}\mathcal{E}(t) & = & \beta^{\nu}\mathcal{S}\mathcal{I} - (d^{\nu} + \gamma^{\nu})\mathcal{E}, \\ {}_0^C\mathbb{D}^{\nu}\mathcal{I}(t) & = & \gamma^{\nu}\mathcal{E} - (d^{\nu} + a^{\nu} + \phi^{\nu})\mathcal{I}, \\ {}_0^C\mathbb{D}^{\nu}\mathcal{H}(t) & = & \phi^{\nu}\mathcal{I} - (\sigma^{\nu} + a^{\nu} + d^{\nu})\mathcal{H}, \\ {}_0^C\mathbb{D}^{\nu}\mathcal{R}(t) & = & \sigma^{\nu}\mathcal{H} - d^{\nu}\mathcal{R}, \end{cases} \tag{9} $$

For this model, the corresponding basic reproduction number is given by

$$ R_0 = \frac{\beta^{\nu}\gamma^{\nu}(d^{\nu} + \alpha^{\nu})\Lambda^{\nu}}{(d^{\nu} + \gamma^{\nu})(d^{\nu} + a^{\nu} + \phi^{\nu})(d^{\nu} + \alpha^{\nu} + c^{\nu})d^{\nu}}. \tag{10} $$

From Equation (10), we note that for $\nu = 1$, the basic reproduction number coincides for both models.

As in the case of the model with the integer derivative Equation (2), the fractional model Equation (4) only has one endemic equilibrium $\mathbb{E}_1 = (\mathcal{S}_1, \mathcal{V}_1, \mathcal{E}_1, \mathcal{I}_1, \mathcal{H}_1, \mathcal{R}_1)'$, where

$$\begin{cases} \mathcal{I}_1 &= \dfrac{\Lambda^\nu \gamma^\nu}{R_0 k_3 k_4}(R_0 - 1),\ \mathcal{S}_1 = \dfrac{(d^\nu + \alpha^\nu)\Lambda^\nu}{c\alpha + k_2(k_1 + \beta^\nu \mathcal{I}_1)}, \\ \mathcal{V}_1 &= \dfrac{c^\nu}{k_2}\mathcal{S}_1,\ \mathcal{E}_1 = \dfrac{\beta^\nu}{k_3}\mathcal{S}_1\mathcal{I}_1,\ \mathcal{H}_1 = \dfrac{\phi^\nu}{k_5}\mathcal{I}_1,\ \mathcal{R}_1 = \dfrac{\sigma^\nu}{d^\nu}\mathcal{H}_1, \end{cases} \tag{11}$$

with Setting $k_1 = c^\nu + d^\nu$, $k_2 = d^\nu + \alpha^\nu$, $k_3 = d^\nu + \gamma^\nu$, $k_4 = d^\nu + a^\nu + \phi^\nu$, and $k_5 = \sigma^\nu + a^\nu + d^\nu$.

4.2.1. Asymptotic Stability of the Disease-Free Equilibrium

The following theorems are proved in the same way as in [2].

Theorem 4. *The disease-free equilibrium point $\mathbb{E}_0$ is locally asymptotically stable if $R_0 < 1$; otherwise, it is unstable.*

Theorem 5. *The disease-free equilibrium point $(\mathbb{E}_0)$ is globally asymptotically stable whenever $R_0 < 1$.*

Theorem 6. *The endemic equilibrium point $(\mathbb{E}_1)$ is locally stable, when $R_0 > 1$.*

4.2.2. Existence and Uniqueness of Solution

In this section, we present the results of the existence and uniqueness of the solution of the fractional differential Equation (9).

For this purpose, let $\eta = \mathcal{C}([0;a], \mathbb{R})$, a Banach space of the continuous function from $[0;a]$ to $\mathbb{R}$, endowed with the norm $\| \mathcal{Z} \|_\eta = \sup_{t\in[0,a]} \{|\mathcal{Z}|\}$.

The system Equation (9) can be rewritten in the following compact form:

$$^C_t\mathbb{D}\mathcal{X}(t) = \xi(t, \mathcal{X}),\ \ \mathcal{X}(0) = \mathcal{X}_0 \geq 0,\ \ t \in [0, a],\ \ \nu \in (0;1). \tag{12}$$

where $\mathcal{X} = (\mathcal{S}, \mathcal{V}, \mathcal{E}, \mathcal{I}, \mathcal{R})'$ and $\xi = (\xi_1, \xi_2, \xi_3, \xi_4, \xi_5, \xi_6)'$ with

$$\begin{cases} \xi_1 = \Lambda^\nu + \alpha^\nu \mathcal{V} - \beta^\nu \mathcal{S}\mathcal{I} - (c^\nu + d^\nu)\mathcal{S}, \\ \xi_2 = c^\nu \mathcal{S} - (d^\nu + \alpha^\nu)\mathcal{V}, \\ \xi_3 = \beta^\nu \mathcal{S}\mathcal{I} - (d^\nu + \gamma^\nu)\mathcal{E}, \\ \xi_4 = \gamma^\nu \mathcal{E} - (d^\nu + a^\nu + \phi^\nu)\mathcal{I}, \\ \xi_5 = \phi^\nu \mathcal{I} - (\sigma^\nu + a^\nu + d^\nu)\mathcal{H}, \\ \xi_6 = \sigma^\nu \mathcal{H} - d^\nu \mathcal{R}, \end{cases}$$

Since $\xi_i \in \eta$, for all of $i \in [0;6] \cap \mathbb{N}$, then the following operator

$$(\mathcal{P}\mathcal{X})(t) = \mathcal{X}_0 + \frac{1}{\Gamma(\nu)} \int_0^t (t - \alpha)^{\nu - 1} \xi(\alpha, \mathcal{X}(\alpha)) d\alpha, \tag{13}$$

is well-defined. The following result is valid:

Lemma 1. *Let $\overline{\mathcal{X}} = (\overline{\mathcal{S}}, \overline{\mathcal{V}}, \overline{\mathcal{E}}, \overline{\mathcal{I}}, \overline{\mathcal{H}}, \overline{\mathcal{R}})'$. The function $\xi = (\xi_i)'$ defined above satisfies $\| \xi(t, \mathcal{X}(t)) - \xi(t, \overline{\mathcal{X}}(t)) \|_\eta \leq \mathbf{N}_\xi \| \mathcal{X} - \overline{\mathcal{X}} \|_\eta$ for some $\mathbf{N}_\xi > 0$.*

Proof. We proceed as follows for the first component of ξ:

$$\begin{aligned} &\left|\xi_1(t, \mathcal{X}(t)) - \xi_1(t, \overline{\mathcal{X}}(t))\right| \\ &= \left|\alpha^\nu(\mathcal{V}(t) - \overline{\mathcal{V}}(t)) - \beta^\nu(\mathcal{S}(t)\mathcal{I}(t) - \overline{\mathcal{S}}(t)\overline{\mathcal{I}}(t)) - h_1(\mathcal{S}(t) - \overline{\mathcal{S}}(t))\right|, \\ &\leq h_1\left|\mathcal{S}(t) - \overline{\mathcal{S}}(t)\right| + \alpha^\nu\left|\mathcal{V}(t) - \overline{\mathcal{V}}(t)\right| + \beta^\nu\left|\mathcal{S}(t)\mathcal{I}(t) - \overline{\mathcal{S}}(t)\overline{\mathcal{I}}(t)\right|. \end{aligned} \tag{14}$$

However,

$$\begin{aligned}|\mathcal{S}(t)\mathcal{I}(t)-\overline{\mathcal{S}}(t)\overline{\mathcal{I}}(t)| &= |\mathcal{I}(t)(\mathcal{S}(t)-\overline{\mathcal{S}}(t))+\overline{\mathcal{S}}(t)(\mathcal{I}(t)-\overline{\mathcal{I}}(t))| \\ &\leq |\mathcal{I}(t)||\mathcal{S}(t)-\overline{\mathcal{S}}(t)|+|\overline{\mathcal{S}}(t)||\mathcal{I}(t)-\overline{\mathcal{I}}(t)|.\end{aligned}$$

Thus, we obtain

$$\begin{aligned}&|\xi_1(t,\mathcal{X}(t))-\xi_1(t,\overline{\mathcal{X}}(t))| \\ &= |\alpha^\nu(\mathcal{V}(t)-\overline{\mathcal{V}}(t))-\beta^\nu(\mathcal{S}(t)\mathcal{I}(t)-\overline{\mathcal{S}}(t)\overline{\mathcal{I}}(t))-h_1(\mathcal{S}(t)-\overline{\mathcal{S}}(t))|, \\ &\leq h_1|\mathcal{S}(t)-\overline{\mathcal{S}}(t)|+\alpha^\nu|\mathcal{V}(t)-\overline{\mathcal{V}}(t)| \\ &+\beta^\nu(|\mathcal{I}(t)||\mathcal{S}(t)-\overline{\mathcal{S}}(t)|+|\overline{\mathcal{S}}(t)||\mathcal{I}(t)-\overline{\mathcal{I}}(t)|) \\ &= (h_1+\beta^\nu|\mathcal{I}(t)|)|\mathcal{S}(t)-\overline{\mathcal{S}}(t)|+\alpha^\nu|\mathcal{V}(t)-\overline{\mathcal{V}}(t)| \\ &+\beta^\nu|\overline{\mathcal{S}}(t)||\mathcal{I}(t)-\overline{\mathcal{I}}(t)| \\ &\leq \mathbf{N}_1(|\mathcal{S}(t)-\overline{\mathcal{S}}(t)|+|\mathcal{V}(t)-\overline{\mathcal{V}}(t)|+|\mathcal{I}(t)-\overline{\mathcal{I}}(t)|),\end{aligned} \tag{15}$$

where $\mathbf{N}_1 = h_1+\alpha^\nu+\max_{t\in[0,a]}\{\beta^\nu|\mathcal{I}(t)|+\beta^\nu|\overline{\mathcal{S}}(t)|\}$, with $h_1 = c^\nu+d^\nu$.

Similarly, with ξ_2, we also have :

$$\begin{aligned}|\xi_2(t,\mathcal{X}(t))-\xi_2(t,\overline{\mathcal{X}}(t))| &= |c^\nu(\mathcal{S}(t)-\overline{\mathcal{S}}(t))-h_2(\mathcal{V}(t)-\overline{\mathcal{V}}(t))|, \\ &\leq c^\nu|(\mathcal{S}(t)-\overline{\mathcal{S}}(t))|+h_2|(\mathcal{V}(t)-\overline{\mathcal{V}}(t))| \\ &\leq \mathbf{N}_2(|\mathcal{S}(t)-\overline{\mathcal{S}}(t)|+|\mathcal{V}(t)-\overline{\mathcal{V}}(t)|),\end{aligned} \tag{16}$$

where $\mathbf{N}_2 = c^\nu+h_2$, where $h_2 = d^\nu+\alpha^\nu$.

Based on the analogous reasoning, we obtain ξ_i, $i\in[0;6]\cap\mathcal{N}$:

$$\begin{aligned}|\xi_3(t,\mathcal{X}(t))-\xi_3(t,\overline{\mathcal{X}}(t))| &\leq \mathbf{N}_3(|\mathcal{S}(t)-\overline{\mathcal{S}}(t)|+|\mathcal{I}(t)-\overline{\mathcal{I}}(t)|+|\mathcal{E}(t)-\overline{\mathcal{E}}(t)|), \\ |\xi_4(t,\mathcal{X}(t))-\xi_4(t,\overline{\mathcal{X}}(t)| &\leq \mathbf{N}_4(|\mathcal{E}(t)-\overline{\mathcal{E}}(t)|+|\mathcal{I}(t)-\overline{\mathcal{I}}(t)|), \\ |\xi_5(t,\mathcal{X}(t)-\xi_5(t,\overline{\mathcal{X}}(t)| &\leq \mathbf{N}_5(|\mathcal{I}(t)-\overline{\mathcal{I}}(t|+|\mathcal{H}(t)-\overline{\mathcal{H}}(t)|), \\ |\xi_6(t,\mathcal{X}(t))-\xi_6(t,\overline{\mathcal{XY}}(t))| &\leq \mathbf{N}_6(|\mathcal{H}(t)-\overline{\mathcal{H}}(t)|+|\mathcal{R}(t)-\overline{\mathcal{R}}(t)|),\end{aligned} \tag{17}$$

where $\mathbf{N}_3 = h_3+\max_{t\in[0,a]}\{\beta^\nu(|\mathcal{I}|+|\overline{\mathcal{S}}|)\}$, $\mathbf{N}_4 = \gamma^\nu+h_4$, $\mathbf{N}_5 = \phi^\nu+h_5$, $\mathbf{N}_6 = \phi^\nu+d^\nu$ with $h_3 = d^\nu-\gamma^\nu$,$h_4 = d^\nu+a^\nu+\phi^\nu$, $h_5 = \sigma^\nu+a^\nu+d^\nu$.

Therefore, we finally obtain

$$\begin{aligned}\parallel \xi(t,\mathcal{X}(t))-\xi(t,\overline{\mathcal{X}}(t)) \parallel_\eta &= sup_{t\in[0;a]}\sum_{i=1}^{6}|\xi_i(t,\mathcal{X}(t))-\xi_i(t,\overline{\mathcal{X}}(t))|, \\ &\leq \overbrace{(\mathbf{N}_1+\mathbf{N}_2+\mathbf{N}_3+\mathbf{N}_4+\mathbf{N}_5+\mathbf{N}_6)}^{\mathbf{N}_\eta} \parallel \mathcal{X}-\overline{\mathcal{X}} \parallel_\eta .\end{aligned}$$

□

Theorem 7. *Let the result of Lemma 1 hold and* $\varpi = \dfrac{a^\nu}{\Gamma(\nu+1)}$. *If* $\varpi\mathbf{N}_\eta < 1$, *then there exists a unique solution of the model Equation (12) on* $[0,a]$, *which is uniformly Lyapunov-stable.*

Proof. The function $\xi : [0,a]\times\mathbb{R}^6 \to \mathbb{R}^6_+$ is clearly continuous in its domain. Thus, the existence of the solutions to Equations (9) and (12) follows from (Theorem 3.1, [58]).

The Banach contraction mapping principle on operator $\mathcal{P}$ (see Equation (13)) will be used in the following to prove the uniqueness of the solution of the fractional model

Equation (12). By definition, $\sup_{t\in[0,a]} \parallel \xi(t,0) \parallel = \Lambda^{\nu}$. Let us now define $\zeta > \dfrac{\parallel \mathcal{X}_0 \parallel + \varpi\Lambda^{\nu}}{1-\varpi\mathbf{N}_{\eta}}$ and a closed convex set $\mathcal{Z}_{\zeta} = \{\mathcal{X} \in \eta : \parallel \mathcal{X} \parallel_{\eta} \leq \zeta\}$. Thus, for the sel- map property, it suffices to show that $\mathcal{P}\mathcal{Z}_{\zeta} \subseteq \mathcal{Z}_{\zeta}$. Let $\mathcal{X} \in \mathcal{Z}_{\zeta}$, we have

$$\begin{aligned}
\parallel \mathcal{P}\mathcal{X} \parallel_{\eta} &= \sup_{t\in[0,a]} \left\{ \left| \mathcal{X}_0 + \frac{1}{\Gamma(\nu)} \int_0^t (t-\varepsilon)^{\nu-1} \xi(\varepsilon, \mathcal{X}(\varepsilon)) d\varepsilon \right| \right\}, \\
&\leq |\mathcal{X}_0| + \frac{1}{\Gamma(\nu)} \sup_{t\in[0,a]} \left\{ \int_0^t (t-\varepsilon)^{\nu-1} (|\xi(\varepsilon, \mathcal{X}(\varepsilon)) - \xi(\varepsilon,0)| + |\xi(\varepsilon,0)|) d\varepsilon \right\}, \\
&\leq |\mathcal{X}_0| + \frac{1}{\Gamma(\nu)} \sup_{t\in[0,a]} \left\{ \int_0^t (t-\varepsilon)^{\nu-1} (\parallel \xi(\varepsilon, \mathcal{X}(\varepsilon)) - \xi(\varepsilon,0) \parallel_{\eta} + \parallel \xi(\varepsilon,0) \parallel_{\eta}) d\varepsilon \right\}, \\
&\leq |\mathcal{X}_0| + \frac{\mathbf{N}_{\eta} \parallel \mathcal{X} \parallel_{\eta} + \Lambda^{\nu}}{\Gamma(\nu)} \sup_{t\in[0,a]} \left\{ \int_0^t (t-\varepsilon)^{\nu-1} d\varepsilon \right\}, \\
&\leq |\mathcal{X}_0| + \frac{\mathbf{N}_{\eta}\zeta + \Lambda^{\nu}}{\Gamma(\nu)} \sup_{t\in[0,a]} \left\{ \int_0^t (t-\varepsilon)^{\nu-1} d\varepsilon \right\}, \\
&= |\mathcal{X}_0| + \frac{a^{\nu}}{\Gamma(\nu+1)} (\mathbf{N}_{\eta}\zeta + \Lambda^{\nu}), \\
&= |\mathcal{X}_0| + \varpi(\mathbf{N}_{\eta}\zeta + \Lambda^{\nu}), \\
&\leq \zeta.
\end{aligned}$$

□

Then, $\mathcal{P}\mathcal{X} \subseteq \mathcal{Z}_{\zeta}$ and $\mathcal{P}$ are indeed the self-map. It remains to show that $\mathcal{P}$ is a contraction. Let $\mathcal{X}$ and $\overline{\mathcal{X}}$ two solutions of Equation (12). Using the result of Lemma 1, we obtain

$$\begin{aligned}
\parallel \mathcal{P}\mathcal{X} - \mathcal{P}\overline{\mathcal{X}} \parallel_{\eta} &= \sup_{t\in[0,a]} \{ |(\mathcal{P}\mathcal{X})(t) - (\mathcal{P}\overline{\mathcal{X}})(t)| \}, \\
&= \frac{1}{\Gamma(\nu)} \sup_{t\in[0,a]} \left\{ \int_0^t (t-\varepsilon)^{\nu-1} |\xi(\varepsilon, \mathcal{X}(\varepsilon)) - \xi(\varepsilon, \overline{\mathcal{X}}(\varepsilon))| d\varepsilon \right\}, \\
&\leq \frac{\mathbf{N}_{\eta}}{\Gamma(\nu)} \sup_{t\in[0,a]} \left\{ \int_0^t (t-\varepsilon)^{\nu-1} (|\mathcal{X}(\varepsilon) - \overline{\mathcal{X}}(\varepsilon)| d\varepsilon \right\}, \\
&\leq \varpi\mathbf{N}_{\eta} \parallel \mathcal{X} - \overline{\mathcal{X}} \parallel_{\eta}.
\end{aligned}$$

The condition $\varpi N_{\eta} < 1$ ensures that $\mathcal{P}$ is a contraction mapping. Thus, by the Banach contraction mapping principle, $\mathcal{P}$ has a unique fixed point on $[0,a]$, which is the solution of Equation (12). Theorem 3.2 in [58] ensures the uniformly Lyapunov stability of the solution.

4.3. Global Stability of the Fractional Model

In what follows, we will perform the global stability of the fractional-order model Equation (9) in the sense of Ulam–Hyers [59,60]. To this aim, we introduce the following inequality:

$$\left| {}_0^{C}\mathbb{D}^{\nu} \mathcal{X}(t) - \xi(t, \mathcal{X}(t)) \right| \leq b, \quad t \in [0;a]. \tag{18}$$

A function $\overline{\mathcal{X}} \in \eta$ is a solution of Equation (18) if there exists $\aleph \in \eta$ satisfying

1. $|\aleph(t)| \leq b$;
2. ${}_0^{C}\mathbb{D}^{\nu} \overline{\mathcal{X}}(t) = \xi(t, \overline{\mathcal{X}}(t)) + \aleph(t), \quad t \in [0;a]$.

Since $\overline{\mathcal{X}} \in \eta$ is a solution of Equation (18), then $\overline{\mathcal{X}} \in \eta$ is also a solution of the following integral inequality

$$\left|\overline{\mathcal{X}}(t) - \overline{\mathcal{X}}(0) - \frac{1}{\Gamma(\nu)}\int_0^t (t-\epsilon)^{\nu-1}\xi(\epsilon, \overline{\mathcal{X}}(\epsilon))d\epsilon\right| \leq \varpi b. \tag{19}$$

We claim the following result:

Theorem 8. *Assume that the result of Lemma 1 holds, and $1 - \varpi \mathbf{N}_\eta > 0$ with $\varpi = \dfrac{a^\nu}{\Gamma(\nu+1)}$. The fractional order model Equation (9) (and equivalently Equation (12)) is Ulam–Hyers-stable and, consequently, generalized Ulam–Hyers-stable.*

Proof. Let $\mathcal{X}$ be a unique solution of Equation (12), and $\overline{\mathcal{X}}$ satisfies Equation (18). For $b > 0$, $t \in [0, a]$, we have

$$\begin{aligned}
\| \overline{\mathcal{X}}(t) - \mathcal{X}(t) \|_\eta &= \sup_{t\in[0,a]} |\overline{\mathcal{X}}(t) - \mathcal{X}(t)| \\
&= \sup_{t\in[0,a]} \left|\overline{\mathcal{X}}(t) - \mathcal{X}(0) - \frac{1}{\Gamma(\nu)}\int_0^t (t-\epsilon)^{\nu-1}\xi(\epsilon, \mathcal{X}(\epsilon))d\epsilon\right| \\
&\leq \sup_{t\in[0,a]} \left|\overline{\mathcal{X}}(t) - \overline{\mathcal{X}}(0) - \frac{1}{\Gamma(\nu)}\int_0^t (t-\epsilon)^{\nu-1}\xi(\epsilon, \overline{\mathcal{X}}(\epsilon))d\epsilon\right| \\
&+ \sup_{t\in[0,a]} \frac{1}{\Gamma(\nu)}\int_0^t (t-\epsilon)^{\nu-1}|\xi(\epsilon, \overline{\mathcal{X}}(\epsilon)) - \xi(\epsilon, \mathcal{X}(\epsilon))|d\epsilon \\
&\leq \varpi b + \frac{\mathbf{N}_\eta}{\Gamma(\nu)} \sup_{t\in[0,a]} \int_0^t (t-\epsilon)^{\nu-1}|\overline{\mathcal{X}}(\epsilon) - \mathcal{X}(\epsilon)|d\epsilon \\
&\leq \varpi b + \varpi \mathcal{N}_\eta \| \overline{\mathcal{X}}(t) - \mathcal{X}(t) \|_\eta,
\end{aligned}$$

which implies that $\| \overline{\mathcal{X}}(t) - \mathcal{X}(t) \|_\eta \leq b \overbrace{\left(\dfrac{\varpi}{1 - \varpi \mathbf{N}_\eta}\right)}^{\mathbf{C}_\eta}$. Thus, from (Definitions 4.5 & 4.6, [59]), we conclude that The fractional order model Equation (9) (and, equivalently, Equation (12)) is Ulam–Hyers-stable and, consequently, generalized Ulam–Hyers-stable. This ends the proof. □

4.4. Numerical Scheme

In this section, we provide the numerical solution of the nonlinear mathematical model using an appropriate iterative scheme, which is very important in mathematical modeling. We use the Adams-type predictor–corrector iterative scheme [50,51] to numerically solve our fractional order model, Equation (9). Let us consider a uniform discretization of $[0, a]$ given by $t_n = n\chi$, $n = 0, 1, 2, \ldots, N$, where $\chi = \dfrac{a}{n}$ denotes the step size. Now, given any approximation,

$$\begin{aligned}
\mathcal{Y}_\chi(t_i) &= (\mathcal{S}_\chi(t_i), \mathcal{V}_\chi(t_i), \mathcal{E}_\chi(t_i), \mathcal{I}_\chi(t_i), \mathcal{H}_\chi(t_i), \mathcal{R}_\chi(t_i)) \\
&\approx \mathcal{Y}(t_i) = (\mathcal{S}(t_i), \mathcal{V}(t_i), \mathcal{E}(t_i), \mathcal{I}(t_i), \mathcal{H}(t_i), \mathcal{R}(t_i)),
\end{aligned}$$

we obtain the next approximation $\mathcal{Y}_\chi(t_{i+1})$ using the Adams-type predictor–corrector iterative scheme, as follows :

$$\begin{aligned}\mathcal{S}_{n+1} &= \mathcal{S}_0 + \frac{\chi^\nu}{\Gamma(\nu+2)}\left(\Lambda^\nu + \alpha^\nu \mathcal{V}^p_{n+1} - \beta^\nu \mathcal{S}^p_{n+1}\mathcal{I}^p_{n+1} - k_1\mathcal{S}^p_{n+1}\right) \\ &+ \frac{\chi^\nu}{\Gamma(\nu+2)}\sum_{j=0}^{n}\theta_{j,n+1}\left(\Lambda^\nu + \alpha^\nu \mathcal{V}_j - \beta^\nu \mathcal{S}_j\mathcal{I}_j - k_1\mathcal{S}_j\right),\end{aligned} \tag{20}$$

$$\mathcal{V}_{n+1} = \mathcal{V}_0 + \frac{\chi^\nu}{\Gamma(\nu+2)}\left\{c^\nu \mathcal{S}^p_{n+1} - k_2\mathcal{V}^p_{n+1} + \sum_{j=0}^{n}\theta_{j,n+1}(c^\nu\mathcal{S}_j - k_2\mathcal{V}_j)\right\}, \tag{21}$$

$$\mathcal{E}_{n+1} = \mathcal{E}_0 + \frac{\chi^\nu}{\Gamma(\nu+2)}\left\{\beta^\nu \mathcal{S}^p_{n+1}\mathcal{I}^p_{n+1} - k_3\mathcal{E}^p_{n+1} + \sum_{j=0}^{n}\theta_{j,n+1}(\beta^\nu\mathcal{S}_j\mathcal{I}_j - k_3\mathcal{E}_j)\right\}, \tag{22}$$

$$\mathcal{I}_{n+1} = \mathcal{I}_0 + \frac{\chi^\nu}{\Gamma(\nu+2)}\left\{\gamma^\nu \mathcal{E}^p_{n+1} - k_4\mathcal{I}^p_{n+1} + \sum_{j=0}^{n}\theta_{j,n+1}(\gamma^\nu\mathcal{E}_j - k_4\mathcal{I}_j)\right\}, \tag{23}$$

$$\mathcal{H}_{n+1} = \mathcal{H}_0 + \frac{\chi^\nu}{\Gamma(\nu+2)}\left\{\phi^\nu \mathcal{I}^p_{n+1} - k_5\mathcal{H}^p_{n+1} + \sum_{j=0}^{n}\theta_{j,n+1}(\phi^\nu\mathcal{I}_j - k_5\mathcal{H}_j)\right\}, \tag{24}$$

$$\mathcal{R}_{n+1} = \mathcal{R}_0 + \frac{\chi^\nu}{\Gamma(\nu+2)}\left\{\sigma^\nu \mathcal{H}^p_{n+1} - d^\nu\mathcal{R}^p_{n+1} + \sum_{j=0}^{n}\theta_{j,n+1}(\sigma^\nu\mathcal{H}_j - d^\nu\mathcal{R}_j)\right\}, \tag{25}$$

where

$$\begin{aligned}\mathcal{S}^p_{n+1} &= \mathcal{S}_0 + \frac{1}{\Gamma(\nu)}\sum_{j=0}^{n} b_{j,n+1}\left(\Lambda^\nu + \alpha^\nu\mathcal{V}_j - \beta^\nu\mathcal{S}_j\mathcal{I}_j - k_1\mathcal{S}_j\right),\\ \mathcal{V}^p_{n+1} &= \mathcal{V}_0 + \frac{1}{\Gamma(\nu)}\sum_{j=0}^{n} b_{j,n+1}(c^\nu\mathcal{S}_j - k_2\mathcal{V}_j),\\ \mathcal{E}^p_{n+1} &= \mathcal{E}_0 + \frac{1}{\Gamma(\nu)}\sum_{j=0}^{n} b_{j,n+1}(\beta^\nu\mathcal{S}_j\mathcal{I}_j - k_3\mathcal{E}_j),\\ \mathcal{I}^p_{n+1} &= \mathcal{I}_0 + \frac{1}{\Gamma(\nu)}\sum_{j=0}^{n} b_{j,n+1}(\gamma^\nu\mathcal{E}_j - k_4\mathcal{I}_j),\\ \mathcal{H}^p_{n+1} &= \mathcal{H}_0 + \frac{1}{\Gamma(\nu+2)}\sum_{j=0}^{n} b_{j,n+1}(\phi^\nu\mathcal{I}_j - k_5\mathcal{H}_j),\\ \mathcal{R}^p_{n+1} &= \mathcal{R}_0 + \frac{1}{\Gamma(\nu+2)}\sum_{j=0}^{n} b_{j,n+1}(\sigma^\nu\mathcal{H}_j - d^\nu\mathcal{R}_j),\end{aligned} \tag{26}$$

with

$$\theta_{i,n+1} = \begin{cases} n^{\nu+1} - (n-\nu)(n+1), & if \quad i = 0,\\ (n-i+2)^{\nu+1} - 2(n-i+1)^{\nu+1} + (n-i)^{\nu+1}, & if \quad 1 \le i \le n,\\ 1, & if \quad i = n+1.\end{cases}$$

and

$$b_{i,n+1} = \frac{\chi^\nu}{\nu}((n-i+1)^\nu - (n-i)^\nu).$$

Theorem 9. *The numerical scheme given by Equations (20)–(26) is stable.*

Proof. The proof follows the proof of Theorem 3.3 in [61] (see also Theorem 5.1, [62]). □

5. Numerical Simulations

In this section, we present numerical simulations of the model using the parameter values listed in Table 2, which were estimated using real data from the measles outbreak in Nigeria [2]. From Equation (10), it is clear that the basic reproduction number varies according to the value of the fractional order ν. Thus, for fractional orders $\nu \in 1, 0.9, 0.8, 0.7, 0.6$, the corresponding values of the basic reproduction number R_0 are, respectively, 3.13, 2.60, 2.15, 1.76, and 1.44. From Figure 5, it is clear that R_0 is an increasing function of the fractional order ν.

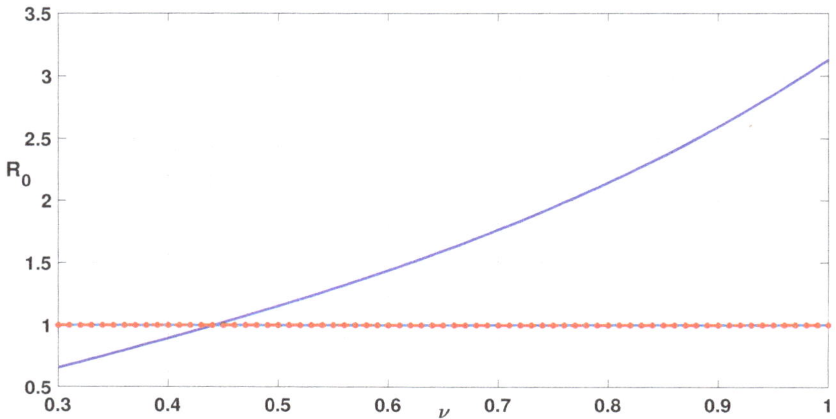

Figure 5. The basic reproduction number R_0 versus the fractional order ν. All other parameter values are listed in Table 2.

Figure 6a shows the combined effect of the fractional order and vaccination on the basic reproduction number, while Figure 6b shows the combined effect of the fractional order and hospitalization on the basic reproduction number. We see that R_0 decreases according to the increases in the vaccination rate c. Note that for $\nu = 1$ and $c \in \{10^{-6}, 10^{-3}, 10^{-2}, 10^{-1}\}$, the basic reproduction number is $R_0 \in \{3.13, 2.45, 0.83, 0.11\}$, for $\nu = 0.9$ and $c \in \{10^{-6}, 10^{-3}, 10^{-2}, 10^{-1}\}$, the basic reproduction number is $R_0 \in \{2.60, 1.99, 0.76, 0.13\}$, for $\nu = 0.8$ and $c \in \{10^{-6}, 10^{-3}, 10^{-2}, 10^{-1}\}$, the basic reproduction number is $R_0 \in \{2.15, 1.61, 0.69, 0.15\}$. This proves that the measles epidemic can be controllable if the vaccination rate coverage is very high (indeed, from $c = 0.01$, $\mathcal{R}_0$ is less than one). Moreover, effectively caring for the sick makes it possible to reduce the number of infected. Indeed, $\mathcal{R}_0$ decreases when the hospitalized rate ϕ increases. Figure 7 combines mass vaccination with the efficient treatment in the basic reproduction number, while the combining effect of individual protection with effective healthcare is depicted in Figure 8. It is clear that when these two control strategies are at their high levels, the basic reproduction number is less than one, which implies the end of the epidemic.

The effect of the fractional order on the dynamics of the measles model is depicted in Figures 9 and 10. It is clear that susceptible (as well as vaccinated) individuals decrease according to the decrease of the fractional order ν, while recovered individuals increase (Figure 9). The populations of all infected classes ($\mathcal{E}$, $\mathcal{I}$, and $\mathcal{H}$) increase when the fractional order decreases (Figure 10). Nevertheless, we can observe in Figure 10 that the epidemic 'pick' is backward-delayed according to the decrease of the fractional order ν. For example, the total number of infected individuals in the latent stage E tended to its maximum value (11,000,000) at the 160th week after the beginning of the epidemic for $\nu = 1$, while this time was backward-delayed to the 15th week.

The effects of the two control measures, namely, vaccination and effective medical care, are depicted in Figures 11 and 12. As in the case of the basic reproduction number, it is clear from Figure 11 that a high level of vaccination coverage can decrease the number of new cases of measles. Similarly, we see in Figure 12 that better care for patients (consisting of systematic hospitalization of new cases from the first symptoms of the disease) will allow for better control of the epidemic. Thus, the combination of these two control measures will help in the fight against the disease spread.

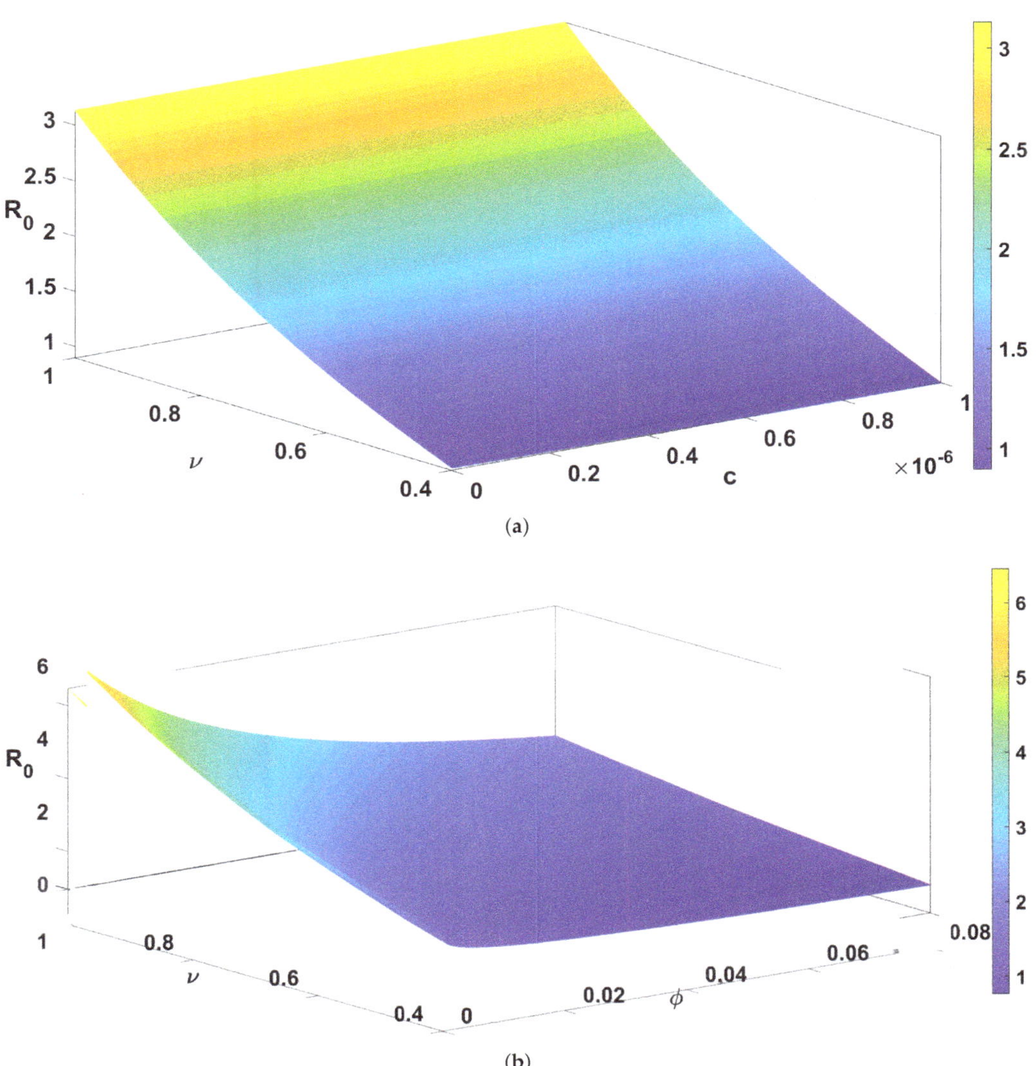

Figure 6. The basic reproduction number R_0 versus (**a**) the vaccination rate c and the fractional order ν; (**b**) the hospitalized rate ϕ and the fractional order ν.

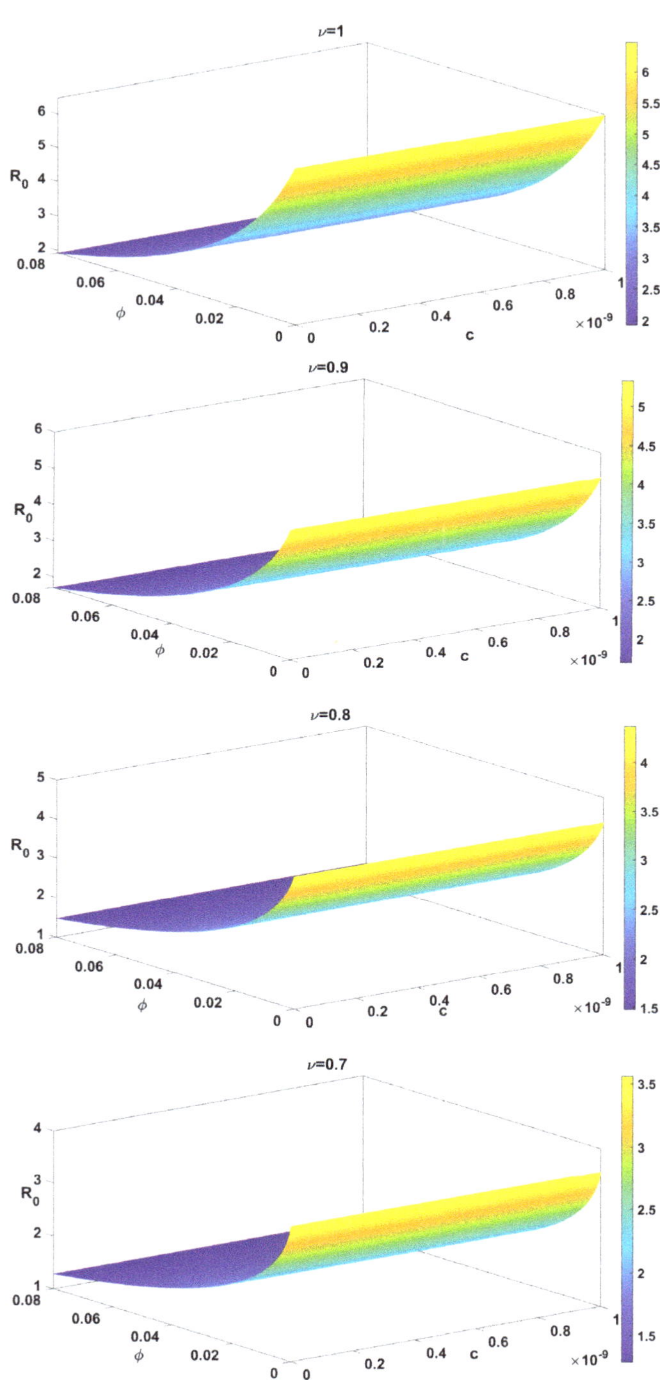

Figure 7. The basic reproduction number R_0 versus the vaccination rate c and the hospitalization rate ϕ.

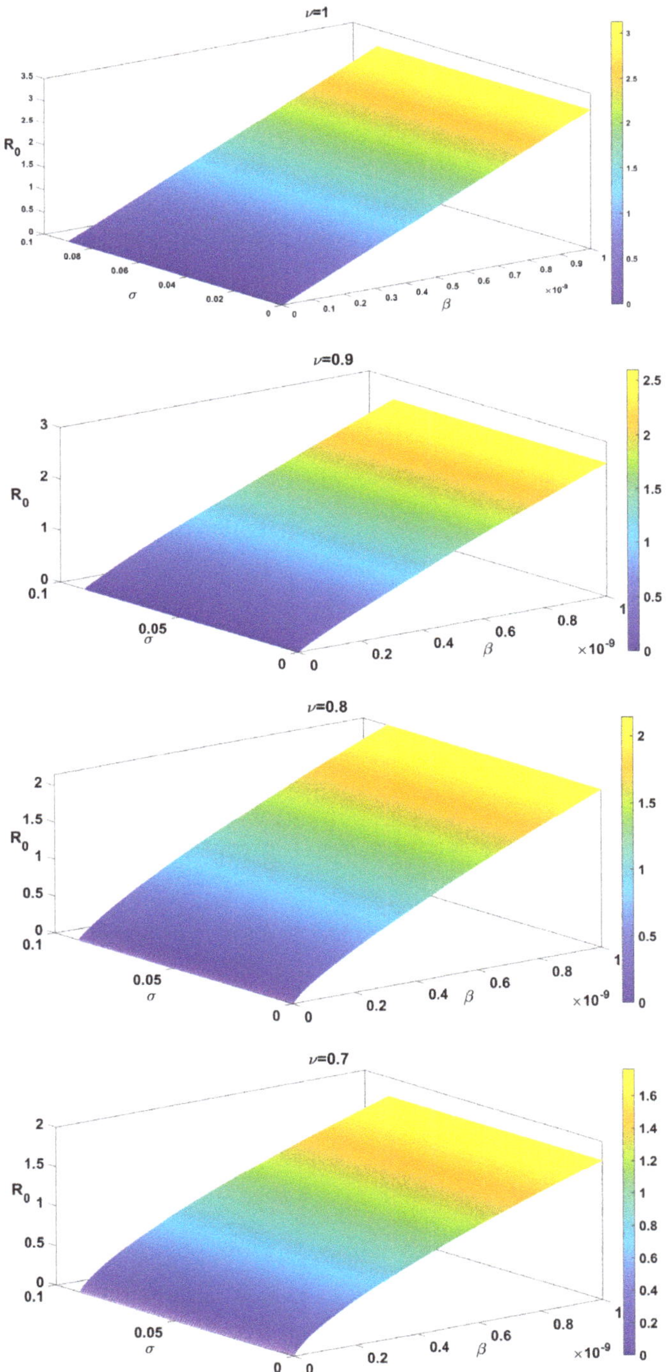

Figure 8. The basic reproduction number R_0 versus the transmission rate β and the recovered rate σ.

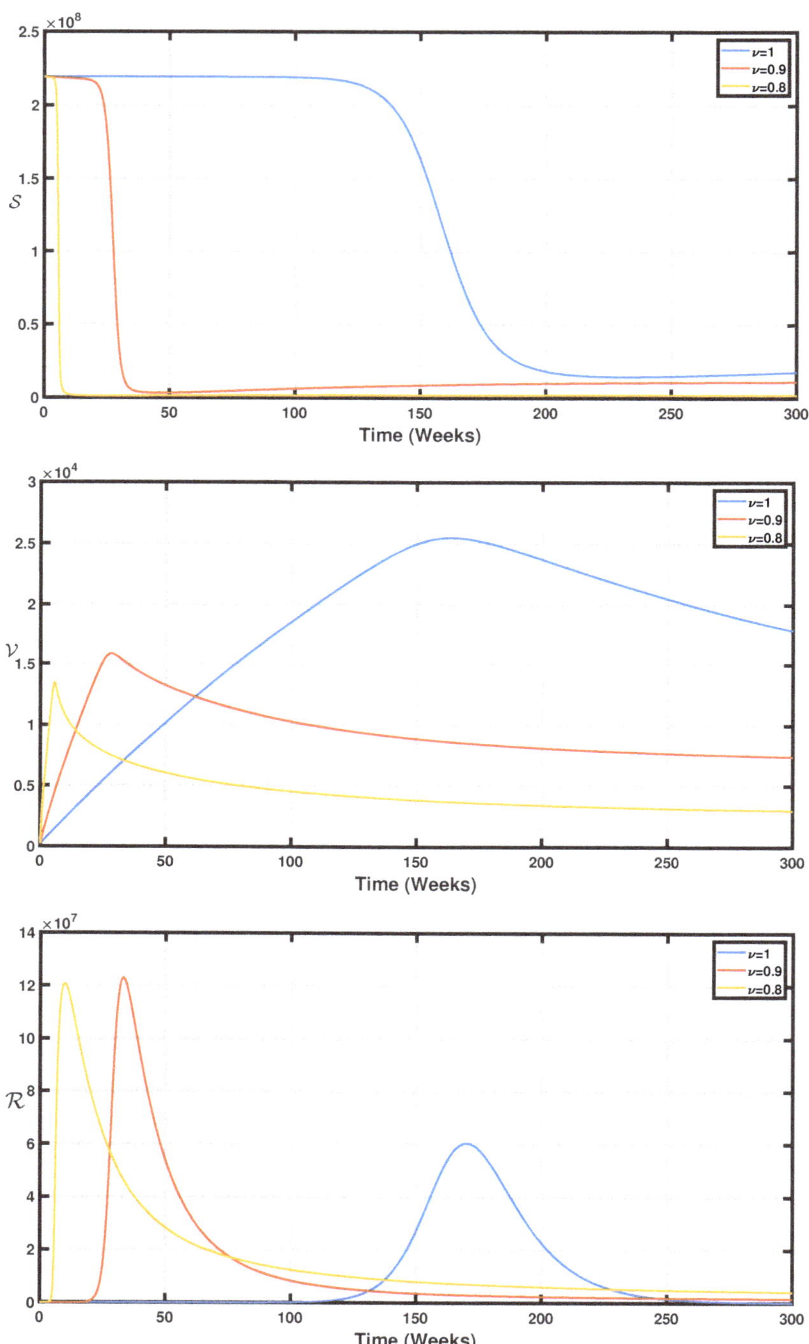

Figure 9. The time series of non-infected state variables of the model Equation (9) with the parameter values listed in Table 2 and different values for the fractional order derivatives ν.

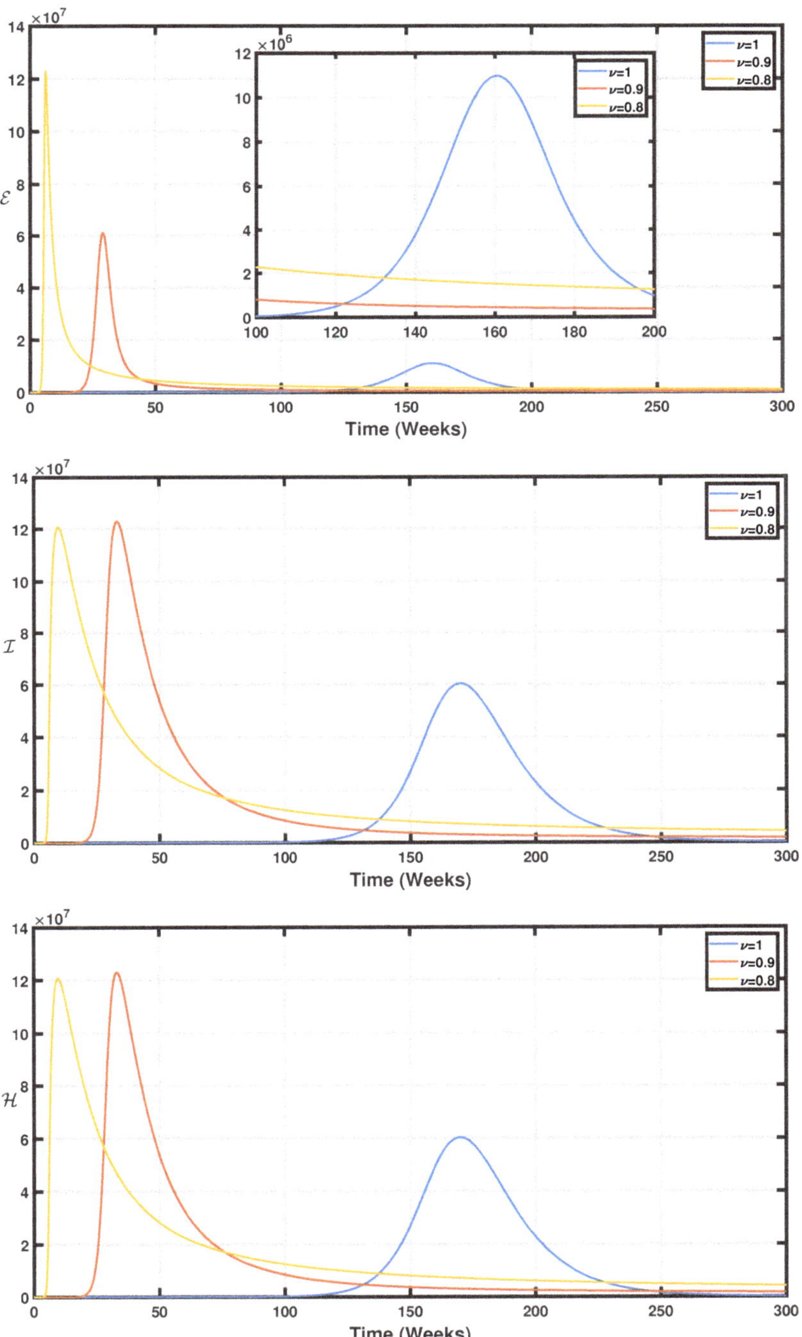

Figure 10. The time series of infected state variables of the model Equation (9) with the parameter values listed in Table 2 and different values for the fractional order derivatives ν.

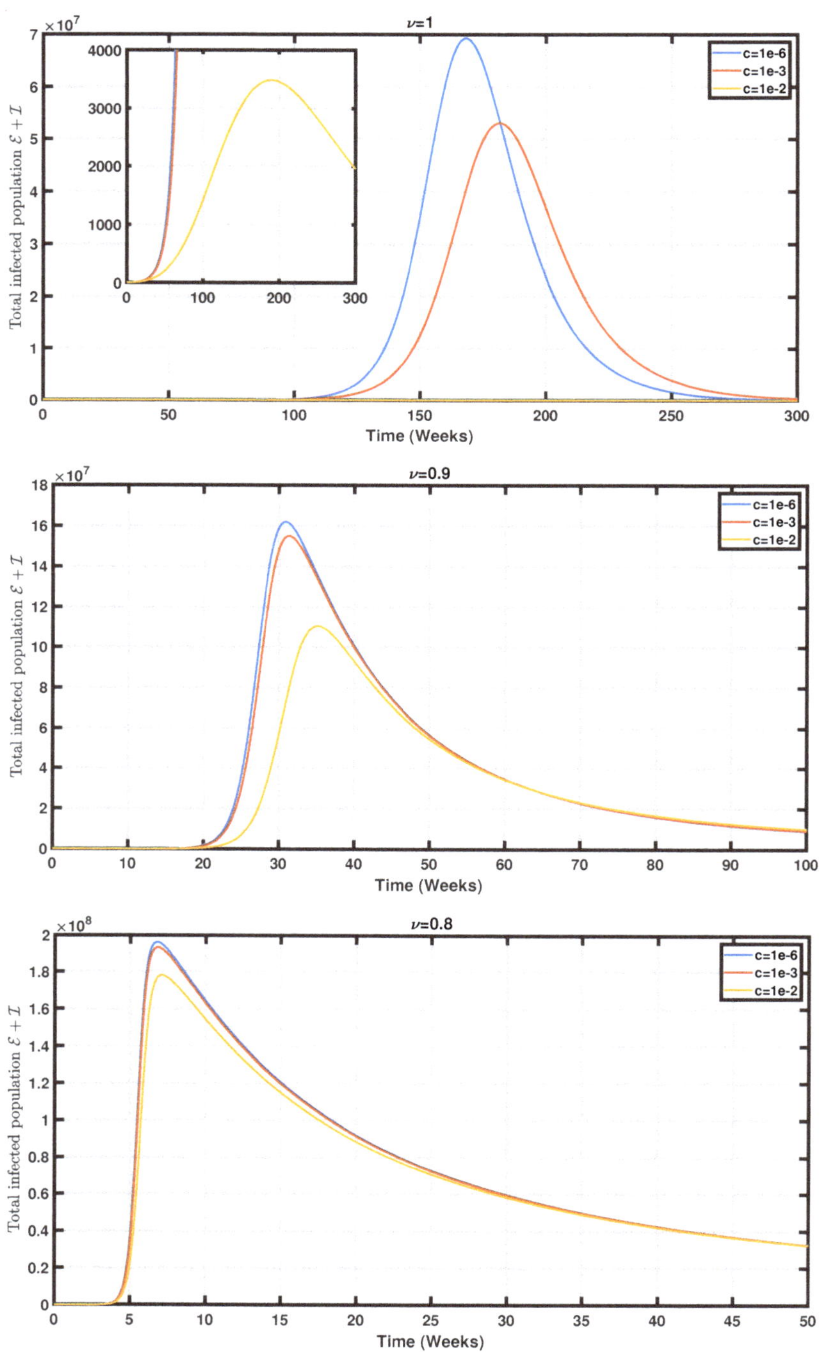

Figure 11. The time series of the total infected population $E(t) + I(t)$ with the parameter values listed in Table 2, except the vaccination rate c and the fractional order derivative ν, which vary.

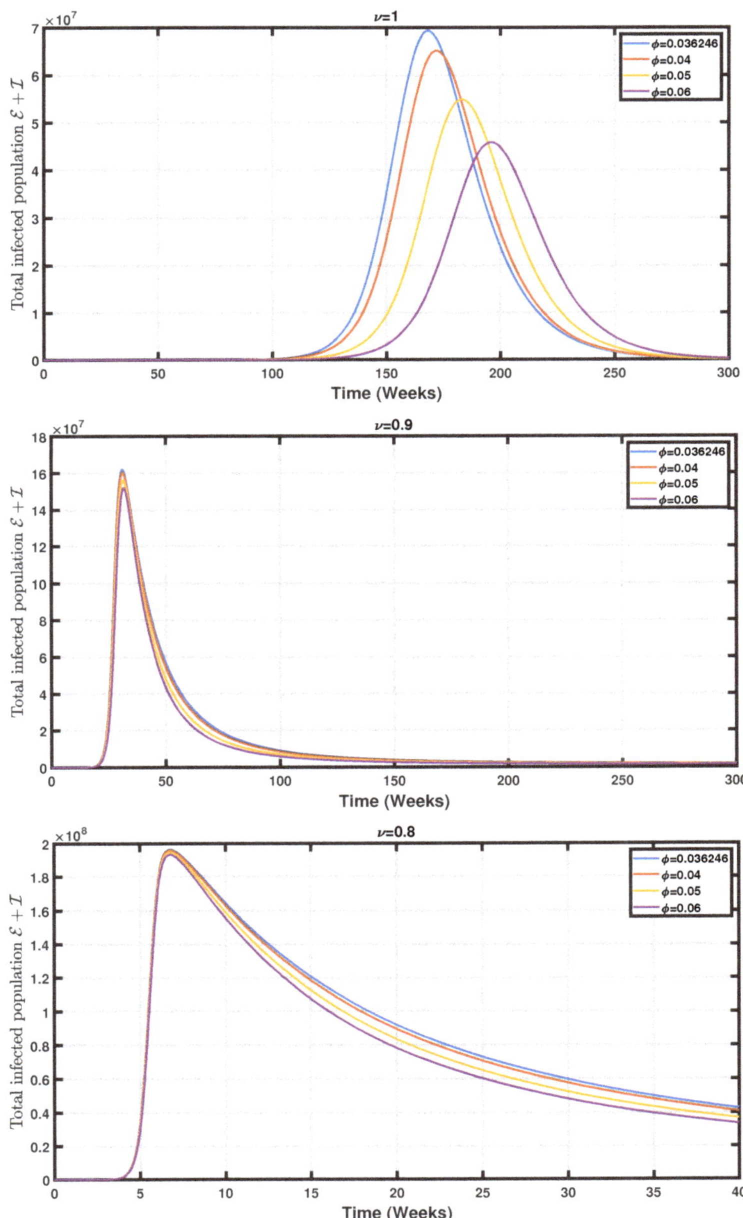

Figure 12. The time series of the total infected population $\mathcal{E}(t) + \mathcal{I}(t)$ with the parameter values listed in Table 2, except the hospitalized rate ϕ and the fractional order derivative ν, which vary.

6. Conclusions and Perspectives

The main objective of this work was to compare the quantitative dynamics of an epidemic model of measles with the integer derivative and fractional derivative (in the sense of Caputo). Thus, we replaced the integer derivative with the Caputo derivative in an existing measles compartmental model, which took into account vaccinations and hospitalized individuals. We performed a global sensitivity analysis by computing the partial rank correlation coefficient between the model parameters, the basic reproduction number, and each state variable of the model. For the fractional model, we computed the basic reproduction number, which is a function of the model parameters and fractional-order parameter. We proved the local and global stability of the disease-free equilibrium as well as the local stability of the endemic equilibrium point. The existence, uniqueness of the solutions, and global stability of the fractional model were also conducted using the Ulam–Hyers stability method. Simulations via the Adams-type predictor–corrector iterative scheme, were conducted to validate our theoretical results and to see the impact of the variation of the fractional order on the disease dynamics. Indeed, the simulation results reveal that the model with the Caputo fractional derivative has a different quantitative behavior than the model with the integer derivative. This suggests that depending on the fractional order, we can forecast the number of total cases in a given interval of time.

It is important to note that some other aspects, such as limited medical resources, mass vaccination, and other fractional derivatives operator (Caputo–Fabrizio, Atangana–Baleanu, piecewise derivatives) can represent direct perspectives to this work. Moreover, there is evidence that the measles vaccine is flawed [63]. Thus, another future direction of this work would be to consider that some individuals who received the measles vaccine may be infected.

Author Contributions: Conceptualization, H.A., R.F. and B.S.S.; methodology, H.A., R.F. and B.S.S.; software, H.A. and R.F.; validation, H.A., R.F., B.S.S. and H.P.E.F.; formal analysis, H.A. and R.F.; investigation, H.A. and R.F.; resources, H.A.; data curation, H.A., R.F. and B.S.S.; writing—original draft preparation, H.A.; writing—review and editing, H.A., R.F., B.S.S. and H.P.E.F.; visualization, H.A.; supervision, H.A. and H.P.E.F.; project administration, H.A. and H.P.E.F.; funding acquisition, H.A. All authors have read and agreed to the published version of the manuscript.

Funding: This research received no external funding.

Institutional Review Board Statement: Not applicable.

Informed Consent Statement: Not applicable.

Data Availability Statement: Not applicable.

Acknowledgments: The authors thank the Editor and the reviewers for their comments and suggestions, which helped us improve the work.

Conflicts of Interest: The authors declare no conflict of interest.

References

1. Griffin, D.E. The immune response in measles: Virus control, clearance and protective immunity. *Viruses* **2016**, *8*, 282. [CrossRef] [PubMed]
2. James Peter, O.; Ojo, M.M.; Viriyapong, R.; Abiodun Oguntolu, F. Mathematical model of measles transmission dynamics using real data from Nigeria. *J. Differ. Equ. Appl.* **2022**, 1–18. [CrossRef]
3. Roberts, M.; Tobias, M. Predicting and preventing measles epidemics in New Zealand: Application of a mathematical model. *Epidemiol. Infect.* **2000**, *124*, 279–287. [CrossRef] [PubMed]
4. Subaiya, S.; Tabu, C.; N'ganga, J.; Awes, A.A.; Sergon, K.; Cosmas, L.; Styczynski, A.; Thuo, S.; Lebo, E.; Kaiser, R.; et al. Use of the revised World Health Organization cluster survey methodology to classify measles-rubella vaccination campaign coverage in 47 counties in Kenya, 2016. *PLoS ONE* **2018**, *13*, e0199786. [CrossRef]
5. Nabi, K.N.; Abboubakar, H.; Kumar, P. Forecasting of COVID-19 pandemic: From integer derivatives to fractional derivatives. *Chaos Solitons Fractals* **2020**, *141*, 110283. [CrossRef]
6. Inaba, H. Kermack and McKendrick revisited: The variable susceptibility model for infectious diseases. *Jpn. J. Ind. Appl. Math.* **2001**, *18*, 273–292. [CrossRef]
7. Kermack, W.O.; McKendrick, A.G. Contributions to the mathematical theory of epidemics. II.—The problem of endemicity. *Proc. R. Soc. Lond. Ser. Contain. Pap. Math. Phys. Character* **1932**, *138*, 55–83.
8. Raza, A.; Rafiq, M.; Baleanu, D.; Shoaib Arif, M.; Naveed, M.; Ashraf, K. Competitive numerical analysis for stochastic HIV/AIDS epidemic model in a two-sex population. *IET Syst. Biol.* **2019**, *13*, 305–315. [CrossRef] [PubMed]
9. Sánchez, Y.G.; Sabir, Z.; Guirao, J.L. Design of a nonlinear SITR fractal model based on the dynamics of a novel coronavirus (COVID-19). *Fractals* **2020**, *28*, 2040026. [CrossRef]

10. Shoaib, M.; Anwar, N.; Ahmad, I.; Naz, S.; Kiani, A.K.; Raja, M.A.Z. Intelligent networks knacks for numerical treatment of nonlinear multi-delays SVEIR epidemic systems with vaccination. *Int. J. Mod. Phys. B* **2022**, 2250100. [CrossRef]
11. Umar, M.; Sabir, Z.; Zahoor Raja, M.A.; Gupta, M.; Le, D.N.; Aly, A.A.; Guerrero-Sánchez, Y. Computational intelligent paradigms to solve the nonlinear SIR system for spreading infection and treatment using Levenberg–Marquardt backpropagation. *Symmetry* **2021**, *13*, 618. [CrossRef]
12. Umar, M.; Kusen; Raja, M.A.Z.; Sabir, Z.; Al-Mdallal, Q. A computational framework to solve the nonlinear dengue fever SIR system. *Comput. Methods Biomech. Biomed. Eng.* **2022**, 1–14. [CrossRef] [PubMed]
13. Arsal, S.R.; Aldila, D.; Handari, B.D. Short review of mathematical model of measles. *AIP Conf. Proc.* **2020**, *2264*, 02003.
14. Aldila, D.; Asrianti, D. A deterministic model of measles with imperfect vaccination and quarantine intervention. *J. Phys. Conf. Ser.* **2019**, *1218*, 012044. [CrossRef]
15. Bashir, A.; Mushtaq, M.; Zafar, Z.U.A.; Rehan, K.; Muntazir, R.M.A. Comparison of fractional order techniques for measles dynamics. *Adv. Differ. Equ.* **2019**, *2019*, 334. [CrossRef]
16. Berhe, H.W.; Makinde, O.D. Computational modelling and optimal control of measles epidemic in human population. *Biosystems* **2020**, *190*, 104102. [CrossRef] [PubMed]
17. Memon, Z.; Qureshi, S.; Memon, B.R. Mathematical analysis for a new nonlinear measles epidemiological system using real incidence data from Pakistan. *Eur. Phys. J. Plus* **2020**, *135*, 378. [CrossRef]
18. Momoh, A.; Ibrahim, M.; Uwanta, I.; Manga, S. Mathematical model for control of measles epidemiology. *Int. J. Pure Appl. Math.* **2013**, *87*, 707–718. [CrossRef]
19. Mossong, J.; Muller, C.P. Modelling measles re-emergence as a result of waning of immunity in vaccinated populations. *Vaccine* **2003**, *21*, 4597–4603. [CrossRef]
20. Obumneke, C.; Adamu, I.I.; Ado, S.T. Mathematical model for the dynamics of measles under the combined effect of vaccination and measles therapy. *Int. J. Sci. Technol.* **2017**, *6*, 862–874.
21. Okyere-Siabouh, S.; Adetunde, I. Mathematical model for the study of measles in cape coast metropolis. *Int. J. Mod. Biol. Med.* **2013**, *4*, 110–133.
22. Qureshi, S. Real life application of Caputo fractional derivative for measles epidemiological autonomous dynamical system. *Chaos Solitons Fractals* **2020**, *134*, 109744. [CrossRef]
23. Center of Disease Control and Prevention. Global Measles Outbreaks. Available online: https://www.cdc.gov/globalhealth/measles/data/global-measles-outbreaks.html (accessed on 12 June 2022).
24. Relief, D. The Global Measles Epidemic Isn't (Just) About Measles. Available online: https://reliefweb.int/report/world/global-measles-epidemic-isn-t-just-about-measles (accessed on 21 April 2022).
25. Sun, H.; Zhang, Y.; Baleanu, D.; Chen, W.; Chen, Y. A new collection of real world applications of fractional calculus in science and engineering. *Commun. Nonlinear Sci. Numer. Simul.* **2018**, *64*, 213–231. [CrossRef]
26. Abboubakar, H.; Kom Regonne, R.; Sooppy Nisar, K. Fractional Dynamics of Typhoid Fever Transmission Models with Mass Vaccination Perspectives. *Fractal Fract.* **2021**, *5*, 149. [CrossRef]
27. Neirameh, A. New fractional calculus and application to the fractional-order of extended biological population model. *Bol. Soc. Parana. Matemática* **2018**, *36*, 115–128. [CrossRef]
28. He, J.H.; Ji, F.Y. Two-scale mathematics and fractional calculus for thermodynamics. *Therm. Sci.* **2019**, *23*, 2131–2133. [CrossRef]
29. Tarasov, V.E. On history of mathematical economics: Application of fractional calculus. *Mathematics* **2019**, *7*, 509. [CrossRef]
30. Caputo, M. Linear models of dissipation whose Q is almost frequency independent—II. *Geophys. J. Int.* **1967**, *13*, 529–539. [CrossRef]
31. Derbazi, C.; Baitiche, Z.; Abdo, M.S.; Shah, K.; Abdalla, B.; Abdeljawad, T. Extremal Solutions of Generalized Caputo-Type Fractional-Order Boundary Value Problems Using Monotone Iterative Method. *Fractal Fract.* **2022**, *6*, 146. [CrossRef]
32. Haidong, Q.; Arfan, M. Fractional model of smoking with relapse and harmonic mean type incidence rate under Caputo operator. *J. Appl. Math. Comput.* **2022**, 1–18. [CrossRef]
33. Jarad, F.; Abdeljawad, T.; Baleanu, D. On the generalized fractional derivatives and their Caputo modification. *Int. Sci. Res. Publ.* **2017**. [CrossRef]
34. Li, L.; Liu, J.G. A generalized definition of Caputo derivatives and its application to fractional ODEs. *SIAM J. Math. Anal.* **2018**, *50*, 2867–2900. [CrossRef]
35. Odibat, Z.; Baleanu, D. Numerical simulation of initial value problems with generalized Caputo-type fractional derivatives. *Appl. Numer. Math.* **2020**, *156*, 94–105. [CrossRef]
36. Xu, C.; Liao, M.; Li, P.; Guo, Y.; Liu, Z. Bifurcation properties for fractional order delayed BAM neural networks. *Cogn. Comput.* **2021**, *13*, 322–356. [CrossRef]
37. Xu, C.; Zhang, W.; Aouiti, C.; Liu, Z.; Yao, L. Further analysis on dynamical properties of fractional-order bi-directional associative memory neural networks involving double delays. *Math. Methods Appl. Sci.* **2022**. [CrossRef]
38. Xu, C.; Liao, M.; Li, P.; Yuan, S. Impact of leakage delay on bifurcation in fractional-order complex-valued neural networks. *Chaos Solitons Fractals* **2021**, *142*, 110535. [CrossRef]
39. Caputo, M.; Fabrizio, M. A new definition of fractional derivative without singular kernel. *Progr. Fract. Differ. Appl.* **2015**, *1*, 73–85.
40. Atangana, A.; Baleanu, D. New fractional derivatives with non-local and non-singular kernel: Theory and Application to Heat Transfer Model. *arXiv* **2016**, arXiv:1602.03408.

41. Liu, X.; Arfan, M.; Ur Rahman, M.; Fatima, B. Analysis of SIQR type mathematical model under Atangana-Baleanu fractional differential operator. *Comput. Methods Biomech. Biomed. Eng.* **2022**, *3*, 1–15. [CrossRef] [PubMed]
42. Zifan, A.; Saberi, S.; Moradi, M.H.; Towhidkhah, F. Automated ECG segmentation using piecewise derivative dynamic time warping. *Int. J. Biol. Med. Sci.* **2006**, *1*. Available online: https://www.researchgate.net/publication/228734344_Automated_ECG_Segmentation_Using_Piecewise_Derivative_Dynamic_Time_Warping (accessed on 26 June 2022).
43. Abboubakar, H.; Kumar, P.; Rangaig, N.A.; Kumar, S. A malaria model with Caputo–Fabrizio and Atangana–Baleanu derivatives. *Int. J. Model. Simulation, Sci. Comput.* **2021**, *12*, 2150013. [CrossRef]
44. Atangana, A. Mathematical model of survival of fractional calculus, critics and their impact: How singular is our world? *Adv. Differ. Equ.* **2021**, *2021*, 403. [CrossRef]
45. Abdulazeez, S.T.; Modanli, M. Solutions of fractional order pseudo-hyperbolic telegraph partial differential equations using finite difference method. *Alex. Eng. J.* **2022**, *61*, 12443–12451. [CrossRef]
46. Farman, M.; Saleem, M.U.; Ahmad, A.; Ahmad, M. Analysis and numerical solution of SEIR epidemic model of measles with non-integer time fractional derivatives by using Laplace Adomian Decomposition Method. *Ain Shams Eng. J.* **2018**, *9*, 3391–3397. [CrossRef]
47. Ogunmiloro, O.M.; Idowu, A.S.; Ogunlade, T.O.; Akindutire, R.O. On the Mathematical Modeling of Measles Disease Dynamics with Encephalitis and Relapse Under the Atangana–Baleanu–Caputo Fractional Operator and Real Measles Data of Nigeria. *Int. J. Appl. Comput. Math.* **2021**, *7*, 185. [CrossRef]
48. Qureshi, S. Monotonically decreasing behavior of measles epidemic well captured by Atangana–Baleanu–Caputo fractional operator under real measles data of Pakistan. *Chaos Solitons Fractals* **2020**, *131*, 109478. [CrossRef]
49. Qureshi, S.; Jan, R. Modeling of measles epidemic with optimized fractional order under Caputo differential operator. *Chaos Solitons Fractals* **2021**, *145*, 110766. [CrossRef]
50. Diethelm, K. An algorithm for the numerical solution of differential equations of fractional order. *Electron. Trans. Numer. Anal* **1997**, *5*, 1–6.
51. Diethelm, K.; Ford, N.J.; Freed, A.D. A predictor-corrector approach for the numerical solution of fractional differential equations. *Nonlinear Dyn.* **2002**, *29*, 3–22. [CrossRef]
52. Marino, S.; Hogue, I.B.; Ray, C.J.; Kirschner, D.E. A methodology for performing global uncertainty and sensitivity analysis in systems biology. *J. Theor. Biol.* **2008**, *254*, 178–196. [CrossRef]
53. Wu, J.; Dhingra, R.; Gambhir, M.; Remais, J.V. Sensitivity analysis of infectious disease models: Methods, advances and their application. *J. R. Soc. Interface* **2013**, *10*, 20121018. [CrossRef] [PubMed]
54. Stein, M. Large sample properties of simulations using Latin hypercube sampling. *Technometrics* **1987**, *29*, 143–151. [CrossRef]
55. Bozkurt, F.; Yousef, A.; Abdeljawad, T.; Kalinli, A.; Al Mdallal, Q. A fractional-order model of COVID-19 considering the fear effect of the media and social networks on the community. *Chaos Solitons Fractals* **2021**, *152*, 111403. [CrossRef] [PubMed]
56. Podlubny, I. *Fractional Differential Equations: An Introduction to Fractional Derivatives, Fractional Differential Equations, to Methods of Their Solution and Some of Their Applications*; Elsevier: Amsterdam, The Netherlands, 1998.
57. Diethelm, K. *The Analysis of Fractional Differential Equations: An Application-Oriented Exposition Using Differential Operators of Caputo Type*; Springer: Berlin/Heidelberg, Germany, 2010.
58. Lin, W. Global existence theory and chaos control of fractional differential equations. *J. Math. Anal. Appl.* **2007**, *332*, 709–726. [CrossRef]
59. Akindeinde, S.O.; Okyere, E.; Adewumi, A.O.; Lebelo, R.S.; Fabelurin, O.O.; Moore, S.E. Caputo fractional-order SEIRP model for COVID-19 Pandemic. *Alex. Eng. J.* **2022**, *61*, 829–845. [CrossRef]
60. Jung, S.M. *Hyers-Ulam-Rassias Stability of Functional Equations in Nonlinear Analysis*; Springer: Berlin/Heidelberg, Germany, 2011; Volume 48.
61. Li, C.; Zeng, F. The finite difference methods for fractional ordinary differential equations. *Numer. Funct. Anal. Optim.* **2013**, *34*, 149–179. [CrossRef]
62. Kumar, P.; Rangaig, N.A.; Abboubakar, H.; Kumar, A.; Manickam, A. Prediction studies of the epidemic peak of coronavirus disease in Japan: From Caputo derivatives to Atangana–Baleanu derivatives. *Int. J. Model. Simul. Sci. Comput.* **2022**, *13*, 2250012. [CrossRef]
63. van Boven, M.; Kretzschmar, M.; Wallinga, J.; O'Neill, P.D.; Wichmann, O.; Hahné, S. Estimation of measles vaccine efficacy and critical vaccination coverage in a highly vaccinated population. *J. R. Soc. Interface* **2010**, *7*, 1537–1544. [CrossRef]

 MDPI

Article

Local Structure of Convex Surfaces near Regular and Conical Points

Alexander Plakhov [1,2]

1 Center for R & D in Mathematics and Applications, Department of Mathematics, University of Aveiro, 3810-193 Aveiro, Portugal; plakhov@ua.pt
2 Institute for Information Transmission Problems, 127051 Moscow, Russia

Abstract: Consider a point on a convex surface in $\mathbb{R}^d$, $d \geq 2$ and a plane of support Π to the surface at this point. Draw a plane parallel with Π cutting a part of the surface. We study the limiting behavior of this part of the surface when the plane approaches the point, being always parallel with Π. More precisely, we study the limiting behavior of the normalized surface area measure in S^{d-1} induced by this part of the surface. In this paper, we consider two cases: (a) when the point is regular and (b) when it is singular conical, that is the tangent cone at the point does not contain straight lines. In Case (a), the limit is the atom located at the outward normal vector to Π, and in Case (b), the limit is equal to the measure induced by the part of the tangent cone cut off by a plane.

Keywords: convex surfaces; surface area measure of a convex body; Newton's problem of minimal resistance

MSC: 52A20; 26B25

1. Introduction

Consider a convex compact set C with a nonempty interior in Euclidean space $\mathbb{R}^d$, $d \geq 2$. Let $r_0 \in \partial C$ be a point on its boundary, and let Π be a plane of support to C at r_0. Consider the part of the boundary ∂C containing r_0 and bounded by a plane parallel with Π. We are interested in studying the limiting properties of this part of the boundary when the bounding plane approaches Π.

In what follows, a convex compact set with a nonempty interior will be called a convex body.

The point $r_0 \in \partial C$ is called *regular* if the plane of support at this point is unique and *singular* otherwise. It is well known that regular points form a full-measure set in ∂C.

Let e denote the outward unit normal vector to Π. Take $t > 0$, and let Π_t be the plane parallel with Π at the distance t from it, on the side opposite to the normal vector. Thus, the plane $\Pi = \Pi_0$ is given by the equation $\langle r - r_0, e\rangle = 0$ and Π_t by the equation $\langle r - r_0, e\rangle = -t$. The body C is contained in the closed half-space $\{r : \langle r - r_0, e\rangle \leq 0\}$. Here and in what follows, $\langle \cdot\,, \cdot\rangle$ means the scalar product.

Consider the convex body:

$$C_t = C \cap \{r : \langle r - r_0, e\rangle \geq -t\}.$$

In other words, C_t is the part of C cut off by the plane Π_t. The boundary of C_t is the union of the convex set of codimension 1:

$$B_t = C \cap \{r : \langle r - r_0, e\rangle = -t\} \tag{1}$$

and the convex surface:

$$S_t = \partial C \cap \{r : \langle r - r_0, e\rangle \geq -t\}; \tag{2}$$

Citation: Plakhov, A. Local Structure of Convex Surfaces near Regular and Conical Points. *Axioms* **2022**, *11*, 356. https://doi.org/10.3390/axioms11080356

Academic Editor: Hans J. Haubold

Received: 20 June 2022
Accepted: 19 July 2022
Published: 23 July 2022

Publisher's Note: MDPI stays neutral with regard to jurisdictional claims in published maps and institutional affiliations.

thus, $\partial C_t = B_t \cup S_t$.

In what follows, we will denote as $|\mathcal{A}|_m$ the m-dimensional Hausdorff measure of the Borel set $\mathcal{A} \subset \mathbb{R}^d$. By default, $|\cdot|$ means $|\cdot|_{d-1}$.

Let n_r denote the outward unit normal to C at a regular point $r \in \partial C$, and let S be a Borel subset of ∂C. *The surface area measure induced by S* is the Borel measure ν_S defined in S^{d-1} satisfying

$$\nu_S(\mathcal{A}) := |\{r \in S : n_r \in \mathcal{A}\}|$$

for any Borel subset $\mathcal{A} \subset S^{d-1}$. In the case when S coincides with ∂C, we obtain the well-known measure $\nu_{\partial C}$ called the *surface area measure of the convex body C*. For this measure, the following well-known relation takes place:

$$\int_{S^{d-1}} n \, \nu_{\partial C}(dn) = \vec{0}. \tag{3}$$

Denote by ν_t the normalized measure induced by the surface S_t; more precisely,

$$\nu_t := \frac{1}{|B_t|} \nu_{S_t}.$$

That is, for any Borel set $\mathcal{A} \subset S^{d-1}$, it holds

$$\nu_t(\mathcal{A}) = \frac{1}{|B_t|} |\{r \in S_t : n_r \in \mathcal{A}\}|.$$

The surface area measure of ∂C_t equals $\nu_{\partial C_t} = |B_t|\delta_{-e} + |B_t|\nu_t$; hence,

$$\int_{S^{d-1}} n \, d\nu_{\partial C_t}(n) = |B_t|(-e + \int_{S^{d-1}} n \, \nu_t(dn)).$$

Here and in what follows, δ_e means the unit atom supported at e. Applying Formula (3) to ∂C_t, one obtains

$$\int_{S^{d-1}} n \, \nu_t(dn) = e. \tag{4}$$

We say that ν_t *weakly converges* to ν_* as $t \to 0$ and denote $\lim_{t\to 0} \nu_t = \nu_*$, if for any continuous function f on S^{d-1}, it holds

$$\lim_{t\to 0} \int_{S^{d-1}} f(n) \, \nu_t(dn) = \int_{S^{d-1}} f(n) \, \nu_*(dn).$$

Similarly, we say that ν_* is a *weak partial limit* of the measure ν_t, if there exists a sequence of positive numbers t_i, $i \in \mathbb{N}$ converging to 0 such that, for any continuous function f on S^{d-1}, it holds

$$\lim_{i\to\infty} \int_{S^{d-1}} f(n) \, \nu_{t_i}(dn) = \int_{S^{d-1}} f(n) \, \nu_*(dn).$$

In this article, we are going to study the limiting properties of the measure ν_t as $t \to 0$.

One such property is derived immediately. Let ν_* be a weak limit or a weak partial limit of ν_t. Passing to the limit $t \to 0$ or to the limit $t_i \to 0$ in Formula (4), one obtains

$$\int_{S^{d-1}} n \, \nu_*(dn) = e. \tag{5}$$

The *tangent cone* to C at $r_0 \in \partial C$ is the closure of the union of all rays with vertex at r_0 that intersect $C \setminus r_0$. Equivalently, the tangent cone at r_0 is the smallest closed cone with the vertex at r_0 that contains C; see Figure 1.

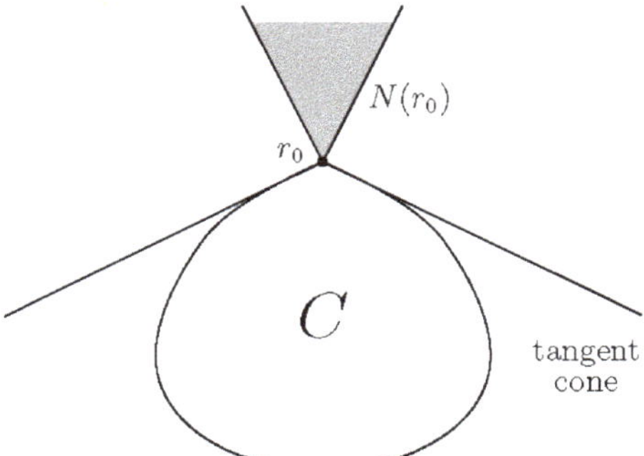

Figure 1. The tangent cone and the normal cone to a convex body C.

If the tangent cone at r_0 is a half-space, then the point r_0 is regular, and vice versa.

The *normal cone* to C at r_0 is the union of all rays with vertex at r_0 whose director vector is the outward normal to a plane of support at r_0. It is denoted as $N(r_0)$. An equivalent definition is the following: the normal cone at r_0 is the set of points r that satisfy $\langle r - r_0, r' - r_0 \rangle \leq 0$ for all $r' \in C$. The normal cone to a convex body does not contain straight lines. Both tangent and normal cones are, of course, convex sets.

If the dimension of $N(r_0)$ equals d (equivalently, if the tangent cone does not contain straight lines), then r_0 is called a *conical point* of C. If the dimension of $N(r_0)$ equals 1, then r_0 is regular, and vice versa. In the intermediate case, that is if the dimension of $N(r_0)$ is greater than 1, but smaller than d, r_0 is called a *ridge point*. This notation goes back to Pogorelov [1].

The motivation for this study comes, to a great extent, from extremal problems in classes of convex bodies and, in particular, from Newton's problem of least resistance for convex bodies [2]. It is natural to try to develop a geometric method of the small variation of convex bodies for such problems, and perhaps, the simplest way would be cutting a small part of the body by a plane. This method proved itself to be effective in the case of Newton's problem. Let us describe this problem in some detail.

The problem in a class of radially symmetric bodies was first stated and solved by Newton himself in 1687 in [3]. The more general version of the problem was posed by Buttazzo and Kawohl in 1993 in [2]. This general problem can be formulated in the functional form as follows:

Find the smallest value of the functional

$$\iint_{\Omega} \frac{1}{1 + |\nabla u(x,y)|^2}\, dxdy \tag{6}$$

in the class of convex functions $u : \Omega \to \mathbb{R}$ satisfying $0 \leq u \leq M$, where $\Omega \subset \mathbb{R}^2$ is a planar convex body and $M > 0$.

The physical meaning of this problem is as follows: find the optimal streamlined shape of a convex body moving downwards through an extremely rarefied medium, provided that the body–particle collisions are perfectly elastic.

Problem (6) (along with its further generalizations) has been studied in various papers including [4–13], but has not been solved completely until now.

It was conjectured in 1995 in [6] that the slope of the graph of an optimal function near the zero level set $L_0 = \{(x,y) : u(x,y) = 0\}$ equals 1. This conjecture was numerically disproven by Wachsmuth (personal communication) in the case when L_0 has an empty interior and, therefore, is a line segment. Moreover, numerical simulation shows that the infimum of $|\nabla u|$ in the complement of L_0 is strictly greater than 1.

On the other hand, this conjecture was proven by the author in [14] in the case when L_0 has a nonempty interior. More precisely, it was proven that if u minimizes functional (6), then for almost all $(x,y) \in \partial L_0$, it holds

$$\lim_{(x',y')\,(\notin L_0)\to(x,y)} |\nabla u(x',y')| = 1.$$

The proof is based on the results concerning local properties of convex surfaces near ridge points in the case $d = 3$. These results were formulated, with the proofs being briefly outlined, in [14].

Remark 1. *The limiting behavior of ν_t in the case $d = 2$ is quite simple. In this case, the tangent cone is an angle, which degenerates to a half-plane if the point is regular. We will call it the* tangent angle. *Let the tangent angle to $C \subset \mathbb{R}^2$ at r_0 be given by*

$$\langle r - r_0,\, e_1\rangle \le 0, \quad \langle r - r_0,\, e_2\rangle \le 0, \quad |e_1| = 1,\ |e_2| = 1,$$

and e be given by

$$e = \lambda_1 e_1 + \lambda_2 e_2, \quad \lambda_1 \ge 0,\ \lambda_2 \ge 0, \quad |e| = 1.$$

Thus, e_1 and e_2 are the outward unit normals to the sides of the angle, and e is the outward unit normal to a line of support at r_0. Then, the limiting measure is the sum of two atoms:

$$\lim_{t\to 0} \nu_t = \lambda_1 \delta_{e_1} + \lambda_2 \delta_{e_2}.$$

The proof of this relation is simple and is left to the reader.

Note that if the point r_0 is regular, then $e_1 = e_2 = e$. It may also happen that the point is singular, that is $e_1 \neq e_2$, and e coincides with one of the vectors e_1 and e_2. In both cases, the limiting measure is an atom:

$$\lim_{t\to 0} \nu_t = \delta_e.$$

The limiting behavior of ν_t is different for different kinds of points:

(a) If the point r_0 is regular, then the limiting measure is an atom.

(b) If r is a conical point, then the limiting measure coincides with the measure induced by the part of the boundary of the tangent cone cut off by a plane Π_t, $t = \sigma$ (note that all the induced measures with $t > 0$ are proportional).

(c) The case of ridge points is the most interesting. In this case, the limiting measure may not exist, and the characterization of all possible partial limits is a difficult task.

Still, the study is nontrivial also in Cases (a) and (b). In this paper, we restrict ourselves to these cases, while Case (c) is postponed to the future. The main results of the paper are contained in the following Theorems 1 and 2.

Theorem 1. *If r_0 is a regular point of ∂C, then*

$$\lim_{t\to 0} \nu_t = \delta_e. \tag{7}$$

Let r_0 be a conical point, K be the tangent cone at r_0, $\hat{S}_t$ be the part of ∂K containing r_0 cut off by the plane Π_t, and $\hat{B}_t$ be the intersection of the cone with the cutting plane Π_t, $t > 0$, that is

$$\hat{S}_t = \partial K \cap \{r : \langle r - r_0,\, e\rangle \ge -t\} \quad \text{and} \quad \hat{B}_t = K \cap \{r : \langle r - r_0,\, e\rangle = -t\}.$$

Let $K_t = K \cap \{r : \langle r - r_0,\, e\rangle \ge -t\}$ be the part of the cone cut off by the plane Π_t; its boundary is $\partial K_t = \hat{S}_t \cup \hat{B}_t$.

All measures induced by $\hat{S}_t$ are proportional, that is the measure:

$$\nu_\star := \frac{1}{|\hat{B}_t|} \nu_{\hat{S}_t}$$

does not depend on t.

Theorem 2. *If r_0 is a conical point of ∂C, then*

$$\lim_{t\to 0} \nu_t = \nu_\star. \tag{8}$$

2. Proof of Theorem 1

The proof is based on several propositions.

Consider a convex set $D \subset \mathbb{R}^{d-1}$, and let $A = |D|$ be its $(d-1)$-dimensional volume and $P = |\partial D|_{d-2}$ be the $(d-2)$-dimensional volume of its boundary.

Proposition 1. *If D contains a circle of radius a, then*

$$P \le \frac{d-1}{a} A. \tag{9}$$

Proof. Let $ds \subset \partial D$ be an infinitesimal element of the boundary of D and denote by $p(ds)$ its $(d-2)$-dimensional volume. Consider the pyramid with the vertex at the center O of the circle and with the base ds, that is the union of line segments joining O with the points of ds. Let $A(ds)$ be the element of the $(d-1)$-dimensional volume of this pyramid; see Figure 2. Then, we have

$$A(ds) \ge \frac{a}{d-1} p(ds),$$

and therefore,

$$A = \int_{\partial D} A(ds) \ge \frac{a}{d-1} \int_{\partial D} p(ds) = \frac{a}{d-1} P.$$

From here follows Inequality (9). □

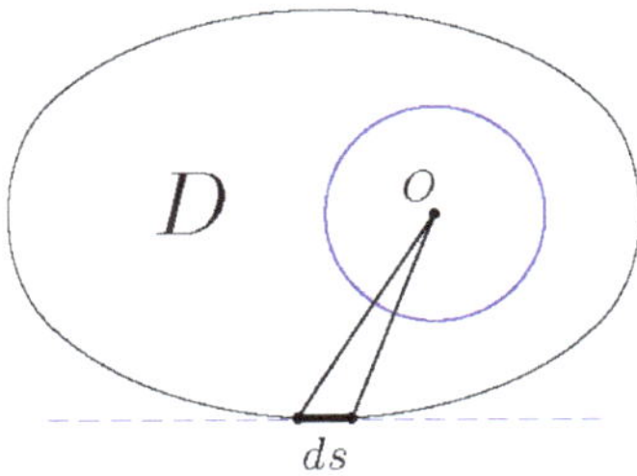

Figure 2. The convex set D containing a circle of radius a.

Consider Euclidean space $\mathbb{R}^d$ with the coordinates (x, z), $x = (x_1, \dots, x_{d-1})$, and fix $t > 0$ and $0 < \varphi < \pi/2$.

Proposition 2. *Let a convex body $C \subset \mathbb{R}^d$ be contained between the planes $z = 0$ and $z = t$, $t > 0$. Let D be the image of C under the natural projection of $\mathbb{R}^d$ on the x-plane, $(x, z) \mapsto x$, and let $P = |\partial D|_{d-2}$ be the $(d-2)$-dimensional volume of ∂D. Let a domain $\mathcal{U} \subset \partial C$ be such that the outward normal $n_r = (n_{r,1}, \dots n_{r,d-1}, n_{r,d})$ at each regular point $r \in \mathcal{U}$ satisfies $|n_{r,d}| \le \cos\varphi$. (In other words, the angles between n_r for $r \in \mathcal{U}$ and the vectors $\pm(0, \dots, 0, 1)$ are $\ge \varphi$.) Then,*

$$|\mathcal{U}| \le \frac{2tP}{\sin\varphi}.$$

Proof. The body C is bounded below by the graph of a convex function, say u_1, and above by the graph of a concave function, say u_2; see Figure 3. Both functions are defined on D. That is, we have

$$C = \{(x,z) : x \in D,\ u_1(x) \le z \le u_2(x)\}.$$

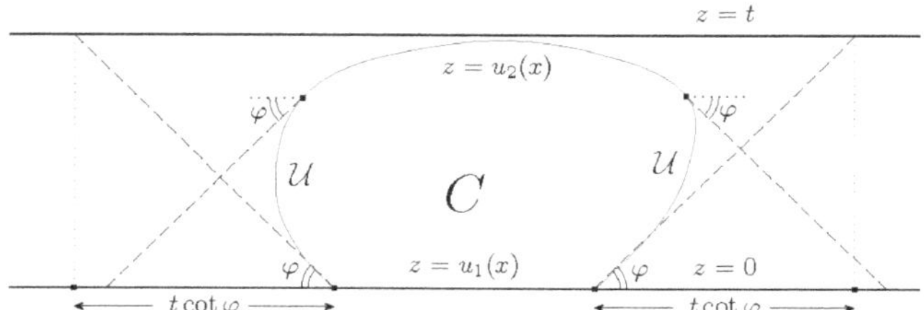

Figure 3. The body C between two parallel planes $z = 0$ and $z = t$ is shown. Here, $\mathcal{U}$ is represented by the union of two curves bounded by the points.

Let $\mathcal{U}_i$ $(i = 1, 2)$ denote the intersection of $\mathcal{U}$ with the graph of u_i. Clearly, if a point $(x, u_i(x))$ is regular and belongs to $\mathcal{U}_i$, then $|\nabla u_i(x)| \ge \tan\varphi$.

For $0 \le z \le t$, denote by P_z the $(d-2)$-dimensional volume of the set:

$$L_z = \{x : u_1(x) = z \text{ and } (x, u_1(x)) \in \mathcal{U}_1\}.$$

One clearly has $P_z \le P$. Let s be the $(d-2)$-dimensional parameter in L_z, and let ds be the element of the $(d-2)$-dimensional volume in L_z. Denote by $x(z,s)$ the point in L_z corresponding to the parameter s. Then, the $(d-1)$-dimensional volume of $\mathcal{U}_1$ equals

$$|\mathcal{U}_1| = \int_0^t dz \int_{L_z} \sqrt{1 + \frac{1}{|\nabla u_1(x(z,s))|^2}}\, ds \le \int_0^t P_z \sqrt{1 + \cot^2\varphi}\, dz \le \frac{tP}{\sin\varphi}.$$

The same argument holds for $\mathcal{U}_2$. It follows that $|\mathcal{U}| = |\mathcal{U}_1| + |\mathcal{U}_2| \le 2tP/\sin\varphi$. □

Proposition 3. *If a convex set in $\mathbb{R}^{d-1}$ contains $d-1$ mutually orthogonal line segments of length 1, then it also contains a ball of radius $c = 1/2(d-1)$.*

Proof. Denote the convex set by D and the segments by $[A_i^0, A_i^1]$, $i = 1, \dots, d-1$. Since all points A_i^j lie in D, each convex combination of the form $P_J = \frac{1}{d-1}\sum_{i=1}^{d-1} A_i^{J(i)}$, where J denotes a map $\{1, \dots, d-1\} \mapsto \{0,1\}$, also lies in D. The convex combination of the set of points P_J is a hypercube with the size of length $1/(d-1)$ and contains the ball of radius $1/2(d-1)$ with the center at the hypercube's center. □

Proposition 4. *B_t contains $d-1$ mutually orthogonal line segments of length β_t, where $\beta_t/t \to \infty$ as $t \to 0$.*

Proof. Take a unit vector e' orthogonal to e, and consider the 2-dimensional plane Π' through r_0 parallel with e and e'. The intersection $\Pi' \cap C =: C'$ is a 2-dimensional convex body, and r_0 is a regular point on its boundary; the intersection $\Pi' \cap \Pi_t = l_t'$ is a line orthogonal to e at the distance t from r_0; the intersection $\Pi' \cap B_t$ is a line segment (maybe degenerating to a point or the empty set). Equivalently, this segment is the intersection of the body C_t' with the line l_t'. Since the point $r_0 \in \partial C'$ is regular, we conclude that the length of this segment β_t' satisfies $\beta_t'/t \to \infty$ as $t \to 0$.

Now, choose unit vectors $e_1, \dots, e_{d-1}$ in such a way that the set of vectors $e_1, \dots, e_{d-1}, e$ forms an orthonormal system in $\mathbb{R}^d$. For each $i = 1, \dots, d-1$, draw the 2-dimensional plane Π^i through r_0 parallel with e and e_i. The intersections $\Pi^i \cap B_t$ are line segments parallel with e_i, and therefore, they are mutually orthogonal. The lengths of these segments

β_t^i satisfy $\beta_t^i/t \to \infty$ as $t \to 0$. Taking $\beta_t = \min_{1\le i\le d-1} \beta_t^i$, one comes to the statement of the proposition. □

Recall that S_t is the intersection of ∂C with the half-space $\{r : \langle r - r_0,\, e\rangle \ge -t\}$ and Π_t is the plane of the equation $\langle r - r_0,\, e\rangle = -t$. For $\varphi \in (0,\, \pi/2)$, denote by $S_{t,\varphi}$ the part of S_t containing the regular points r satisfying $\langle n_r,\, e\rangle \le \cos\varphi$. In other words, $S_{t,\varphi}$ is the set of regular points r in S_t such that the angle between e and n_r is greater than or equal to φ.

Proposition 5. *We have*

$$\frac{|S_{t,\varphi}|}{|B_t|} \to 0 \quad \textit{as } t \to 0.$$

Proof. Consider a coordinate system (x, z), $x = (x_1, \ldots, x_{d-1})$ such that the x-plane coincides with Π_t and the z-axis is directed toward the vector e. For $t_0 > 0$ sufficiently small, the intersection of Π_t and the interior of C is nonempty for all $t \le t_0$. The angle between $-e$ and the outward normal at each regular point of S_t, $t \le t_0$ is greater than a positive value φ_0. That is, for any regular point $r \in S_t$, it holds $\langle n_r,\, e\rangle \ge -\cos\varphi_0$. Without loss of generality, one can take $\varphi < \varphi_0$, and then, for all regular points $r \in S_{t,\varphi}$, it holds $|\langle n_r,\, e\rangle| \le \cos\varphi$.

In the chosen coordinate system, C_t is contained between the planes $z = 0$ and $z = t$. Denote by D_t the image of C_t under the natural projection $(x, z) \mapsto x$. The domain D_t contains B_t and is contained in the $(t\cot\varphi)$-neighborhood of B_t; hence, its $(d-2)$-dimensional volume does not exceed $P_t = |\partial B_t|_{d-2} + s_{d-2}(t\cot\varphi)^{d-2}$, where $s_{d-2} = |S^{d-2}|_{d-2}$ means the area of the $(d-2)$-dimensional unit sphere.

Applying Proposition 2 to the body $C = C_t$ and the domain $\mathcal{U} = S_{t,\varphi}$, one obtains

$$|S_{t,\varphi}| \le \frac{2tP_t}{\sin\varphi} = 2t\,\frac{|\partial B_t|_{d-2} + s_{d-2}(t\cot\varphi)^{d-2}}{\sin\varphi}.$$

By Propositions 3 and 4, B_t contains a ball of radius $c\beta_t$, and therefore, by Proposition 1,

$$|\partial B_t|_{d-2} \le \frac{d-1}{c\beta_t}\,|B_t|$$

and additionally, $|B_t| \ge b_{d-1}(c\beta_t)^{d-1}$, where b_{d-1} means the volume of the unit ball in $\mathbb{R}^{d-1}$. Hence,

$$\frac{|S_{t,\varphi}|}{|B_t|} \le 2t\,\frac{\frac{d-1}{c\beta_t}\,|B_t| + s_{d-2}(t\cot\varphi)^{d-2}}{\sin\varphi|B_t|} \le \frac{2(d-1)}{c\sin\varphi}\,\frac{t}{\beta_t} + \frac{2s_{d-2}}{c\sin\varphi\, b_{d-1}}\,\frac{t}{\beta_t} \to 0 \quad \text{as } t \to 0.$$

□

Let us now finish the proof of Theorem 1.

Recall that ν_S is the surface area measure induced by S. For all $\varphi \in (0,\, \pi/2)$, one has

$$\nu_t = \frac{1}{|B_t|}\,\nu_{S_{t,\varphi}} + \frac{1}{|B_t|}\,\nu_{S_t\setminus S_{t,\varphi}}.$$

Proposition 5 implies that the measure $\frac{1}{|B_t|}\,\nu_{S_{t,\varphi}}$ converges to 0 as $t \to 0$. Indeed, for any continuous function f on S^{d-1},

$$\int_{S^{d-1}} f(n)\,\frac{1}{|B_t|}\,\nu_{S_{t,\varphi}}(dn) \le \max|f|\,\frac{|S_{t,\varphi}|}{|B_t|} \to 0 \quad \text{as } t \to 0.$$

On the other hand, the measure $\frac{1}{|B_t|}\,\nu_{S_t\setminus S_{t,\varphi}}$ is supported in the set in S^{d-1} containing all points whose radius vector forms the angle $\le \varphi$ with e. It follows that each partial limit of $\frac{1}{|B_t|} \nu_{S_t \setminus S_{t,\varphi}}$ and, therefore, each partial limit of ν_t are supported in this set. Since $\varphi > 0$ can be made arbitrary small, one concludes that each partial limit of ν_t is proportional to δ_e. Finally, utilizing Equality (5) true for each partial limit ν_*, one concludes that the limit of ν_t exists and is equal to δ_e.

3. Proof of Theorem 2

In the proof, we will use the well-known fact that the surface area measure is continuous with respect to the Hausdorff topology in the space of convex bodies.

More precisely, we say that a family of convex bodies C_t, $t > 0$ in $\mathbb{R}^d$ converges to a convex body $C \subset \mathbb{R}^d$ as $t \to 0$ in the sense of Hausdorff and write $C_t \underset{t\to 0}{\longrightarrow} C$, if for any $\varepsilon > 0$, there exists $t_0 > 0$ such that for all $t \le t_0$, C_t is contained in the ε-neighborhood of C and C is contained in the ε-neighborhood of C_t.

It is well known that if $C_t \underset{t\to 0}{\longrightarrow} C$, then $\nu_{\partial C_t} \to \nu_{\partial C}$ as $t \to 0$.

Choose $\sigma > 0$ so $|\hat{B}_\sigma| = 1$, and therefore,

$$\nu_{\hat{S}_\sigma} = \nu_\star. \tag{10}$$

Let the origin coincide with the point r_0, that is $r_0 = \vec{0}$; then, the homothety of a set $\mathcal{A}$ with the center at r_0 and ratio k is $k\mathcal{A}$. See Figure 4.

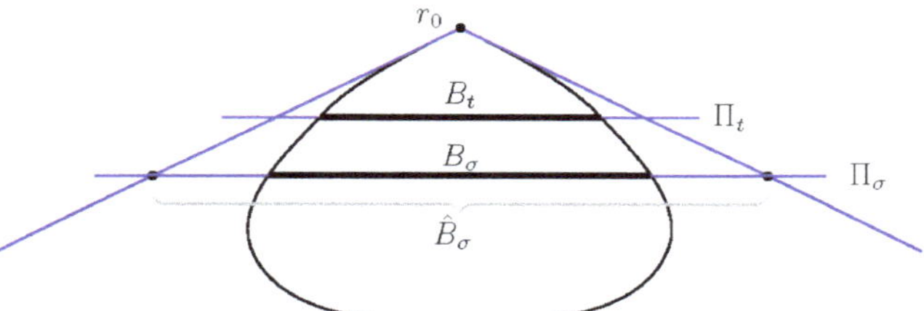

Figure 4. The tangent cone at r_0, the cutting planes Π_t and Π_σ, and the sets B_t, B_σ, and $\hat{B}_\sigma$ in the case when the point r_0 is conical.

Proposition 6. $\frac{\sigma}{t} B_t \underset{t\to 0}{\longrightarrow} \hat{B}_\sigma$.

Proof. Note that for all positive t_1 and t_2,

$$\frac{t_1}{t_2}\,\Pi_{t_2} = \Pi_{t_1} \quad \text{and} \quad \frac{t_1}{t_2}\,\hat{B}_{t_2} = \hat{B}_{t_1}.$$

Additionally, since the tangent cone K contains C, then $\hat{B}_t$ contains B_t, and so,

$$\frac{\sigma}{t}\,B_t \subset \frac{\sigma}{t}\,\hat{B}_t = \hat{B}_\sigma.$$

Let now $0 < t_1 \le t_2$. Since $\vec{0}$ and B_{t_2} belong to C, so does their linear combination,

$$\frac{t_1}{t_2}\,B_{t_2} = \Big(1 - \frac{t_1}{t_2}\Big)\vec{0} + \frac{t_1}{t_2}\,B_{t_2} \subset C.$$

On the other hand, $\frac{t_1}{t_2}\,B_{t_2} \subset \frac{t_1}{t_2}\,\Pi_{t_2} = \Pi_{t_1}$. It follows that $\frac{t_1}{t_2}\,B_{t_2} \subset C \cap \Pi_{t_1} = B_{t_1}$. We conclude that

$$\frac{\sigma}{t_2}\,B_{t_2} \subset \frac{\sigma}{t_1}\,B_{t_1},$$

that is $\frac{\sigma}{t}B_t$, $t > 0$ form a nested family of sets contained in $\hat{B}_\sigma$.

Suppose that $\frac{\sigma}{t} B_t$ does not converge to $\hat{B}_\sigma$. This implies that the closure of the union

$$\overline{\bigcup_{t>0} \frac{\sigma}{t} B_t} =: \tilde{B}_\sigma$$

is contained in, but does not coincide with, $\hat{B}_\sigma$.

The union:

$$\bigcup_{t>0} \frac{t}{\sigma} \tilde{B}_\sigma =: \tilde{K}$$

is a cone with the vertex at r_0; it is contained in the tangent cone K, but does not coincide with it. On the other hand,

$$C = \bigcup_{t\geq 0} B_t \subset \bigcup_{t\geq 0} \frac{t}{\sigma} \tilde{B}_\sigma = \tilde{K};$$

that is C is contained in the cone $\tilde{K}$, which is smaller than the tangent cone K. This contradiction proves our proposition. □

From Proposition 6, it follows, in particular, that

$$\lim_{t\to 0} \left| \frac{\sigma}{t} B_t \right| = |\hat{B}_\sigma| = 1, \tag{11}$$

and therefore,

$$\nu_{\frac{\sigma}{t} B_t} = \left| \frac{\sigma}{t} B_t \right| \delta_{-e} \longrightarrow |\hat{B}_\sigma| \delta_{-e} = \nu_{\hat{B}_\sigma} \quad \text{as } t \to 0. \tag{12}$$

Denote

$$\Sigma_\sigma^t := \text{conv}\Big(\frac{\sigma}{t} B_t \cup r_0 \Big).$$

Since the convex body $\frac{\sigma}{t} C_t$ contains both r_0 and $\frac{\sigma}{t} B_t$, we have $\Sigma_\sigma^t \subset \frac{\sigma}{t} C_t$.

Recall that K_σ is the part of the tangent cone cut off by the plane Π_σ. We have $K_\sigma = \text{conv}(\hat{B}_\sigma \cup r_0)$. Since by Proposition 6, $\frac{\sigma}{t} B_t \underset{t\to 0}{\longrightarrow} \hat{B}_\sigma$, we conclude that $\text{conv}(\frac{\sigma}{t} B_t \cup r_0) \underset{t\to 0}{\longrightarrow} \text{conv}(\hat{B}_\sigma \cup r_0)$, that is

$$\Sigma_\sigma^t \underset{t\to 0}{\longrightarrow} K_\sigma.$$

Using this relation and the double inclusion:

$$\Sigma_\sigma^t \subset \frac{\sigma}{t} C_t \subset K_\sigma,$$

one concludes that $\frac{\sigma}{t} C_t$ converges to K_σ in the sense of Hausdorff, and therefore,

$$\nu_{\frac{\sigma}{t}\partial C_t} \to \nu_{\partial K_\sigma} \quad \text{as} \quad t \to 0.$$

Using that $\frac{\sigma}{t}\partial C_t = \frac{\sigma}{t} S_t \cup \frac{\sigma}{t} B_t$ and $\partial K_\sigma = \hat{S}_\sigma \cup \hat{B}_\sigma$ and using (12), one obtains

$$\nu_{\frac{\sigma}{t} S_t} \to \nu_{\hat{S}_\sigma} \quad \text{as} \quad t \to 0,$$

and taking account of (11), one obtains

$$\lim_{t\to 0} \nu_t = \lim_{t\to 0} \frac{1}{|B_t|} \nu_{S_t} = \frac{1}{\lim_{t\to 0} |\frac{\sigma}{t} B_t|} \lim_{t\to 0} \nu_{\frac{\sigma}{t} S_t} = \nu_{\hat{S}_\sigma} = \nu_\star.$$

Theorem 2 is proven.

Funding: This research received no external funding.

Institutional Review Board Statement: Not applicable.

Informed Consent Statement: Not applicable.

Data Availability Statement: Not applicable.

Acknowledgments: This work was supported by the Center for Research and Development in Mathematics and Applications (CIDMA) through the Portuguese Foundation for Science and Technology (FCT), within Projects UIDB/04106/2020 and UIDP/04106/2020.

Conflicts of Interest: The author declares no conflict of interest.

References

1. Pogorelov, A.V. *Extrinsic Geometry of Convex Surfaces*; American Mathematical Society (AMS): Providence, RI, USA, 1973.
2. Buttazzo, G.; Kawohl, B. On Newton's problem of minimal resistance. *Math. Intell.* **1993**, *15*, 7–12. [CrossRef]
3. Newton, I. *Philosophiae Naturalis Principia Mathematica*; Streater: London, UK, 1687.
4. Belloni, M.; Kawohl, B. A paper of Legendre revisited. *Forum Math.* **1997**, *9*, 655–668. [CrossRef]
5. Brock, F.; Ferone, V.; Kawohl, B. A symmetry problem in the calculus of variations. *Calc. Var.* **1996**, *4*, 593–599. [CrossRef]
6. Buttazzo, G.; Ferone, V.; Kawohl, B. Minimum problems over sets of concave functions and related questions. *Math. Nachr.* **1995**, *173*, 71–89. [CrossRef]
7. Buttazzo, G.; Guasoni, P. Shape optimization problems over classes of convex domains. *J. Convex Anal.* **1997**, *4*, 343–351.
8. Comte, M.; Lachand-Robert, T. Newton's problem of the body of minimal resistance under a single-impact assumption. *Calc. Var. Partial Differ. Equ.* **2001**, *12*, 173–211. [CrossRef]
9. Lachand-Robert, T.; Oudet, E. Minimizing within convex bodies using a convex hull method. *SIAM J. Optim.* **2006**, *16*, 368–379. [CrossRef]
10. Lachand-Robert, T.; Peletier, M.A. Newton's problem of the body of minimal resistance in the class of convex developable functions. *Math. Nachr.* **2001**, *226*, 153–176. [CrossRef]
11. Plakhov, A. Billiards and two-dimensional problems of optimal resistance. *Arch. Ration. Mech. Anal.* **2009**, *194*, 349–382. [CrossRef]
12. Plakhov, A.; Torres, D. Newton's aerodynamic problem in media of chaotically moving particles. *Sbornik Math.* **2005**, *196*, 885–933. [CrossRef]
13. Wachsmuth, G. The numerical solution of Newton's problem of least resistance. *Math. Program.* **2014**, *147*, 331–350. [CrossRef]
14. Plakhov, A. On generalized Newton's aerodynamic problem. *Trans. Mosc. Math. Soc.* **2021**, *82*, 217–226. [CrossRef]

 MDPI

Article

Pattern Formation Induced by Fuzzy Fractional-Order Model of COVID-19

Abeer S. Alnahdi [1,†], Ramsha Shafqat [2,*,†], Azmat Ullah Khan Niazi [2,†] and Mdi Begum Jeelani [1,*,†]

1 Department of Mathematics and Statistics, College of Science, Imam Mohammad Ibn Saud Islamic University, Riyadh 11564, Saudi Arabia; asalnahdi@imamu.edu.sa
2 Department of Mathematics and Statistics, The University of Lahore, Sargodha 40100, Pakistan; azmatullah.khan@math.uol.edu.pk
* Correspondence: ramshawarriach@gmail.com (R.S.); mbshaikh@imamu.edu.sa (M.B.J.)
† These authors contributed equally to this work.

Abstract: A novel coronavirus infection system is established for the analytical and computational aspects of this study, using a fuzzy fractional evolution equation (FFEE) stated in Caputo's sense for order (1,2). It is constructed using the FFEE formulated in Caputo's meaning. The model consist of six components illustrating the coronavirus outbreak, involving the susceptible people $\mathbf{K}_{\ell}(\omega)$, the exposed population $\mathbf{L}_{\ell}(\omega)$, total infected strength $\mathbf{C}_{\ell}(\omega)$, asymptotically infected population $\mathbf{M}_{\ell}(\omega)$, total number of humans recovered $\mathbf{E}_{\ell}(\omega)$, and reservoir $\mathbf{Q}_{\ell}(\omega)$. Numerical results using the fuzzy Laplace approach in combination with the Adomian decomposition transform are developed to better understand the dynamical structures of the physical behavior of COVID-19. For the controlling model, such behavior on the generic characteristics of RNA in COVID-19 is also examined. The findings show that the proposed technique of addressing the uncertainty issue in a pandemic situation is effective.

Keywords: approximation solution; fuzzy number; fuzzy fractional order derivative; coronavirus infection system; Adomian decomposition method

MSC: 26A33; 34K37

Citation: Alnahdi, A.S.; Shafqat, R.; Niazi, A.U.K.; Jeelani, M.B. Pattern Formation Induced by Fuzzy Fractional-Order Model of COVID-19. *Axioms* **2022**, *11*, 313. https://doi.org/10.3390/axioms11070313

Academic Editors: Natália Martins, Ricardo Almeida, Cristiana João Soares da Silva and Moulay Rchid Sidi Ammi

Received: 30 May 2022
Accepted: 23 June 2022
Published: 27 June 2022

1. Introduction

Recently, the entire world has been afflicted by a novel coronavirus pandemic known as the "novel coronavirus 2019", abbreviated as "nCOVID-19", which was initially reported in Wuhan, central China [1]. It has been discovered that nCOVID-19 is spread from animal to human; several afflicted people claimed to have contracted the virus after visiting a local fish and wild animal market in Wuhan on 28 November [2]. Following that, other researchers confirmed that transmission can also occur from one person to another [3]. According to World Health Organization data, the number of reported laboratory-confirmed human infections in 187 countries, territories, or places around the world reached 292,142 on 21 March 2020, with 12,784 mortality cases [4]. The death rate was as high as 0.0666 in some nations, such as Italy and Spain. This confirms the severity and high infectivity of nCOVID-19. Most patients infected with nCOVID-19 will have mild to moderate respiratory symptoms, such as shortness of breath, low fever, nausea, cough, and other symptoms. Other symptoms have been described, including gastroenteritis and neurological illnesses of varying severity [5]. nCOVID-19 is primarily transmitted by droplets from the nose when an infected person coughs or sneezes. A person is in danger of catching the virus if he or she inhales droplets from infected people in the air. As a result, avoiding meetings and contacting other individuals is the greatest approach to avoid contracting the virus.

To manage people flow and movement, Wuhan has been shut down by the Chinese government, and they have decreased or restricted the transportation system of the country, including airplanes, trains, buses, and private cars, among other things. People have had to stay at home and have their body temperature taken every day. If they have to go outside, they are advised to wear respirators. With the spread of nCOVID-19 around the world, more governments have entered the antivirus fight in the footsteps of the Chinese government. It was reported that an increasing number of governments have begun to issue restrictions prohibiting international travel, as well as closing schools, shopping malls, and businesses. The nCOVID-19 pandemic has caused significant economic loss throughout the world, as well as huge hardship for country administrations and even all human beings. A large number of doctors and researchers also dedicated themselves to the anti-pandemic fight and conducted research based on their knowledge. They studied nCOVID-19 from a variety of perspectives, including microbiology, virology, sociology, veterinary sciences, infectious diseases, public environmental occupational health, political economy, media studies, and so on. The main countries in nCOVID-19 research include China, the United States, and Korea, as a result of the virus's early epidemic, which prompted them to begin relevant research right away. The origins of nCOVID-19 were investigated by a group of scientists. Initially, bats were thought to be the source of nCOVID-19, which is comparable to SARS (severe acute respiratory syndrome), a worldwide epidemic that began in China and other parts of the world in 2003 [6,7].

Following that, some studies linked nCOVID-19 to the 2012 pandemics of SARS and MERS (Middle East respiratory syndrome) to show that there are lessons to be learned from the two pandemics. SARS-COVID, MERS-COVID, and nCOVID-19 all belong to SARS-COVID, according to Lu [8] and belong to the same family of Betacoronaviruses. Previous studies, according to Zhou, suggest that nCOVID-19 has a significant degree of resemblance to SARS-COVID, and, as evidenced by [9], has the same predicted cell entry mechanism and human cell receptor use based on full-length genome phylogenetic research. Xiaolong and Mose also analyzed the high RBD (receptor binding domain) identity between nCOVID-19 and SARS-COVID, and proposed that the SARS-COVID specific human antibody, CR3022, might bind potently to the virus. nCOVID-19 RBD has a KD of 6.3 nM, indicating that the difference within the RBD is 6.3 nM. SARS-COVID and nCOVID-19 have a significant impact on neutralizing antibody cross-reactivity, which is still needed to produce novel monoclonal antibodies that bind selectively to nCOVID-19 RBD [10]. Syed et al. determined SARS-COVID-derived B lymphocyte epitopes and T-cell epitopes experimentally based on previous studies on SARS-COVID immunological systems and structures and discovered that they are similar and contain no mutation within the available nCOVID-19 sequences, which is critical for narrowing down the hunt for potent targets for an efficient vaccine against the nCOVID-19. Some researchers are concentrating on the transmission and identification of the nCOVID-19 virus in people. Human-to-human transmission is widely acknowledged as a factor in the rapid spread of illnesses. Ahmed said that viral strains from the area's affected persons had been analyzed, but that there was a limited genetic difference, meaning that they all descended from a single ancestor [11]. Zhou, on the other hand, claimed that the sequences of the seven conserved viral replicase domains in ORF 1ab in nCOVID-19 and SARS-COVID [9] are 94.6 percent similar. Chaudhury et al. demonstrated that using accurate, physics-based energy functions, computational protein–protein docking can disclose the native-like, low-energy protein–protein complex from the unbound structures of two separate, interacting protein components [12]. In this paper, we attempt to mathematically examine the nCOVID-19 infection mechanism. The numerical findings are obtained using the fuzzy Laplace transform based on Adomian decomposition, which can be useful in understanding the dynamical structures of the physical behavior of nCOVID-19. In the form of nonlinear fractional order differential equations (FODEs), we define the system of six equations illustrating the coronavirus outbreak, involving susceptible people $\mathbf{A}_{\ell}(\omega)$, the exposed population $\mathbf{B}_{\ell}(\omega)$, total infected strength $\mathbf{C}_{\ell}(\omega)$, asymptotically infected population $\mathbf{D}_{\ell}(\omega)$, total number of humans recovered $\mathbf{E}_{\ell}(\omega)$, and reservoir $\mathbf{F}_{\ell}(\omega)$, which are listed in the following order [13]:

$$
\begin{aligned}
&D^{\vartheta}_{\omega}\mathbf{A}_{\ell}(\omega) = r_{\ell} - \imath_{\ell}\mathbf{A}_{\ell} - s_{\ell}\mathbf{A}_{\ell}(C_{\ell} + \lambda\mathbf{D}_{\ell}) - s_{\ell}\mathbf{S}_{\ell}\mathbf{F}_{\ell},\\
&D^{\vartheta}_{\omega}\mathbf{B}_{\ell}(\omega) = s_{\ell}\mathbf{A}_{\ell}(C_{\ell} + \varpi\mathbf{D}_{\ell}) + s_{l}\mathbf{A}_{\ell}\mathbf{F} - (1-\varpi_{\ell})\mu_{\ell}\mathbf{B}_{\ell} - \varpi_{\ell}\mu'_{\ell}\mathbf{B}_{\ell} - \imath_{\ell}\mathbf{B}_{\ell},\\
&D^{\vartheta}_{\omega}\mathbf{C}_{\ell}(\omega) = (1-\varpi_{\ell})\mu_{\ell}\mathbf{B}_{\ell} - (\vartheta_{\ell} + \imath_{\ell})\mathbf{C}_{\ell},\\
&D^{\vartheta}_{\omega}\mathbf{D}_{\ell}(\omega) = \varpi_{\ell}\mu_{\ell'}\mathbf{B}_{\ell} - (\gamma'_{\ell} + \imath_{\ell})\mathbf{C}_{\ell},\\
&D^{\vartheta}_{\ell}\mathbf{E}_{\ell}(\omega) = \gamma_{\ell}\mathbf{C}_{\ell} + \gamma'_{\ell}\mathbf{D}_{\ell} - \imath_{\ell}\mathbf{E}_{\ell},\\
&D^{\vartheta}_{\ell}\mathbf{F}_{\ell}(\ell) = \varsigma\mathbf{C}_{\ell} + \psi\mathbf{D}_{\ell} - \varrho\mathbf{F}_{\ell},
\end{aligned}
\tag{1}
$$

where r_{ℓ} is the birth rate, $\imath_{\ell}$ is the death rate of infected people, s_{ℓ} is the transmission coefficient, s_{l} is the disease transmission coefficient, and λ is the transmissibility multiple. The symbols μ_{ℓ} and μ_{ℓ} signify the incubation period. The recovery rates of $\mathbf{C}_{\ell}$ and $\mathbf{D}_{\ell}$ are represented by ϑ_{ℓ} and ϑ'_{ℓ}, respectively. ς and ψ reflect the virus's influence from $\mathbf{C}_{\ell}$ and $\mathbf{D}_{\ell}$ to $\mathbf{F}_{\ell}$, respectively, and ϱ indicates the virus's elimination rate from $\mathbf{F}_{\ell}$. The parameters are included in Table 1 for convenience. In recent years, fuzzy calculus and FODEs [14–17] have been added to current calculus and DEs, respectively. Then, FODEs were expanded to fuzzy FODEs [18–20]. Many academics have researched FODEs and fuzzy integral equations in order to establish the existence–uniqueness theory of solutions [21–26]. It is extremely time consuming to compute more precise solutions to each fuzzy FODE when dealing with fuzzy FODEs. Mathematicians have put forth a lot of effort to solve fuzzy FODEs using various approaches, such as perturbation methods, integral transform methods, and spectral techniques [27–32]. Some researchers examined the stability of fuzzy DEs [33]. Niazi et al. [34], Iqbal et al. [35], Shafqat et al. [36] and Abuasbeh et al. [37] investigated the existence and uniqueness of the FFEE. Ahmad et al. [38] worked on Model (2), the fuzzy fractional-order model of the novel coronavirus as

$$
\begin{aligned}
&D^{\vartheta}_{\omega}\mathbf{L}_{\ell}(\omega) = \hat{r}_{\ell} - \hat{\imath}_{\ell}\mathbf{L}_{\ell} - \hat{s}_{\ell}\mathbf{L}_{\ell}(\mathbf{N}_{\ell} + \hat{\lambda}\mathbf{P}_{\ell}) - \hat{s}_{\ell}\mathbf{L}_{\ell}\mathbf{R}_{\ell},\\
&D^{\vartheta}_{\omega}\mathbf{M}_{\ell}(\omega) = \hat{s}_{\ell}\mathbf{L}_{\ell}(N_{\ell} + \varpi\mathbf{P}_{\ell}) + \hat{s}_{\ell}\mathbf{L}_{\ell}\mathbf{R} - (1-\hat{\varpi}_{\ell})\hat{\mu}_{\ell}\mathbf{M}_{\ell} - \hat{\varpi}_{\ell}\hat{\mu}_{\ell'}\mathbf{M}_{\ell} - \hat{\imath}_{\ell}\mathbf{M}_{\ell},\\
&D^{\vartheta}_{\omega}\mathbf{N}_{\ell}(\omega) = (1-\hat{\varpi}_{\ell})\hat{\mu}_{\ell}\mathbf{M}_{\ell} - (\hat{\vartheta}_{\ell} + \hat{\imath}_{\ell})\mathbf{N}_{\ell},\\
&D^{\vartheta}_{\omega}\mathbf{P}_{\ell}(\omega) = \hat{\varpi}_{\ell}\hat{\mu}_{\ell'}\mathbf{K}_{\ell} - (\hat{\vartheta}_{\ell'} + \hat{\imath}_{\ell})\mathbf{P}_{\ell},\\
&D^{\vartheta}_{\omega}\mathbf{Q}_{\ell}(\omega) = \hat{\vartheta}_{\ell}\mathbf{N}_{\ell} + \hat{\vartheta}_{\ell'}\mathbf{P}_{\ell} - \imath_{\ell}\mathbf{Q}_{\ell},\\
&D^{\vartheta}_{\omega}\mathbf{R}_{\ell}(\omega) = \hat{\varsigma}\mathbf{N}_{\ell} + \hat{\psi}\mathbf{P}_{\ell} - \hat{\varrho}\mathbf{R}_{\ell}.
\end{aligned}
\tag{2}
$$

Inspired by the above, Model (3) with a fuzzy fractional-order derivative by using mild solution is investigated here, with the uncertainty in initial data. For $1 < \vartheta \leq 2$,

$$
\begin{aligned}
&{}^{c}_{0}D^{\vartheta}_{\omega}\mathbf{K}_{\ell}(\omega) = \hat{r}_{\ell} - \hat{\imath}_{\ell}\mathbf{K}_{\ell} - \hat{s}_{\ell}\mathbf{K}_{\ell}(\mathbf{C}_{\ell} + \hat{\lambda}\mathbf{M}_{\ell}) - \hat{s}_{\ell}\mathbf{K}_{\ell}\mathbf{Q}_{\ell},\\
&{}^{c}_{0}D^{\vartheta}_{\omega}\mathbf{L}_{\ell}(\omega) = \hat{s}_{\ell}\mathbf{K}_{\ell}(\mathbf{C}_{\ell} + \varpi\mathbf{M}_{\ell}) + \hat{s}_{\ell}\mathbf{K}_{\ell}\mathbf{Q} - (1-\hat{\varpi}_{\ell})\hat{\mu}_{\ell}\mathbf{L}_{\ell} - \hat{\varpi}_{\ell}\hat{\mu}'_{\ell}\mathbf{L}_{\ell} - \hat{\imath}_{\ell}\mathbf{L}_{\ell},\\
&{}^{c}_{0}D^{\vartheta}_{\omega}\mathbf{C}_{\ell}(\omega) = (1-\hat{\varpi}_{\ell})\hat{\mu}_{\ell}\mathbf{L}_{\ell} - (\hat{\vartheta}_{\ell} + \hat{\imath}_{\ell})\mathbf{C}_{\ell},\\
&{}^{c}_{0}D^{\vartheta}_{\omega}\mathbf{M}_{\ell}(\omega) = \hat{\varpi}_{\ell}\hat{\mu}'_{\ell}\mathbf{L}_{\ell} - (\hat{\vartheta}'_{\ell} + \hat{\imath}_{\ell})\mathbf{M}_{\ell},\\
&{}^{c}_{0}D^{\vartheta}_{\omega}\mathbf{E}_{\ell}(\omega) = \hat{\vartheta}_{\ell}\mathbf{C}_{\ell} + \hat{\vartheta}'_{\ell}\mathbf{M}_{\ell} - \imath_{\ell}\mathbf{E}_{\ell},\\
&{}^{c}_{0}D^{\vartheta}_{\omega}\mathbf{Q}_{\ell}(\omega) = \hat{\varsigma}\mathbf{C}_{\ell} + \hat{\psi}\mathbf{M}_{\ell} - \hat{\varrho}\mathbf{Q}_{\ell},
\end{aligned}
\tag{3}
$$

associated to fuzzy initial condition, for $\Xi \in (0,1)$,

$$
\begin{aligned}
&\mathbf{K}(0,\Xi) = (\underline{\mathbf{K}}(0,\Xi), \overline{\mathbf{K}}(0,\Xi)) + (-1)g(\mathbf{K}), \quad \mathbf{K}'(0,\Xi) = (\underline{\mathbf{K}}(0,\Xi), \overline{\mathbf{K}}(0,\Xi)),\\
&\mathbf{L}(0,\Xi) = (\underline{\mathbf{L}}(0,\Xi), \overline{\mathbf{L}}(0,\Xi)) + (-1)g(\mathbf{L}), \quad \mathbf{L}'(0,\Xi) = (\underline{\mathbf{L}}(0,\Xi), \overline{\mathbf{L}}(0,\Xi)),\\
&\mathbf{C}(0,\Xi) = (\underline{\mathbf{C}}(0,\Xi), \overline{\mathbf{C}}(0,\Xi)) + (-1)g(\mathbf{C}), \quad \mathbf{C}'(0,\Xi) = (\underline{\mathbf{C}}(0,\Xi), \overline{\mathbf{C}}(0,\Xi)),\\
&\mathbf{M}(0,\Xi) = (\underline{\mathbf{M}}(0,\Xi), \overline{\mathbf{M}}(0,\Xi)) + (-1)g(\mathbf{M}), \quad \mathbf{M}'(0,\Xi) = (\underline{\mathbf{M}}(0,\Xi), \overline{\mathbf{M}}(0,\Xi)),\\
&\mathbf{E}(0,\Xi) = (\underline{\mathbf{E}}(0,\Xi), \overline{\mathbf{E}}(0,\Xi)) + (-1)g(\mathbf{E}), \quad \mathbf{E}'(0,\Xi) = (\underline{\mathbf{E}}(0,\Xi), \overline{\mathbf{E}}(0,\Xi)),\\
&\mathbf{Q}(0,\Xi) = (\underline{\mathbf{Q}}(0,\Xi), \overline{\mathbf{Q}}(0,\Xi)) + (-1)g(\mathbf{Q}), \quad \mathbf{Q}'(0,\Xi) = (\underline{\mathbf{Q}}(0,\Xi), \overline{\mathbf{Q}}(0,\Xi)),
\end{aligned}
$$

where r_{ℓ} is birth rate, $\imath_{\ell}$ is death rate of infected people, s_{ℓ} is transmission coefficient, s_{l} is the disease transmission coefficient, and λ is the transmissibility multiple. The symbols μ_{ℓ} and μ'_{ℓ} signify the incubation period. The recovery rates of $\mathbf{C}_{\ell}$ and $\mathbf{M}_{\ell}$ are represented by ϑ_{ℓ} and ϑ_{ℓ}, respectively. ς and ψ reflect the virus's influence from $\mathbf{C}_{\ell}$ and $\mathbf{M}_{\ell}$ to $\mathbf{Q}_{\ell}$, respectively, and ϱ indicates the virus's elimination rate from $\mathbf{Q}_{\ell}$.

Table 1. Description of the model's parameters (3).

Notation	Parameters Description	Numerical Value
$\overline{r_\ell}$	Birth rate	1
$\overline{\iota_\ell}$	Death rate of infected population	$\frac{1}{(76.79\times 365)}$
$\overline{s_\ell}$	Transmission coefficient	0.05
$\overline{s_l}$	Disease transmission coefficient	0.001231
$\overline{\mu_\ell}, \overline{\mu'_\ell}$	Signified incubation period	0.001234, 0.05
$\overline{\vartheta_\ell}, \overline{\vartheta'_\ell}$	Recovery rate of $\mathbf{C}_\ell, \mathbf{M}_\ell$	0.1389
$\overline{\varsigma}, \overline{\psi}$	Influence of virus from $\mathbf{C}_\ell$ and $\mathbf{M}_\ell$ to $\mathbf{Q}_\ell$	0.0398, 0.01
$\hat{\varrho}$	Amount of asymptotic infection	0.1243
$\hat{\lambda}$	Transmissibility multiple	0.02
$\hat{\psi}$	Elimination rate of virus from $\mathbf{Q}_\ell$	0.01
$\mathbf{K}_0$	Initial value of susceptible	220 million
$\mathbf{C}_0$	Initial value of infected	0.015 million
$\mathbf{L}_0$	Initial value of exposed	100 million
$\mathbf{M}_0$	Initial value of asymptotically infected	0.60 million
$\mathbf{E}_0$	Initial value of recovered	0 million
$\mathbf{Q}_0$	Initial value of reservoir	0.1 million

We use fuzzy fractional-order model of order (1, 2) by using a fuzzy mild solution of the novel coronavirus with nonlocal conditions. Owing to it, our model's graphs are more accurate. The theory of fuzzy sets continues to gain scholars' attention because of its huge range of applications in domains such as engineering, mechanics, robotics, electrical, control, thermal systems, and signal processing. In light of the foregoing arguments and to meet the current uncertain scenario, based on fuzzy fractional calculus, we suggested a new coronavirus infection strategy. We ensure that the proposed model is closer to the true behavior of a system growing the basic properties of RNA in COVID-19 by researching it, which also improves the physical behavior of such an infection system. Our major goal is to obtain the existence–uniqueness result of a COVID-19 model of order (1, 2). Applying mild solutions to this COVID-19 model becomes more complicated. However, via rigorous analysis, we show that the suggested function deduces a novel representation of solution operators and then provides a new idea of mild solutions. As a result, the research of system (3) differs greatly from earlier studies on the COVID-19 model. The other section of this paper is as follows. Section 2 discusses the definitions. Section 3 introduces the existence–uniqueness of the solution to the succeeding fuzzy model. A general method is also shown here for using fuzzy Laplace transform to determine the solution of the examined system. In Section 4, numerical results and discussion are presented. Finally, in Section 5, a conclusion is given.

2. Preliminaries

Definition 1 ([39,40])**.** *Take $\rho : \mathbf{R}^{\mathrm{m}} \to [0,1]$ to be a fuzzy real line set meeting the below properties:*

(i) *ρ is normal.*
(ii) *ρ is upper semicontinuous on $\mathbf{R}^{\mathrm{m}}$.*
(iii) *ρ is convex.*
(iv) *$cl\{a \in \mathbf{R}^{\mathrm{m}}, \rho(a) > 0\}$ is compact.*

Then, it is known as a fuzzy number.

Definition 2 ([39])**.** *On a fuzzy number ρ, the $\wp$-level set is defined by*

$$[u]^{\wp} = \{x \in \mathbf{R}^{\mathrm{m}} : \mu(x) \leq \wp\},$$

where $\wp \in (0,1]$ and $x \in \mathbf{R}^{\mathrm{m}}$.

Definition 3 ([39,40]). *Suppose $[\underline{\rho}(\theta), \overline{\rho}(\theta)]$ is the parametric form of a fuzzy number ρ, where $0 \leq \theta \leq 1$ and the below properties are satisfied:*

(a) *$\underline{\rho}(\theta)$ is left continuous, bounded, and increasing function over $(0, 1]$ and right continuous at 0.*

(b) *$\overline{\rho}(\theta)$ is left continuous, bounded, and increasing function over $(0, 1]$ and right continuous at 0.*

(c) *$\underline{\rho}(\theta) \leq \overline{\rho}(\theta)$.*
Additionally, if $\underline{\rho}(\theta) = \overline{\rho}(\theta) = 0$, then θ is called a crisp number.

Definition 4 ([30]). *If a mapping $v : \mathbf{E}^{\mathrm{m}} \times \mathbf{E}^{\mathrm{m}} \to \mathbf{R}^{\mathrm{m}}$ and let $\varrho = (\underline{\varrho}(\theta) \leq \overline{\varrho}(\theta))$ and $\mu = (\underline{\mu}(\theta) \leq \overline{\mu}(\theta))$ be two fuzzy numbers in their parametric form. The Hausdorff distance between ϱ and μ is defined as*

$$v(\varrho, \mu) = \max\{\sup_{a \in A} \inf_{b \in B} ||a - b||, \sup_{b \in B} \inf_{a \in A} ||a - b||\}.$$

In $\mathbf{E}^{\mathrm{m}}$, the metric v has the below properties:

(i) $v(v + \varrho, w + \varrho) = v(v, w)$ for all $v, \varrho, w \in \mathbf{E}^{\mathrm{m}}$;
(ii) $v(v\sigma, w\sigma) = |\sigma v(v, w)$ for all $v, w \in \mathbf{E}^{\mathrm{m}}, \varrho \in \mathbf{R}^{\mathrm{m}}$;
(iii) $v(v + \varsigma, w + \zeta) \leq v(v, w) + v(\xi, \zeta)$ for all $v, w, \varsigma, \zeta \in \mathbf{E}^{\mathrm{m}}$;
(iv) $(\mathbf{E}^{\mathrm{m}}, v)$ is a complete metric space.

Definition 5 ([30]). *Let $\tau_1, \tau_2 \in \mathbf{E}^{\mathrm{m}}$. If there exist $\tau_3 \in \mathbf{E}^{\mathrm{m}}$ such that $\tau_1 = \tau_2 + \tau_3$ then τ_3 is said to be the H-difference of τ_1 and τ_2, denoted by $\tau_1 \ominus \tau_2$.*

Definition 6 ([30]). *If $\Theta : \mathbf{R}^{\mathrm{m}} \to \mathbf{E}^{\mathrm{m}}$ be a fuzzy mapping. Then Θ is called continuous if for any $\epsilon > 0 \ \exists \ \delta > 0$ and a fixed value of $\lambda_0 \in [\zeta_1, \zeta_2]$, we have*

$$v(\Theta(\lambda), \Theta(\lambda_0)) < \epsilon,$$

whenever

$$|\lambda - \lambda_0| < \varpi.$$

Definition 7 ([27,30]). *Let φ be a continuous fuzzy function on $[0, b] \subseteq \mathbf{R}^{\mathrm{m}}$, a fuzzy fractional integral in RL sense corresponding to ω is defined by*

$$I^{\lambda}\phi(\omega) = \frac{1}{\Gamma(\lambda)} \int_0^{\omega} (\omega - \zeta)^{\lambda - 1} \phi(\zeta) d\zeta, \text{ where } \lambda, \zeta \in (0, \infty).$$

If $\phi \in C^F[0, b] \bigcap L^F[0, b]$, where $C^F[0, b]$ and $L^F[0, b]$ are the spaces of fuzzy continuous fractions and fuzzy Lebesgue integrable functions, respectively, then the fuzzy fractional integral is defined as

$$[I^{\lambda}\phi(\omega)]_{\wp} = [I^{\kappa}\underline{\phi}_{\wp}(\omega), I^{\lambda}\overline{\phi}_{\wp}(\omega)], 0 \leq \wp \leq 1,$$

where

$$\begin{aligned} I^{\lambda}\underline{\phi}_{\wp}(\omega) &= \frac{1}{\Gamma(\lambda)} \int_0^{\omega} (\omega - \zeta)^{\lambda - 1} \underline{\phi}_{\wp}(\omega) d\zeta, \ \lambda, \zeta \in (0, \infty), \\ I^{\lambda}\overline{\phi}_{\wp}(\omega) &= \frac{1}{\Gamma(\lambda)} \int_0^{\omega} (\omega - \zeta)^{\lambda - 1} \overline{\phi}_{\wp}(\omega) d\zeta, \ \lambda, \zeta \in (0, \infty). \end{aligned}$$

Definition 8 ([30]). *If a fuzzy fraction $\phi \in C^F[0, b] \bigcap L^F[0, b]$ is such that $\phi = [\underline{\phi}_{\wp}(\omega), \overline{\phi}_{\wp}(\omega)]$, $0 \leq \wp \leq 1$ and $\omega_1 \in (0, b)$, then the fuzzy fractional Caputo's derivative is defined as*

$$[D^{\beta}\underline{\phi}_{\wp}(\omega_0), D^{\beta}\overline{\phi}_{\wp}(\omega_0)], 0 \leq \beta \leq 1,$$

where

$$\begin{aligned} D^{\beta}\underline{\phi}_{\wp}(\omega_0) &= \frac{1}{\Gamma(n-\beta)}\left[\int_0^{\omega}(\omega-\zeta)^{n-\beta-1}\frac{d^n}{d\zeta^n}\underline{\phi}_{\wp}(\omega)d\zeta\right]_{\omega=0}, \\ D^{\beta}\overline{\phi}_{\wp}(\omega_0) &= \frac{1}{\Gamma(n-\beta)}\left[\int_0^{\omega}(\omega-\zeta)^{n-\beta-1}\frac{d^n}{d\zeta^n}\overline{\phi}_{\wp}(\omega)d\zeta\right]_{\omega=\omega_0}, \end{aligned}$$

whenever the integrals on the right-hand sides converge and $n = [\beta]$.

Definition 9 ([29–31]). *Suppose* ϕ *is a continuous fuzzy-valued function. Assume that* $\phi(\chi).e^{-s\chi}$ *is an improper fuzzy Riemann-integrable on* $[0,\infty)$*, then its fuzzy Laplace transform is*

$$L[\phi(\chi)] = \int_0^{\infty}\phi(\chi).e^{-s\chi}d\chi.$$

For $0 \le r \le 1$*, the parametric form of* $\phi(\chi)$ *is represented by*

$$\int_0^{\infty}\phi(\chi,r).e^{-s\chi}d\chi = \left[\int_0^{\infty}\underline{\phi}(\chi,r).e^{-s\chi}d\chi, \int_0^{\infty}\overline{\phi}(\chi,r).e^{-s\chi}d\chi\right].$$

Hence,

$$L[\phi(\chi,r)] = [L\underline{\phi}(\chi,r), L\overline{\phi}(\chi,r)].$$

Theorem 1 ([31]). *If* $\phi \in C^F[0,b]\bigcap L^F[0,b]$*, then the Laplace transform of the fuzzy fractional derivative in Caputo's form is given for* $0 \le \wp \le 1$ *and* $0 < \beta \le 1$ *by*

$$L[(D^{\beta}\phi(\omega))_{\wp}] = s^{\beta}L[\phi(\omega)] - s^{\beta-1}[\phi(0)].$$

Theorem 2 (Schauder Fixed Point Theorem). *Let* $(X, \|.\|)$ *be a Banach space and* $S \subset X$ *be compact, convex and nonempty. Any continuous operator* $A : S \to S$ *has at least one fixed point.*

3. Main Result

The existence–uniqueness of solutions to succeeding fuzzy fractional model (FFM) are examined in the following section, and we show how to discover a semi-analytic solution to model (3) using the fuzzy Laplace transform by using mild the solution.

3.1. Existence–Uniqueness

The existence–uniqueness of the succeeding FFM are addressed in this section using fixed point theory. Consider the right-hand side of Model (3):

$$\begin{aligned} &\digamma(\omega,\mathbf{K}_\ell(\omega),\mathbf{L}_\ell(\omega),\mathbf{C}_\ell(\omega),\mathbf{M}_\ell(\omega),\mathbf{E}_\ell(\omega),\mathbf{Q}_\ell(\omega)) = \hat{r}_\ell - \hat{\imath}_\ell\mathbf{K}_\ell - \hat{s}_\ell\mathbf{K}_\ell(\mathbf{C}_\ell + \hat{\lambda}\mathbf{M}_\ell) - \hat{s}_\ell\mathbf{K}_\ell\mathbf{Q}_\ell, \\ &\daleth(\omega,\mathbf{K}_\ell(\omega),\mathbf{L}_\ell(\omega),\mathbf{C}_\ell(\omega),\mathbf{M}_\ell(\omega),\mathbf{E}_\ell(\omega),\mathbf{Q}_\ell(\omega)) = \hat{s}_\ell\mathbf{K}_\ell(\mathbf{C}_\ell + \varpi\mathbf{M}_\ell) + \hat{s}_\ell\mathbf{K}_\ell\mathbf{Q} - (1-\hat{\varpi}_\ell)\hat{\mu}_\ell\mathbf{L}_\ell - \hat{\varpi}_\ell\hat{\mu}'_\ell\mathbf{L}_\ell - \hat{\imath}_\ell\mathbf{L}_\ell, \\ &\Phi(\omega,\mathbf{K}_\ell(\omega),\mathbf{L}_\ell(\omega),\mathbf{C}_\ell(\omega),\mathbf{M}_\ell(\omega),\mathbf{E}_\ell(\omega),\mathbf{Q}_\ell(\omega)) = (1-\hat{\varpi}_\ell)\hat{\mu}_\ell\mathbf{L}_\ell - (\hat{\vartheta}_\ell + \hat{\imath}_\ell)\mathbf{C}_\ell, \\ &\Psi(\omega,\mathbf{K}_\ell(\omega),\mathbf{L}_\ell(\omega),\mathbf{C}_\ell(\omega),\mathbf{M}_\ell(\omega),\mathbf{E}_\ell(\omega),\mathbf{Q}_\ell(\omega)) = \hat{\varpi}_\ell\hat{\mu}_{\ell'}\mathbf{L}_\ell - (\hat{\vartheta}'_\ell + \hat{\imath}_\ell)\mathbf{M}_\ell, \\ &\Upsilon(\omega,\mathbf{K}_\ell(\omega),\mathbf{L}_\ell(\omega),\mathbf{C}_\ell(\omega),\mathbf{M}_\ell(\omega),\mathbf{E}_\ell(\omega),\mathbf{Q}_\ell(\omega)) = \hat{\vartheta}_\ell\mathbf{C}_\ell + \hat{\vartheta}'_\ell\mathbf{M}_\ell - \iota_\ell\mathbf{E}_\ell, \\ &\Lambda(\omega,\mathbf{K}_\ell(\omega),\mathbf{L}_\ell(\omega),\mathbf{C}_\ell(\omega),\mathbf{M}_\ell(\omega),\mathbf{E}_\ell(\omega),\mathbf{Q}_\ell(\omega)) = \hat{\varsigma}\mathbf{C}_\ell + \hat{\psi}\mathbf{M}_\ell - \hat{\varrho}\mathbf{Q}_\ell, \end{aligned} \tag{4}$$

where $\digamma, \daleth, \Phi, \Psi, \Upsilon$ and Λ are fuzzy functions. Thus, for $1 < \gamma \le 2$, the model (3) is

$$\begin{aligned} D^{\vartheta}_{\omega}\mathbf{K}_\ell(\omega) &= \psi(\digamma(\omega,\mathbf{K}_\ell(\omega),\mathbf{L}_\ell(\omega),\mathbf{C}_\ell(\omega),\mathbf{M}_\ell(\omega),\mathbf{E}_\ell(\omega),\mathbf{Q}_\ell(\omega))), \\ D^{\vartheta}_{\omega}\mathbf{L}_\ell(\omega) &= \daleth(\omega,\mathbf{K}_\ell(\omega),\mathbf{L}_\ell(\omega),\mathbf{C}_\ell(\omega),\mathbf{M}_\ell(\omega),\mathbf{E}_\ell(\omega),\mathbf{Q}_\ell(\omega))), \\ D^{\vartheta}_{\omega}\mathbf{C}_\ell(\omega) &= \Phi(\omega,\mathbf{K}_\ell(\omega),\mathbf{L}_\ell(\omega),\mathbf{C}_\ell(\omega),\mathbf{M}_\ell(\omega),\mathbf{E}_\ell(\omega),\mathbf{Q}_\ell(\omega))), \\ D^{\vartheta}_{\omega}\mathbf{M}_\ell(\omega) &= \Psi(\omega,\mathbf{K}_\ell(\omega),\mathbf{L}_\ell(\omega),\mathbf{C}_\ell(\omega),\mathbf{M}_\ell(\omega),\mathbf{E}_\ell(\omega),\mathbf{Q}_\ell(\omega))), \\ D^{\vartheta}_{\omega}\mathbf{E}_\ell(\omega) &= \Upsilon(\omega,\mathbf{K}_\ell(\omega),\mathbf{L}_\ell(\omega),\mathbf{C}_\ell(\omega),\mathbf{M}_\ell(\omega),\mathbf{E}_\ell(\omega),\mathbf{Q}_\ell(\omega))), \\ D^{\vartheta}_{\omega}\mathbf{Q}_\ell(\omega))) &= \Lambda(\omega,\mathbf{K}_\ell(\omega),\mathbf{L}_\ell(\omega),\mathbf{C}_\ell(\omega),\mathbf{M}_\ell(\omega),\mathbf{E}_\ell(\omega),\mathbf{Q}_\ell(\omega))), \end{aligned} \tag{5}$$

with fuzzy initial conditions

$$\begin{aligned}
&\mathbf{K}(0,\Xi) = (\underline{\mathbf{K}}(0,\Xi), \overline{\mathbf{K}}(0,\Xi)) + (-1)g(\mathbf{K}), \quad \mathbf{K}'(0,\Xi) = (\underline{\mathbf{K}}(0,\Xi), \overline{\mathbf{K}}(0,\Xi)),\\
&\mathbf{L}(0,\Xi) = (\underline{\mathbf{L}}(0,\Xi), \overline{\mathbf{L}}(0,\Xi)) + (-1)g(\mathbf{L}), \quad \mathbf{L}'(0,\Xi) = (\underline{\mathbf{L}}(0,\Xi), \overline{\mathbf{L}}(0,\Xi)),\\
&\mathbf{C}(0,\Xi) = (\underline{\mathbf{C}}(0,\Xi), \overline{\mathbf{C}}(0,\Xi)) + (-1)g(\mathbf{C}), \quad \mathbf{C}'(0,\Xi) = (\underline{\mathbf{C}}(0,\Xi), \overline{\mathbf{C}}(0,\Xi)),\\
&\mathbf{M}(0,\Xi) = (\underline{\mathbf{M}}(0,\Xi), \overline{\mathbf{M}}(0,\Xi)) + (-1)g(\mathbf{M}), \quad \mathbf{M}'(0,\Xi) = (\underline{\mathbf{M}}(0,\Xi), \overline{\mathbf{M}}(0,\Xi)),\\
&\mathbf{E}(0,\Xi) = (\underline{\mathbf{E}}(0,\Xi), \overline{\mathbf{E}}(0,\Xi)) + (-1)g(\mathbf{E}), \quad \mathbf{E}'(0,\Xi) = (\underline{\mathbf{E}}(0,\Xi), \overline{\mathbf{E}}(0,\Xi)),\\
&\mathbf{Q}(0,\Xi) = (\underline{\mathbf{Q}}(0,\Xi), \overline{\mathbf{Q}}(0,\Xi)) + (-1)g(\mathbf{Q}), \quad \mathbf{Q}'(0,\Xi) = (\underline{\mathbf{Q}}(0,\Xi), \overline{\mathbf{Q}}(0,\Xi)).
\end{aligned}$$

Now applying a mild solution and using the initial conditions, we obtain

$$\begin{aligned}
&\mathbf{K}_\ell(\omega) = C_q(\omega)(\mathbf{K}(0,\Xi) + (-1)g(\mathbf{K})) + K_q(\omega)\mathbf{K}'(0,\Xi) +\\
&\tfrac{1}{\Gamma(\vartheta)}\textstyle\int_0^\omega (\omega-\wedge)^{\vartheta-1}P_\vartheta(\omega-\wedge)\digamma(\omega,\mathbf{K}_\ell(\omega),\mathbf{L}_\ell(\omega),\mathbf{C}_\ell(\omega),\mathbf{M}_\ell(\omega),\mathbf{E}_\ell(\omega),\mathbf{Q}_\ell(\omega)))d\wedge,\\
&\mathbf{L}_\ell(\omega) = C_q(\omega)(\mathbf{L}(0,\Xi) + (-1)g(\mathbf{L})) + K_q(\omega)\mathbf{L}'(0,\Xi) +\\
&\tfrac{1}{\Gamma(\vartheta)}\textstyle\int_0^\omega (\omega-\wedge)^{\vartheta-1}P_\vartheta(\omega-\wedge)\daleth(\omega,\mathbf{K}_\ell(\omega),\mathbf{L}_\ell(\omega),\mathbf{C}_\ell(\omega),\mathbf{M}_\ell(\omega),\mathbf{E}_\ell(\omega),\mathbf{Q}_\ell(\omega)))d\wedge,\\
&\mathbf{C}_\ell(\omega) = C_q(\omega)(\mathbf{C}(0,\Xi) + (-1)g(\mathbf{C})) + K_q(\omega)\mathbf{C}'(0,\Xi) +\\
&\tfrac{1}{\Gamma(\vartheta)}\textstyle\int_0^\omega (\omega-\wedge)^{\vartheta-1}P_\vartheta(\omega-\wedge)\Phi(\omega,\mathbf{K}_\ell(\omega),\mathbf{L}_\ell(\omega),\mathbf{C}_\ell(\omega),\mathbf{M}_\ell(\omega),\mathbf{E}_\ell(\omega),\mathbf{Q}_\ell(\omega)))d\wedge),\\
&\mathbf{M}_\ell(\omega) = C_q(\omega)(\mathbf{M}(0,\Xi) + (-1)g(\mathbf{M})) + K_q(\omega)\mathbf{M}'(0,\Xi) +\\
&\tfrac{1}{\Gamma(\vartheta)}\textstyle\int_0^\omega (\omega-\wedge)^{\vartheta-1}P_\vartheta(\omega-\wedge)\Psi(\omega,\mathbf{K}_\ell(\omega),\mathbf{L}_\ell(\omega),\mathbf{C}_\ell(\omega),\mathbf{M}_\ell(\omega),\mathbf{E}_\ell(\omega),\mathbf{Q}_\ell(\omega)))d\wedge,\\
&\mathbf{E}_\ell(\omega)) = C_q(\omega)(\mathbf{E}(0,\Xi) + (-1)g(\mathbf{E})) + K_q(\omega)\mathbf{E}'(0,\Xi) +\\
&\tfrac{1}{\Gamma(\vartheta)}\textstyle\int_0^\omega (\omega-\wedge)^{\vartheta-1}P_\vartheta(\omega-\wedge)\Upsilon(\omega,\mathbf{K}_\ell(\omega),\mathbf{L}_\ell(\omega),\mathbf{C}_\ell(\omega),\mathbf{M}_\ell(\omega),\mathbf{E}_\ell(\omega),\mathbf{Q}_\ell(\omega)))d\wedge,\\
&\mathbf{Q}_\ell(\omega) = C_q(\omega)(\mathbf{Q}(0,\Xi) + (-1)g(\mathbf{Q})) + K_q(\omega)\mathbf{Q}'(0,\Xi) +\\
&\tfrac{1}{\Gamma(\vartheta)}\textstyle\int_0^\omega (\omega-\wedge)^{\vartheta-1}P_\vartheta(\omega-\wedge)\Lambda(\omega,\mathbf{K}_\ell(\omega),\mathbf{L}_\ell(\omega),\mathbf{C}_\ell(\omega),\mathbf{M}_\ell(\omega),\mathbf{E}_\ell(\omega),\mathbf{Q}_\ell(\omega)))d\wedge.
\end{aligned} \tag{6}$$

Let us define a Banach space as $\mathbf{G} = \mathbf{G}_1 \times \mathbf{G}_2$ under the fuzzy norm:

$$||\mathbf{K}_\ell(\omega) + \mathbf{L}_\ell(\omega) + \mathbf{C}_\ell(\omega) + \mathbf{M}_\ell(\omega) + \mathbf{E}_\ell(\omega) + \mathbf{Q}_\ell(\omega)|| = \max_{\omega\in[0,\Im]}[|\mathbf{K}_\ell(\omega) + \mathbf{L}_\ell(\omega) + \mathbf{C}_\ell(\omega) + \mathbf{M}_\ell(\omega) + \mathbf{E}_\ell(\omega) + \mathbf{Q}_\ell(\omega)|].$$

Equation (6) becomes

$$\mathbf{N}(\omega) = C_q(\omega)\mathbf{N}(0,\Xi) + K_q(\omega)\mathbf{N}'(0,\Xi) + \frac{1}{\Gamma(\vartheta)}\int_0^\omega (\omega-\wedge)^{\vartheta-1}P_q(\omega-\wedge)\Theta(\wedge,\mathbf{N}(\wedge))d\wedge, \tag{7}$$

where

$$\Theta(\omega,\mathbf{N}_\ell(\omega)) = \begin{array}{l}
\digamma(\omega,\mathbf{K}_\ell(\omega),\mathbf{L}_\ell(\omega),\mathbf{C}_\ell(\omega),\mathbf{M}_\ell(\omega),\mathbf{E}_\ell(\omega),\mathbf{Q}_\ell(\omega))),\\
\daleth(\omega,\mathbf{K}_\ell(\omega),\mathbf{L}_\ell(\omega),\mathbf{C}_\ell(\omega),\mathbf{M}_\ell(\omega),\mathbf{E}_\ell(\omega),\mathbf{Q}_\ell(\omega))),\\
\Phi(\omega,\mathbf{K}_\ell(\omega),\mathbf{L}_\ell(\omega),\mathbf{C}_\ell(\omega),\mathbf{M}_\ell(\omega),\mathbf{E}_\ell(\omega),\mathbf{Q}_\ell(\omega))),\\
\Psi(\omega,\mathbf{K}_\ell(\omega),\mathbf{L}_\ell(\omega),\mathbf{C}_\ell(\omega),\mathbf{M}_\ell(\omega),\mathbf{E}_\ell(\omega),\mathbf{Q}_\ell(\omega))),\\
\Upsilon(\omega,\mathbf{K}_\ell(\omega),\mathbf{L}_\ell(\omega),\mathbf{C}_\ell(\omega),\mathbf{M}_\ell(\omega),\mathbf{E}_\ell(\omega),\mathbf{Q}_\ell(\omega))),\\
\Lambda(\omega,\mathbf{K}_\ell(\omega),\mathbf{L}_\ell(\omega),\mathbf{C}_\ell(\omega),\mathbf{M}_\ell(\omega),\mathbf{E}_\ell(\omega),\mathbf{Q}_\ell(\omega))).
\end{array}$$

On the nonlinear function $\Theta : \mathbf{G} \to \mathbb{G}$, we add the following assumptions:

Assumption 1 (H_1). *There exists constant $K_\mathbf{N} > 0$ such that for each $\mathbf{N}_{\ell_1}(\omega), \mathbf{N}_{\ell_2}(\omega) \in \mathbf{G}$,*

$$|\Theta(\omega,\mathbf{N}_{\ell_1}(\omega)) - \Theta(\omega,\mathbf{N}_{\ell_2}(\omega))| \le K_\mathbf{N}|\Theta(\omega,\mathbf{N}_{\ell_1}(\omega)) - \Theta(\omega,\mathbf{N}_{\ell_2}(\omega))|.$$

Assumption 2 (H_2). *There exist constants $F_\mathbf{N} > 0$ such that*

$$|\Theta(\omega,\mathbf{N}_\ell(\omega))| \le F_\mathbf{N}|\mathbf{N}_\ell(\omega))| + N_\mathbf{N}.$$

Theorem 3. *Under Assumption 2, the considered Model (5) has at least one solution.*

Proof. Suppose $A = \{\mathbf{N}_\ell(\omega) \in \mathbf{G} : ||\mathbf{N}_\ell(\omega))|| \leq r\} \subset \mathbf{G}$ is a closed and convex fuzzy set, and $\psi : I \to I$ is a mapping defined as

$$\zeta(\mathbf{N}_k(\omega)) = C_q(\omega)\mathbf{N}(0,\Xi) + K_q(\omega)\mathbf{N}'(0,\omega) + \frac{1}{\Gamma(\vartheta)}\int_0^\omega (\omega-\wedge)^{\vartheta-1}P_q(\omega-\wedge)\Theta(\wedge,\mathbf{N}_\ell(\wedge))d\wedge. \quad (8)$$

For any $\mathbf{N}_\ell(\omega) \in I$, we have

$$\begin{aligned}
||\psi(\mathbf{N}_\ell(\omega)|| &= \max_{\omega\in[0,\Im]}\left|C_q(\omega)\mathbf{N}(0,\Xi) + K_q(\omega)\mathbf{N}'(0,\Xi) + \frac{1}{\Gamma(\vartheta)}\int_0^\omega (\omega-\wedge)^{\vartheta-1}P_q(\omega-\wedge)\Theta(\wedge,\mathbf{N}_\ell(\wedge))d\wedge\right| \\
&\leq |C_q(\ell)\mathbf{N}(0,\Xi) + K_q(\omega)\mathbf{N}'(0,\Xi)| + \frac{1}{\Gamma(\vartheta)}\int_0^\omega (\psi-\wedge)^{\vartheta-1}P_q(\omega-\wedge)|\Theta(\wedge,\mathbf{N}_\ell(\wedge))|d\wedge \\
&\leq |C_q(\omega)\mathbf{N}(0,\Xi) + K_q(\omega)\mathbf{N}'(0,\Xi)| + \frac{1}{\Gamma(\vartheta)}\int_0^\omega (\omega-\wedge)^{\vartheta-1}P_q(\omega-\wedge)[M_{\mathbf{N}}|\mathbf{N}_\ell(\omega)| + N_{\mathbf{N}}]d\wedge \\
&\leq |\mathbf{N}(0,\Xi)| + \frac{\tau^\vartheta}{\Gamma(\vartheta+1)}[M_{\mathbf{N}}|\mathbf{N}_k(\omega)| + N_{\mathbf{N}}].
\end{aligned}$$

We obtain $\zeta(I) \subset I$ from the previous inequality, implying that the operator φ is bounded. The operator ϕ is then shown to be completely continuous. Allow $\phi_1, \phi_2 \in [0,\Im]$ to be such that $\phi_1 < \phi_2$, and then

$$\begin{aligned}
||\phi(\mathbf{N}_\ell(\omega)(\phi_2) - \phi(\mathbf{N}_\ell(\omega)(\phi_1)|| &= \left|\frac{1}{\Gamma(\gamma)}\int_0^{\phi_2}(\phi_2-\wedge)^{\vartheta-1}P_q(\omega-\wedge)\Theta(\wedge,\mathbf{N}_\ell(\wedge))d\wedge\right. \\
&\quad \left.-\frac{1}{\Gamma(\gamma)}\int_0^{\phi_1}(\phi_1-\wedge)^{\gamma-1}P_q(\omega-\wedge)\Theta(\wedge,\mathbf{N}_\ell(\wedge))d\wedge\right| \\
&\leq [\phi_2^\gamma - \phi_1^\gamma]\frac{[F_{\mathbf{N}}|\mathbf{N}_\ell(\omega)| + N_{\mathbf{N}}]}{\Gamma(\gamma+1)}.
\end{aligned}$$

We can see that the right-hand side of the inequality goes to zero as $\phi_2 \to \phi_1$. Hence,

$$||\phi(\mathbf{N}_\ell(\omega)(\phi_2) - \phi(\mathbf{N}_\ell(\omega))(\phi_1)|| \to as\ \phi_2 \to \phi_1.$$

As a result, ϕ is an equicontinuous operator. The operator ϕ is entirely continuous according to the Arzela–Ascoli theorem, and it was previously bounded. As a result of Schauder's fixed point theorem, System (5) has at least one solution. □

Theorem 4. *If $\tau^\vartheta K_{\mathbf{N}} < \Gamma(\vartheta+1)$, the investigated system (5) has a unique solution if Assumption 1 holds.*

Proof. Let $\mathbf{N}_{\ell_1}(\omega), \mathbf{N}_{\ell_2}(\omega) \in \mathbf{G}$, then

$$\begin{aligned}
||\phi(\mathbf{N}_{\ell_1}(\omega)) - \phi(\mathbf{N}_{\ell_2}(\omega))|| &= \max_{\omega\in[0,\Im]}\left|\frac{1}{\Gamma(\vartheta)}\int_0^\omega(\omega-\wedge)^{\vartheta-1}P_q(\omega-\wedge)\Theta(\wedge,\mathbf{N}_{\ell_1}(\wedge))d\wedge\right. \\
&\quad \left.-\frac{1}{\Gamma(\vartheta)}\int_0^\omega(\omega-\wedge)^{\vartheta-1}P_q(\omega-\wedge)\Theta(\wedge,\mathbf{N}_{\ell_2}(\wedge))d\wedge\right| \\
&\leq \frac{\tau^\vartheta}{\Gamma(\vartheta+1)}K_{\mathbf{N}}|\mathbf{N}_{\ell_1}(\omega) - \mathbf{N}_{\ell_2}(\omega)|.
\end{aligned}$$

As a result, ζ is a contraction. As a result, according to the Banach contraction theorem, System (5) has a single solution. □

3.2. Procedure for Solution

A general method is shown here for using the fuzzy Laplace transform to determine the solution of the examined system.

We have the fuzzy Laplace transform of (5) when we use beginning conditions:

$$
\begin{aligned}
L[{}_0^cD_\omega^\vartheta \mathbf{K}_\ell(\omega)] &= L[\digamma(\omega, \mathbf{K}_\ell(\omega), \mathbf{L}_\ell(\omega), \mathbf{C}_\ell(\omega), \mathbf{M}_\ell(\omega), \mathbf{E}_\ell(\omega), \mathbf{Q}_\ell(\omega))], \\
L[{}_0^cD_\omega^\vartheta \mathbf{L}_\ell(\omega)] &= L[\beth(\omega, \mathbf{K}_\ell(\omega), \mathbf{L}_\ell(\omega), \mathbf{C}_\ell(\omega), \mathbf{M}_\ell(\omega), \mathbf{E}_\ell(\omega), \mathbf{Q}_\ell(\omega))], \\
L[{}_0^cD_\omega^\vartheta \mathbf{C}_\ell(\omega)] &= L[\Phi(\omega, \mathbf{K}_\ell(\omega), \mathbf{L}_\ell(\omega), \mathbf{C}_\ell(\omega), \mathbf{M}_\ell(\omega), \mathbf{E}_\ell(\omega), \mathbf{Q}_\ell(\omega))], \\
L[{}_0^cD_\omega^\vartheta \mathbf{M}_\ell(\omega)] &= L[\Psi(\omega, \mathbf{K}_\ell(\omega), \mathbf{L}_\ell(\omega), \mathbf{C}_\ell(\omega), \mathbf{M}_\ell(\omega), \mathbf{E}_\ell(\omega), \mathbf{Q}_\ell(\omega))], \\
L[{}_0^cD_\omega^\vartheta \mathbf{E}_\ell(\omega)] &= L[\Upsilon(\omega, \mathbf{K}_\ell(\omega), \mathbf{L}_\ell(\omega), \mathbf{C}_\ell(\omega), \mathbf{M}_\ell(\omega), \mathbf{E}_\ell(\omega), \mathbf{Q}_\ell(\omega))], \\
L[{}_0^cD_\omega^\vartheta \mathbf{Q}_\ell(\omega)] &= L[\Lambda(\omega, \mathbf{K}_\ell(\omega), \mathbf{L}_\ell(\omega), \mathbf{C}_\ell(\omega), \mathbf{M}_\ell(\omega), \mathbf{E}_\ell(\omega), \mathbf{Q}_\ell(\omega))], \\
s^\vartheta L[{}_0^cD_\omega^\vartheta \mathbf{K}_\ell(\omega)] &= s^{\vartheta-1}\mathbf{K}(0,\Xi) + s^{\vartheta-1}\mathbf{K}'(0,\Xi) + L[\digamma(\omega, \mathbf{K}_\ell(\omega), \mathbf{L}_\ell(\omega), \mathbf{C}_\ell(\omega), \mathbf{M}_\ell(\omega), \mathbf{E}_\ell(\omega), \mathbf{Q}_\ell(\omega))], \\
s^\vartheta L[{}_0^cD_\omega^\vartheta \mathbf{L}_\ell(\omega)] &= s^{\vartheta-1}\mathbf{L}(0,\Xi) + s^{\vartheta-1}\mathbf{L}'(0,\Xi) + L[\beth(\omega, \mathbf{K}_\ell(\omega), \mathbf{L}_\ell(\omega), \mathbf{C}_\ell(\omega), \mathbf{MM}_\ell(\omega), \mathbf{E}_\ell(\omega), \mathbf{Q}_\ell(\omega))], \\
s^\vartheta L[{}_0^cD_\omega^\vartheta \mathbf{C}_\ell(\omega)] &= s^{\vartheta-1}\mathbf{C}(0,\Xi) + s^{\vartheta-1}\mathbf{C}'(0,\Xi) + L\Phi(\omega, \mathbf{K}_\ell(\omega), \mathbf{L}_\ell(\omega), \mathbf{C}_\ell(\omega), \mathbf{M}_\ell(\omega), \mathbf{E}_\ell(\omega), \mathbf{Q}_\ell(\omega))], \\
s^\vartheta L[{}_0^cD_\omega^\vartheta \mathbf{M}_\ell(\omega)] &= s^{\vartheta-1}\mathbf{M}(0,\Xi) + s^{\vartheta-1}\mathbf{D}'(0,\Xi) + L[\Psi(\omega, \mathbf{K}_\ell(\omega), \mathbf{L}_\ell(\omega), \mathbf{C}_\ell(\omega), \mathbf{M}_\ell(\omega), \mathbf{E}_\ell(\omega), \mathbf{Q}_\ell(\omega))], \\
s^\vartheta L[{}_0^cD_\omega^\vartheta \mathbf{E}_\ell(\omega)] &= s^{\vartheta-1}\mathbf{E}(0,\Xi) + s^{\vartheta-1}\mathbf{E}'(0,\Xi) + L[\Upsilon(\omega, \mathbf{K}_\ell(\omega), \mathbf{L}_\ell(\omega), \mathbf{C}_\ell(\omega), \mathbf{M}_\ell(\omega), \mathbf{E}_\ell(\omega), \mathbf{Q}_\ell(\omega))], \\
s^\vartheta L[{}_0^cD_\omega^\vartheta \mathbf{Q}_\ell(\omega)] &= s^{\vartheta-1}\mathbf{Q}(0,\Xi) + s^{\vartheta-1}\mathbf{Q}'(0,\Xi) + L[\Lambda(\omega, \mathbf{K}_\ell(\omega), \mathbf{L}_\ell(\omega), \mathbf{C}_\ell(\omega), \mathbf{M}_\ell(\omega), \mathbf{E}_\ell(\omega), \mathbf{Q}_\ell(\omega))], \\
L[{}_0^cD_\omega^\vartheta \mathbf{K}_\ell(\omega)] &= \frac{1}{s}\mathbf{K}(0,\Xi) + \frac{1}{s}\mathbf{K}'(0,\Xi) + \frac{1}{s^\vartheta}L[\digamma(\omega, \mathbf{K}_\ell(\omega), \mathbf{L}_\ell(\omega), \mathbf{C}_\ell(\omega), \mathbf{M}_\ell(\omega), \mathbf{E}_\ell(\omega), \mathbf{Q}_\ell(\omega))], \\
L[{}_0^cD_\omega^\vartheta \mathbf{L}_\ell(\omega)] &= \frac{1}{s}\mathbf{L}(0,\Xi) + \frac{1}{s}\mathbf{L}'(0,\Xi) + \frac{1}{s^\vartheta}L[\beth(\omega, \mathbf{K}_\ell(\omega), \mathbf{L}_\ell(\omega), \mathbf{C}_\ell(\omega), \mathbf{M}_\ell(\omega), \mathbf{E}_\ell(\omega), \mathbf{Q}_\ell(\omega)))], \\
L[{}_0^cD_\omega^\vartheta \mathbf{C}_\ell(\omega)] &= \frac{1}{s}\mathbf{C}(0,\Xi) + \frac{1}{s}\mathbf{C}'(0,\Xi) + \frac{1}{s^\vartheta}L[\Phi(\omega, \mathbf{K}_\ell(\omega), \mathbf{L}_\ell(\omega), \mathbf{C}_\ell(\omega), \mathbf{M}_\ell(\omega), \mathbf{E}_\ell(\omega), \mathbf{Q}_\ell(\omega))], \\
L[{}_0^cD_\omega^\vartheta \mathbf{M}_\ell(\omega)] &= \frac{1}{s}\mathbf{M}(0,\Xi) + \frac{1}{s}\mathbf{M}'(0,\Xi) + \frac{1}{s^\vartheta}L[\Psi(\omega, \mathbf{K}_\ell(\omega), \mathbf{L}_\ell(\omega), \mathbf{C}_\ell(\omega), \mathbf{M}_\ell(\omega), \mathbf{E}_\ell(\omega), \mathbf{Q}_\ell(\omega))], \\
L[{}_0^cD_\omega^\vartheta \mathbf{E}_\ell(\omega)] &= \frac{1}{s}\mathbf{E}(0,\Xi) + \frac{1}{s}\mathbf{E}'(0,\Xi) + \frac{1}{s^\vartheta}L[\Upsilon(\omega, \mathbf{K}_\ell(\omega), \mathbf{L}_\ell(\omega), \mathbf{C}_\ell(\omega), \mathbf{M}_\ell(\omega), \mathbf{E}_\ell(\omega), \mathbf{Q}_\ell(\omega))], \\
L[{}_0^cD_\omega^\vartheta \mathbf{Q}_\ell(\omega)] &= \frac{1}{s}\mathbf{Q}(0,\Xi) + \frac{1}{s}\mathbf{Q}'(0,\Xi) + \frac{1}{s^\vartheta}L[\Lambda(\omega, \mathbf{K}_\ell(\omega), \mathbf{L}_\ell(\omega), \mathbf{C}_\ell(\omega), \mathbf{M}_\ell(\omega), \mathbf{E}_\ell(\omega), \mathbf{Q}_\ell(\omega))].
\end{aligned}
$$

The infinite series solution is

$$
\begin{aligned}
&\mathbf{K}_\ell(\omega) = \sum_{n=0}^{\infty} \mathbf{K}_{\ell_n}(\omega), \quad \mathbf{L}_\ell(\omega) = \sum_{n=0}^{\infty} \mathbf{L}_{\ell_n}(\omega), \quad \mathbf{C}_\ell(\omega) = \sum_{n=0}^{\infty} \mathbf{C}_{\ell_n}(\omega), \\
&\mathbf{M}_\ell(\omega) = \sum_{n=0}^{\infty} \mathbf{M}_{\ell_n}(\omega), \quad \mathbf{E}_\ell(\omega) = \sum_{n=0}^{\infty} \mathbf{E}_{\ell_n}(\omega), \quad \mathbf{Q}_\ell(\omega) = \sum_{n=0}^{\infty} \mathbf{Q}_{\ell_n}(\omega), \\
&\mathbf{K}_\ell(\omega))\mathbf{C}_\ell(\omega) = \sum_{n=0}^{\infty} \mathbf{X}_{1,n}, \quad \mathbf{K}_\ell(\omega)\mathbf{M}_\ell(\omega) = \sum_{n=0}^{\infty} \mathbf{X}_{2,n}, \quad \mathbf{K}_\ell(\omega)\mathbf{Q}_\ell(\omega) = \sum_{n=0}^{\infty} \mathbf{X}_{3,n},
\end{aligned}
$$

where $\mathbf{X}_{1,n}, \mathbf{X}_{2,n}$ and $\mathbf{X}_{3,n}$ are Adomian polynomials that represent nonlinear terms. As a result, the final equation is

$$
L\left[\sum_{n=0}^{\infty} D_\omega^\vartheta \mathbf{K}_\ell(\omega)\right] = \frac{1}{s}\mathbf{K}(0,\Xi) + \frac{1}{s}\mathbf{K}'(0,\Xi) + \frac{1}{s^\vartheta}L[\digamma(\omega, \mathbf{K}_\ell(\omega), \mathbf{L}_\ell(\omega), \mathbf{C}_\ell(\omega), \mathbf{M}_\ell(\omega), \mathbf{E}_\ell(\omega), \mathbf{Q}_\ell(\omega))],
$$

$$
\begin{aligned}
L\left[\sum_{n=0}^{\infty} D_{\omega}^{\vartheta}\mathbf{L}_{\ell}(\omega)\right] &= \frac{1}{s}\mathbf{L}(0,\Xi)+\frac{1}{s}\mathbf{L}'(0,\Xi)+\frac{1}{s^{\vartheta}}L[\beth(\omega,\mathbf{K}_{\ell}(\omega),\mathbf{L}_{\ell}(\omega),\mathbf{C}_{\ell}(\omega),\mathbf{M}_{\ell}(\omega),\mathbf{E}_{\ell}(\omega),\mathbf{Q}_{\ell}(\omega)))],\\
L\left[\sum_{n=0}^{\infty} D_{\omega}^{\vartheta}\mathbf{C}_{\ell}(\omega)\right] &= \frac{1}{s}\mathbf{C}(0,\Xi)+\frac{1}{s}\mathbf{C}'(0,\Xi)+\frac{1}{s^{\vartheta}}L[\Phi(\omega,\mathbf{K}_{\ell}(\omega),\mathbf{L}_{\ell}(\omega),\mathbf{C}_{\ell}(\omega),\mathbf{M}_{\ell}(\omega),\mathbf{E}_{\ell}(\omega),\mathbf{Q}_{\ell}(\omega))],\\
L\left[\sum_{n=0}^{\infty} D_{\omega}^{\vartheta}\mathbf{M}_{\ell}(\omega)\right] &= \frac{1}{s}\mathbf{D}(0,\Xi)+\frac{1}{s}\mathbf{M}'(0,\Xi)+\frac{1}{s^{\vartheta}}L[\Psi(\omega,\mathbf{K}_{\ell}(\omega),\mathbf{L}_{\ell}(\omega),\mathbf{C}_{\ell}(\omega),\mathbf{M}_{\ell}(\omega),\mathbf{E}_{\ell}(\omega),\mathbf{Q}_{\ell}(\omega))],\\
L\left[\sum_{n=0}^{\infty} D_{\omega}^{\vartheta}\mathbf{E}_{\ell}(\omega)\right] &= \frac{1}{s}\mathbf{E}(0,\Xi)+\frac{1}{s}\mathbf{E}'(0,\Xi)+\frac{1}{s^{\vartheta}}L[\Upsilon(\omega,\mathbf{K}_{\ell}(\omega),\mathbf{L}_{\ell}(\omega),\mathbf{C}_{\ell}(\omega),\mathbf{M}_{\ell}(\omega),\mathbf{E}_{\ell}(\omega),\mathbf{Q}_{\ell}(\omega))],\\
L\left[\sum_{n=0}^{\infty} D_{\omega}^{\vartheta}\mathbf{Q}_{\ell}(\omega)\right] &= \frac{1}{s}\mathbf{Q}(0,\Xi)+\frac{1}{s}\mathbf{Q}'(0,\Xi)+\frac{1}{s^{\vartheta}}L[\Lambda(\omega,\mathbf{K}_{\ell}(\omega),\mathbf{L}_{\ell}(\omega),\mathbf{C}_{\ell}(\omega),\mathbf{M}_{\ell}(\omega),\mathbf{E}_{\ell}(\omega),\mathbf{Q}_{\ell}(\omega))].
\end{aligned}
$$

When we use the inverse Laplace transform, we obtain

$$
\begin{aligned}
\sum_{n=0}^{\infty} D_{\omega}^{\vartheta}\mathbf{K}_{\ell}(\omega) &= \mathbf{K}(0,\Xi)+\frac{1}{s}\mathbf{K}'(0,\Xi)+L^{-1}\left[\frac{1}{s^{\vartheta}}L[\digamma(\omega,\mathbf{K}_{\ell}(\omega),\mathbf{L}_{\ell}(\omega),\mathbf{C}_{\ell}(\omega),\mathbf{M}_{\ell}(\omega),\mathbf{E}_{\ell}(\omega),\mathbf{Q}_{\ell}(\omega))]\right],\\
\sum_{n=0}^{\infty} D_{\omega}^{\vartheta}\mathbf{L}_{\ell}(\omega) &= \mathbf{L}(0,\Xi)+\frac{1}{s}\mathbf{L}'(0,\Xi)+L^{-1}\left[\frac{1}{s^{\vartheta}}L[\beth(\omega,\mathbf{K}_{\ell}(\omega),\mathbf{L}_{\ell}(\omega),\mathbf{C}_{\ell}(\omega),\mathbf{M}_{\ell}(\omega),\mathbf{E}_{\ell}(\omega),\mathbf{Q}_{\ell}(\omega))]\right],\\
\sum_{n=0}^{\infty} D_{\omega}^{\vartheta}\mathbf{C}_{\ell}(\omega) &= \mathbf{C}(0,\Xi)+\frac{1}{s}\mathbf{C}'(0,\Xi)+L^{-1}\left[\frac{1}{s^{\vartheta}}L[\Phi(\omega,\mathbf{K}_{\ell}(\omega),\mathbf{L}_{\ell}(\omega),\mathbf{C}_{\ell}(\omega),\mathbf{M}_{\ell}(\omega),\mathbf{E}_{\ell}(\omega),\mathbf{Q}_{\ell}(\omega))]\right],\\
\sum_{n=0}^{\infty} D_{\omega}^{\vartheta}\mathbf{M}_{\ell}(\omega) &= \mathbf{M}(0,\Xi)+\frac{1}{s}\mathbf{M}'(0,\Xi)+L^{-1}\left[\frac{1}{s^{\vartheta}}L[\Psi(\omega,\mathbf{K}_{\ell}(\omega),\mathbf{L}_{\ell}(\omega),\mathbf{C}_{\ell}(\omega),\mathbf{M}_{\ell}(\omega),\mathbf{E}_{\ell}(\omega),\mathbf{Q}_{\ell}(\omega))]\right],\\
\sum_{n=0}^{\infty} D_{\omega}^{\vartheta}\mathbf{E}_{\ell}(\omega) &= \mathbf{E}(0,\Xi)+\frac{1}{s}\mathbf{E}'(0,\Xi)+L^{-1}\left[\frac{1}{s^{\vartheta}}L[\Upsilon(\omega,\mathbf{K}_{\ell}(\omega),\mathbf{L}_{\ell}(\omega),\mathbf{C}_{\ell}(\omega),\mathbf{M}_{\ell}(\omega),\mathbf{E}_{\ell}(\omega),\mathbf{Q}_{\ell}(\omega))]\right],\\
\sum_{n=0}^{\infty} D_{\omega}^{\vartheta}\mathbf{Q}_{\ell}(\omega) &= \mathbf{Q}(0,\Xi)+\frac{1}{s}\mathbf{Q}'(0,\Xi)+L^{-1}\left[\frac{1}{s^{\vartheta}}L[\Lambda(\omega,\mathbf{K}_{\ell}(\omega),\mathbf{L}_{\ell}(\omega),\mathbf{C}_{\ell}(\omega),\mathbf{M}_{\ell}(\omega),\mathbf{E}_{\ell}(\omega),\mathbf{Q}_{\ell}(\omega))]\right],
\end{aligned}
$$

We evaluate the first two terms of two terms of the series when comparing the terms on both sides:

$$
\begin{array}{llll}
\underline{\mathbf{K}}_{\ell_0}(\omega)=\underline{\mathbf{K}}(0,\Xi), & \underline{\mathbf{K}}'_{\ell_0}(\omega)=\underline{\mathbf{K}}'(0,\Xi), & \overline{\mathbf{K}}_{\ell_0}(\omega)=\overline{\mathbf{K}}(0,\Xi), & \overline{\mathbf{K}}'_{\ell_0}(\omega)=\overline{\mathbf{K}}'(0,\Xi),\\
\underline{\mathbf{L}}_{\ell_0}(\omega)=\underline{\mathbf{L}}(0,\Xi), & \underline{\mathbf{L}}'_{\ell_0}(\omega)=\underline{\mathbf{L}}'(0,\Xi), & \overline{\mathbf{L}}_{\ell_0}(\omega)=\overline{\mathbf{L}}(0,\Xi), & \overline{\mathbf{L}}'_{\ell_0}(\omega)=\overline{\mathbf{L}}'(0,\Xi),\\
\underline{\mathbf{C}}_{\ell_0}(\omega)=\underline{\mathbf{C}}(0,\Xi), & \underline{\mathbf{C}}'_{\ell_0}(\omega)=\underline{\mathbf{C}}'(0,\Xi), & \overline{\mathbf{C}}_{\ell_0}(\omega)=\overline{\mathbf{C}}(0,\Xi), & \overline{\mathbf{C}}'_{\ell_0}(\omega)=\overline{\mathbf{C}}'(0,\Xi),\\
\underline{\mathbf{M}}_{\ell_0}(\omega)=\underline{\mathbf{M}}(0,\Xi), & \underline{\mathbf{M}}'_{\ell_0}(\omega)=\underline{\mathbf{M}}'(0,\Xi), & \overline{\mathbf{M}}_{\ell_0}(\omega)=\overline{\mathbf{M}}(0,\Xi), & \overline{\mathbf{M}}'_{\ell_0}(\omega)=\overline{\mathbf{M}}'(0,\Xi),\\
\underline{\mathbf{E}}_{\ell_0}(\omega)=\underline{\mathbf{E}}(0,\Xi), & \underline{\mathbf{E}}'_{\ell_0}(\omega)=\underline{\mathbf{E}}'(0,\Xi), & \overline{\mathbf{E}}_{\ell_0}(\omega)=\overline{\mathbf{E}}(0,\Xi), & \overline{\mathbf{E}}'_{\ell_0}(\omega)=\overline{\mathbf{E}}'(0,\Xi),\\
\underline{\mathbf{Q}}_{\ell_0}(\omega)=\underline{\mathbf{Q}}(0,\Xi), & \underline{\mathbf{Q}}'_{\ell_0}(\omega)=\underline{\mathbf{Q}}'(0,\Xi), & \overline{\mathbf{Q}}_{\ell_0}(\omega)=\overline{\mathbf{Q}}(0,\Xi), & \overline{\mathbf{Q}}'_{\ell_0}(\omega)=\overline{\mathbf{Q}}'(0,\Xi),
\end{array}
\tag{9}
$$

$$
\begin{aligned}
\underline{\mathbf{K}}_{\ell_1}(\omega) &= L^{-1}[\tfrac{1}{s^{\vartheta}}L[\hat{r}_{\ell}-\hat{\imath}_{\ell}\underline{\mathbf{K}}_{\ell_0}-\hat{\imath}_{\ell}\underline{\mathbf{K}}'_{\ell_0}-\hat{s}\underline{\mathbf{K}}_{\ell_0}(\underline{\mathbf{C}}_{\ell_0}+\lambda\underline{\mathbf{M}}_{\ell_0})-\hat{s}\underline{\mathbf{K}}'_{\ell_0}(\underline{\mathbf{C}}'_{\ell_0}+\lambda\underline{\mathbf{M}}'_{\ell_0})-\hat{s}_{\ell}\underline{\mathbf{K}}_{\ell_0}\underline{\hat{\mathbf{Q}}}_{\ell_0}-\hat{s}_{\ell}\underline{\mathbf{K}}'_{\ell_0}\underline{\hat{\mathbf{Q}}}'_{\ell_0}]],\\
\overline{\mathbf{K}}_{\ell_1}(\omega) &= L^{-1}[\tfrac{1}{s^{\vartheta}}L[\hat{r}_{\ell}-\hat{\imath}_{\ell}\overline{\mathbf{K}}_{\ell_0}-\hat{\imath}_{\ell}\overline{\mathbf{K}}'_{\ell_0}-\hat{s}\overline{\mathbf{K}}_{\ell_0}(\overline{\mathbf{C}}_{\ell_0}+\lambda\overline{\mathbf{M}}_{\ell_0})-\hat{s}\overline{\mathbf{K}}'_{\ell_0}(\overline{\mathbf{C}}'_{\ell_0}+\lambda\overline{\mathbf{M}}'_{\ell_0})-\hat{s}_{\ell}\overline{\mathbf{K}}_{\ell_0}\overline{\hat{\mathbf{Q}}}_{\ell_0}-\hat{s}_{\ell}\overline{\mathbf{K}}'_{\ell_0}\overline{\hat{\mathbf{Q}}}'_{\ell_0}]].
\end{aligned}
\tag{10}
$$

We may find the other terms in the same way.

As a result, the examined system's series solution is

$$\begin{aligned}
\underline{\mathbf{K}}_\ell(\omega) &= \underline{\mathbf{K}}_{\ell_0}(\omega) + \underline{\mathbf{K}}_{\ell_0}(\omega)' + \underline{\mathbf{K}}_{\ell_1}(\omega) + \underline{\mathbf{K}}_{\ell_1}(\omega)' + \underline{\mathbf{K}}_{\ell_2}(\omega) + \underline{\mathbf{K}}_{\ell_2}(\omega)' + \dots, \\
\overline{\mathbf{K}}_\ell(\omega) &= \overline{\mathbf{K}}_{\ell_0}(\omega) + \overline{\mathbf{K}}_{\ell_0}(\omega)' + \overline{\mathbf{K}}_{\ell_1}(\omega) + \overline{\mathbf{K}}_{\ell_1}(\omega)' + \underline{\mathbf{K}}_{\ell_2}(\omega) + \underline{\mathbf{K}}_{\ell_2}(\omega)' + \dots, \\
\underline{\mathbf{L}}_\ell(\omega) &= \underline{\mathbf{L}}_{\ell_0}(\omega) + \underline{\mathbf{L}}_{\ell_0}(\omega)' + \underline{\mathbf{L}}_{\ell_1}(\omega) + \underline{\mathbf{L}}_{\ell_1}(\omega)' + \underline{\mathbf{L}}_{\ell_2}(\omega) + \underline{\mathbf{L}}_{\ell_2}(\omega)' + \dots, \\
\overline{\mathbf{L}}_\ell(\omega) &= \overline{\mathbf{L}}_{\ell_0}(\omega) + \overline{\mathbf{L}}_{\ell_0}(\omega)' + \overline{\mathbf{L}}_{\ell_1}(\omega) + \overline{\mathbf{L}}_{\ell_1}(\omega)' + \overline{\mathbf{L}}_{k_2}(\omega) + \overline{\mathbf{L}}_{\ell_2}(\omega)' + \dots, \\
\underline{\mathbf{C}}_\ell(\omega) &= \underline{\mathbf{C}}_{\ell_0}(\omega) + \underline{\mathbf{C}}_{\ell_0}(\omega)' + \underline{\mathbf{C}}_{\ell_1}(\omega) + \underline{\mathbf{C}}_{\ell_1}(\omega)' + \underline{\mathbf{C}}_{\ell_2}(\omega) + \underline{\mathbf{C}}_{\ell_2}(\omega)' + \dots, \\
\overline{\mathbf{C}}_\ell(\omega) &= \overline{\mathbf{C}}_{\ell_0}(\omega) + \overline{\mathbf{C}}_{\ell_0}(\omega)' + \overline{\mathbf{C}}_{\ell_1}(\omega) + \overline{\mathbf{C}}_{\ell_1}(\omega)' + \underline{\mathbf{C}}_{\ell_2}(\omega) + \underline{\mathbf{C}}_{\ell_2}(\omega)' + \dots, \\
\underline{\mathbf{M}}_\ell(\omega) &= \underline{\mathbf{M}}_{\ell_0}(\omega) + \underline{\mathbf{M}}_{\ell_0}(\omega)' + \underline{\mathbf{M}}_{\ell_1}(\omega) + \underline{\mathbf{M}}_{\ell_1}(\omega)' + \underline{\mathbf{M}}_{\ell_2}(\omega) + \underline{\mathbf{M}}_{\ell_2}(\omega)' + \dots, \\
\overline{\mathbf{M}}_\ell(\omega) &= \overline{\mathbf{M}}_{\ell_0}(\omega) + \overline{\mathbf{M}}_{\ell_0}(\omega)' + \overline{\mathbf{M}}_{\ell_1}(\omega) + \overline{\mathbf{M}}_{\ell_1}(\omega)' + \underline{\mathbf{M}}_{\ell_2}(\omega) + \underline{\mathbf{M}}_{\ell_2}(\omega)' + ..., \\
\underline{\mathbf{E}}_\ell(\omega) &= \underline{\mathbf{E}}_{\ell_0}(\omega) + \underline{\mathbf{E}}_{\ell_0}(\omega)' + \underline{\mathbf{E}}_{\ell_1}(\omega) + \underline{\mathbf{E}}_{\ell_1}(\omega)' + \underline{\mathbf{E}}_{\ell_2}(\omega) + \underline{\mathbf{E}}_{\ell_2}(\omega)' + \dots, \\
\overline{\mathbf{E}}_\ell(\omega) &= \overline{\mathbf{E}}_{\ell_0}(\omega) + \overline{\mathbf{E}}_{\ell_0}(\omega)' + \overline{\mathbf{E}}_{\ell_1}(\omega) + \overline{\mathbf{E}}_{\ell_1}(\omega)' + \underline{\mathbf{E}}_{\ell_2}(\omega) + \underline{\mathbf{E}}_{\ell_2}(\omega)' + \dots, \\
\underline{\mathbf{Q}}_\ell(\omega) &= \underline{\mathbf{Q}}_{\ell_0}(\omega) + \underline{\mathbf{Q}}_{\ell_0}(\omega)' + \underline{\mathbf{Q}}_{\ell_1}(\omega) + \underline{\mathbf{Q}}_{\ell_1}(\omega)' + \underline{\mathbf{Q}}_{\ell_2}(\omega) + \underline{\mathbf{Q}}_{\ell_2}(\omega)' + \dots, \\
\overline{\mathbf{Q}}_\ell(\omega) &= \overline{\mathbf{Q}}_{\ell_0}(\omega) + \overline{\mathbf{Q}}_{\ell_0}(\omega)' + \overline{\mathbf{Q}}_{\ell_1}(\omega) + \overline{\mathbf{Q}}_{\ell_1}(\omega)' + \underline{\mathbf{Q}}_{\ell_2}(\omega) + \underline{\mathbf{Q}}_{\ell_2}(\omega)' + \dots
\end{aligned} \tag{11}$$

4. Numerical Results and Discussion

We take a look at Table 1 that corresponds to the model's parameters. Consider initial conditions for the proposed model (3).

We used the initial conditions and applied the proposed procedure to (3):

$$\begin{aligned}
&\underline{\mathbf{K}}_{\ell_0}(\omega,\Xi) = 2\Xi - 1, \quad \overline{\mathbf{K}}_{\hbar_0}(\omega,\Xi) = 1 - 2\Xi, \quad \underline{\mathbf{K}}'_{\ell_0}(\omega,\Xi) = 2\Xi - 1, \quad \overline{\mathbf{K}}'_{\hbar_0}(\omega,\Xi) = 1 - 2\Xi, \\
&\underline{\mathbf{L}}_{\ell_0}(\omega,\Xi) = 2\Xi - 1, \quad \overline{\mathbf{L}}_{\hbar_0}(\omega,\Xi) = 1 - 2\Xi, \quad \underline{\mathbf{L}}'_{\ell_0}(\omega,\Xi) = 2\Xi - 1, \quad \overline{\mathbf{L}}'_{\hbar_0}(\omega,\Xi) = 1 - 2\Xi, \\
&\underline{\mathbf{C}}_{\ell_0}(\omega,\Xi) = 2\Xi - 1, \quad \overline{\mathbf{C}}_{\hbar_0}(\omega,\Xi) = 1 - 2\Xi, \quad \underline{\mathbf{C}}'_{\ell_0}(\omega,\Xi) = 2\Xi - 1, \quad \overline{\mathbf{C}}'_{\hbar_0}(\omega,\Xi) = 1 - 2\Xi, \\
&\underline{\mathbf{M}}_{\ell_0}(\omega,\Xi) = 2\Xi - 1, \quad \overline{\mathbf{M}}_{\hbar_0}(\omega,\Xi) = 1 - 2\Xi, \quad \underline{\mathbf{M}}'_{\ell_0}(\omega,\Xi) = 2\Xi - 1, \quad \overline{\mathbf{M}}'_{\hbar_0}(\omega,\Xi) = 1 - 2\Xi, \\
&\underline{\mathbf{E}}_{\ell_0}(\omega,\Xi) = 2\Xi - 1, \quad \overline{\mathbf{E}}_{\hbar_0}(\ell,\Xi) = 1 - 2\Xi, \quad \underline{\mathbf{E}}'_{\ell_0}(\omega,\Xi) = 2\Xi - 1, \quad \overline{\mathbf{E}}'_{\hbar_0}(\omega,\Xi) = 1 - 2\Xi, \\
&\underline{\mathbf{Q}}_{\ell_0}(\omega,\Xi) = 2\Xi - 1, \quad \overline{\mathbf{Q}}_{\hbar_0}(\omega,\Xi) = 1 - 2\Xi, \quad \underline{\mathbf{Q}}'_{\ell_0}(\omega,\Xi) = 2\Xi - 1, \quad \overline{\mathbf{Q}}'_{\hbar_0}(\omega,\Xi) = 1 - 2\Xi,
\end{aligned}$$

The second term of the series solution is

$$\begin{aligned}
\underline{\mathbf{K}}_{\ell_1}(\omega,\Xi) &= [\hat{r}_\ell - \hat{\imath}_\ell(2\Xi-1) - \hat{s}(2\Xi-1)^2 - \lambda\hat{s}_\ell(2\Xi-1)^2 - \hat{s}_\ell(2\Xi-1)^2]\tfrac{\omega^\Xi}{\Gamma(\vartheta+1)}, \\
\underline{\mathbf{K}}'_{\ell_1}(\omega,\Xi) &= [\hat{r}'_\ell - \hat{\imath}'_\ell(2\Xi-1) - \hat{s}'(2\Xi-1)^2 - \lambda\hat{s}'_\ell(2\Xi-1)^2 - \hat{s}'_\ell(2\Xi-1)^2]\tfrac{\omega^\Xi}{\Gamma(\vartheta+1)}, \\
\overline{\mathbf{K}}_{\ell_1}(\omega,\Xi) &= [\hat{r}_\ell - \hat{\imath}_\ell(1-2\Xi) - \hat{s}(1-2\Xi)^2 - \lambda\hat{s}_\ell(1-2\Xi)^2 - \hat{s}_\ell(1-2\Xi)^2]\tfrac{\omega^\Xi}{\Gamma(\vartheta+1)}, \\
\overline{\mathbf{K}}'_{\ell_1}(\omega,\Xi) &= [\hat{r}'_\ell - \hat{\imath}'_\ell(1-2\Xi) - \hat{s}'(1-2\Xi)^2 - \lambda\hat{s}'_\ell(1-2\Xi)^2 - \hat{s}'_\ell(1-2\Xi)^2]\tfrac{\omega^\Xi}{\Gamma(\vartheta+1)},
\end{aligned} \tag{12}$$

$$\begin{aligned}
\underline{\mathbf{L}}_{\ell_1}(\omega,\Xi) &= [\hat{s}_\ell(2\Xi-1)^2 + \lambda\hat{s}_\ell(2\Xi-1)^2 + \hat{s}_\ell(2\Xi-1)^2 - (1-\hat{\omega})\hat{\mu}_\ell(2\Xi-1) - \hat{\omega}_\ell\hat{\mu}_\ell(2\Xi-1) - \\
&\quad \hat{\imath}_\ell(2\Xi-1)]\tfrac{\omega^\Xi}{\Gamma(\vartheta+1)}, \\
\underline{\mathbf{L}}'_{\ell_1}(\omega,\Xi) &= [\hat{s}'_\ell(2\Xi-1)^2 + \lambda\hat{s}'_\ell(2\Xi-1)^2 + \hat{s}'_\ell(2\Xi-1)^2 - (1-\hat{\omega})\hat{\mu}'_\ell(2\Xi-1) - \hat{\omega}_\ell\hat{\mu}'_\ell(2\Xi-1) - \\
&\quad \hat{\imath}'_\ell(2\Xi-1)]\tfrac{\omega^\Xi}{\Gamma(\vartheta+1)}, \\
\overline{\mathbf{L}}_{\ell_1}(\omega,\Xi) &= [\hat{s}_\ell(1-2\Xi)^2 + \lambda\hat{s}_\ell(1-2\Xi)^2 + \hat{s}_\ell(1-2\Xi)^2 - (1-\hat{\omega})\hat{\mu}_\ell(1-2\Xi) - \hat{\omega}_\ell\hat{\mu}_\ell(1-2\Xi) - \\
&\quad \hat{\imath}_\ell(1-2\Xi)]\tfrac{\omega^\Xi}{\Gamma(\vartheta+1)}, \\
\overline{\mathbf{L}}'_{\ell_1}(\omega,\Xi) &= [\hat{s}'_\ell(2\Xi-1)^2 + \lambda\hat{s}'_\ell(2\Xi-1)^2 + \hat{s}'_\ell(2\Xi-1)^2 - (1-\hat{\omega})\hat{\mu}'_\ell(2\Xi-1) - \hat{\omega}_\ell\hat{\mu}'_\ell(1-2\Xi) - \\
&\quad \hat{\imath}'_\ell(1-2\Xi)]\tfrac{\omega^\Xi}{\Gamma(\vartheta+1)},
\end{aligned} \tag{13}$$

$$\begin{aligned}
\underline{\mathbf{C}}_{\ell_1}(\omega,\Xi) &= [(1-\hat{\$}_\ell)\hat{\mu}'_\ell(2\Xi-1) - (\hat{\vartheta}'_p + \hat{\imath}_\ell)(2\Xi-1)]\tfrac{\omega^\Xi}{\Gamma(\vartheta+1)}, \\
\underline{\mathbf{C}}'_{\ell_1}(\omega,\Xi) &= [(1-\hat{\$}_\ell)\hat{\mu}_k(2\Xi-1) - (\hat{\vartheta}_p + \hat{\imath}_\ell)(2\Xi-1)]\tfrac{\omega^\Xi}{\Gamma(\vartheta+1)}, \\
\overline{\mathbf{C}}_{\ell_1}(\omega,\Xi) &= [(1-\hat{\$}_\ell)\hat{\mu}'_\ell(1-2\Xi) - (\hat{\vartheta}'_p + \hat{\imath}_\ell)(1-2\Xi)]\tfrac{\omega^\Xi}{\Gamma(\vartheta+1)}, \\
\overline{\mathbf{C}}'_{\ell_1}(\omega,\Xi) &= [(1-\hat{\$}_\ell)\hat{\mu}_k(1-2\Xi) - (\hat{\vartheta}_p + \hat{\imath}_\ell)(1-2\Xi)]\tfrac{\omega^\Xi}{\Gamma(\vartheta+1)}
\end{aligned} \tag{14}$$

$$\begin{aligned}
\underline{\mathbf{M}}_{\ell_1}(\omega,\Xi) &= [\hat{\varpi}_\ell\hat{\mu}'_\ell(2\Xi-1) - (\hat{\vartheta}'_\ell - \hat{\imath}_\ell)(2\Xi-1)]\tfrac{\omega^{\Xi}}{(2\Xi-1)}\tfrac{\omega^{\Xi}}{\Gamma(\vartheta+1)},\\
\underline{\mathbf{M}}'_{\ell_1}(\omega,\Xi) &= [\hat{\varpi}_\ell\hat{\mu}'_\ell(2\Xi-1) - (\hat{\vartheta}'_\ell - \hat{\imath}'_\ell)(2\Xi-1)]\tfrac{\omega^{\Xi}}{(2\Xi-1)}\tfrac{\omega^{\Xi}}{\Gamma(\vartheta+1)},\\
\overline{\mathbf{M}}_{\ell_1}(\omega,\Xi) &= [\hat{\varpi}_\ell\hat{\mu}'_\ell(1-2\Xi) - (\hat{\vartheta}'_\ell - \hat{\imath}_\ell)(1-2\Xi)]\tfrac{\omega^{\Xi}}{(1-2\Xi)}\tfrac{\omega^{\Xi}}{\Gamma(\vartheta+1)},\\
\overline{\mathbf{M}}'_{\ell_1}(\omega,\Xi) &= [\hat{\varpi}_\ell\hat{\mu}'_\ell(1-2\Xi) - (\hat{\vartheta}'_\ell - \hat{\imath}'_\ell)(1-2\Xi)]\tfrac{\omega^{\Xi}}{(1-2\Xi)}\tfrac{\omega^{\Xi}}{\Gamma(\vartheta+1)},
\end{aligned} \tag{15}$$

$$\begin{aligned}
\underline{\mathbf{E}}_{\ell_1}(\omega,\Xi) &= (2\Xi-1)[\hat{\vartheta}_\ell + \hat{\vartheta}'_\ell - \hat{\imath}_\ell]\tfrac{\omega^{\Xi}}{\Gamma(\vartheta+1)},\\
\underline{\mathbf{E}}'_{\ell_1}(\omega,\Xi) &= (2\Xi-1)[\hat{\vartheta}'_\ell + \hat{\vartheta}'_\ell - \hat{\imath}'_\ell]\tfrac{\omega^{\Xi}}{\Gamma(\vartheta+1)},\\
\overline{\mathbf{E}}_{\ell_1}(\omega,\Xi) &= (1-2\Xi)[\hat{\vartheta}_\ell + \hat{\vartheta}'_\ell - \hat{\imath}_\ell]\tfrac{\omega^{\Xi}}{\Gamma(\vartheta+1)},\\
\overline{\mathbf{E}}'_{\ell_1}(\omega,\Xi) &= (1-2\Xi)[\hat{\vartheta}'_\ell + \hat{\vartheta}'_\ell - \hat{\imath}'_\ell]\tfrac{\omega^{\Xi}}{\Gamma(\vartheta+1)},
\end{aligned} \tag{16}$$

$$\begin{aligned}
\underline{\mathbf{Q}}_{\ell_1}(\omega,\Xi) &= (2\Xi-1)[\hat{\varsigma}_\ell + \hat{\psi} - \hat{\varrho}]\tfrac{\omega^{\Xi}}{\Gamma(\vartheta+1)},\\
\underline{\mathbf{Q}}'_{\ell_1}(\omega,\Xi) &= (2\Xi-1)[\hat{\varsigma}'_\ell + \hat{\psi}' - \hat{\varrho}']\tfrac{\omega^{\Xi}}{\Gamma(\vartheta+1)},\\
\overline{\mathbf{Q}}_{\ell_1}(\omega,\Xi) &= (1-2\Xi)[\hat{\varsigma}_\ell + \hat{\psi} - \hat{\varrho}]\tfrac{\omega^{\Xi}}{\Gamma(\vartheta+1)},\\
\overline{\mathbf{Q}}'_{\ell_1}(\omega,\Xi) &= (1-2\Xi)[\hat{\varsigma}'_\ell + \hat{\psi}' - \hat{\varrho}']\tfrac{\omega^{\Xi}}{\Gamma(\vartheta+1)},
\end{aligned} \tag{17}$$

assume that

$$\begin{aligned}
\underline{I}_1 &= \hat{r}_\ell - \hat{\imath}_\ell(2\Xi-1) - \hat{s}(2\Xi-1)^2 - \lambda\hat{s}_\ell(2\Xi-1)^2 - \hat{s}_(2\Xi-1)^2,\\
\overline{I}_1 &= \hat{r}_\ell - \hat{\imath}_\ell(1-2\Xi) - \hat{s}(1-2\Xi)^2 - \lambda\hat{s}_\ell(1-2\Xi)^2 - \hat{s}_\ell(1-2\Xi)^2,\\
\underline{I}_2 &= \hat{s}_\ell(2\Xi-1)^2 + \lambda\hat{s}_\ell(2\Xi-1)^2 + \hat{s}_\ell(2\Xi-1)^2 - (1-\hat{\varpi})\hat{\mu}_\ell(2\Xi-1) - \hat{\varpi}_\ell\hat{\mu}'_\ell(2\Xi-1) - \hat{\imath}_\ell(2\Xi-1)],\\
\overline{I}_2 &= \hat{s}_\ell(1-2\Xi)^2 + \lambda\hat{s}_\ell(1-2\Xi)^2 + \hat{s}_\ell(1-2\Xi)^2 - (1-\hat{\varpi})\hat{\mu}_\ell(1-2\Xi) - \hat{\varpi}_\ell\hat{\mu}'_\ell(1-2\Xi) - \hat{\imath}_\ell(1-2\Xi)],\\
\underline{I}_3 &= [(1-\hat{\varpi}_\ell)\hat{\mu}_\ell(2\Xi-1) - (\hat{\vartheta}_p + \hat{\imath}_\ell)(2\Xi-1)],\\
\overline{I}_3 &= [(1-\hat{\varpi}_\ell)\hat{\mu}_\ell(1-2\Xi) - (\hat{\vartheta}_p + \hat{\imath}_\ell)(1-2\Xi)],\\
\underline{I}_4 &= \hat{\varpi}_\ell\hat{\mu}'_\ell(1-2\Xi) - (\hat{\vartheta}_\ell + \hat{\imath}_\ell)(1-2\Xi),\\
\overline{I}_4 &= \hat{\varpi}_\ell\hat{\mu}'_\ell(1-2\Xi) - (\hat{\vartheta}'_\ell - \hat{\imath}_\ell](1-2\Xi),\\
\underline{I}_5 &= (2\Xi-1)[\hat{\vartheta}_\ell + \hat{\vartheta}'_\ell - \hat{\imath}_\ell],
\end{aligned}$$

$$\begin{aligned}
\overline{I}_5 &= (1-2\Xi)[\hat{\vartheta}_\ell + \hat{\vartheta}'_\ell - \hat{\imath}_\ell],\\
\underline{I}_6 &= (2\Xi-1)[\hat{\varsigma}_\ell + \hat{\psi} - \hat{\varrho}],\\
\overline{I}_6 &= (1-2\Xi)[\hat{\varsigma}_\ell + \hat{\psi} - \hat{\varrho}],
\end{aligned}$$

$$\begin{aligned}
\underline{I}'_1 &= \hat{r}'_\ell - \hat{\imath}'_\ell(2\Xi-1) - \hat{s}'(2\Xi-1)^2 - \lambda\hat{s}'_\ell(2\Xi-1)^2 - \hat{s}'_\ell(2\Xi-1)^2,\\
\overline{I}'_1 &= \hat{r}'_\ell - \hat{\imath}'_\ell(1-2\Xi) - \hat{s}'(1-2\Xi)^2 - \lambda\hat{s}'_\ell(1-2\Xi)^2 - \hat{s}'_\ell(1-2\Xi)^2,\\
\underline{I}'_2 &= \hat{s}'_\ell(2\Xi-1)^2 + \lambda\hat{s}'_\ell(2\Xi-1)^2 + \hat{s}'_\ell(2\Xi-1)^2 - (1-\hat{\varpi}')\hat{\mu}'_\ell(2\Xi-1) - \hat{\varpi}'_\ell\hat{\mu}'_\ell(2\Xi-1) - \hat{\imath}'_\ell(2\Xi-1)],\\
\overline{I}_2 &= \hat{s}'_\ell(1-2\Xi)^2 + \lambda\hat{s}'_\ell(1-2\Xi)^2 + \hat{s}'_\ell(1-2\Xi)^2 - (1-\hat{\varpi})\hat{\mu}'_\ell(1-2\Xi) - \hat{\varpi}'_\ell\hat{\mu}'_\ell{\prime}(1-2\Xi) - \hat{\imath}'_\ell(1-2\Xi)],\\
\underline{I}_3 &= [(1-\hat{\varpi}'_\ell)\hat{\mu}'_\ell(2\Xi-1) - (\hat{\vartheta}_p + \hat{\imath}'_\ell)(2\Xi-1)],\\
\overline{I}_3 &= [(1-\hat{\varpi}'_\ell)\hat{\mu}'_\ell(1-2\Xi) - (\hat{\vartheta}_p + \hat{\imath}'_\ell)(1-2\Xi)],\\
\underline{I}_4 &= \hat{\varpi}'_\ell\hat{\mu}'_\ell(1-2\Xi) - (\hat{\vartheta}'_\ell{\prime} + \hat{\imath}'_\ell)(1-2\Xi),\\
\overline{I}_4 &= \hat{\varpi}'_\ell\hat{\mu}'_\ell(1-2\Xi) - (\hat{\vartheta}'_\ell - \hat{\imath}_\ell](1-2\Xi),\\
\underline{I}_5 &= (2\Xi-1)[\hat{\vartheta}'_\ell + \hat{\vartheta}'_\ell{\prime} - \hat{\imath}_\ell],\\
\overline{I}_5 &= (1-2\Xi)[\hat{\vartheta}'_\ell + \hat{\vartheta}'_\ell - \hat{\imath}'_\ell],\\
\underline{I}_6 &= (2\Xi-1)[\hat{\varsigma}'_\ell + \hat{\psi} - \hat{\varrho}],\\
\overline{I}_6 &= (1-2\Xi)[\hat{\varsigma}'_\ell + \hat{\psi} - \hat{\varrho}],
\end{aligned}$$

The series' third term is now underway.

$$
\begin{aligned}
&\underline{\mathbf{K}}_{\ell_2}(\omega,\Xi) = \hat{r}_\ell \tfrac{\omega^\vartheta}{\Gamma(\vartheta+1)} - \hat{\imath}_\ell \underline{I}_1 \tfrac{\omega^{2\vartheta}}{\Gamma(2\vartheta+1)} [\hat{s}_\ell(2\Xi-1)\underline{I}_3 + \underline{I}_1] \tfrac{\omega^{2\vartheta}}{\Gamma(2\vartheta+1)} - [\hat{s}_\ell(2\Xi-1)(\underline{I}_6+\underline{I}_1)] \tfrac{\omega^{2\vartheta}}{\Gamma(2\vartheta+1)},\\
&\overline{\mathbf{K}}_{\ell_2}(\omega,\Xi) = \hat{r}_\ell \tfrac{\omega^\vartheta}{\Gamma(\vartheta+1)} - \hat{\imath}_\ell \overline{I}_1 \tfrac{\omega^{2\vartheta}}{\Gamma(2\vartheta+1)} [\hat{s}_\ell(2\Xi-1)\overline{I}_3 + \overline{I}_1] \tfrac{\omega^{2\vartheta}}{\Gamma(2\vartheta+1)} - [\hat{s}_\ell(2\Xi-1)(\overline{I}_6+\overline{I}_1)] \tfrac{\omega^{2\vartheta}}{\Gamma(2\vartheta+1)},\\
&\underline{\mathbf{L}}_{\ell_2}(\omega,\Xi) = [\hat{s}_\ell(2\Xi-1)(\underline{I}_3+\underline{I}_1)] \tfrac{\omega^{2\Xi}}{\Gamma(2\vartheta+1)} + [\lambda\hat{s}_\ell(2\Xi-1)^2 + \hat{s}_\ell(2\Xi-1)^2\\
&-(1-\hat{\omega})\hat{\mu}_\ell(2\Xi-1) - \hat{\omega}_\ell\hat{\mu}'_\ell(2\Xi-1) - \hat{\imath}_\ell(2\Xi-1)] \tfrac{\omega^{\Xi}}{\Gamma(\vartheta+1)},\\
&\overline{\mathbf{L}}_{\ell_2}(\omega,\Xi) = [\hat{s}_\ell(1-2\Xi)(\overline{I}_3+\overline{I}_1)] \tfrac{\omega^{2\Xi}}{\Gamma(2\vartheta+1)} + [\lambda\hat{s}_\ell(1-2\Xi)^2 + \hat{s}_\ell(1-2\Xi)^2\\
&-(1-\hat{\omega})\hat{\mu}_\ell(1-2\Xi) - \hat{\omega}_\ell\hat{\mu}'_\ell(1-2\Xi) - \hat{\imath}_\ell(2\Xi-1)] \tfrac{\omega^{\Xi}}{\Gamma(\vartheta+1)},\\
&\underline{\mathbf{C}}_{\ell_1}(\omega,\Xi) = (1-\hat{\$}_\ell)\hat{\omega}_\ell(2\Xi-1) - (\hat{\vartheta}_p + \hat{\imath}_\ell)(2\Xi-1)] \tfrac{\omega^{\Xi}}{\Gamma(\vartheta+1)},\\
&\overline{\mathbf{C}}_{\ell_1}(\omega,\Xi) = (1-\hat{\$}_\ell)\hat{\mu}_\ell(1-2\Xi) - (\hat{\vartheta}_p + \hat{\imath}_\ell)(1-2\Xi)] \tfrac{\omega^{\Xi}}{\Gamma(\vartheta+1)},\\
&\underline{\mathbf{M}}_{\ell_1}(\omega,\Xi) = [\hat{\omega}_\ell\hat{\mu}'_\ell(2\Xi-1) - (\hat{\vartheta}'_\ell - \hat{\imath}_\ell)] \tfrac{\omega^{\Xi}}{(} 2\Xi-1) \tfrac{\omega^{\Xi}}{\Gamma(\vartheta+1)},\\
&\overline{\mathbf{M}}_{\ell_1}(\omega,\Xi) = [\hat{\omega}_\ell\hat{\mu}'_\ell(1-2\Xi) - (\hat{\vartheta}'_\ell - \hat{\imath}_\ell)] \tfrac{\omega^{\Xi}}{(} 1-2\Xi) \tfrac{\omega^{\Xi}}{\Gamma(\vartheta+1)},\\
&\underline{\mathbf{E}}_{\ell_2}(\omega,\Xi) = [\hat{\vartheta}_\ell\underline{I}_3 - \hat{\vartheta}'_\ell\underline{I}_4 - \hat{\imath}_\ell\underline{I}_5] \tfrac{\omega^{2\Xi}}{\Gamma(2\Xi+1)},\\
&\overline{\mathbf{E}}_{\ell_2}(\omega,\Xi) = [\hat{\vartheta}_\ell\overline{I}_3 - \hat{\vartheta}'_\ell\overline{I}_4 - \hat{\imath}_\ell\overline{I}_5] \tfrac{\omega^{2\Xi}}{\Gamma(2\vartheta+1)},\\
&\underline{\mathbf{Q}}_{\ell_2}(\omega,\Xi) = [\hat{\varsigma}_\ell\underline{I}_3 + \hat{\psi}\underline{I}_4 - \hat{\varrho}\underline{I}_6] \tfrac{\omega^{\Xi}}{\Gamma(\vartheta+1)},\\
&\overline{\mathbf{Q}}_{\ell_2}(\omega,\Xi) = (1-2\Xi)[\hat{\varsigma}_\ell\underline{I}_3 + \hat{\psi}\underline{I}_4 - \hat{\varrho}\underline{I}_6] \tfrac{\omega^{\Xi}}{\Gamma(\vartheta+1)},
\end{aligned} \tag{18}
$$

$$
\begin{aligned}
&\underline{\mathbf{K}}'_{\ell_2}(\omega,\Xi) = \hat{r}'_\ell \tfrac{\omega^\vartheta}{\Gamma(\vartheta+1)} - \hat{\imath}'_\ell \underline{I}'_1 \tfrac{\omega^{2\vartheta}}{\Gamma(2\vartheta+1)} [\hat{s}'_\ell(2\Xi-1)\underline{I}'_3 + \underline{I}'_1] \tfrac{\omega^{2\vartheta}}{\Gamma(2\vartheta+1)} - [\hat{s}'_\ell(2\Xi-1)(\underline{I}'_6+\underline{I}'_1)] \tfrac{\omega^{2\vartheta}}{\Gamma(2\vartheta+1)},\\
&\overline{\mathbf{K}}'_{\ell_2}(\omega,\Xi) = \hat{r}'_\ell \tfrac{\omega^\vartheta}{\Gamma(\vartheta+1)} - \hat{\imath}'_\ell \overline{I}'_1 \tfrac{\omega^{2\vartheta}}{\Gamma(2\vartheta+1)} [\hat{s}'_\ell(2\Xi-1)\overline{I}'_3 + \overline{I}'_1] \tfrac{\omega^{2\vartheta}}{\Gamma(2\vartheta+1)} - [\hat{s}'_\ell(2\Xi-1)(\overline{I}'_6+\overline{I}'_1)] \tfrac{\omega^{2\vartheta}}{\Gamma(2\vartheta+1)},\\
&\underline{\mathbf{L}}'_{\ell_2}(\omega,\Xi) = [\hat{s}'_\ell(2\Xi-1)(\underline{I}'_3+\underline{I}'_1)] \tfrac{\omega^{2\Xi}}{\Gamma(2\vartheta+1)} + [\lambda\hat{s}'_\ell(2\Xi-1)^2 + \hat{s}'_\ell(2\Xi-1)^2\\
&-(1-\hat{\omega})\hat{\mu}'_\ell(2\Xi-1) - \hat{\omega}'_\ell\hat{\mu}'_\ell(2\Xi-1) - \hat{\imath}'_\ell(2\Xi-1)] \tfrac{\omega^{\Xi}}{\Gamma(\vartheta+1)},\\
&\overline{\mathbf{L}}'_{\ell_2}(\omega,\Xi) = [\hat{s}'_\ell(1-2\Xi)(\overline{I}'_3+\overline{I}_1)'] \tfrac{\omega^{2\Xi}}{\Gamma(2\vartheta+1)} + [\lambda\hat{s}'_\ell(1-2\Xi)^2 + \hat{s}'_\ell(1-2\Xi)^2\\
&-(1-\hat{\omega})\hat{\mu}'_\ell(1-2\Xi) - \hat{\omega}_\ell\hat{\mu}'_\ell(1-2\Xi) - \hat{\imath}'_\ell(2\Xi-1)] \tfrac{\omega^{\Xi}}{\Gamma(\vartheta+1)},\\
&\underline{\mathbf{C}}'_{\ell_1}(\omega,\Xi) = (1-\hat{\$}'_\ell)\hat{\omega}'_\ell(2\Xi-1) - (\hat{\vartheta}'_p + \hat{\imath}'_\ell)(2\Xi-1)] \tfrac{\omega^{\Xi}}{\Gamma(\vartheta+1)},\\
&\overline{\mathbf{C}}'_{\ell_1}(\omega,\Xi) = (1-\hat{\$}'_\ell)\hat{\mu}'_\ell(1-2\Xi) - (\hat{\vartheta}'_p + \hat{\imath}'_\ell)(1-2\Xi)] \tfrac{\omega^{\Xi}}{\Gamma(\vartheta+1)},\\
&\underline{\mathbf{M}}'_{\ell_1}(\omega,\Xi) = [\hat{\omega}'_\ell\hat{\mu}'_\ell{}'(2\Xi-1) - (\hat{\vartheta}'_\ell{}' - \hat{\imath}'_\ell)] \tfrac{\omega^{\Xi}}{(} 2\Xi-1) \tfrac{\omega^{\Xi}}{\Gamma(\vartheta+1)},\\
&\overline{\mathbf{M}}'_{\ell_1}(\omega,\Xi) = [\hat{\omega}'_\ell\hat{\mu}'_\ell{}'(1-2\Xi) - (\hat{\vartheta}'_\ell{}' - \hat{\imath}_\ell)'] \tfrac{\omega^{\Xi}}{(} 1-2\Xi) \tfrac{\omega^{\Xi}}{\Gamma(\vartheta+1)},\\
&\underline{\mathbf{E}}'_{\ell_2}(\omega,\Xi) = [\hat{\vartheta}'_\ell\underline{I}'_3 - \hat{\vartheta}'_\ell{}'\underline{I}'_4 - \hat{\imath}'_\ell\underline{I}'_5] \tfrac{\omega^{2\Xi}}{\Gamma(2\Xi+1)},\\
&\overline{\mathbf{E}}'_{\ell_2}(\omega,\Xi) = [\hat{\vartheta}'_\ell\overline{I}'_3 - \hat{\vartheta}'_\ell{}'\overline{I}'_4 - \hat{\imath}'_\ell\overline{I}'_5] \tfrac{\omega^{2\Xi}}{\Gamma(2\vartheta+1)},\\
&\underline{\mathbf{Q}}'_{\ell_2}(\omega,\Xi) = [\hat{\varsigma}'_\ell\underline{I}'_3 + \hat{\psi}'\underline{I}'_4 - \hat{\varrho}'\underline{I}'_6] \tfrac{\omega^{\Xi}}{\Gamma(\vartheta+1)},\\
&\overline{\mathbf{Q}}'_{\ell_2}(\omega,\Xi) = (1-2\Xi)[\hat{\varsigma}'_\ell\underline{I}'_3 + \hat{\psi}'\underline{I}'_4 - \hat{\varrho}'\underline{I}'_6] \tfrac{\omega^{\Xi}}{\Gamma(\vartheta+1)}.
\end{aligned} \tag{19}
$$

Here, we present the computational results based on the numerical scheme discussed above. Figures 1–6 show a comparison of approximate fuzzy and approximate normal solutions for the considered model at various fractional orders for the given uncertainty. Figure 1 presents the susceptible population versus days; in Figure 2, the exposed population versus days; in Figure 3 the infected population versus days; in Figure 4, the recovered population versus days; in Figure 5, the asymptotically infected population versus days; and in Figure 6, the reservoir population versus days for a fuzzy and normal solution at 1.25, 1.50 and 1.75. Figure 7 shows the recovered population versus days and in Figure 8, the asymptotically infected population versus days for a fuzzy and normal solution at 1.00, 1.25, 1.50, 1.75 and 2.00. As the susceptible class value decreases, the exposed population grows, and the infection spreads at a different rate due to changing fractional orders. Similarly, when the number of death cases rises, the recovered class expands, the asymptotically infected class expands, and the virus population in the reservoir expands. We can see from the pictures that fuzziness, in combination with fractional order $(1,2)$, provides global dynamics to nonlinear problems where the data are uncertain.

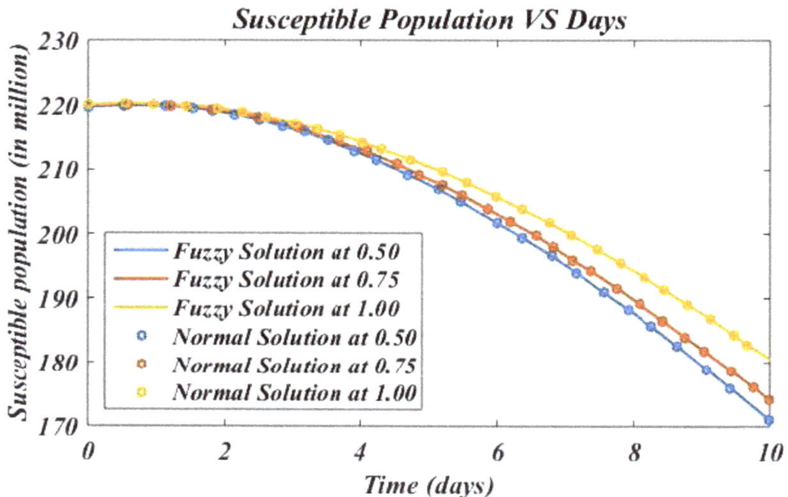

Figure 1. Illustration of approximate fuzzy and normal susceptible compartment solutions for three terms at given uncertainty levels $\vartheta \in [1, 2]$ versus fractional order.

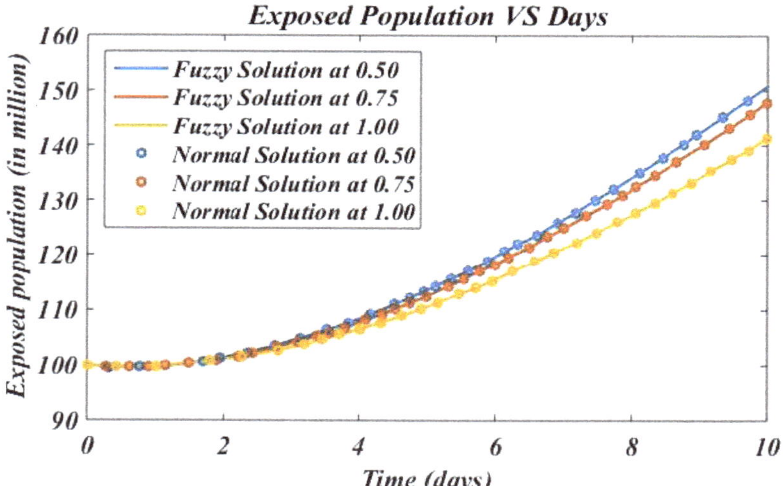

Figure 2. Illustration of approximate fuzzy and normal exposed compartment solutions for three terms at given uncertainty levels $\vartheta \in [1, 2]$ versus fractional order.

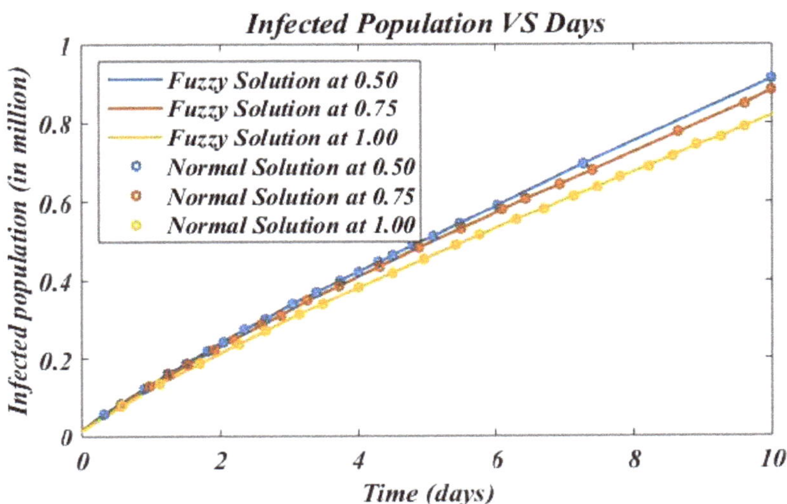

Figure 3. Illustration of approximate fuzzy and normal infected compartment solutions for three terms at given uncertainty levels $\vartheta \in [1,2]$ versus fractional order

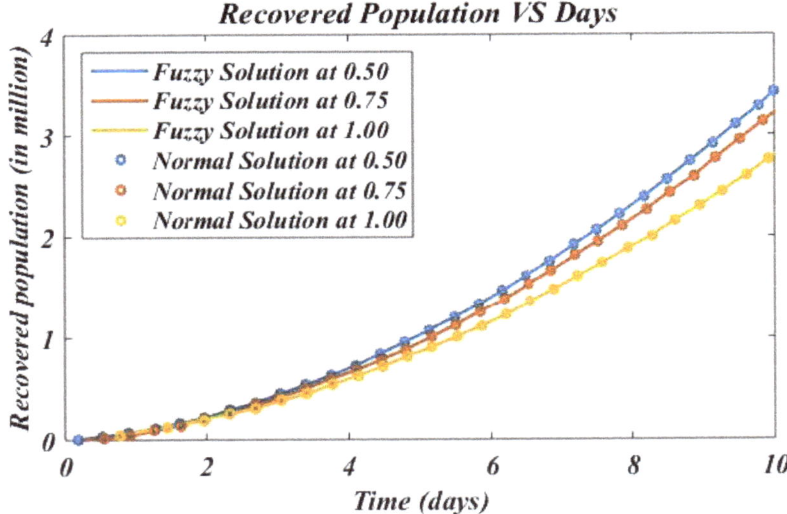

Figure 4. Illustration of approximate fuzzy and normal recovered compartment solutions for three terms at given uncertainty levels $\vartheta \in [1,2]$ versus fractional order.

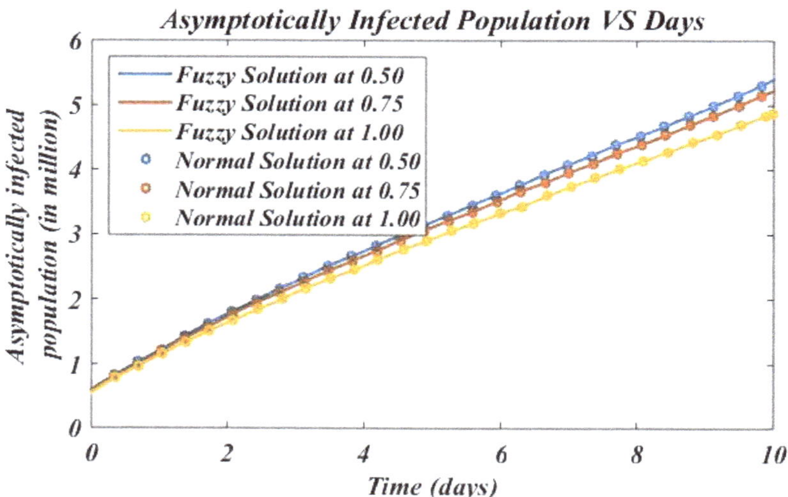

Figure 5. Illustration of approximate fuzzy and normal asymptotically infected compartment solutions for three terms at given uncertainty levels $\vartheta \in [1,2]$ versus fractional order.

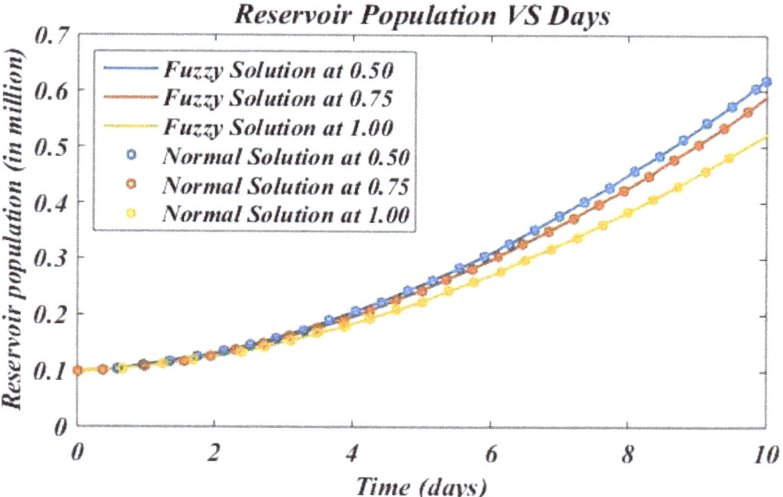

Figure 6. Illustration of approximate fuzzy and normal reservoir compartment solutions for three terms at given uncertainty levels $\vartheta \in [1,2]$ versus fractional order.

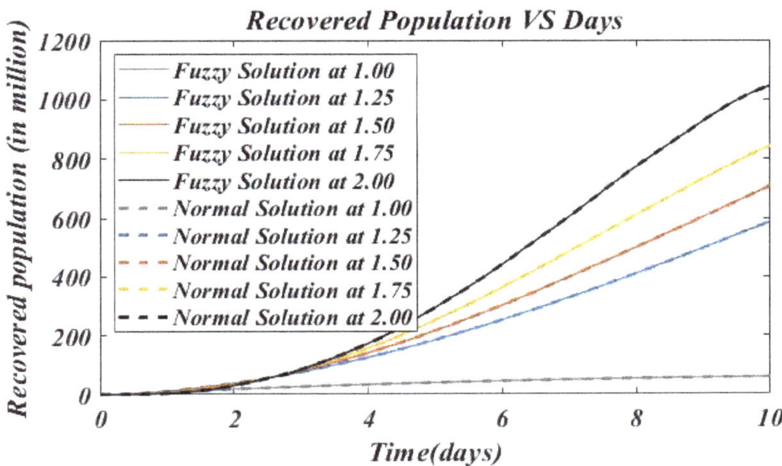

Figure 7. Illustration of approximate fuzzy and normal recovered compartment solutions for five terms at given uncertainty levels $\vartheta \in [1,2]$ versus fractional order.

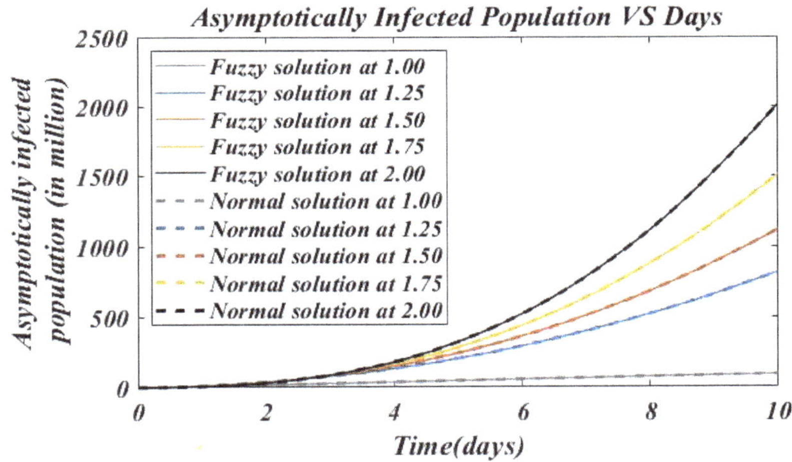

Figure 8. Illustration of approximate fuzzy and normal asymptotically infected compartment solutions for five terms at given uncertainty levels $\vartheta \in [1,2]$ versus fractional order.

Remark 1. *According to the results, the lower bounded is a growing set-valued function, while the upper bounded is a decreasing one, indicating that the solutions are fuzzy numbers. It is also worth noting that under fuzzy differentiability, identical findings can be derived in general circumstances.*

Remark 2. *Given that stochastic and random parameters are more difficult to address, and that uncertainty might contribute to an increase in computation costs, using fuzzy notions to model such real-world systems may be the best option.*

5. Conclusions

In this paper, we present an analytical investigation of the fractional order COVID-19 model (3) for existence–uniqueness results, as well as numerical simulations. We used the Schauder's fixed point theorem to prove that the solution to the fuzzy fractional-order model of COVID-19 exists and is unique. Using the fuzzy Laplace transform and the

Adomian decomposition approach, we established a reasonable strategy for obtaining an approximate solution for the recommended model. We compared fuzzy and normal results for up to three phrases to demonstrate the utility of this method. We discovered that fuzziness combined with a fractional calculus technique yielded outstanding global dynamics in instances where data uncertainty exists. For future research, we recommend that readers revisit the problem for stochastic differential operators, optimal control and sensitivity analysis. Furthermore, the provided results can be compared to simulations for different fractional derivatives.

Author Contributions: Formal analysis, A.U.K.N.; Project administration, M.B.J.; Supervision, A.S.A. and A.U.K.N.; Writing–original draft, R.S.; Writing–review & editing, M.B.J. All authors have read and agreed to the published version of the manuscript.

Funding: This research was funded by [Imam Mohammad Ibn Saud Islamic University] grant number [21-13-18-069] and the APC was funded by [Imam Mohammad Ibn Saud Islamic University].

Institutional Review Board Statement: Not applicable.

Informed Consent Statement: Not applicable.

Data Availability Statement: No new data were created this study.

Acknowledgments: The authors extend their appreciation to the Deanship of Scientific Research at Imam Mohammad Ibn Saud Islamic University for funding this work through Research Group no. 21-13-18-069.

Conflicts of Interest: The authors declare that they have no known competing financial interest or personal relationships that could have appeared to influence the work reported in this paper.

References

1. Chan, J.F.W.; Kok, K.H.; Zhu, Z.; Chu, H.; To, K.K.W.; Yuan, S.; Yuen, K.Y. Genomic characterization of the 2019 novel human-pathogenic coronavirus isolated from a patient with atypical pneumonia after visiting Wuhan. *Emerg. Microbes Infect.* **2020**, *9*, 221–236. [CrossRef] [PubMed]
2. Lu, H.; Stratton, C.W.; Tang, Y.W. Outbreak of pneumonia of unknown etiology in Wuhan, China: The mystery and the miracle. *J. Med. Virol.* **2020**, *92*, 401. [CrossRef] [PubMed]
3. Ji, W.; Wang, W.; Zhao, X.; Zai, J.; Li, X. Homologous recombination within the spike glycoprotein of the newly identified coronavirus may boost cross-species transmission from snake to human. *J. Med. Virol.* **2020**, *92*, 433–440. [CrossRef]
4. Fahmi, I. *World Health Organization Coronavirus Disease 2019 (COVID-19) Situation Report*; WHO: Geneva, Switzerland, 2019.
5. Chen, Y.; Guo, D. Molecular mechanisms of coronavirus RNA capping and methylation. *Virol. Sin.* **2016**, *31*, 3–11. [CrossRef] [PubMed]
6. Wang, L.F.; Shi, Z.; Zhang, S.; Field, H.; Daszak, P.; Eaton, B.T. Review of bats and SARS. *Emerg. Infect. Dis.* **2006**, *12*, 1834. [CrossRef]
7. Ge, X.Y.; Li, J.L.; Yang, X.L.; Chmura, A.A.; Zhu, G.; Epstein, J.H.; Mazet, J.K.; Hu, B.; Zhang, W.; Peng, C.; et al. Isolation and characterization of a bat SARS-like coronavirus that uses the ACE2 receptor. *Nature* **2013**, *503*, 535–538. [CrossRef]
8. Lu, R.; Zhao, X.; Li, J.; Niu, P.; Yang, B.; Wu, H.; Wang, W.; Song, H.; Huang, B.; Zhu, N.; et al. Genomic characterisation and epidemiology of 2019 novel coronavirus: Implications for virus origins and receptor binding. *Lancet* **2020**, *395*, 565–574. [CrossRef]
9. Zhou, P.; Yang, X.L.; Wang, X.G.; Hu, B.; Zhang, L.; Zhang, W.; Si, H.-R.; Zhu, Y.; Li, B.; Huang, C.-L.; et al. A pneumonia outbreak associated with a new coronavirus of probable bat origin. *Nature* **2020**, *579*, 270–273. [CrossRef]
10. Tian, X.; Li, C.; Huang, A.; Xia, S.; Lu, S.; Shi, Z.; Lu, L.; Jiang, S.; Yang, Z.; Wu, Y.; et al. Potent binding of 2019 novel coronavirus spike protein by a SARS coronavirus-specific human monoclonal antibody. *Emerg. Microbes Infect.* **2020**, *9*, 382–385. [CrossRef]
11. Ahmed, S.F.; Quadeer, A.A.; McKay, M.R. Preliminary identification of potential vaccine targets for 2019-nCoV based on SARS-CoV immunological studies. *Viruses* **2020**, *12*, 254. [CrossRef]
12. Chaudhury, S.; Berrondo, M.; Weitzner, B.D.; Muthu, P.; Bergman, H.; Gray, J.J. Benchmarking and analysis of protein docking performance in Rosetta v3. 2. *PLoS ONE* **2011**, *6*, e22477. [CrossRef] [PubMed]
13. Khan, M.A.; Atangana, A. Modeling the dynamics of novel coronavirus (2019-nCov) with fractional derivative. *Alex. Eng. J.* **2020**, *59*, 2379–2389. [CrossRef]
14. Agarwal, R.P.; Lakshmikantham, V.; Nieto, J.J. On the concept of solution for fractional differential equations with uncertainty. *Nonlinear Anal. Theory Methods Appl.* **2010**, *72*, 2859–2862. [CrossRef]
15. Asjad, M.I.; Aleem, M.; Ahmadian, A.; Salahshour, S.; Ferrara, M. New trends of fractional modeling and heat and mass transfer investigation of (SWCNTs and MWCNTs)-CMC based nanofluids flow over inclined plate with generalized boundary conditions. *Chin. J. Phys.* **2020**, *66*, 497–516. [CrossRef]

16. Abbas, A.; Shafqat, R.; Jeelani, M.B.; Alharthi, N.H. Significance of Chemical Reaction and Lorentz Force on Third-Grade Fluid Flow and Heat Transfer with Darcy-Forchheimer Law over an Inclined Exponentially Stretching Sheet Embedded in a Porous Medium. *Symmetry* **2022**, *14*, 779. [CrossRef]
17. Abbas, A.; Shafqat, R.; Jeelani, M.B.; Alharthi, N.H. Convective Heat and Mass Transfer in Third-Grade Fluid with Darcy–Forchheimer Relation in the Presence of Thermal-Diffusion and Diffusion-Thermo Effects over an Exponentially Inclined Stretching Sheet Surrounded by a Porous Medium: A CFD Study. *Processes* **2022**, *10*, 776. [CrossRef]
18. Kaleva, O. Fuzzy differential equations. *Fuzzy Sets Syst.* **1987**, *24*, 301–317. [CrossRef]
19. Lupulescu, V. Fractional calculus for interval-valued functions. *Fuzzy Sets Syst.* **2015**, *265*, 63–85. [CrossRef]
20. Arshad, S.; Lupulescu, V. Fractional differential equation with the fuzzy initial condition. *Electron. J. Differ. Equ.* **2011**, *34*, 1–8.
21. Benchohra, M.; Cabada, A.; Seba, D. An existence result for nonlinear fractional differential equations on Banach spaces. *Bound. Value Probl.* **2009**, *2009*, 628916 . [CrossRef]
22. Belmekki, M.; Nieto, J.; Rodriguez-Lopez, R. Existence of periodic solution for a nonlinear fractional differential equation. *Bound. Value Probl.* **2009**, *2009*, 324561. [CrossRef]
23. Park, J.Y.; Han, H.K. Existence and uniqueness theorem for a solution of fuzzy Volterra integral equations. *Fuzzy Sets Syst.* **1999**, *105*, 481–488. [CrossRef]
24. Ali, N.; Khan, R.A. Existence of positive solution to a class of fractional differential equations with three point boundary conditions. *Math. Sci. Lett.* **2016**, *5*, 291–296. [CrossRef]
25. Khan, R.A.; Shah, K. Existence and uniqueness of solutions to fractional order multi-point boundary value problems. *Commun. Appl. Anal.* **2015**, *19*, 515–525.
26. Lakshmikantham, V.; Leela, S. Nagumo-type uniqueness result for fractional differential equations. *Nonlinear Anal.* **2009**, *71*, 2886–2889. [CrossRef]
27. Miller, K.S.; Ross, B. *An Introduction to the Fractional Calculus and Fractional Differential Equations*; Wiley: Hoboken, NJ, USA, 1993.
28. Lakshmikantham, V.; Vatsala, A.S. Basic theory of fractional differential equations. *Nonlinear Anal. Theory Methods Appl.* **2008**, *69*, 2677–2682. [CrossRef]
29. Perfilieva, I. Fuzzy transforms: Theory and applications. *Fuzzy Sets Syst.* **2006**, *157*, 993–1023. [CrossRef]
30. Salahshour, S.; Allahviranloo, T. Application of fuzzy differential transform method for solving fuzzy Volterra integral equations. *Appl. Math. Model.* **2013**, *37*, 1016–1027. [CrossRef]
31. Allahviranloo, T.; Salahshour, S.; Abbasbandy, S. Explicit solutions of fractional differential equations with uncertainty. *Soft Comput.* **2012**, *16*, 297–302. [CrossRef]
32. Allahviranloo, T.; Ahmadi, M.B. Fuzzy laplace transforms. *Soft Comput.* **2010**, *14*, 235. [CrossRef]
33. Zhu, Y. Stability analysis of fuzzy linear differential equations. *Fuzzy Optim. Decis. Mak.* **2010**, *9*, 169–186. [CrossRef]
34. Niazi, A.U.K.; He, J.; Shafqat, R.; Ahmed, B. Existence, Uniqueness, and Eq–Ulam-Type Stability of Fuzzy Fractional Differential Equation. *Fractal Fract.* **2021**, *5*, 66. [CrossRef]
35. Iqbal, N.; Niazi, A.U.K.; Shafqat, R.; Zaland, S. Existence and Uniqueness of Mild Solution for Fractional-Order Controlled Fuzzy Evolution Equation. *J. Funct. Spaces* **2021**, *2021*, 5795065. [CrossRef]
36. Shafqat, R.; Niazi, A.U.K.; Jeelani, M.B.; Alharthi, N.H. Existence and Uniqueness of Mild Solution Where $\alpha \in (1,2)$ for Fuzzy Fractional Evolution Equations with Uncertainty. *Fractal Fract.* **2022**, *6*, 65. [CrossRef]
37. Abuasbeh, K.; Shafqat, R.; Niazi, A.U.K.; Awadalla, M. Local and Global Existence and Uniqueness of Solution for Time-Fractional Fuzzy Navier–Stokes Equations. *Fractal Fract.* **2022**, *6*, 330. [CrossRef]
38. Ahmad, S.; Ullah, A.; Shah, K.; Salahshour, S.; Ahmadian, A.; Ciano, T. Fuzzy fractional-order model of the novel coronavirus. *Adv. Differ. Equations* **2020**, *2020*, 472. [CrossRef]
39. Gottwald, S. *Fuzzy Set Theory and Its Applications*; Kluwer Academic Publishers: Dordrecht, The Netherlands, 1991; ISBN 0-7923-9075-X.
40. Zadeh, L.A. Fuzzy sets. In *Fuzzy Sets, Fuzzy Logic, and Fuzzy Systems*; World Scientific: Singapore, 1996; pp. 394–432.

MDPI

Article

Existence Results for a Multipoint Fractional Boundary Value Problem in the Fractional Derivative Banach Space

Djalal Boucenna [1,†], Amar Chidouh [2,†] and Delfim F. M. Torres [3,*,†]

1 High School of Technological Teaching, Enset, Skikda 21001, Algeria; djalal.boucenna@enset-skikda.dz
2 Department of Mathematics, Chadli Bendjedid University, Eltarf 36000, Algeria; m2ma.chidouh@gmail.com
3 Center for Research and Development in Mathematics and Applications (CIDMA), Department of Mathematics, University of Aveiro, 3810-193 Aveiro, Portugal
* Correspondence: delfim@ua.pt
† These authors contributed equally to this work.

Abstract: We study a class of nonlinear implicit fractional differential equations subject to nonlocal boundary conditions expressed in terms of nonlinear integro-differential equations. Using the Krasnosel'skii fixed-point theorem we prove, via the Kolmogorov–Riesz criteria, the existence of solutions. The existence results are established in a specific fractional derivative Banach space and they are illustrated by two numerical examples.

Keywords: fractional differential equations; boundary value problems; Kolmogorov–Riesz theorem; fixed point theorems; Nemytskii operator

MSC: 34B10; 34K37; 45J05

Citation: Boucenna, D.; Chidouh, A.; Torres, D.F.M. Existence Results for a Multipoint Fractional Boundary Value Problem in the Fractional Derivative Banach Space. *Axioms* **2022**, *11*, 295. https://doi.org/10.3390/axioms11060295

Academic Editor: Chris Goodrich

Received: 2 May 2022
Accepted: 14 June 2022
Published: 16 June 2022

1. Introduction

It is noticeable, in recent years, that the field of fractional calculus has been swept for research by many mathematicians, due to its effectiveness in describing many physical phenomena, see, e.g., [1–7].

A fractional derivative is a generalization of the ordinary one. Despite the emergence of several definitions of fractional derivative, the content is one that depends entirely on Volterra integral equations and their kernel, which facilitates the description of each phenomenon as a temporal lag, such as rheological phenomena [8–10].

The study of differential equations is considered of primary importance in mathematics. In applications, differential equations serve as mathematical models for all natural phenomena. Regardless of their type (ordinary, partial, or fractional), the study of differential equations is developed in three directions: existence, uniqueness, and stability of solutions. Therefore, to investigate boundary value problems is always a central question in mathematics [11–16].

Often, it is of central importance to know the behavior of any solution, of the equation under study, at the boundary of the domain, because that makes it easier to find the solution. In 2009, Ahmad and Nieto considered the following boundary value problem [17]:

$$\begin{aligned} {}^{C}D^{\alpha}y(t) &= f\left(t, y(t), \int_0^t \varphi(t,s)y(s)ds\right), \quad 1<\alpha<2, \\ ay(0)+by'(0) &= \int_0^1 q_1(y(s))ds, \\ ay(1)+by'(1) &= \int_0^1 q_2(y(s))ds. \end{aligned} \tag{1}$$

Their results of existence are obtained via Krasnosel'skii fixed-point theorem in the space of continuous functions. For that, they apply Ascoli's theorem in order to provide the compactness of the first part of the Krasnosel'skii operator.

The pioneering work of Ahmad and Nieto of 2009 [17] gave rise to several different investigations. These include: inverse source problems for fractional integrodifferential equations [18]; the study of positive solutions for singular fractional boundary value problems with coupled integral boundary conditions [19]; the expression and properties of Green's function for nonlinear boundary value problems of fractional order with impulsive differential equations [20]; existence of solutions to several kinds of differential equations using the coincidence theory [21]; existence and uniqueness of solution for fractional differential equations with Riemann–Liouville fractional integral boundary conditions [22]; sufficient conditions for the existence and uniqueness of solutions for a class of terminal value problems of fractional order on an infinite interval [23]; existence of solutions, and stability, for fractional integro-differential equations involving a general form of Hilfer fractional derivative with respect to another function [24]; existence of solutions for a boundary value problem involving mixed generalized fractional derivatives of Riemann–Liouville and Caputo, supplemented with nonlocal multipoint boundary conditions [25]; existence conditions to fractional order hybrid differential equations [26]; and an existence analysis for a nonlinear implicit fractional differential equation with integral boundary conditions [27]. Motivated by all these existence results, we consider here a more general multipoint fractional boundary value problem in the fractional derivative Banach space.

Let $1 < p < \infty$ and $1 \geq \gamma > \frac{1}{p}$ and consider the following fractional boundary value problem:

$$\begin{aligned} {}^{C}D^{\alpha}y(t) &= f\Big(t, y(t), {}^{C}D^{\gamma}y(t)\Big) + {}^{C}D^{\alpha-2}g(t, y(t)), \quad 2 < \alpha < 3, \\ y(0) + y'(0) &= \int_0^1 q_1(y(s))ds, \\ y(1) + y'(1) &= \int_0^1 q_2(y(s))ds, \\ y''(0) &= 0, \end{aligned} \tag{2}$$

where ${}^{C}D^{\alpha}$ is the standard Caputo derivative, $f : [0,1] \times \mathbb{R} \times \mathbb{R} \to \mathbb{R}$, and $g : [0,1] \times \mathbb{R} \to \mathbb{R}$ and $q_1, q_2 : \mathbb{R} \to \mathbb{R}$ are given functions such that $g(t,0) = g(0,y) = q_1(0) = q_2(0) = 0$ for any $(t,y) \in [0,1] \times \mathbb{R}$. Our problem (2) generalizes (1) and finds applications in viscoelasticity, where the fractional operators are associated with delay kernels that make the fractional differential equations the best models for several rheological Maxwell phenomena. In particular, for $\alpha \in (1,2)$, we can model oscillatory processes with fractional damping [28].

We prove existence of a solution to problem (2) in the special Banach space $E^{\gamma,p}$ that is known in the literature as the fractional derivative space [29]. This Banach space is equipped with the norm

$$\|u\|_{\gamma,p} = \left(\int_0^T |u(t)|^p + \int_0^T \left|{}^{C}D_0^{\gamma}u(t)\right|^p\right)^{\frac{1}{p}}. \tag{3}$$

The paper is organized as follows. In Section 2, we recall some useful definitions and lemmas to prove our main results. The original contributions are then given in Section 3. The main result is Theorem 1, which establishes the existence of solutions to the fractional boundary value problem (2) using Krasnosel'skii fixed point theorem. Two illustrative examples are given. We end with Section 4, discussing the obtained existence result.

2. Preliminaries

For the convenience of the reader, and to facilitate the analysis of problem (2), we begin by recalling the necessary background from the theory of fractional calculus [30,31].

Definition 1. *The Riemann–Liouville fractional integral of order $\alpha > 0$ of a function $f : (0, +\infty) \to \mathbb{R}$ is given by*

$$I_0^\alpha f(t) = \frac{1}{\Gamma(\alpha)} \int_0^t (t-s)^{\alpha-1} f(s) ds.$$

Definition 2. *The Caputo fractional derivative of order $\alpha > 0$ of a function $f : (0, +\infty) \to \mathbb{R}$ is given by*

$${}^C D_0^\alpha f(t) = \frac{1}{\Gamma(n-\alpha)} \int_0^t \frac{f^{(n)}(s)}{(t-s)^{\alpha-n+1}} ds = I_0^{n-\alpha} f^{(n)}(t),$$

where $n = [\alpha] + 1$, with $[\alpha]$ denoting the integer part of α.

Lemma 1 (See [17]). *Let $\alpha > 0$. Then, the fractional differential equation ${}^C D_{0^+}^\alpha u(t) = 0$ has*

$$u(t) = c_0 + c_1 t + c_2 t^2 + \cdots + c_{n-1} t^{n-1}, \quad c_i \in \mathbb{R}, \quad i = 1, 2, \ldots, n-1,$$

as solution.

Definition 3. *A map $f : [0,1] \times \mathbb{R} \times \mathbb{R} \to \mathbb{R}$ is said to be Carathéodory if*

(a) $t \to f(t, u; v)$ *is measurable for each* $u, v \in \mathbb{R}$;
(b) $(u, v) \to f(t, u; v)$ *is continuous for almost all* $t \in [0,1]$.

Definition 4. *Let J be a measurable subset of $\mathbb{R}$ and $f : J \times \mathbb{R}^{d_1} \to \mathbb{R}^{d_2}$ satisfies the Carathéodory condition. By a generalized Nemytskii operator we mean the mapping N_f taking a (measurable) vector function $u = (u_1, \ldots, u_{d_1})$ to the function $N_f u(t) = f(t, u(t))$, $t \in J$.*

The following lemma is concerned with the continuity of the operator N_f with $d_1 = 2$ and $d_2 = 1$.

Lemma 2 (See [32]). *Consider the same data of Definition 4. Assume there exists $w \in L^1([0,1])$ with $1 \le p < \infty$ and a constant $c > 0$ such that $|f(t,u,v)| \le w(t) + c(|u|^p + |v|^p)$ for almost all $t \in [0,1]$ and $u, v \in \mathbb{R}$. Then, the Nemytskii operator*

$$N_f u(t) = f(t, u(t)), \quad u = (u_1, u_2) \in L^p(0,1) \times L^p(0,1), \quad t \in [0,1] \text{ a.e.},$$

is continuous from $L^p([0,1]) \times L^p([0,1])$ to $L^1(0,1)$.

Lemma 3 (See [33]). *Let $\mathcal{F}$ be a bounded set in $L^p([0,1])$ with $1 \le p < \infty$. Assume that*

$$\lim_{|h| \to 0} \|\tau_h f - f\|_p = 0 \text{ uniformly on } \mathcal{F}.$$

Then, $\mathcal{F}$ is relatively compact in $L^p([0,1])$.

For any $1 \le p < \infty$, we denote

$$\|u\|_{L^p[0,T]} := \left(\int_0^T |u(t)|^p \right)^{\frac{1}{p}}, \quad \|u\|_\infty := \max_{t \in [0,T]} |u(t)|.$$

Now, we give the definition and some properties of $E^{\gamma,p}$. For more details about the following lemmas, see [29,34] and references therein.

Definition 5. *Let $0 < \gamma \leq 1$ and $1 < p < \infty$. The fractional derivative space $E^{\gamma,p}$ is defined by the closure of $C^\infty([0,T])$ with respect to the norm*

$$\|u\|_{\gamma,p} = \left(\int_0^T |u(t)|^p + \int_0^T \left|{}^C D_0^\gamma u(t)\right|^p\right)^{\frac{1}{p}}. \tag{4}$$

Lemma 4 (See [29,34]). *Let $0 < \gamma \leq 1$ and $1 < p < \infty$. The fractional derivative space $E^{\gamma,p}$ is a reflexive and separable Banach space.*

Lemma 5 (See [29,34]). *Let $0 < \gamma \leq 1$ and $1 < p < \infty$. For all $u \in E^{\gamma,p}$, we have*

$$\|u\|_{L^p} \leq \frac{T^\alpha}{\Gamma(\gamma+1)} \left\|{}^C D_0^\gamma u\right\|_{L^p}. \tag{5}$$

Moreover, if $\gamma > \frac{1}{p}$ and $\frac{1}{p} + \frac{1}{q} = 1$, then

$$\|u\|_\infty \leq \frac{T^{\alpha - \frac{1}{p}}}{\Gamma(\gamma)((\gamma-1)q+1)^{\frac{1}{q}}} \left\|{}^C D_0^\gamma u\right\|_{L^p}. \tag{6}$$

According to the inequality (5), we can also consider the space $E^{\gamma,p}$ with respect to the equivalent norm

$$\|u\|_{\gamma,p} = \left\|{}^C D_0^\gamma u\right\|_{L^p} = \left(\int_0^T \left|{}^C D_0^\gamma u(t)\right|^p\right)^{\frac{1}{p}}, \quad u \in E^{\gamma,p}.$$

3. Main Results

We begin by considering a linear problem and obtain its solution in terms of a Green function.

Lemma 6. *Assume $h, k \in C([0,1])$, $k(0) = 0$ and $\alpha \in (2,3)$. Then, the solution to the boundary value problem*

$$\begin{aligned} {}^C D^\alpha y(t) &= h(t) + {}^C D^{\alpha-2} k(t), \quad t \in (0,1), \\ y''(0) &= 0, \\ y(0) + y'(0) &= \int_0^1 \eta_1(s) ds, \\ y(1) + y'(1) &= \int_0^1 \eta_2(s) ds, \end{aligned} \tag{7}$$

is given by

$$y(t) = \int_0^1 G(t,s)h(s)ds + \int_0^1 H(t,s)k(s)ds + (2-t)\int_0^1 \eta_1(s)ds + (t-1)\int_0^1 \eta_2(s)ds,$$

where

$$G(t,s) = \begin{cases} \frac{(t-s)^{\alpha-1} + (1-t)(1-s)^{\alpha-1}}{\Gamma(\alpha)} + \frac{(1-t)(1-s)^{\alpha-2}}{\Gamma(\alpha-1)}, & 0 \leq s \leq t \leq 1, \\ \frac{(1-t)(1-s)^{\alpha-1}}{\Gamma(\alpha)} + \frac{(1-t)(1-s)^{\alpha-2}}{\Gamma(\alpha-1)}, & 0 \leq t \leq s \leq 1, \end{cases} \tag{8}$$

and

$$H(t,s) = \begin{cases} (t-s) + (1-t)(2-s), & 0 \leq s \leq t \leq 1, \\ (1-t)(2-s), & 0 \leq t \leq s \leq 1. \end{cases} \tag{9}$$

Proof. Let y be a solution of problem (7). By Lemma 1, we have

$$y(t) = c_0 + c_1 t + c_2 t^2 + \frac{1}{\Gamma(\alpha)} \int_0^t (t-s)^{\alpha-1} h(s) ds + I_0^2 k(t).$$

Taking the conditions (7) into account, it follows that

$$c_2 = 0,$$

$$y(0) + y'(0) = c_0 + c_1 = \int_0^1 \eta_1(s) ds,$$

and

$$\begin{aligned} y(1) + y'(1) &= c_0 + 2c_1 + \frac{1}{\Gamma(\alpha)} \int_0^1 (1-s)^{\alpha-1} h(s) ds + \int_0^1 (1-s) k(s) ds \\ &+ \frac{1}{\Gamma(\alpha-1)} \int_0^1 (1-s)^{\alpha-2} h(s) ds + \int_0^1 k(s) ds \\ &= \int_0^1 \eta_2(s) ds, \end{aligned}$$

which implies

$$\begin{aligned} c_0 = &\frac{1}{\Gamma(\alpha)} \int_0^1 (1-s)^{\alpha-1} h(s) ds + \frac{1}{\Gamma(\alpha-1)} \int_0^1 (1-s)^{\alpha-2} h(s) ds \\ &+ \int_0^1 (2-s) k(s) ds + 2 \int_0^1 \eta_1(s) ds - \int_0^1 \eta_2(s) ds, \end{aligned}$$

and

$$\begin{aligned} c_1 = &-\frac{1}{\Gamma(\alpha)} \int_0^1 (1-s)^{\alpha-1} h(s) ds - \frac{1}{\Gamma(\alpha-1)} \int_0^1 (1-s)^{\alpha-2} h(s) ds \\ &- \int_0^1 (2-s) k(s) ds + \int_0^1 \eta_2(s) ds - \int_0^1 \eta_1(s) ds. \end{aligned}$$

Hence, the solution of problem (7) is

$$\begin{aligned} y(t) = &\int_0^t \left(\frac{(t-s)^{\alpha-1} + (1-t)(1-s)^{\alpha-1}}{\Gamma(\alpha)} + \frac{(1-t)(1-s)^{\alpha-2}}{\Gamma(\alpha-1)} \right) h(s) ds \\ &+ \int_t^1 \left(\frac{(1-t)(1-s)^{\alpha-1}}{\Gamma(\alpha)} + \frac{(1-t)(1-s)^{\alpha-2}}{\Gamma(\alpha-1)} \right) h(s) ds \\ &+ \int_0^t ((t-s) + (1-t)(2-s)) k(s) ds + \int_t^1 (1-t)(2-s) k(s) ds \\ &+ (2-t) \int_0^1 \eta_1(s) ds + (t-1) \int_0^1 \eta_2(s) ds \\ = &\int_0^1 G(t,s) h(s) ds + \int_0^1 H(t,s) k(s) ds \\ &+ (2-t) \int_0^1 \eta_1(s) ds + (t-1) \int_0^1 \eta_2(s) ds. \end{aligned}$$

The proof is complete. □

Lemma 7. *Functions $G, H, \frac{\partial^\gamma}{\partial t} G$ and $\frac{\partial^\gamma}{\partial t} H$ are continuous on $[0,1] \times [0,1]$ and satisfy for all $t, s \in [0,1]$:*

1. $|G(t,s)| \leq \frac{3}{\Gamma(\alpha-1)}$, $\quad |H(t,s)| \leq 3$.

2. $\left|\frac{\partial^\gamma}{\partial t}G(t,s)\right| \le \frac{\Gamma(\alpha)}{\Gamma(\alpha-\gamma)} + \frac{2}{\Gamma(2-\gamma)\Gamma(\alpha-1)}$, $\left|\frac{\partial^\gamma}{\partial t}H(t,s)\right| \le \frac{3}{\Gamma(2-\gamma)}$.

Proof. We have

$$^{C}D_0^\gamma(1-t) = I_0^{1-\gamma}(1-t)' = -\frac{1}{\Gamma(2-\gamma)}t^{1-\gamma}, \tag{10}$$

and

$$\frac{\partial^\gamma}{\partial t}(t-s)^{\alpha-1} = I_0^{1-\gamma}\frac{\partial}{\partial t}(t-s)^{\alpha-1} = (\alpha-1)I_0^{1-\gamma}(t-s)^{\alpha-2}.$$

Thus, for $0 \le s \le t \le 1$, we get $\frac{\partial^\gamma}{\partial t}(t-s)^{\alpha-1} \ge 0$ and

$$\frac{\partial^\gamma}{\partial t}(t-s)^{\alpha-1} \le^C D_0^\gamma t^{\alpha-1} = \frac{\Gamma(\alpha)}{\Gamma(\alpha-\gamma)}t^{\alpha-\gamma-1}. \tag{11}$$

On the other hand, we have $\Gamma(\alpha-1) \le \Gamma(\alpha)$ for $2 \le \alpha \le 3$. Now, we give the bound of functions $|G(t,s)|$ and $\left|\frac{\partial^\gamma}{\partial t}G(t,s)\right|$. From the definition of function G and (10) and (11), we obtain:

- For $0 \le s \le t \le 1$,

$$\begin{aligned} |G(t,s)| &= \frac{(t-s)^{\alpha-1}+(1-t)(1-s)^{\alpha-1}}{\Gamma(\alpha)} + \frac{(1-t)(1-s)^{\alpha-2}}{\Gamma(\alpha-1)} \\ &\le \frac{(1-s)^{\alpha-1}(1+(1-t))}{\Gamma(\alpha)} + \frac{(1-t)(1-s)^{\alpha-2}}{\Gamma(\alpha-1)} \\ &\le \frac{1+(1-t)}{\Gamma(\alpha)} + \frac{(1-t)}{\Gamma(\alpha-1)} \\ &\le \frac{3}{\Gamma(\alpha-1)}, \end{aligned}$$

  and

$$\begin{aligned} \left|\frac{\partial^\gamma}{\partial t}G(t,s)\right| &\le \left|\frac{\Gamma(\alpha)}{\Gamma(\alpha-\gamma)}t^{\alpha-\gamma-1}\right| + \left|\frac{t^{1-\gamma}}{\Gamma(2-\gamma)}\left(\frac{(1-s)^{\alpha-1}}{\Gamma(\alpha)} + \frac{(1-s)^{\alpha-2}}{\Gamma(\alpha-1)}\right)\right| \\ &\le \frac{\Gamma(\alpha)}{\Gamma(\alpha-\gamma)} + \frac{1}{\Gamma(2-\gamma)}\left(\frac{1}{\Gamma(\alpha)} + \frac{1}{\Gamma(\alpha-1)}\right) \\ &\le \frac{\Gamma(\alpha)}{\Gamma(\alpha-\gamma)} + \frac{2}{\Gamma(2-\gamma)\Gamma(\alpha-1)}. \end{aligned}$$

- For $0 \le t \le s \le 1$,

$$\begin{aligned} |G(t,s)| &= \frac{(1-t)(1-s)^{\alpha-1}}{\Gamma(\alpha)} + \frac{(1-t)(1-s)^{\alpha-2}}{\Gamma(\alpha-1)} \\ &\le \frac{(1-t)}{\Gamma(\alpha)} + \frac{(1-t)}{\Gamma(\alpha-1)} \\ &\le \frac{2}{\Gamma(\alpha-1)}, \end{aligned}$$

  and

$$\begin{aligned} \left|\frac{\partial^\gamma}{\partial t}G(t,s)\right| &= \left|-\frac{t^{1-\gamma}}{\Gamma(2-\gamma)}\left(\frac{(1-s)^{\alpha-1}}{\Gamma(\alpha)} + \frac{(1-s)^{\alpha-2}}{\Gamma(\alpha-1)}\right)\right| \\ &\le \frac{1}{\Gamma(2-\gamma)}\left(\frac{1}{\Gamma(\alpha)} + \frac{1}{\Gamma(\alpha-1)}\right) \\ &\le \frac{2}{\Gamma(2-\gamma)\Gamma(\alpha-1)}. \end{aligned}$$

By using the same above calculation, we obtain the estimation of $|H(t,s)|$ and $\left|\frac{\partial^\gamma}{\partial t}H(t,s)\right|$. The proof is complete. □

In the sequel, we denote

$$G_\gamma(t,s) := \frac{\partial^\gamma}{\partial t}G(t,s), \quad t,s \in [0,1] \times [0,1].$$

Moreover, we also use the following notations: $G^* := \max_{t,s\in[0,1]\times[0,1]}|G(t,s)|$ and

$$G_\gamma^* := \max_{t,s\in[0,1]\times[0,1]}|G_\gamma(t,s)|.$$

Theorem 1. *Assume that the following four hypotheses hold:*

(H1) $f : [0,1] \times \mathbb{R} \times \mathbb{R} \to \mathbb{R}$ *satisfies the Carathéodory condition.*

(H2) There exist $w \in L^1(0,1)$ *and* $c > 0$ *such that*

$$|f(t,u,v)| \leq w(t) + c\big(|u|^p + |v|^p\big) \text{ for } t \in (0,1) \text{ and } u,v \in \mathbb{R}. \tag{12}$$

(H3) There exist two strictly positive constants k_1 *and* k_2 *and a function* $\varphi_1 \in L^q((0,1),\mathbb{R}_+)$, $\frac{1}{p} + \frac{1}{q} = 1$, *such that for all* $t \in [0,1]$ *and* $x, y \in \mathbb{R}$, *we have*

$$\begin{aligned} |g(t,x) - g(t,y)| &\leq \varphi_1(t)|x - y|, \\ |q_1(x) - q_1(y)| &\leq k_1|x - y|, \\ |q_2(x) - q_2(y)| &\leq k_2|x - y|. \end{aligned}$$

(H4) There exists a real number $R > 0$ *such that*

$$\frac{R\Big[3\|\varphi_1\|_q + k_1 + k_2\Big]}{\Gamma(2-\gamma)\Gamma(1+\gamma)} + G^*_\gamma\left(\|w\|_1 + c\left(1 + \left(\frac{1}{\Gamma(\gamma+1)}\right)^p\right)R^p\right) \leq R. \tag{13}$$

Then, if

$$\frac{3\|\varphi_1\|_q + k_1 + k_2}{\Gamma(2-\gamma)\Gamma(1+\gamma)} < 1, \tag{14}$$

the boundary value problem (2) has a solution in $E^{\gamma,p}$.

Proof. We transform problem (2) into a fixed-point problem. Define two operators $F, L : E^{\gamma,p} \to E^{\gamma,p}$ by

$$Fy(t) = \int_0^1 G(t,s)f(s,y(s),D^\gamma y(s))ds,$$

and

$$Ly(t) = \int_0^1 H(t,s)g(s,y(s))ds + (2-t)\int_0^1 q_1(y(s))ds + (t-1)\int_0^1 q_2(y(s))ds.$$

Then, y is a solution of problem (2) if, and only if, y is a fixed point of $F + L$. We define the set B_R as follows:

$$B_R := \{u \in E^{\gamma,p}, \|u\|_{E^{\gamma,p}} \leq R\},$$

where R is the same constant defined in (H_3). It is clear that B_R is convex, closed, and a bounded subset of $E^{\gamma,p}$. We shall show that F, G satisfy the assumptions of Krasnosel'skii fixed-point theorem. The proof is given in several steps.

(i) We prove that F is continuous. Let $(y_n)_{n\in\mathbb{N}}$ be a sequence such that $y_n \to y$ in $E^{\gamma,p}$. From (12) and Lemma 2, and for each $t \in [0,1]$, we obtain

$$\begin{aligned} &\left|\left({}^C D_0^\gamma F y_n\right)(t) - \left({}^C D_0^\gamma F y\right)(t)\right| \\ &\leq \int_0^1 |G_\gamma(t,s)|\,|f(s,y_n(s),D^\gamma y_n(s)) - f(s,y(s),D^\gamma y(s))|ds \\ &\leq G^*_\gamma \Big\|N_f y_n - N_f y\Big\|_1. \end{aligned}$$

Applying the L^p norm, we obtain that $\|Fy_n - Fy\|_{E^{\gamma,p}} \to 0$ when $y_n \to y$ in$E^{\gamma,p}$. Thus, the operator F is continuous.

(ii) Now, we prove that $Fx + Ly \in B_R$ for $x, y \in B_R$. Let $x, y \in B_R$, $t \in (0,1)$. In view of hypothesis $(H3)$, we obtain

$$\begin{aligned}\left|{}^C D_0^\gamma Fy(t)\right| &\leq \int_0^1 |G_\gamma(t,s)||f(s, y(s), D^\gamma y(s))|ds \\ &\leq G_\gamma^*\left(\|w\|_1 + c\left(\|y\|_p^p + \left\|{}^C D_0^\gamma y\right\|_p^p\right)\right) \\ &\leq G_\gamma^*\left(\|w\|_1 + c\left(1 + \left(\frac{1}{\Gamma(\gamma+1)}\right)^p\right)R^p\right).\end{aligned}$$

Applying the L^p norm, we obtain that

$$\|Fy\|_{E^{\gamma,p}} \leq G_\gamma^*\left(\|w\|_1 + c\left(1 + \left(\frac{1}{\Gamma(\gamma+1)}\right)^p\right)R^p\right). \tag{15}$$

Also,

$$\begin{aligned}\left|{}^C D_0^\gamma L(x)(t)\right| &\leq \frac{3}{\Gamma(2-\gamma)}\int_0^1 |g(s, x(s))|ds \\ &+ \frac{1}{\Gamma(2-\gamma)}\int_0^1 |q_1(x(s))|ds + \frac{1}{\Gamma(2-\gamma)}\int_0^1 |q_2(x(s))|ds \\ &\leq \frac{3}{\Gamma(2-\gamma)}\int_0^1 \varphi_1(s)|x(s)|ds + \frac{1}{\Gamma(2-\gamma)}\int_0^1 k_1|x(s)|ds \\ &+ \frac{1}{\Gamma(2-\gamma)}\int_0^1 k_2|x(s)|ds.\end{aligned}$$

Applying again the L^p norm, we obtain from Holder's inequality that

$$\|L(x)\|_{E^{\gamma,p}} \leq \frac{3}{\Gamma(2-\gamma)}\left(\|\varphi_1\|_q\|x\|_p\right) + \frac{k_1}{\Gamma(2-\gamma)}\|x\|_p + \frac{k_2}{\Gamma(2-\gamma)}\|x\|_p.$$

In view of (5), we obtain

$$\|L(x)\|_{E^{\gamma,p}} \leq \left[\frac{3\|\varphi_1\|_q}{\Gamma(2-\gamma)\Gamma(1+\gamma)} + \frac{k_1}{\Gamma(2-\gamma)\Gamma(1+\gamma)} + \frac{k_2}{\Gamma(2-\gamma)\Gamma(1+\gamma)}\right]\|x\|_{E^{\gamma,p}}.$$

Then,

$$\|L(x)\|_{E^{\gamma,p}} \leq \frac{R\left[3\|\varphi_1\|_q + k_1 + k_2\right]}{\Gamma(2-\gamma)\Gamma(1+\gamma)}. \tag{16}$$

From (13), (15) and (16), we conclude that $Fx + Ly \in B_R$ whenever $x, y \in B_R$.

(iii) Let us prove that $F(B_R) = \{F(u) : u \in B_R\}$ is relatively compact in $E^{\gamma,p}$. Let $t \in (0,1)$ and $h > 0$, where $t + h \leq 1$, and let $u \in D_R$. From (12), we obtain that

$$\begin{aligned}&\left|{}^C D_0^\gamma Fy(t+h) -{}^C D_0^\gamma Fy(t)\right| \\ &\leq \int_0^1 |G_\gamma(t+h,s) - G_\gamma(t,s)||f(s, y(s), D^\gamma y(s))|ds \\ &\leq \int_0^1 |G_\gamma(t+h,s) - G_\gamma(t,s)|\left[w(s) + c(|y(s)|^p + |D^\gamma y(s)|^p)\right]ds \\ &\leq \sup_{t\in[0,1]}\left[\sup_{s\in[0,1]}|G_\gamma(t+h,s) - G_\gamma(t,s)|\right]\left(\|w\|_1 + c\left(1 + \left(\frac{1}{\Gamma(\gamma+1)}\right)^p\right)R^p\right).\end{aligned}$$

Therefore,

$$\frac{\|Fu(\cdot+h)-Fu(\cdot)\|_{E^{\gamma,p}}}{\left(\|w\|_1+c\left(1+\left(\frac{1}{\Gamma(\gamma+1)}\right)^p\right)R^p\right)} \leq \sup_{t\in[0,1]}\left[\sup_{s\in[0,1]}|G_\gamma(t+h,s)-G_\gamma(t,s)|\right]. \quad (17)$$

Then, $\|Fu(\cdot+h)-Fu(\cdot)\|_{E^{\gamma,p}} \to 0$ as $h \to 0$ for any $u \in B_R$, since G_γ is a continuous function on $[0,1]\times[0,1]$. From Lemma 3, we conclude that $F : B_R \to B_R$ is compact.

(iv) Finally, we prove that L is a contraction. Let $x, y \in D_R$ and $t \in (0,1)$. Then,

$$\begin{aligned}\left|{}^C D_0^\gamma L(x)(t) - {}^C D_0^\gamma L(y)(t)\right| &\leq \frac{3}{\Gamma(2-\gamma)}\int_0^1 |g(s,x(s))-g(s,y(s))|ds \\ &+ \frac{1}{\Gamma(2-\gamma)}\int_0^1 |q_1(x(s))-q_1(x(s))|ds \\ &+ \frac{1}{\Gamma(2-\gamma)}\int_0^1 |q_2(x(s))-q_2(x(s))|ds \\ &\leq \frac{3}{\Gamma(2-\gamma)}\int_0^1 \varphi_1(s)|x(s)-y(s)|ds \\ &+ \frac{k_1}{\Gamma(2-\gamma)}\int_0^1 |x(s)-x(s)|ds \\ &+ \frac{k_2}{\Gamma(2-\gamma)}\int_0^1 |x(s)-x(s)|ds.\end{aligned}$$

Applying the L^p norm and Holder's inequality, we obtain that

$$\begin{aligned}\|L(x)-L(y)\|_{E^{\gamma,p}} \leq \frac{3}{\Gamma(2-\gamma)}\left(\|\varphi_1\|_q\|x-y\|_p\right) + \frac{k_1}{\Gamma(2-\gamma)}\left(\|x-y\|_p\right) \\ + \frac{k_2}{\Gamma(2-\gamma)}\left(\|x-y\|_p\right).\end{aligned}$$

Then, from (5), we obtain

$$\|L(x)-L(y)\|_{E^{\gamma,p}} \leq \frac{3\|\varphi_1\|_q + k_1 + k_2}{\Gamma(2-\gamma)\Gamma(1+\gamma)}\|x-y\|_{E^{\gamma,p}}.$$

From (14), the operator L is a contraction.

As a consequence of (i)–(iv), we conclude that $F : B_R \to B_R$ is continuous and compact. As a consequence of Krasnosel'skii fixed point theorem, we deduce that $F+G$ has a fixed point $y \in B_R \subset E^{\gamma,p}$, which is a solution to problem (2). □

We now illustrate our Theorem 1 with two examples.

Example 1. *Consider the fractional boundary value problem* (2) *with*

$$\begin{aligned}\alpha = 2.5, \quad \gamma = 0.5, \quad p = 3, \quad q = \frac{3}{2}, \\ f(t,x,y) = \frac{\exp(-t)}{5} - \frac{1}{164\pi}\arctan\left(x^3+y^3\right), \\ g(t,x) = \frac{1}{10}t^{\frac{2}{3}}x, \\ q_1(x) = q_2(x) = \frac{1}{20}x,\end{aligned}$$

which we denote by (P_1). *Hypotheses* $(H1)$ *and* $(H2)$ *are satisfied for*

$$w(t) = \frac{\exp(-t)}{5} \in L^1(0,1), \quad c = \frac{1}{164\pi}, \quad \varphi_1(t) = \frac{t^{\frac{2}{3}}}{10}, \quad \textit{and} \quad k_1 = k_2 = \frac{1}{20}.$$

Moreover, we have

$$\frac{\left[3\|\varphi_1\|_q + k_1 + k_2\right]}{\Gamma(2-\gamma)\Gamma(1+\gamma)} = \frac{\left[\frac{3}{10}\left(\frac{1}{2}\right)^{\frac{2}{3}} + \frac{1}{10}\right]}{\left(\Gamma\left(\frac{3}{2}\right)\right)^2} \simeq 0.368 < 1.$$

If we choose $R = 2$, *then we obtain*

$$\begin{aligned} &\frac{R\left[3\|\varphi_1\|_q + k_1 + k_2\right]}{\Gamma(2-\gamma)\Gamma(1+\gamma)} + G_\gamma^*\left(\|w\|_1 + c\left(1 + \left(\frac{1}{\Gamma(\gamma+1)}\right)^p\right)R^p\right) - R \\ &\leq \frac{2\left[\frac{3}{10}\left(\frac{1}{2}\right)^{\frac{2}{3}} + \frac{1}{10}\right]}{\left(\Gamma\left(\frac{3}{2}\right)\right)^2} + 4.047\left(\frac{1}{5} + \frac{1}{164\pi}\left(1 + \left(\frac{1}{\Gamma\left(\frac{3}{2}\right)}\right)^3\right)2^3\right) - 2 \\ &\simeq -0.301. \end{aligned}$$

Since all conditions of our Theorem 1 are satisfied, we conclude that the fractional boundary value problem (P_1) *has a solution in* $E^{\gamma,p}$.

Example 2. *Consider the fractional boundary value problem (2) with*

$$\begin{aligned} \alpha = 2.7, \quad \gamma &= 0.7, \quad p = 4, \quad q = \frac{4}{3}, \\ f(t,x,y) &= \frac{1}{10}\sin(t) + \frac{1}{200}\cos\left(x^4 + y^4\right), \\ g(t,x) &= \frac{1}{9\pi}t^{\frac{3}{4}}\arctan(x), \\ q_1(x) = q_2(x) &= \frac{1}{10}\sin(x), \end{aligned}$$

which we denote by (P_2). *Hypotheses* $(H1)$ *and* $(H2)$ *are satisfied for*

$$w(t) = \frac{1}{10}\sin(t) \in L^1(0,1), \quad c = \frac{1}{200}, \quad \varphi_1(t) = \frac{t^{\frac{3}{4}}}{9\pi} \quad \textit{and} \quad k_1 = k_2 = \frac{1}{10}.$$

Moreover, we have

$$\frac{\left[3\|\varphi_1\|_q + k_1 + k_2\right]}{\Gamma(2-\gamma)\Gamma(1+\gamma)} = \frac{\left[\frac{1}{3\pi}\left(\frac{1}{2}\right)^{\frac{3}{4}} + \frac{1}{5}\right]}{\Gamma(1.3)\Gamma(1.7)} \simeq 0.323 < 1.$$

If we choose $R = 2$, *then we obtain*

$$\begin{aligned} &\frac{R\left[3\|\varphi_1\|_q + k_1 + k_2\right]}{\Gamma(2-\gamma)\Gamma(1+\gamma)} + G_\gamma^*\left(\|w\|_1 + c\left(1 + \left(\frac{1}{\Gamma(\gamma+1)}\right)^p\right)R^p\right) - R \\ &\leq \frac{2\left[\frac{1}{3\pi}\left(\frac{1}{2}\right)^{\frac{3}{4}} + \frac{1}{5}\right]}{\Gamma(1.3)\Gamma(1.7)} + 3.9995\left(\frac{1}{10} + \frac{1}{200}\left(1 + \left(\frac{1}{\Gamma(1.7)}\right)^4\right)2^4\right) - 2 \\ &\simeq -0.163. \end{aligned}$$

Since all conditions of our Theorem 1 are satisfied, we conclude that the fractional boundary value problem (P_2) *has a solution in* $E^{\gamma,p}$.

4. Discussion

The celebrated existence result of Ahmad and Nieto [17] for problem (1) is obtained via Krasnosel'skii fixed-point theorem in the space of continuous functions. For that, they needed to apply Ascoli's theorem in order to provide the compactness of the first part of the Krasnosel'skii operator. Here, we proved existence for the more general problem (2) in the fractional derivative Banach space $E^{\gamma,p}$, equipped with the norm (3). From norm (3), it is natural to deal with a subspace of $L^p \times L^p$. Since Ascoli's theorem is limited to the space of continuous functions for the compactness, we had to make use of a different approach to ensure existence of solution in the fractional derivative space $E^{\gamma,p}$. Our tool was the Kolmogorov–Riesz compactness theorem, which turned out to be a powerful tool to address the problem. To the best of our knowledge, the use of the Kolmogorov–Riesz compactness theorem to prove existence results for boundary value problems involving nonlinear integrodifferential equations of fractional order with integral boundary conditions is a completely new approach. In this direction, we are only aware of the work [35], where a necessary and sufficient condition of pre-compactness in variable exponent Lebesgue spaces is established and, as an application, the existence of solutions to a fractional Cauchy problem is obtained in the Lebesgue space of variable exponent. As future work, we intend to generalize our existence result to the variable-order case [36].

Author Contributions: Conceptualization, D.B., A.C. and D.F.M.T.; validation, D.B., A.C. and D.F.M.T.; formal analysis, D.B., A.C. and D.F.M.T.; investigation, D.B., A.C. and D.F.M.T.; writing—original draft preparation, D. B., A.C. and D.F.M.T.; writing—review and editing, D.B., A.C. and D.F.M.T. All authors have read and agreed to the published version of the manuscript.

Funding: This research was partially funded by FCT, grant number UIDB/04106/2020 (CIDMA).

Institutional Review Board Statement: Not applicable.

Informed Consent Statement: Not applicable.

Data Availability Statement: Data sharing not applicable.

Acknowledgments: The authors are grateful to the referees for their comments and remarks.

Conflicts of Interest: The authors declare no conflict of interest.

References

1. Dhar, B.; Gupta, P.K.; Sajid, M. Solution of a dynamical memory effect COVID-19 infection system with leaky vaccination efficacy by non-singular kernel fractional derivatives. *Math. Biosci. Eng.* **2022**, *19*, 4341–4367. [CrossRef] [PubMed]
2. Kumar, A.; Malik, M.; Sajid, M.; Baleanu, D. Existence of local and global solutions to fractional order fuzzy delay differential equation with non-instantaneous impulses. *AIMS Math.* **2022**, *7*, 2348–2369. [CrossRef]
3. Failla, G.; Zingales, M. Advanced materials modelling via fractionalcalculus: Challenges and perspectives. *Philos. Trans. R. Soc.* **2020**, *A378*, 20200050. [CrossRef] [PubMed]
4. Fang, C.; Shen, X.; He, K.; Yin, C.; Li, S.; Chen, X.; Sun, H. Application of fractional calculus methods to viscoelastic behaviours of solid propellants. *Philos. Trans. R Soc. A* **2020**, *378*, 20190291. [CrossRef]
5. Wei, E.; Hu, B.; Li, J.; Cui, K.; Zhang, Z.; Cui, A.; Ma, L. Nonlinear viscoelastic-plastic creep model of rock based on fractional calculus. *Adv. Civ. Eng.* **2022**, *2022*, 3063972. [CrossRef]
6. Marin, M.; Othman, M.I.; Vlase, S.; Codarcea-Munteanu, L. Thermoelasticity of initially stressed bodies with voids: A domain of influence. *Symmetry* **2019**, *11*, 573. [CrossRef]
7. Ndaïrou, F.; Area, I.; Nieto, J.J.; Silva, C.J.; Torres, D.F.M. Fractional model of COVID-19 applied to Galicia, Spain and Portugal. *Chaos Solitons Fractals* **2021**, *144*, 110652. [CrossRef]
8. Mainardi, F.; Spada, G. Creep, relaxation and viscosity properties for basic fractional models in rheology. *Eur. Phys. J. Spec. Top.* **2011**, *193*, 133–160. [CrossRef]
9. Chidouh, A.; Guezane-Lakoud, A.; Bebbouchi, R.; Bouaricha, A.; Torres, D.F.M. Linear and Nonlinear Fractional Voigt Models. In *Theory and Applications of Non-Integer Order Systems*; Springer: Berlin/Heidelberg, Germany, 2017; pp. 157–167. [CrossRef]
10. Mainardi, F.; Gorenflo, R. Time-fractional derivatives in relaxation processes: A tutorial survey. *Fract. Calc. Appl. Anal.* **2007**, *10*, 269–308.

11. Keten, A.; Yavuz, M.; Baleanu, D. Nonlocal Cauchy problem via a fractional operator involving power kernel in Banach spaces. *Fractal Fract.* **2019**, *3*, 27. [CrossRef]
12. Wang, Y.; Liang, S.; Wang, Q. Existence results for fractional differential equations with integral and multi-point boundary conditions. *Bound. Value Probl.* **2018**, *2018*, 7129796. [CrossRef]
13. Ahmad, B. Existence results for multi-point nonlinear boundary value problems for fractional differential equations. *Mem. Differ. Equ. Math. Phys.* **2010**, *49*, 83–94.
14. Area, I.; Cabada, A.; Cid, J.A.; Franco, D.; Liz, E.; Pouso, R.L.; Rodríguez-López, R. (Eds.) *Nonlinear Analysis and Boundary Value Problems*; Springer: Cham, Switzerland, 2019. [CrossRef]
15. Behrndt, J.; Hassi, S.; de Snoo, H. *Boundary Value Problems, Weyl Functions, and Differential Operators*; Monographs in Mathematics; Birkhäuser/Springer: Cham, Switzerland, 2020; Volume 108. [CrossRef]
16. Kusraev, A.G.; Totieva, Z.D. (Eds.) *Operator Theory and Differential Equations*; Trends in Mathematics; Birkhäuser/Springer: Cham, Switzerland, 2021. [CrossRef]
17. Ahmad, B.; Nieto, J.J. Existence results for nonlinear boundary value problems of fractional integrodifferential equations with integral boundary conditions. *Bound. Value Probl.* **2009**, *2009*, 708576. [CrossRef]
18. Wu, B.; Wu, S. Existence and uniqueness of an inverse source problem for a fractional integrodifferential equation. *Comput. Math. Appl.* **2014**, *68*, 1123–1136. [CrossRef]
19. Wang, Y.; Liu, L.; Zhang, X.; Wu, Y. Positive solutions for $(n-1,1)$-type singular fractional differential system with coupled integral boundary conditions. *Abstr. Appl. Anal.* **2014**, *2014*, 142391. [CrossRef]
20. Zhou, J.; Feng, M. Green's function for Sturm-Liouville-type boundary value problems of fractional order impulsive differential equations and its application. *Bound. Value Probl.* **2014**, *2014*, 69. [CrossRef]
21. Ariza-Ruiz, D.; Garcia-Falset, J. Existence and uniqueness of solution to several kinds of differential equations using the coincidence theory. *Taiwan. J. Math.* **2015**, *19*, 1661–1692. [CrossRef]
22. Abbas, M.I. Existence and uniqueness results for fractional differential equations with Riemann-Liouville fractional integral boundary conditions. *Abstr. Appl. Anal.* **2015**, *2015*, 290674. [CrossRef]
23. Hussain Shah, S.A.; Rehman, M.U. A note on terminal value problems for fractional differential equations on infinite interval. *Appl. Math. Lett.* **2016**, *52*, 118–125. [CrossRef]
24. Abdo, M.S.; Panchal, S.K. Fractional integro-differential equations involving ψ-Hilfer fractional derivative. *Adv. Appl. Math. Mech.* **2019**, *11*, 338–359. [CrossRef]
25. Shammakh, W.; Alzumi, H.Z.; AlQahtani, B.A. On more general fractional differential equations involving mixed generalized derivatives with nonlocal multipoint and generalized fractional integral boundary conditions. *J. Funct. Spaces* **2020**, *2020*, 3102142. [CrossRef]
26. Shah, A.; Khan, R.A.; Khan, A.; Khan, H.; Gómez-Aguilar, J.F. Investigation of a system of nonlinear fractional order hybrid differential equations under usual boundary conditions for existence of solution. *Math. Methods Appl. Sci.* **2021**, *44*, 1628–1638. [CrossRef]
27. Ali, A.; Shah, K.; Abdeljawad, T.; Mahariq, I.; Rashdan, M. Mathematical analysis of nonlinear integral boundary value problem of proportional delay implicit fractional differential equations with impulsive conditions. *Bound. Value Probl.* **2021**, *2021*, 7. [CrossRef]
28. Bagley, R.L.; Torvik, P.J. On the appearence of the fractional derivative in the behavior of real materials. *J. Appl. Mech.* **1984**, *51*, 294–298. [CrossRef]
29. Jiao, F.; Zhou, Y. Existence of solutions for a class of fractional boundary value problems via critical point theory. *Comput. Math. Appl.* **2011**, *62*, 1181–1199. [CrossRef]
30. Kilbas, A.A.; Srivastava, H.M.; Trujillo, J.J. *Theory and Applications of Fractional Differential Equations*; North-Holland Mathematics Studies; Elsevier Science B.V.: Amsterdam, The Netherlands, 2006; Volume 204.
31. Podlubny, I. *Fractional Differential Equations*; Mathematics in Science and Engineering; Academic Press, Inc.: San Diego, CA, USA, 1999; Volume 198.
32. Precup, R. *Methods in Nonlinear Integral Equations*; Kluwer Academic Publishers: Dordrecht, The Netherlands, 2002. [CrossRef]
33. Brezis, H. *Functional Analysis, Sobolev Spaces and Partial Differential Equations*; Universitext; Springer: New York, NY, USA, 2011.
34. Nyamoradi, N.; Rodríguez-López, R. On boundary value problems for impulsive fractional differential equations. *Appl. Math. Comput.* **2015**, *271*, 874–892. [CrossRef]
35. Dong, B.; Fu, Z.; Xu, J. Riesz-Kolmogorov theorem in variable exponent Lebesgue spaces and its applications to Riemann-Liouville fractional differential equations. *Sci. China Math.* **2018**, *61*, 1807–1824. [CrossRef]
36. Almeida, R.; Tavares, D.; Torres, D.F.M. *The Variable-Order Fractional Calculus of Variations*; Springer Briefs in Applied Sciences and Technology; Springer: Cham, Switzerland, 2019. [CrossRef]

Article

Weighted Generalized Fractional Integration by Parts and the Euler–Lagrange Equation

Houssine Zine [1,†], El Mehdi Lotfi [2,†], Delfim F. M. Torres [1,*,†] and Noura Yousfi [2,†]

1 Center for Research and Development in Mathematics and Applications (CIDMA), Department of Mathematics, University of Aveiro, 3810-193 Aveiro, Portugal; zinehoussine@ua.pt
2 Laboratory of Analysis, Modeling and Simulation (LAMS), Faculty of Sciences Ben M'sik, Hassan II University of Casablanca, P.O. Box 7955, Sidi Othman, Casablanca 20000, Morocco; lotfiimehdi@gmail.com (E.M.L.); nourayousfi.fsb@gmail.com (N.Y.)
* Correspondence: delfim@ua.pt
† These authors contributed equally to this work.

Abstract: Integration by parts plays a crucial role in mathematical analysis, e.g., during the proof of necessary optimality conditions in the calculus of variations and optimal control. Motivated by this fact, we construct a new, right-weighted generalized fractional derivative in the Riemann–Liouville sense with its associated integral for the recently introduced weighted generalized fractional derivative with Mittag–Leffler kernel. We rewrite these operators equivalently in effective series, proving some interesting properties relating to the left and the right fractional operators. These results permit us to obtain the corresponding integration by parts formula. With the new general formula, we obtain an appropriate weighted Euler–Lagrange equation for dynamic optimization, extending those existing in the literature. We end with the application of an optimization variational problem to the quantum mechanics framework.

Keywords: weighted generalized fractional calculus; integration by parts formula; Euler–Lagrange equation; quantum mechanics; calculus of variations

MSC: 26A33; 49K05

Citation: Zine, H.; Lotfi, E.M.; Torres, D.F.M.; Yousfi, N. Weighted Generalized Fractional Integration by Parts and the Euler–Lagrange Equation. *Axioms* **2022**, *11*, 178. https://doi.org/10.3390/axioms11040178

Academic Editor: Chris Goodrich

Received: 21 March 2022
Accepted: 13 April 2022
Published: 15 April 2022

1. Introduction

In the last decade, fractional calculus played an important role in the theoretical study of dynamical systems by showing significant results in many natural fields and engineering domains [1,2]. For this reason, mathematicians are paying more attention to the generalization of several important formulas in the integral theory of Mathematical Analysis, namely, the Newton–Leibniz formula, the Green formula, and the Gauss and Stokes formulas [3,4]. Some are central tools that enable mathematicians to extend other theories, such as the integration by parts formula, Taylor's formula, the Euler–Lagrange equation, Grönwall's inequality, Lyapunov theorems and LaSalle's invariance principle [5,6].

Often, memory effects are fractionally modeled with Riemann–Liouville and Caputo derivatives [7,8]. However, the fact that the Mittag–Leffler function is a generalization of the exponential function naturally gives rise to new definitions for fractional operators [9,10]. In 2020, Hattaf [11] has proposed a new left-weighted generalized fractional derivative for both Caputo and Riemann–Liouville senses and their associated integral operator, see also [12]. Motivated by their applications in mechanics, where the introduction of the correct operator is needed [8,13], here, we introduce the right-weighted generalized fractional derivative and its associated integral operator, proving their main properties and, in particular, their integration by parts formula.

It is worth emphasizing that integration by parts is of great interest in integral calculus and mathematical analysis. For example, it represents a strong tool to develop the calculus of variations through the so-called Euler–Lagrange equation, which is the central result of dynamic optimization [8]. In recent years, the development of some theoretical practices using fractional derivatives has drawn the attention of several researchers. In 2012, Almeida, Malinowska and Torres [14] reviewed some recent results of fractional variational calculus and discussed the necessary optimality conditions of Euler–Lagrange type for functionals with a Lagrangian containing left and right Caputo derivatives. In 2017, Abdeljawad and Baleanu obtained an adequate integration by parts formula and the corresponding Euler–Lagrange equations using the nonlocal fractional derivative with Mittag–Leffler kernel. In 2019, Abdeljawad et al. [15] developed a fractional integration by parts formula for Riemann–Liouville, Liouville–Caputo, Caputo–Fabrizio and Atangana–Baleanu fractional derivatives. In 2020, Zine and Torres [16] introduced a stochastic fractional calculus, and obtained a stochastic fractional Euler–Lagrange equation. Motivated by these works, particularly [14–17], and with the help of our weighted generalized fundamental integration by parts formula, we extend the available Euler–Lagrange equations.

The main purpose of our work is to compute a new integration by parts formula for the weighted generalized fractional derivative and to discuss the associated necessary optimality conditions of Euler–Lagrange type. To do this, we organize the paper as follows. In Section 2, we recall some necessary results from the literature. We proceed with Section 3, introducing the right-weighted generalized fractional derivative and its associated integral and studying their well-posedness. Integration by parts is investigated in Section 4, followed by Section 5, where the weighted generalized fractional Euler–Lagrange equation is rigorously proved. We end with Section 6, illustrating the obtained theoretical results with their application in the quantum mechanics framework.

2. Preliminaries

In this section, we present some definitions and properties from the fractional calculus literature, which will help us to prove our main results. In the text, $f \in H^1(a,b)$ is a sufficiently smooth function on $[a,b]$ with $a,b \in \mathbb{R}$. In addition, we adopt the following notations:

$$\phi(\alpha) := \frac{1-\alpha}{B(\alpha)}, \quad \psi(\alpha) := \frac{\alpha}{B(\alpha)},$$

where $0 \leq \alpha < 1$ and $B(\alpha)$ is a normalization function obeying $B(0) = B(1) = 1$. In the paper, we denote

$$\mu_\alpha := \frac{\alpha}{1-\alpha}.$$

Lemma 1 (See [18]). *Let $\alpha > 0$, $p \geq 1$, $q \geq 1$ and $\frac{1}{p} + \frac{1}{q} \leq 1 + \alpha$ ($p \neq 1$ and $q \neq 1$ in the case $\frac{1}{p} + \frac{1}{q} = 1 + \alpha$). If $f \in L_p(a,b)$ and $g \in L_q(a,b)$, then*

$$\int_a^b f(x)\, {}^{RL}_{a,1}I^\alpha g(x)dx = \int_a^b g(x)\, {}^{RL}I^\alpha_{b,1} f(x)dx,$$

where ${}^{RL}_{a,1}I^\alpha$ is the left standard Riemann–Liouville fractional integral of order α given by

$$ {}^{RL}_{a,1}I^\alpha f(x) = \frac{1}{\Gamma(\alpha)} \int_a^x (x-s)^{\alpha-1} f(s)ds, \quad x > a, \tag{1}$$

and ${}^{RL}I^\alpha_{b,1}$ is the right standard Riemann–Liouville fractional integral of order α given by

$$ {}^{RL}I^\alpha_{b,1} f(x) = \frac{1}{\Gamma(\alpha)} \int_x^b (s-x)^{\alpha-1} f(s)ds, \quad x < b. \tag{2}$$

Definition 1 (See [11]). *Let $0 \leq \alpha < 1$ and $\beta > 0$. The left-weighted generalized fractional derivative of order α of function f, in the Riemann–Liouville sense, is defined by*

$$ {}^{R}_{a,w}D^{\alpha,\beta}f(x) = \frac{1}{\phi(\alpha)}\frac{1}{w(x)}\frac{d}{dx}\int_a^x (wf)(s)E_\beta\left[-\mu_\alpha(x-s)^\beta\right]ds, \tag{3} $$

where E_β denotes the Mittag–Leffler function of parameter β defined by

$$ E_\beta(z) = \sum_{j=0}^{\infty}\frac{z^j}{\Gamma(\beta j+1)}, \quad z \in \mathbb{C}, \tag{4} $$

and $w \in C^1([a,b])$ with $w, w' > 0$. The corresponding fractional integral is defined by

$$ {}_{a,w}I^{\alpha,\beta}f(x) = \phi(\alpha)f(x) + \psi(\alpha)\,{}^{RL}_{a,w}I^\beta f(x), \tag{5} $$

where ${}^{RL}_{a,w}I^\beta$ is the standard weighted Riemann–Liouville fractional integral of order β given by

$$ {}^{RL}_{a,w}I^\beta f(x) = \frac{1}{\Gamma(\beta)}\frac{1}{w(x)}\int_a^x (x-s)^{\beta-1}w(s)f(s)ds, \quad x > a. \tag{6} $$

3. Well-Posedness of the Right-Weighted Fractional Operators

We denote the right-weighted generalized fractional derivative of order α in the Riemann–Liouville sense by ${}^{R}D^{\alpha,\beta}_{b,w}$, and we define this so that the following identity occurs:

$$ Q\left({}^{R}_{a,w}D^{\alpha,\beta}f\right)(x) = \left({}^{R}D^{\alpha,\beta}_{b,w}Qf\right)(x) $$

with Q being the *reflection operator*, that is, $(Qf)(x) = f(a+b-x)$ with function f defined on the interval $[a,b]$.

Definition 2 (right-weighted generalized fractional derivative). *Let $0 \leq \alpha < 1$ and $\beta > 0$. The right-weighted generalized fractional derivative of order α of function f, in the Riemann–Liouville sense, is defined by*

$$ {}^{R}D^{\alpha,\beta}_{b,w}f(x) = \frac{-1}{\phi(\alpha)}\frac{1}{w(x)}\frac{d}{dx}\int_x^b (wf)(s)E_\beta\left[-\mu_\alpha(s-x)^\beta\right]ds, \tag{7} $$

where $w \in C^1([a,b])$ with $w, w' > 0$.

To properly define the new right-weighted fractional integral, we need to solve the equation ${}^{R}D^{\alpha,\beta}_{b,w}f(x) = u(x)$. We have

$$ {}^{R}D^{\alpha,\beta}_{b,w}f(x) = {}^{R}D^{\alpha,\beta}_{b,w}QQf(x) = Q\,{}^{R}_{a,w}D^{\alpha,\beta}Qf(x) = u(t). $$

Then,

$$ {}^{R}_{a,w}D^{\alpha,\beta}Qf(x) = Qu(x) $$

and thus,

$$ Qf(x) = \phi(\alpha)Qu(x) + \psi(\alpha)\,{}^{RL}_{a,w}I^\beta Qu(x) = \phi(\alpha)Qu(x) + \psi(\alpha)\,Q\,{}^{RL}I^\beta_{b,w}u(x), $$

where ${}^{RL}I^\beta_{b,w}$ is the right-weighted standard Riemann–Liouville fractional integral of order β given by

$$ {}^{RL}I^\beta_{b,w}f(x) = \frac{1}{\Gamma(\beta)}\frac{1}{w(x)}\int_x^b (s-x)^{\beta-1}w(s)f(s)ds, \quad x < b. \tag{8} $$

Applying Q to both sides of (8), we obtain

$$f(t) = \phi(\alpha)u(x) + \psi(\alpha)\, {}^{RL}I^{\beta}_{b,w}u(x).$$

Moreover,

$$\begin{aligned} {}_{a,w}I^{\alpha,\beta}Qf(x) &= \phi(\alpha)Qf(x) + \psi(\alpha)\, {}^{RL}_{a,w}I^{\beta}Qf(x) \\ &= \phi(\alpha)Qf(x) + \psi(\alpha)\, Q^{RL}I^{\beta}_{b,w}f(x) \\ &= Q\big[\phi(\alpha)f(x) + \psi(\alpha)\, {}^{RL}I^{\beta}_{b,w}f(x)\big]. \end{aligned}$$

We are now in the position to introduce the concept of the right-weighted generalized fractional integral.

Definition 3 (right-weighted generalized fractional integral). *Let $0 \le \alpha < 1$ and $\beta > 0$. The right-weighted generalized fractional integral of order α of function f is given by*

$$I^{\alpha,\beta}_{b,w}f(x) = \phi(\alpha)f(x) + \psi(\alpha)\, {}^{RL}I^{\beta}_{b,w}f(x), \tag{9}$$

where $w \in C^1([a,b])$ with $w, w' > 0$.

Our next result provides a series representation to the left- and right-weighted generalized fractional derivatives.

Theorem 1. *Let $0 \le \alpha < 1$ and $\beta > 0$. The left- and right-weighted generalized fractional derivatives of order α of function f can be written, respectively, as*

$${}^{R}_{a,w}D^{\alpha,\beta}f(x) = \frac{1}{\phi(\alpha)}\sum_{j=0}^{\infty}(-\mu_\alpha)^j\, {}^{RL}_{a,w}I^{\beta j}f(x) \tag{10}$$

and

$${}^{R}D^{\alpha,\beta}_{b,w}f(x) = \frac{-1}{\phi(\alpha)}\sum_{j=0}^{\infty}(-\mu_\alpha)^j\, {}^{RL}_{b,w}I^{\beta j}f(x). \tag{11}$$

Proof. The Mittag–Leffler function $E_\beta(x)$ is an entire series of x. Since the series (4) locally and uniformly converges in the whole complex plane, the left-weighted generalized fractional derivative can be rewritten as

$$\begin{aligned} {}^{R}_{a,w}D^{\alpha,\beta}f(x) &= \frac{1}{\phi(\alpha)}\frac{1}{w(x)}\frac{d}{dx}\int_a^x (wf)(s)\sum_{j=0}^{\infty}(-\mu_\alpha)^j\frac{(x-s)^{\beta j}}{\Gamma(\beta j+1)}ds \\ &= \frac{1}{\phi(\alpha)}\frac{1}{w(x)}\sum_{j=0}^{\infty}(-\mu_\alpha)^j\frac{1}{\Gamma(\beta j+1)}\frac{d}{dx}\int_a^x (wf)(s)(x-s)^{\beta j}ds \\ &= \frac{1}{\phi(\alpha)}\frac{1}{w(x)}\sum_{j=0}^{\infty}(-\mu_\alpha)^j\frac{1}{\Gamma(\beta j)}\int_a^x (wf)(s)(x-s)^{\beta j-1}ds \\ &= \frac{1}{\phi(\alpha)}\sum_{j=0}^{\infty}(-\mu_\alpha)^j\big({}^{RL}_{a,w}I^{\beta j}f(x)\big). \end{aligned}$$

From Definition 2, and using the same steps that were used before, one can easily rewrite the new right-weighted generalized fractional derivative as equality (11). The proof of (10) is similar. □

Theorem 2. *Let $0 \le \alpha < 1$ and $\beta > 0$. The left- and right-weighted generalized fractional derivative and their associated integrals satisfy the following formulas:*

$${}_{a,w}I^{\alpha,\beta}\left({}_{a,w}^{R}D^{\alpha,\beta}f\right)(x) = {}_{a,w}^{R}D^{\alpha,\beta}\left({}_{a,w}I^{\alpha,\beta}f\right)(x) = f(x) \tag{12}$$

and

$$I_{b,w}^{\alpha,\beta}\left({}^{R}D_{b,w}^{\alpha,\beta}f\right)(x) = {}^{R}D_{b,w}^{\alpha,\beta}\left({}^{\alpha,\beta}I_{b,w}f\right)(x) = -f(x). \tag{13}$$

Proof. We note that

$$\begin{aligned}
{}_{a,w}I^{\alpha,\beta}\left({}_{a,w}^{R}D^{\alpha,\beta}f\right)(x) &= \phi(\alpha)\left({}_{a,w}^{R}D^{\alpha,\beta}f\right)(x) + \psi(\alpha){}_{a,w}^{RL}I^{\beta}\left({}_{a,w}^{R}D^{\alpha,\beta}f\right)(x) \\
&= \sum_{j=0}^{\infty}(-\mu_\alpha)^j\,{}_{a,w}^{RL}I^{\beta j}f(x) + \mu_\alpha\,{}_{a,w}^{RL}I^{\beta}\Big(\sum_{j=0}^{\infty}(-\mu_\alpha)^j\,{}_{a,w}^{RL}I^{\beta j}f\Big)(x) \\
&= \sum_{j=0}^{\infty}(-\mu_\alpha)^j\,{}_{a,w}^{RL}I^{\beta j}f(x) - \sum_{j=0}^{\infty}(-\mu_\alpha)^{j+1}\,{}_{a,w}^{RL}I^{\beta+\beta j}f(x) \\
&= f(t).
\end{aligned}$$

Then,

$$\begin{aligned}
{}_{a,w}^{R}D^{\alpha,\beta}\left({}_{a,w}I^{\alpha,\beta}f\right)(x) &= \frac{1}{\phi(\alpha)}\sum_{j=0}^{\infty}(-\mu_\alpha)^j\,{}_{a,w}^{RL}I^{\beta j}\left({}_{a,w}I^{\alpha,\beta}f\right)(x), \\
&= \frac{1}{\phi(\alpha)}\sum_{j=0}^{\infty}(-\mu_\alpha)^j\,{}_{a,w}^{RL}I^{\beta j}\Big[\phi(\alpha)f(x) + \psi(\alpha)\,{}_{a,w}^{RL}I^{\beta}f(x)\Big] \\
&= \sum_{j=0}^{\infty}(-\mu_\alpha)^j\,{}_{a,w}^{RL}I^{\beta j}f(x) + \mu_\alpha\sum_{j=0}^{\infty}(-\mu_\alpha)^j\,{}_{a,w}^{RL}I^{\beta j+\beta}f(x) \\
&= \sum_{j=0}^{\infty}(-\mu_\alpha)^j\,{}_{a,w}^{RL}I^{\beta j}f(x) - \sum_{j=0}^{\infty}(-\mu_\alpha)^{j+1}\,{}_{a,w}^{RL}I^{\beta j+\beta}f(x) \\
&= f(x)
\end{aligned}$$

and equality (12) holds true. The proof of equality (13) is similar. □

4. Integration by Parts

Our formulas of integration by parts are proved in suitable function spaces.

Definition 4 (See [19]). *For $\alpha > 0$, $\beta > 0$ and $1 \le p \le \infty$, the following function spaces are defined:*

$${}_{a,w}I^{\alpha,\beta}(L_p) := \left\{f : f = {}_{a,w}I^{\alpha,\beta}(\eta), \eta \in L_p(a,b)\right\}$$

and

$$I_{b,w}^{\alpha,\beta}(L_p) := \left\{f : f = I_{b,w}^{\alpha,\beta}(\theta), \theta \in L_p(a,b)\right\}.$$

Theorem 3 (integration by parts without the weighted function). *Let $0 \le \alpha < 1$, $\beta > 0$, $p \ge 1$, $q \ge 1$ and $\frac{1}{p} + \frac{1}{q} \le 1 + \alpha$ ($p \ne 1$ and $q \ne 1$ in the case $\frac{1}{p} + \frac{1}{q} = 1 + \alpha$).*

- *If $f \in L_p(a,b)$ and $g \in L_q(a,b)$, then*

$$\int_a^b f(x)({}_{a,1}I^{\alpha,\beta}g)(x)dx = \int_a^b g(x)(I_{b,1}^{\alpha,\beta}f)(x)dx. \tag{14}$$

- *If* $f \in I^{\alpha,\beta}_{b,w}(L_p)$ *and* $g \in {}_{a,w}I^{\alpha,\beta}(L_q)$, *then*

$$\int_a^b f(x)({}^{R}_{a,1}D^{\alpha,\beta}g)(x)dx = \int_a^b g(x)({}^{R}D^{\alpha,\beta}_{b,1}f)(x)dx. \tag{15}$$

Proof. First, we prove equality (14). Since,

$$\begin{aligned}\int_a^b f(x)({}_{a,1}I^{\alpha,\beta}g)(x)dx &= \int_a^b f(x)\Big[\phi(\alpha)g(x) + \psi(\alpha)\,{}^{RL}_{a,1}I^{\beta}g(x)\Big] \\ &= \phi(\alpha)\int_a^b f(x)g(x)dx + \psi(\alpha)\int_a^b f(x)\,{}^{RL}_{a,1}I^{\beta}g(x)dx,\end{aligned}$$

it follows from Lemma 1 that

$$\begin{aligned}\int_a^b f(x)({}_{a,1}I^{\alpha,\beta}g)(x)dx &= \phi(\alpha)\int_a^b f(x)g(x)dx + \psi(\alpha)\int_a^b g(x)\,{}^{RL}I^{\beta}_{b,1}f(x)dx \\ &= \int_a^b g(x)\Big[\phi(\alpha)f(x) + \psi(\alpha)\,{}^{RL}I^{\beta}_{b,1}f(x)\Big] \\ &= \int_a^b g(x)(I^{\alpha,\beta}_{b,1}f)(x)dx.\end{aligned}$$

Now, we prove (15):

$$\begin{aligned}\int_a^b f(x)({}^{R}_{a,1}D^{\alpha,\beta}g)(x)dx &= \int_a^b I^{\alpha,\beta}_{b,1}\theta(x)\Big({}^{R}_{a,1}D^{\alpha,\beta}({}_{a,1}I^{\alpha,\beta}\eta)\Big)(x)dx \\ &= \int_a^b \eta(x)I^{\alpha,\beta}_{b,1}\theta(x)dx \ \text{ (from Theorem 2)} \\ &= \int_a^b \theta(x)\,{}_{a,1}I^{\alpha,\beta}\eta(x)dx \ \text{ (from equality (14))} \\ &= \int_a^b g(x)({}^{R}D^{\alpha,\beta}_{b,1}f)(x)dx \ \text{ (from Theorem 2)}.\end{aligned}$$

The proof is complete. □

Theorem 4 (weighted generalized integration by parts). *Let* $0 \le \alpha < 1$, $\beta > 0$, $p \ge 1$, $q \ge 1$ *and* $\frac{1}{p} + \frac{1}{q} \le 1 + \alpha$ ($p \ne 1$ *and* $q \ne 1$ *in the case* $\frac{1}{p} + \frac{1}{q} = 1 + \alpha$). *If* $f \in L_p(a,b)$ *and* $g \in L_q(a,b)$, *then*

$$\int_a^b f(x)({}_{a,w}I^{\alpha,\beta}g)(x)dx = \int_a^b w(x)^2 g(x)\left(I^{\alpha,\beta}_{b,w}\left(\frac{f}{w^2}\right)\right)(x)dx, \tag{16}$$

$$\int_a^b f(x)({}^{R}_{a,w}D^{\alpha,\beta}g)(x)dx = \int_a^b w(x)^2 g(x)\left({}^{R}D^{\alpha,\beta}_{b,w}\left(\frac{f}{w^2}\right)\right)(x)dx. \tag{17}$$

Proof. We have

$$\begin{aligned}\int_a^b f(x)({}_{a,w}I^{\alpha,\beta}g)(x)dx &= \int_a^b w(x)\frac{f(x)}{w(x)}\left({}_{a,w}I^{\alpha,\beta}\left(\frac{gw}{w}\right)\right)(x)dx \\ &= \int_a^b \frac{f(x)}{w(x)}\left({}_{a,1}I^{\alpha,\beta}(gw)\right)(x)dx \\ &= \int_a^b w(x)g(x)\left(I^{\alpha,\beta}_{b,1}\left(\frac{f}{w}\right)\right)(x)dx \ \text{(from Theorem 3)} \\ &= \int_a^b g(x)w(x)^2\left(I^{\alpha,\beta}_{b,w}\left(\frac{f}{w^2}\right)\right)(x)dx.\end{aligned}$$

Therefore, equality (16) is true. Similarly, we have

$$\begin{aligned}\int_a^b f(x)({}^R_{a,w}D^{\alpha,\beta}g)(x)dx &= \int_a^b w(x)\frac{f(x)}{w(x)}\left({}^R_{a,w}D^{\alpha,\beta}\left(\frac{gw}{w}\right)\right)(x)dx\\ &= \int_a^b \frac{f(x)}{w(x)}\left({}^R_{a,1}D^{\alpha,\beta}(gw)\right)(x)dx\\ &= \int_a^b w(x)g(x)\left(D^{\alpha,\beta}_{b,1}\left(\frac{f}{w}\right)\right)(x)dx \text{ (from Theorem 3)}\\ &= \int_a^b g(x)w(x)^2\left(D^{\alpha,\beta}_{b,w}\left(\frac{f}{w^2}\right)\right)(x)dx\end{aligned}$$

and equality (17) holds. □

Remark 1. *When $w(t) = 1$ and $\alpha = \beta$, then we can obtain from our Theorem 4 the integration by parts formula [17] associated with Atangana–Baleanu derivatives:*

$$\begin{aligned}\int_a^b f(x)\left({}^{AB}_{a}I^{\alpha}g\right)(x)dx &= \int_a^b g(x)\left({}^{AB}I^{\alpha}_{b}f\right)(x)dx,\\ \int_a^b f(x)\left({}^{ABR}_{a}D^{\alpha}g\right)(x)dx &= \int_a^b g(x)\left({}^{ABR}D^{\alpha}_{b}f\right)(x)dx.\end{aligned}$$

From (16) and (17), we obtain the following consequence.

Corollary 1. *Let $0 \le \alpha < 1$, $\beta > 0$, $p \ge 1$, $q \ge 1$ and $\frac{1}{p} + \frac{1}{q} \le 1 + \alpha$ ($p \neq 1$ and $q \neq 1$ in the case $\frac{1}{p} + \frac{1}{q} = 1 + \alpha$). If $f \in L_p(a,b)$ and $g \in L_q(a,b)$, then*

$$\begin{aligned}\int_a^b f(x)\left(I^{\alpha,\beta}_{b,w}g\right)(x)dx &= \int_a^b w(x)^2 g(x)\left({}_{a,w}I^{\alpha,\beta}\left(\frac{f}{w^2}\right)\right)(x)dx,\\ \int_a^b f(x)\left({}^{R}D^{\alpha,\beta}_{b,w}g\right)(x)dx &= \int_a^b w(x)^2 g(x)\left({}^R_{a,w}D^{\alpha,\beta}\left(\frac{f}{w^2}\right)\right)(x)dx.\end{aligned}$$

For a symmetric view of weighted generalized integration by parts, we propose the following corollary of Theorem 4.

Corollary 2. *Let $0 \le \alpha < 1$, $\beta > 0$, $p \ge 1$, $q \ge 1$ and $\frac{1}{p} + \frac{1}{q} \le 1 + \alpha$ ($p \neq 1$ and $q \neq 1$ in the case $\frac{1}{p} + \frac{1}{q} = 1 + \alpha$). If $f \in L_p(a,b)$ and $g \in L_q(a,b)$, then*

$$\int_a^b w(x)f(x)\left({}_{a,w}I^{\alpha,\beta}\frac{g}{w}\right)(x)dx = \int_a^b w(x)g(x)\left(I^{\alpha,\beta}_{b,w}\frac{f}{w}\right)(x)dx, \tag{18}$$

$$\int_a^b w(x)f(x)\left({}^R_{a,w}D^{\alpha,\beta}\frac{g}{w}\right)(x)dx = \int_a^b w(x)g(x)\left({}^{R}D^{\alpha,\beta}_{b,w}\frac{f}{w}\right)(x)dx. \tag{19}$$

5. The Weighted Generalized Fractional Euler–Lagrange Equation

Let us denote by $AC(I \to \mathbb{R})$ the set of absolutely continuous functions X, where $I = [a,b]$, such that the left and right Riemann–Liouville-weighted generalized fractional derivatives of X exist, endowed with the norm

$$\|X\| = \sup_{t\in I}\left(|\, X(t) \,| + |\, {}^{RL}_{a,w}D^{\alpha,\beta}X(t) \,| + |{}^{RL}\, D^{\alpha,\beta}_{b,w}X(t) \,|\right).$$

Let $L \in C^1(I \times \mathbb{R} \times \mathbb{R} \times \mathbb{R} \to \mathbb{R})$ and consider the following minimization problem:

$$J[X] = \left(\int_a^b L\left(t, X(t), {}^{RL}D_{a,w}^{\alpha,\beta}X(t), {}^{RL}D_{b,w}^{\alpha,\beta}X(t)\right)dt \right) \longrightarrow \min \tag{20}$$

subject to the boundary conditions

$$X(a) = X_a, \quad X(b) = X_b. \tag{21}$$

Under appropriate general conditions, one can prove that the minimum of $J[\cdot]$ exists [20]. Here, we are interested in showing the usefulness of our Theorem 4 to prove the necessary optimality conditions for problem (20) and (21). With the help of weighted generalized fractional integration by parts, we obtain the following Euler–Lagrange necessary optimality condition for the fundamental weighted generalized fractional problem of the calculus of variations (20) and (21).

Theorem 5 (the weighted generalized fractional Euler–Lagrange equation)**.** *If $L \in C^1(I \times \mathbb{R} \times \mathbb{R} \times \mathbb{R} \to \mathbb{R})$ and $X \in AC([a,b] \to \mathbb{R})$ is a minimizer of (20) subject to the fixed end points (21); then, X satisfies the following weighted generalized fractional Euler–Lagrange equation:*

$$\partial_2 L + w(t)^2\, {}^{R}D_{b,w}^{\alpha,\beta}\left(\frac{\partial_3 L}{w(t)^2}\right) + w(t)^2\, {}_{a,w}^{R}D^{\alpha,\beta}\left(\frac{\partial_4 L}{w(t)^2}\right) = 0,$$

where $\partial_i L$ denotes the partial derivative of the Lagrangian L with respect to its ith argument evaluated at $\left(t, X(t), {}^{RL}D_{a,w}^{\alpha,\beta}X(t), {}^{RL}D_{b,w}^{\alpha,\beta}X(t)\right)$.

Proof. Let $J[X] = \int_a^b L\left(t, X(t), {}_{a,w}^{R}D^{\alpha,\beta}X(t), {}^{R}D_{b,w}^{\alpha,\beta}X(t)\right)dt$ and assume that X^* is the optimal solution of problem (20) and (21). Set

$$X = X^* + \varepsilon\eta,$$

where $\eta, X \in AC([a,b] \to \mathbb{R})$ and ε is a small, real parameter. By linearity of the weighted generalized fractional derivative, we obtain

$${}_{a,w}^{R}D^{\alpha,\beta}X(t) = {}_{a,w}^{R}D^{\alpha,\beta}X^* + \varepsilon\left({}_{a,w}^{R}D^{\alpha,\beta}\eta(t)\right)$$

and

$${}^{R}D_{b,w}^{\alpha,\beta}X(t) = {}^{R}D_{b,w}^{\alpha,\beta}X^* + \varepsilon\left({}^{R}D_{b,w}^{\alpha,\beta}\eta(t)\right).$$

Now, consider the following function:

$$J(\varepsilon) = \int_a^b L\Big(t, X^*(t) + \varepsilon\eta(t),\ {}_{a,w}^{R}D^{\alpha,\beta}X^*(t) + \varepsilon\left({}_{a,w}^{R}D^{\alpha,\beta}\eta(t)\right), \\ {}^{R}D_{b,w}^{\alpha,\beta}X^*(t) + \varepsilon\left({}^{R}D_{b,w}^{\alpha,\beta}\eta(t)\right)\Big)dt.$$

Fermat's theorem asserts that $\left.\frac{d}{d\varepsilon}J(\varepsilon)\right|_{\varepsilon=0} = 0$ and we deduce, by the chain rule, that

$$\int_a^b \left(\partial_2 L \cdot \eta + \partial_3 L \cdot {}_{a,w}^{R}D^{\alpha,\beta}\eta + \partial_4 L \cdot {}^{R}D_{b,w}^{\alpha,\beta}\eta\right)dt = 0.$$

Using Theorem 4 of weighted fractional integration by parts, we obtain that

$$\int_a^b \left(\partial_2 L \cdot \eta + w(t)^2 \cdot \eta \cdot {}^{R}D_{b,w}^{\alpha,\beta}\left(\frac{\partial_3 L}{w(t)^2}\right) + w(t)^2 \cdot \eta \cdot {}_{a,w}^{R}D^{\alpha,\beta}\left(\frac{\partial_4 L}{w(t)^2}\right)\right)dt = 0.$$

The result follows by the fundamental theorem of the calculus of variations. □

6. An Application

Let us consider the weighted generalized fractional variational problem (20) and (21) with

$$L\left(t, X(t), {}^{R}_{a,w}D^{\alpha,\beta}X(t), {}^{R}D^{\alpha,\beta}_{b,w}X(t)\right) = \frac{1}{2}\left(\frac{1}{2}m \left|{}^{R}_{a,w}D^{\alpha,\beta}X(t)\right|^2 + \frac{1}{2}m \left|{}^{R}D^{\alpha,\beta}_{b,w}X(t)\right|^2\right) - V(X(t)),$$

where X is an absolutely continuous function on $[a,b]$ and V maps $C^1(I \to \mathbb{R})$ to $\mathbb{R}$. Note that

$$\frac{1}{2}\left(\frac{1}{2}m \left|{}^{R}_{a,w}D^{\alpha,\beta}X(t)\right|^2 + \frac{1}{2}m \left|{}^{R}D^{\alpha,\beta}_{b,w}X(t)\right|^2\right)$$

can be viewed as a weighted generalized kinetic energy in the quantum mechanics framework. By applying our Theorem 5 to the current variational problem, we obtain that

$$\frac{1}{2}m\left[w(t)^2\, {}^{R}D^{\alpha,\beta}_{b,w}\left(\frac{{}^{R}_{a,w}D^{\alpha,\beta}X(t)}{w(t)^2}\right) + w(t)^2\, {}^{R}_{a,w}D^{\alpha,\beta}\left(\frac{{}^{R}D^{\alpha,\beta}_{b,w}X(t)}{w(t)^2}\right)\right] = V'(X(t)), \quad (22)$$

where V' is the derivative of the potential energy of the system. We observe that relation (22) generalizes Newton's dynamical law $m\ddot{X}(t) = V'(X(t))$.

7. Conclusions

In this work, some definitions and properties of a recent class of fractional operators defined by general integral operators, with and without singular kernels, are recalled. A new definition of a right-weighted generalized fractional operator in the Riemann–Liouville sense is then posed, serving as a prerequisite for the establishment of a new weighted generalized integration by parts formula, which shows a duality relation with the existing left weighted generalized fractional operator in the Riemann–Liouville–Hattaf sense [11]. In the context of the fractional calculus of variations, we have investigated weighted generalized Euler–Lagrange equations, which were then used to produce an effective application in the quantum mechanics setting, after a proper definition of kinetic energy.

Author Contributions: Conceptualization, H.Z., E.M.L., D.F.M.T. and N.Y.; validation, H.Z., E.M.L., D.F.M.T. and N.Y.; formal analysis, H.Z., E.M.L., D.F.M.T. and N.Y.; investigation, H.Z., E.M.L., D.F.M.T. and N.Y.; writing—original draft preparation, H.Z., E.M.L., D.F.M.T. and N.Y.; writing—review and editing, H.Z., E.M.L., D.F.M.T. and N.Y.; supervision, D.F.M.T. All authors have read and agreed to the published version of the manuscript.

Funding: This research was funded by Fundação para a Ciência e a Tecnologia (FCT) grant number UIDB/04106/2020 (CIDMA).

Data Availability Statement: Not applicable.

Acknowledgments: The authors would like to express their gratitude to two anonymous reviewers, for their constructive comments and suggestions, which helped them to enrich the paper.

Conflicts of Interest: The authors declare no conflict of interest. The funders had no role in the design of the study; in the collection, analyses, or interpretation of data; in the writing of the manuscript, or in the decision to publish the results.

References

1. Anastassiou, G.A. *Generalized Fractional Calculus—New Advancements and Applications*; Studies in Systems, Decision and Control; Springer: Cham, Switzerland, 2021; p. 305.
2. Mahrouf, M.; Boukhouima, A.; Zine, H.; Lotfi, E.M.; Torres, D.F.M.; Yousfi, N. Modeling and forecasting of COVID-19 spreading by delayed stochastic differential equations. *Axioms* **2021**, *10*, 18. [CrossRef]

3. Jiagui, P.; Qing, C. *Differential Geometry*; Higher Education Press: Beijing, China, 1983.
4. Shengshen, C.; Weihuan, C. *Lecture Notes on Differential Geometry*; Peking University Press: Beijing, China, 1983.
5. LaSalle, J.P. *The Stability of Dynamical Systems*; Regional Conference Series in Applied Mathematics; Society for Industrial and Applied Mathematics: Philadelphia, PA, USA, 1976.
6. Lyapunov, A.M. The general problem of the stability of motion. *Int. J. Control* **1992**, *55*, 521–790. [CrossRef]
7. Balint, A.M.; Balint, S.; Neculae, A. On the objectivity of mathematical description of ion transport processes using general temporal Caputo and Riemann-Liouville fractional partial derivatives. *Chaos Solitons Fractals* **2022**, *156*, 111802. [CrossRef]
8. Malinowska, A.B.; Torres, D.F.M. *Introduction to the Fractional Calculus of Variations*; Imperial College Press: London, UK, 2012.
9. Ghanim, F.; Bendak, S.; Hawarneh, A.A. Certain implementations in fractional calculus operators involving Mittag-Leffler-confluent hypergeometric functions. *Proc. A* **2022**, *478*, 839. [CrossRef]
10. Set, E.; Choi, J.; Demİrbaş, S. Some new Chebyshev type inequalities for fractional integral operator containing a further extension of Mittag-Leffler function in the kernel. *Afr. Mat.* **2022**, *33*, 42. [CrossRef]
11. Hattaf, K. A new generalized definition of fractional derivative with non-singular kernel. *Computation* **2020**, *8*, 49. [CrossRef]
12. Hattaf, K. On some properties of the new generalized fractional derivative with non-singular kernel. *Math. Probl. Eng.* **2021**, *2021*, 1580396. [CrossRef]
13. Riewe, F. Mechanics with fractional derivatives. *Phys. Rev. E* **1997**, *55*, 3581–3592. [CrossRef]
14. Almeida, R.; Malinowska, A.B.; Torres, D.F.M. Fractional Euler-Lagrange differential equations via Caputo derivatives. In *Fractional Dynamics and Control*; Springer: New York, NY, USA, 2012; pp. 109–118.
15. Abdeljawad, T.; Atangana, A.; Gómez-Aguilar, J.F.; Jarad, F. On a more general fractional integration by parts formulae and applications. *Phys. A* **2019**, *536*, 122494. [CrossRef]
16. Zine, H.; Torres, D.F.M. A stochastic fractional calculus with applications to variational principles. *Fractal Fract.* **2020**, *4*, 38. [CrossRef]
17. Abdeljawad, T.; Baleanu, D. Integration by parts and its applications of a new nonlocal fractional derivative with Mittag-Leffler nonsingular kernel. *J. Nonlinear Sci. Appl.* **2017**, *10*, 1098–1107. [CrossRef]
18. Samko, S.G.; Kilbas, A.A.; Marichev, O.I. *Fractional Integrals and Derivatives*; Gordon and Breach Science Publishers: Yverdon, Switzerland, 1993.
19. Kilbas, A.A.; Srivastava, H.M.; Trujillo, J.J. *Theory and Applications of Fractional Differential Equations*; North-Holland Mathematics Studies; Elsevier Science B.V.: Amsterdam, The Netherlands, 2006; p. 204.
20. Bourdin, L.; Odzijewicz, T.; Torres, D.F.M. Existence of minimizers for generalized Lagrangian functionals and a necessary optimality condition—Application to fractional variational problems. *Differ. Integral Equ.* **2014**, *27*, 743–766.

Article

Modeling the Impact of the Imperfect Vaccination of the COVID-19 with Optimal Containment Strategy

Lahbib Benahmadi [1], Mustapha Lhous [1], Abdessamad Tridane [2,*], Omar Zakary [3] and Mostafa Rachik [3]

1 Fundamental and Applied Mathematics Laboratory, Department of Mathematics and Computer Science, Faculty of Sciences Ain Chock, Hassan II University of Casablanca, Maarif B.P 5366, Casablanca 20100, Morocco; 900210781@uaeu.ac.ae (L.B.); mlhous17@gmail.com (M.L.)
2 Department of Mathematical Sciences, United Arab Emirates University, Al Ain P.O. Box 15551, United Arab Emirates
3 Laboratory of Analysis Modelling and Simulation, Department of Mathematics and Computer Science, Faculty of Sciences Ben M'sik, Hassan II University of Casablanca, Sidi Othman Casablanca B.P 7955, Casablanca 20100, Morocco; zakaryma@gmail.com (O.Z.); m_rachik@yahoo.fr (M.R.)
* Correspondence: a-tridane@uaeu.ac.ae

Abstract: Since the beginning of the COVID-19 pandemic, vaccination has been the main strategy to contain the spread of the coronavirus. However, with the administration of many types of vaccines and the constant mutation of viruses, the issue of how effective these vaccines are in protecting the population is raised. This work aimed to present a mathematical model that investigates the imperfect vaccine and finds the additional measures needed to help reduce the burden of disease. We determine the $\mathcal{R}_0$ threshold of disease spread and use stability analysis to determine the condition that will result in disease eradication. We also fitted our model to COVID-19 data from Morocco to estimate the parameters of the model. The sensitivity analysis of the basic reproduction number, with respect to the parameters of the model, is simulated for the four possible scenarios of the disease progress. Finally, we investigate the optimal containment measures that could be implemented with vaccination. To illustrate our results, we perform the numerical simulations of optimal control.

Keywords: COVID-19; vaccination; basic reproduction number; stability; Lyapunov function; optimal control

Citation: Benahmadi, L.; Lhous, M.; Tridane, A.; Zakary, O.; Rachik, M. Modeling the Impact of the Imperfect Vaccination of the COVID-19 with Optimal Containment Strategy. *Axioms* **2022**, *11*, 124. https://doi.org/10.3390/axioms11030124

Academic Editors: Natalia Martins, Cristiana João Soares da Silva, Moulay Rchid Sidi Ammi and Ricardo Almeida

Received: 16 January 2022
Accepted: 25 February 2022
Published: 10 March 2022

1. Introduction

Since the beginning of the ongoing COVID-19 pandemic, the world has been racing to develop a vaccination that helps protect the populations around the world and bring human life to a normal status. This race to find a vaccine was not only challenged by the fast spread of the disease, but also the high rate of mutation of COVID-19. As a result, we witness many vaccination types with different biotechnological approaches and different efficacy [1]. These efficacies are based on clinical trials that might have some limitations as their samples do not necessarily cover a wide population from different parts of the world. These facts make the question of the efficacy of vaccines legitimate and need to be investigated.

This problem was investigated using mathematical modeling to study the possible measure that needed to be implemented to reduce the impact of an imperfect vaccine.

The mathematical model of the imperfect vaccination of infectious disease was started by the work of Arino et al. [2]. Many papers have followed up on this work, looking into the various spreads of imperfect vaccination for various diseases. For example, the work of Abu-Raddad [3] studied the mathematical model of a possible HIV vaccination where the authors investigated the impact of the defectiveness of vaccination on the progress of the disease. Liu et al. [4] studied a general SIR with an added vaccination compartment and imperfect vaccination. This study showed how vaccination reduced the infected population but could not eradicate the infection. In fact, the eradication required an additional necessary condition. If vaccination efficacy improves, this condition may be alleviated. A mathematical model with the imperfect vaccination of birds in the case of avian influenza was studied in [5]. This model considered age-since-vaccination structure and symptomatically infected birds. This study showed that the only way eradicate the disease was by the full coverage of the bird population or by full efficacy. A time delay model of imperfect vaccination was studied in [6] with a possible loss of immunity. The study showed the existence of the critical vaccination coverage needed to eliminate the infection. In the case of imperfect vaccination, the authors showed that a critical proportion of the population needed vaccination. Another delay model with distributed delay [7] and the delay model with a generalized incidence function were studied in [8].

Regarding the ongoing pandemic, there are some studies that have investigated imperfect vaccination in the USA ([9,10]) and the UK [11] but without finding optimal measures that could help contain the pandemic, as the use of an imperfect vaccine cannot achieve the low endemicity of COVID-19. On the other hand, many studies (see [12–22]) used optimal control to find the optimal way to allocate vaccination and the best strategy to vaccinate the population, depending on the age or comorbidity of the population. The goal of this paper was to investigate a mathematical model of the imperfect vaccination of COVID-19. The aim was to study the dynamics of this model and present the possible control measures that need to be implemented in order to reduce the impact of the vaccine's imperfection.

To our knowledge, our work is the only one to date to have studied the potential dynamics of imperfect vaccination and the optimal use of other public health measures that help to reduce the effect of administering imperfect vaccination. The only work that combined these two problems was used in the case of possible malaria vaccination [23].

The structure of this paper is summarized as follows. In Section 2, the mathematical model is formed and the existence conditions of the system are verified. Section 3 takes into account the basic reproduction number. Sections 4 and 5 analyze the local and global stability at the disease-free equilibrium point, respectively. The optimal control problem of vaccination and additional measures to reduce the disease spread are presented in Section 6. In Section 7, we fit our model to data from Morocco to estimate the parameters of the model. We also discussed four possible scenarios of the dynamic of the model via the elasticity of the basic reproduction number and we give the illustration of the optimal solution via numerical simulations. The conclusion and discussion of the results are given in Section 8.

2. The Mathematical Model

Nine epidemiological compartments were recognized in the population: susceptible (S); vaccinated (V); exposed (E) (asymptomatic); infected with mild symptoms (I); quarantined in a home with mild symptoms (Q); hospitalized (H); quarantined in a hospital with complications (C) (i.e., isolated in a hospital with breathing assistance); and mortality due to disease (D):

$$
\begin{aligned}
\frac{dS}{dt} &= -\beta_1 \tfrac{SE}{N} - \beta_2 \tfrac{SI}{N} - \lambda_1 S \\
\frac{dV}{dt} &= \lambda_1 S - \beta_3 \tfrac{VE}{N} - \beta_4 \tfrac{VI}{N} - \lambda_2 V \\
\frac{dE}{dt} &= \beta_1 \tfrac{SE}{N} + \beta_2 \tfrac{SI}{N} + \beta_3 \tfrac{VE}{N} + \beta_4 \tfrac{VI}{N} - \theta E \\
\frac{dI}{dt} &= \theta E - (\gamma_1 + \gamma_2 + \gamma_3) I \\
\frac{dQ}{dt} &= \gamma_1 I - (\sigma_1 + \delta_1) Q \\
\frac{dH}{dt} &= \gamma_2 I + \sigma_1 Q - (\sigma_2 + \delta_2) H \\
\frac{dC}{dt} &= \gamma_3 I + \sigma_2 H - (\mu + \delta_3) C \\
\frac{dR}{dt} &= \delta_1 Q + \delta_2 H + \delta_3 C + \lambda_2 V \\
\frac{dD}{dt} &= \mu C
\end{aligned}
\tag{1}
$$

In this model, we assume that vaccination does not provide complete protection against COVID-19. As a result, the rate of being vaccinated is λ_1, and λ_2 is the rate of vaccinated persons who have recovered and developed immunity. We suppose that each infectious sub-population (E and I) infected the healthy population at varying densities, where β_i is the infection density per capita with $i = 1,2$. In reality, our major assumption about vaccination's imperfection is that some vaccinated persons may only receive partial protection and may become sick if they are exposed to multiple infections. The imperfection can be due to the mutation of the virus. In fact, when people are vaccinated, they tend to relax their guard and take fewer protection measures again the virus. θ is the proportion of infected individuals. Some people show mild symptoms with the per capita rate γ_1 and can stay at home with treatment; whilst others develop hard symptoms and must be monitored in the hospital with the per capita rate γ_2 and another critical condition that requires penetrating breathing with per capita rate γ_3. The parameter δ_i, with $i = 1,2,3$, represents the recovery rates of quarantined, hospitalized and critical infected persons (respectively). Average quarantine and hospitalization times are $1/\sigma_1$ and $1/\sigma_2$, respectively. Finally, μ represents the mortality rate due to disease. The flow chart of the model is given in the Figure 1. When the nine equations in (1) are combined together, the total population size N remains constant.

2.1. Positivity and Boundedness

This part is dedicated to establishing the positivity and boundedness of the model (1) solutions under non-negative conditions.

Theorem 1. *If $S(0) \geq 0$, $V(0) \geq 0$, $E(0) \geq 0$, $I(0) \geq 0$ $Q(0) \geq 0$, $H(0) \geq 0$, $C(0) \geq 0$, $R(0) \geq 0$ and $D(0) \geq 0$, then the solution $S(t)$, $V(t)$, $E(t)$, $I(t)$ $Q(t)$, $H(t)$, $C(t)$, $R(t)$, $D(t)$ of system (1) is non-negative and the solutions exist in Ω for all $t \geq 0$.*

The proof of positivity follows the standard argument, as can be seen, for example, in [24].

The solution of the model (1) exists in the positively invariant region:

$$
\begin{aligned}
\Omega &= \{(S(t), V(t), E(t), I(t), Q(t), H(t), C(t), R(t), D(t)) \in R^9_+ \\
&: S(t) + V(t) + E(t) + I(t) + Q(t) + C(t) + R(t) + D(t) = N \leq N(0)\}.
\end{aligned}
$$

Since all the parameters and sub populations of the system are non-negative and $N(t)$ = constant $\forall t \geq 0$.

Straightforwardly, we obtain $S(t) \leq N(t) \leq N(0)$ for all $t \geq 0$. The other variables yielded the same result. As a result, the overall population is finitely upper bounded. For system (1), the region Ω is positively invariant.

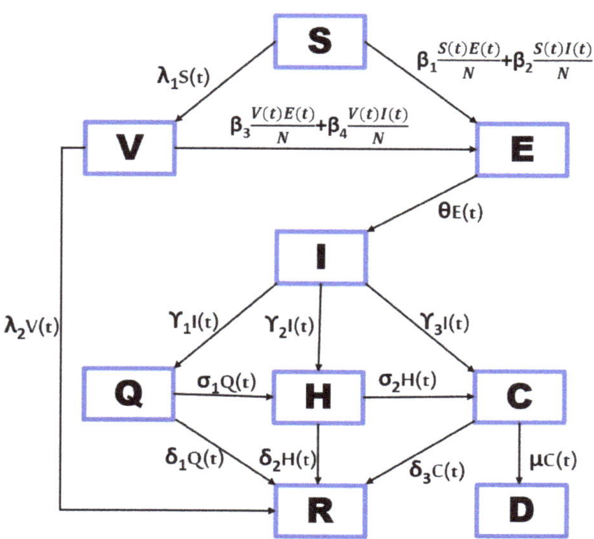

Figure 1. Flow chart of the model.

2.2. Existence and Uniqueness of Solutions

Theorem 2. *The system (1) that fulfills a given initial condition* $(S(0), V(0), E(0), I(0)\ Q(0), H(0), C(0), R(0), D(0))$ *has a unique solution.*

Proof. The system (1) may be expressed as follows:

$$\dot{X} = AX + B(X) \tag{2}$$

where $X(t) = [S(t),\ V(t),\ E(t),\ I(t),\ Q(t),\ H(t),\ C(t),\ R(t),\ D(t)]^\top$

$$A = \begin{bmatrix} -\lambda_1 & 0 & 0 & 0 & 0 & 0 & 0 & 0 & 0 \\ \lambda_1 & -\lambda_2 & 0 & 0 & 0 & 0 & 0 & 0 & 0 \\ 0 & 0 & -\theta & 0 & 0 & 0 & 0 & 0 & 0 \\ 0 & 0 & \theta & -\gamma_1-\gamma_2-\gamma_3 & 0 & 0 & 0 & 0 & 0 \\ 0 & 0 & 0 & \gamma_1 & -\sigma_1-\delta_1 & 0 & 0 & 0 & 0 \\ 0 & 0 & 0 & \gamma_2 & \sigma_1 & -\sigma_2-\delta_2 & 0 & 0 & 0 \\ 0 & 0 & 0 & \gamma_3 & 0 & \sigma_2 & -\mu-\delta_3 & 0 & 0 \\ 0 & \lambda_2 & 0 & 0 & \delta_1 & \delta_2 & \delta_3 & 0 & 0 \\ 0 & 0 & 0 & 0 & 0 & 0 & \mu & 0 & 0 \end{bmatrix}, \tag{3}$$

$$B(X) = \begin{bmatrix} -\beta_1 \frac{SE}{N} - \beta_2 \frac{SI}{N} \\ -\beta_3 \frac{VE}{N} - \beta_4 \frac{VI}{N} \\ \beta_1 \frac{SE}{N} + \beta_2 \frac{SI}{N} + \beta_3 \frac{VE}{N} + \beta_4 \frac{VI}{N} \\ 0 \\ 0 \\ 0 \\ 0 \\ 0 \\ 0 \end{bmatrix}.$$

Equation (2) is a non-linear system that can be written as

$$\Sigma(X) = AX + B(X).$$

We have:

$$\begin{aligned} |B(X_1) - B(X_2)| &\leq \tfrac{2\beta_1}{N}|S_2E_2 - S_1E_1| + \tfrac{2\beta_2}{N}|S_2I_2 - S_1I_1| + \tfrac{2\beta_3}{N}|V_2E_2 - V_1E_1| + \tfrac{2\beta_4}{N}|V_2I_2 - V_1I_1| \\ &\leq \tfrac{2\beta_1}{N}|S_1(E_2 - E_1) + E_2(S_2 - S_1)| + \tfrac{2\beta_2}{N}|S_1(I_2 - I_1) + I_2(S_2 - S_1)| + \\ &\quad \tfrac{2\beta_3}{N}|V_1(E_2 - E_1) + E_2(V_2 - V_1)| + \tfrac{2\beta_4}{N}|V_1(I_2 - I_1) + I_2(V_2 - V_1)| \\ &\leq 2\beta_1(|E_2 - E_1| + |S_2 - S_1|) + 2\beta_2(|I_2 - I_1| + |S_2 - S_1|) + \\ &\quad 2\beta_3(|E_2 - E_1| + |V_2 - V_1|) + 2\beta_4(|I_2 - I_1| + |V_2 - V_1|) \\ &\leq 2(\beta_1 + \beta_3)|E_2 - E_1| + 2(\beta_1 + \beta_2)|S_2 - S_1| + 2(\beta_2 + \beta_4)|I_2 - I_1| + 2(\beta_3 + \beta_4)|V_2 - V_1| \\ &\leq 4M(|S_2 - S_1| + |V_2 - V_1| + |E_2 - E_1| + |I_2 - I_1|), \text{ where } M = \max\{\beta_1, \beta_2, \beta_3, \beta_4\} \\ &\leq 4M\|X_2 - X_1\| \end{aligned}$$

then, we obtain $|\Sigma(X_1) - \Sigma(X_2)| \leq \overline{M}\|X_1 - X_2\|$, where $\overline{M} = \max\{M, \|A\|\} < \infty$. As a result, the function $\Sigma(t)$ is uniformly Lipschitz continuous. The restriction on $S(t) \geq 0$, $V(t) \geq 0$, $E(t) \geq 0$, $I(t) \geq 0$, $Q(t) \geq 0$, $H(t) \geq 0$, $C(t) \geq 0$, $R(t) \geq 0$, and $D(t) \geq 0$. Thus, a solution to the system (2) exists [25]. □

3. The Basic Reproduction Number

The basic reproduction number $\mathcal{R}_0$ is the average number of persons in a susceptible population that one person infected with COVID-19 is expected to infect, and it is calculated using the next generation matrix approach [26]. The disease compartments are thus E and I. Therefore, the all-time disease-free equilibrium point $\mathcal{E}_0 = (N_0, 0, 0, 0, 0, 0, 0)$.

$$\mathcal{F} = \begin{pmatrix} \beta_1 \frac{S^*E^*}{N} + \beta_2 \frac{S^*I^*}{N} + \beta_3 \frac{V^*E^*}{N} + \beta_4 \frac{V^*I^*}{N} \\ 0 \end{pmatrix}, \ \mathcal{V} = \mathcal{V}^- - \mathcal{V}^+ = \begin{pmatrix} \theta E \\ (\gamma_1 + \gamma_2 + \gamma_3)I^* - \theta E^* \end{pmatrix}$$

The Jacobian matrices of $\mathcal{F}$ and $\mathcal{V}$ computed at $\mathcal{E}_0$ are provided by F and V, respectively, such that:

$$F = \begin{pmatrix} \beta_1 & \beta_2 \\ 0 & 0 \end{pmatrix}, \ V = \begin{pmatrix} \theta & 0 \\ -\theta & (\gamma_1 + \gamma_2 + \gamma_3) \end{pmatrix}$$

The inverse of V is given by

$$V^{-1} = \begin{pmatrix} \frac{1}{\theta} & 0 \\ \frac{1}{(\gamma_1+\gamma_2+\gamma_3)} & \frac{1}{(\gamma_1+\gamma_2+\gamma_3)} \end{pmatrix} \text{ and } FV^{-1} = \begin{pmatrix} \frac{\beta_1}{\theta} + \frac{\beta_2}{\gamma_1+\gamma_2+\gamma_3} & \frac{\beta_2}{\gamma_1+\gamma_2+\gamma_3} \\ 0 & 0 \end{pmatrix}$$

Therefore, the domainant eigenvalue of FV^{-1}:

$$\mathcal{R}_0 = \rho(FV^{-1}) = \frac{\beta_1}{\theta} + \frac{\beta_2}{(\gamma_1 + \gamma_2 + \gamma_3)} \tag{4}$$

4. Local Stability Analysis at Disease-Free Equilibrium (DFE) $\mathcal{E}_0$

The DFE's local stability is investigated as follows.

The Jacobian matrix of the system (1) at $\mathcal{E}_0 = (N, 0, 0, 0, 0, 0, 0)$ is:

$$J^*(\mathcal{E}_0) = \begin{pmatrix} -\lambda_1 & 0 & -\beta_1 & -\beta_2 & 0 & 0 & 0 \\ \lambda_1 & -\lambda_2 & 0 & 0 & 0 & 0 & 0 \\ 0 & 0 & \beta_1-\theta & \beta_2 & 0 & 0 & 0 \\ 0 & 0 & \theta & -(\gamma_1+\gamma_2+\gamma_3) & 0 & 0 & 0 \\ 0 & 0 & 0 & \gamma_1 & -(\sigma_1+\delta_1) & 0 & 0 \\ 0 & 0 & 0 & \gamma_2 & \sigma_1 & -(\sigma_2+\delta_2) & 0 \\ 0 & 0 & 0 & \gamma_3 & 0 & \sigma_2 & -(\mu+\delta_3) \end{pmatrix}. \quad (5)$$

The eigenvalues of the Jacobian matrix $J^*(\mathcal{E}_0)$ are the roots of the following characteristic equation:

$$(-\lambda_1-\lambda)(-\lambda_2-\lambda)(-\sigma_1-\delta_1-\lambda)(-\sigma_2-\delta_2-\lambda)(-\mu-\delta_3-\lambda)(\lambda^2+a_1\lambda+a_0)=0 \quad (6)$$

where:

$$\begin{aligned} a_0 &= (\theta-\delta_1)(\gamma_1+\gamma_2+\gamma_3)-\theta\beta_2 \\ a_1 &= (\theta-\delta_1)+(\gamma_1+\gamma_2+\gamma_3) \end{aligned} \quad (7)$$

The roots of $(\lambda^2+a_1\lambda+a_0)=0$ are given by

$$\begin{aligned} \lambda_1 &= \frac{\beta_1-\theta-(\gamma_1+\gamma_2+\gamma_3)-\sqrt{(\beta_1-\theta+\gamma_1+\gamma_2+\gamma_3)^2+4\theta\beta_2}}{2} &= \frac{\alpha_1-\gamma-\sqrt{(\alpha_1+\gamma)^2+4\theta\beta_2}}{2} \\ \lambda_2 &= \frac{\beta_1-\theta-(\gamma_1+\gamma_2+\gamma_3)+\sqrt{(\beta_1-\theta+\gamma_1+\gamma_2+\gamma_3)^2+4\theta\beta_2}}{2} &= \frac{\alpha_1-\gamma+\sqrt{(\alpha_1+\gamma)^2+4\theta\beta_2}}{2} \end{aligned}$$

with $\alpha_1 = \beta_1 - \theta$ and $\gamma = (\gamma_1+\gamma_2+\gamma_3)$.

It is straightforward that if $\alpha_1 - \gamma < 0 \Rightarrow \mathcal{R}^* = \frac{\beta_1}{\theta+\gamma} < 1$ and $\lambda_1 < 0$.

In the case for λ_2, we write the equation:

$$1-\mathcal{R}_0 = \frac{-\alpha_1\gamma-\theta\beta_2}{\gamma\theta} \quad (8)$$

From the previous Equation (8), if $\mathcal{R}_0 < 1$, then:

$$\begin{aligned} & \alpha_1\gamma+\theta\beta_2 < 0 \\ \Rightarrow\ & 4\alpha_1\gamma+4\theta\beta_2 < 0 \\ \Rightarrow\ & 2\alpha_1\gamma+4\theta\beta_2 < -2\alpha_1\gamma \\ \Rightarrow\ & \alpha_1^2+2\alpha_1\gamma+4\theta\beta_2+\gamma^2 < \alpha_1^2-2\alpha_1\gamma+\gamma^2 \\ \Rightarrow\ & (\alpha_1+\gamma)^2+4\theta\beta_2 < (\gamma-\alpha_1)^2 \\ \Rightarrow\ & \sqrt{(\alpha_1+\gamma)^2+4\theta\beta_2} < \gamma-\alpha_1 \\ \Rightarrow\ & \alpha_1-\gamma+\sqrt{(\alpha_1+\gamma)^2+4\theta\beta_2} < 0 \\ \Rightarrow\ & \lambda_2 < 0 \end{aligned}$$

Using the same steps as in Equation (8), if $\mathcal{R}_0 > 1$, then:

$$\begin{aligned} & \alpha_1\gamma+\theta\beta_2 > 0 \\ \Rightarrow\ & 4\alpha_1\gamma+4\theta\beta_2 > 0 \\ \Rightarrow\ & 2\alpha_1\gamma+4\theta\beta_2 > -2\alpha_1\gamma \\ \Rightarrow\ & \alpha_1^2+2\alpha_1\gamma+4\theta\beta_2+\gamma^2 > \alpha_1^2-2\alpha_1\gamma+\gamma^2 \\ \Rightarrow\ & (\alpha_1+\gamma)^2+4\theta\beta_2 > (\gamma-\alpha_1)^2 \\ \Rightarrow\ & \sqrt{(\alpha_1+\gamma)^2+4\theta\beta_2} > \gamma-\alpha_1 \\ \Rightarrow\ & \alpha_1-\gamma+\sqrt{(\alpha_1+\gamma)^2+4\theta\beta_2} > 0 \\ \Rightarrow\ & \lambda_2 > 0 \end{aligned}$$

Notice that the fact that $\mathcal{R}_0 < 1$ implies that $\frac{\beta_1}{\theta} < 1$ and since $\mathcal{R}^* < \frac{\beta_1}{\theta}$. Then, we conclude that it is enough to have $\mathcal{R}_0 < 1$ to ensure that the roots of $(\lambda^2+a_1\lambda+a_0)=0$

are negative. On the other hand, if $\mathcal{R}^* > 1$, then $\mathcal{R}_0 > 1$. As a result, we have just proven the following theorem:

Theorem 3. *If $\mathcal{R}_0 < 1$, the disease-free equilibrium $\mathcal{E}_0$ of the system (1) is locally asymptotically stable, but unstable if $\mathcal{R}^* > 1$, where: $\mathcal{R}^* = \frac{\beta_1}{\theta+\gamma}$.*

5. Global Stability Analysis at Disease-Free Equilibrium

The global stability of the disease-free equilibrium point $\mathcal{E}_0$ was found in this part by creating the Lyapunov function as follows:

Theorem 4. *The disease-free equilibrium $\mathcal{E}_0$ of the model (1) is globally asymptotically stable whenever ${\mathcal{R}_0}^{Total} \leq 1$, where ${\mathcal{R}_0}^{Total} = \mathcal{R}_0 + {\mathcal{R}_0}^{v}$.*

Proof. Consider the Lyapunov function L in the trivial equilibrium point $\mathcal{E}_0$, which has non-negative coefficients g_1 and g_2:

$$L = g_1 E + g_2 I \tag{9}$$

Differentiating Equation (9) with respect to time t, and substituting both $\frac{dE}{dt}$ and $\frac{dI}{dt}$ from Equation (1) yields the result:

$$\begin{aligned} \dot{L} &= g_1\dot{E} + g_2\dot{I} \\ &= g_1\beta_1\tfrac{SE}{N} + g_1\beta_2\tfrac{SI}{N} + g_1\beta_3\tfrac{VE}{N} + g_1\beta_4\tfrac{VI}{N} - g_1\theta E \\ &\quad + g_2\theta E - g_2(\gamma_1+\gamma_2+\gamma_3)I \end{aligned} \tag{10}$$

By simplifying Equation (10) by collecting similar terms of E and I, then by solving for coefficient g_1 and g_2, this yields:

$$\begin{aligned} \dot{L} &\leq g_1\beta_1 E + g_1\beta_3 E + g_2\theta E - g_1\theta E + g_1\beta_2 I + g_1\beta_4 I - g_2(\gamma_1+\gamma_2+\gamma_3)I \\ &\leq g_1\theta\left(\frac{\beta_1+\beta_3}{\theta} + \frac{g_2}{g_1} - 1\right)E + g_1(\beta_2+\beta_4 - g_2(\gamma_1+\gamma_2+\gamma_3))I \\ &\leq g_1\left(\beta_2+\beta_4 - g_1(1-\frac{\beta_1+\beta_3}{\theta})(\gamma_1+\gamma_2+\gamma_3)\right)I \\ &\leq g_1^2\left(\frac{\beta_2+\beta_4}{g_1} - (\gamma_1+\gamma_2+\gamma_3) + \frac{\beta_1+\beta_3}{\theta}(\gamma_1+\gamma_2+\gamma_3)\right)I \\ &\leq g_1^2(\gamma_1+\gamma_2+\gamma_3)\left(\frac{\beta_2}{g_1(\gamma_1+\gamma_2+\gamma_3)} + \frac{\beta_1}{\theta} - 1 + \frac{\beta_4}{g_1(\gamma_1+\gamma_2+\gamma_3)} + \frac{\beta_3}{\theta}\right)I \\ &\leq (\gamma_1+\gamma_2+\gamma_3)(\mathcal{R}_0 + {\mathcal{R}_0}^{v} - 1)I \end{aligned} \tag{11}$$

where $g_1 = 1$, $\frac{g_2}{g_1} = 1 - \frac{\beta_1+\beta_3}{\theta}$.

$${\mathcal{R}_0}^{Total} = \left(\frac{\beta_2}{(\gamma_1+\gamma_2+\gamma_3)} + \frac{\beta_1}{\theta} + \frac{\beta_4}{(\gamma_1+\gamma_2+\gamma_3)} + \frac{\beta_3}{\theta}\right) = \mathcal{R}_0 + {\mathcal{R}_0}^{v}. \tag{12}$$

Since $\frac{g_2}{g_1} > 0$, then $\frac{\beta_1+\beta_3}{\theta} < 1$. Therefore, $\dot{L}$ is negative if ${\mathcal{R}_0}^{Total} < 1$. Furthermore, $\dot{L} = 0$ if and only if $I = 0$. It can thus be investigated whether singleton $\mathcal{E}_0$ is the highest compact invariant set for the model (1). Thus, by LaSalle's invariance principle [27], the DFE is globally asymptotically stable in a region Ω around $\mathcal{E}_0$. □

Remark 1. *The above result shows that ${\mathcal{R}_0}^{Total}$ and $\mathcal{R}_0$ can be reduced to less than a unit so that the disease disappears. Obviously, $\mathcal{R}_0 < {\mathcal{R}_0}^{Total}$ means that if $\mathcal{R}_0 < 1$, then the complete eradication of disease is guaranteed.*

6. The Optimal Imperfect Vaccination

When imperfect vaccination is administered to a population, there is a need to find the optimal approach to use it in order to reduce the burden of the disease in the population. The goal of this section is to implement the best control strategy possible in the situation of an imperfect vaccination. Three types of control are used for this purpose. First, the control u_1 represents the awareness of taking the vaccine via media, as well as creating knowledge of the positive effects of vaccination to gain herd immunity in the population. The second control u_2 is the movement restrictions for susceptible and vaccinated individuals by adhering to a preventative protocol, avoiding the exposure of the vaccinated people to the coronavirus via non-pharmaceutical measures. The third one u_3 seeks to improve the efficacy of the vaccine.

Therefore, the model with control strategies is given by

$$\begin{aligned}
\frac{dS}{dt} &= -\tfrac{S}{N}(1-u_2)(\beta_1 E+\beta_2 I)-(\lambda_1+u_1)S \\
\frac{dV}{dt} &= (\lambda_1+u_1)S-\tfrac{V}{N}(1-u_2)(\beta_3 E+\beta_4 I)-(\lambda_2+u_3)V \\
\frac{dE}{dt} &= \tfrac{S}{N}(1-u_2)(\beta_1 E+\beta_2 I)+\tfrac{V}{N}(1-u_2)(\beta_3 E+\beta_4 I)-\theta E \\
\frac{dI}{dt} &= \theta E-(\gamma_1+\gamma_2+\gamma_3)I \\
\frac{dQ}{dt} &= \gamma_1 I-(\sigma_1+\delta_1)Q \\
\frac{dH}{dt} &= \gamma_2 I+\sigma_1 Q-(\sigma_2+\delta_2)H \\
\frac{dC}{dt} &= \gamma_3 I+\sigma_2 H-(\mu+\delta_3)C \\
\frac{dR}{dt} &= \delta_1 Q+\delta_2 H+\delta_3 C+(\lambda_2+u_3)V \\
\frac{dD}{dt} &= \mu C
\end{aligned} \tag{13}$$

With:

$$(u_1(t),u_2(t),u_3(t)) \in \mathcal{U}_{ad}^T \tag{14}$$

and $\mathcal{U}_{ad}^T$ is a set of admissible controls defined by

$$\mathcal{U}_{ad}^T = \left\{ \begin{array}{ll} u \mid (u_1(t),u_2(t),u_3(t)) \text{ are measurable,} & 0 \le u_1(t) \le 1-\lambda_1,\ 0 \le u_2(t) \le 1, \\ & 0 \le u_3(t) \le 1-\lambda_2,\ t \in [0,T] \end{array} \right\} \tag{15}$$

The objective function to minimize is:

$$J(u_1(t),u_2(t),u_3(t)) = \int_0^T [-A_1 V(t)+A_2 I(t)-A_3 R(t)+\frac{1}{2}\tau_1 u_1^2(t)+\frac{1}{2}\tau_2 u_2^2(t)+\frac{1}{2}\tau_3 u_3^2(t)]dt \tag{16}$$

The positive weight constants A_1, A_2 and A_3, respectively, keep the sizes of $V(t)$, $I(t)$, and $R(t)$ in balance. Positive weight parameters: τ_1, τ_2, and τ_3 are related with the controls $u_1(t)$, $u_2(t)$, and $u_3(t)$ in the objective functional.

To solve the problem, we first compute the Lagrangian and Hamiltonian. Equation (16) in order to identify an optimal solution. The optimal problem's Lagrangian is:

$$L = -A_1 V(t)+A_2 I(t)-A_3 R(t)+\frac{1}{2}\tau_1 u_1^2(t)+\frac{1}{2}\tau_2 u_2^2(t)+\frac{1}{2}\tau_3 u_3^2(t). \tag{17}$$

For the control problem, we may define the Hamiltonian $\overline{H}$ as follows:

$$\overline{H} = L+\zeta_1(t)\frac{dS}{dt}+\zeta_2(t)\frac{dV}{dt}+\zeta_3(t)\frac{dE}{dt}+\zeta_4(t)\frac{dI}{dt}+\zeta_5(t)\frac{dQ}{dt}+\zeta_6(t)\frac{dH}{dt}+\zeta_7(t)\frac{dC}{dt} \\ +\zeta_8(t)\frac{dR}{dt}+\zeta_9(t)\frac{dD}{dt} \tag{18}$$

where $\zeta_1,\ldots,\zeta_9$ are the adjoint functions to be found.

We have the existence result:

Theorem 5. *The optimal control problem, defined by Equations (13)–(16), has a solution (u_1^*,u_2^*,u_3^*) that satisfies*

$$J(u_1^*,u_2^*,u_3^*) = \min_{(u_1,u_2,u_3)\in\mathcal{U}_{ad}^T} J(u_1,u_2,u_3)$$

Proof. We use the result [28] to show that an optimal control exists. The control and the state variables are both non-negative. This minimization problem satisfies the convexity requirement of the objective functional.

The control space previously defined as (15) is both convex and closed by definition. In order for the optimal control to exist, it is necessary for the optimal system to be compact. The boundedness of the optimal system determines the compactness needed. Additionally, an integrand throughout the functional (16) is convex on the control $(u_1(t),u_2(t),u_3(t))$. It can be concluded that the constant $\rho>1$ exists, as do positive integers w_1, w_2 and w_3 such that $J(u_1,u_2,u_3)\geq -w_2+w_1(\|(u_1,u_2,u_3)\|^2)^{\frac{\varrho}{2}}$. This leads us to conclude that optimal control exists. □

Characterization of the Optimal Control

We then investigate the necessary optimal control conditions. For this purpose, the maximum principle of Pontryagin to Hamiltonian [29] can be applied.

Theorem 6. *Let $S^*(t)$, $V^*(t)$, $E^*(t)$, $I^*(t)$, $Q^*(t)$, $H^*(t)$, $C^*(t)$, $R^*(t)$ and $D^*(t)$ represent optimal state solutions with optimal control variables $(u_1^*(t),u_2^*(t),u_3^*(t))$ for the optimal control problem (16). $\zeta_1,\ldots,\zeta_9$ are thus adjoint variables that satisfy:*

$$\begin{aligned}
\dot{\zeta}_1(t) &= (1-u_2)(\beta_1\tfrac{E(t)}{N}+\beta_2\tfrac{I(t)}{N})(\zeta_1(t)-\zeta_3(t))+(\lambda_1+u_1)(\zeta_1(t)-\zeta_2(t))\\
\dot{\zeta}_2(t) &= A_1+(1-u_2)(\beta_3\tfrac{E(t)}{N}+\beta_4\tfrac{I(t)}{N})(\zeta_2(t)-\zeta_3(t))+(\lambda_2+u_3)(\zeta_2(t)-\zeta_8(t))\\
\dot{\zeta}_3(t) &= (1-u_2)\beta_1\tfrac{S(t)}{N}(\zeta_1(t)-\zeta_3(t))+(1-u_2)\beta_3\tfrac{V(t)}{N}(\zeta_2(t)-\zeta_3(t))+\theta(\zeta_3(t)-\zeta_4(t))\\
\dot{\zeta}_4(t) &= -A_2+(1-u_2)\beta_2\tfrac{S(t)}{N}(\zeta_1(t)-\zeta_3(t))+(1-u_2)\beta_4\tfrac{V(t)}{N}(\zeta_2(t)-\zeta_3(t))\\
&\quad+\gamma_1(\zeta_4(t)-\zeta_5(t))+\gamma_2(\zeta_4(t)-\zeta_6(t))+\gamma_3(\zeta_4(t)-\zeta_7(t))\\
\dot{\zeta}_5(t) &= \sigma_1(\zeta_5(t)-\zeta_6(t))+\delta_1(\zeta_5(t)-\zeta_8(t))\\
\dot{\zeta}_6(t) &= \sigma_1(\zeta_6(t)-\zeta_7(t))+\delta_2(\zeta_6(t)-\zeta_8(t))\\
\dot{\zeta}_7(t) &= \mu(\zeta_7(t)-\zeta_9(t))+\delta_3(\zeta_7(t)-\zeta_8(t))\\
\dot{\zeta}_8(t) &= A_3\\
\dot{\zeta}_9(t) &= 0
\end{aligned} \tag{19}$$

Conditions of transversality.

$$\begin{aligned}
&\zeta_1(T)=\zeta_3(T)=\zeta_5(T)=\zeta_6(T)=\zeta_7(T)=\zeta_9(T)=0,\\
&\zeta_2(T)=-A_1,\ \zeta_4(T)=A_2,\ \zeta_8(T)=-A_3,
\end{aligned} \tag{20}$$

Moreover, the optimal control (u_1^*,u_2^*,u_3^*) is provided by

$$\begin{aligned} u_1^*(t) &= \max\{\min\{\tfrac{S(t)}{\tau_1}(\zeta_1(t)-\zeta_2(t)),1-\lambda_1\},0\}. \\ u_2^*(t) &= \max\{\min\{\tfrac{S(t)}{\tau_2 N}(\beta_1 E(t)+\beta_2 I(t))(\zeta_3(t)-\zeta_1(t))+\tfrac{V(t)}{\tau_2 N}(\beta_3 E(t)+\beta_4 I(t))(\zeta_3(t)-\zeta_2(t)),1\},0\}. \\ u_3^*(t) &= \max\{\min\{\tfrac{V(t)}{\tau_3}(\zeta_2(t)-\zeta_8(t)),1-\lambda_2\},0\} \end{aligned} \tag{21}$$

Proof. By using Hamiltonian (18), Pontryagin's maximum principle and setting $S(t) = S^*(t), V(t) = V^*(t), E(t) = E^*(t), I(t) = I^*(t), Q(t) = Q^*(t), H(t) = H^*(t), C(t) = C^*(t), R(t) = R^*(t)$ and $D(t) = D^*(t)$ to obtain the following:

$$\begin{aligned} \frac{d\zeta_1}{dt} &= (1-u_2)(\beta_1\tfrac{E(t)}{N}+\beta_2\tfrac{I(t)}{N})(\zeta_1(t)-\zeta_3(t))+(\lambda_1+u_1)(\zeta_1(t)-\zeta_2(t)) \\ \frac{d\zeta_2}{dt} &= A_1+(1-u_2)(\beta_3\tfrac{E(t)}{N}+\beta_4\tfrac{I(t)}{N})(\zeta_2(t)-\zeta_3(t))+(\lambda_2+u_3)(\zeta_2(t)-\zeta_8(t)) \\ \frac{d\zeta_3}{dt} &= (1-u_2)\beta_1\tfrac{S(t)}{N}(\zeta_1(t)-\zeta_3(t))+(1-u_2)\beta_3\tfrac{V(t)}{N}(\zeta_2(t)-\zeta_3(t))+\theta(\zeta_3(t)-\zeta_4(t)) \\ \frac{d\zeta_4}{dt} &= -A_2+(1-u_2)\beta_2\tfrac{S(t)}{N}(\zeta_1(t)-\zeta_3(t))+(1-u_2)\beta_4\tfrac{V(t)}{N}(\zeta_2(t)-\zeta_3(t))+\gamma_1(\zeta_4(t)-\zeta_5(t)) \\ &+ \gamma_2(\zeta_4(t)-\zeta_6(t))+\gamma_3(\zeta_4(t)-\zeta_7(t)) \end{aligned}$$

$$\begin{aligned} \frac{d\zeta_5}{dt} &= \sigma_1(\zeta_5(t)-\zeta_6(t))+\delta_1(\zeta_5(t)-\zeta_8(t)) \\ \frac{d\zeta_6}{dt} &= \sigma_1(\zeta_6(t)-\zeta_7(t))+\delta_2(\zeta_6(t)-\zeta_8(t)) \\ \frac{d\zeta_7}{dt} &= \mu(\zeta_7(t)-\zeta_9(t))+\delta_3(\zeta_7(t)-\zeta_8(t)) \\ \frac{d\zeta_8}{dt} &= A_3 \\ \frac{d\zeta_9}{dt} &= 0 \end{aligned}$$

Using optimality conditions, we conclude that:

$$\begin{aligned} \frac{d\overline{H}(t)}{du_1(t)} &= \tau_1 u_1^*(t)+S(t)(\zeta_2(t)-\zeta_1(t)) \\ \frac{d\overline{H}(t)}{du_2(t)} &= \tau_2 u_2^*(t)+\tfrac{S(t)}{N}(\beta_1 E(t)+\beta_2 I(t))(\zeta_1(t)-\zeta_3(t))+\tfrac{V(t)}{N}(\beta_3 E(t)+\beta_4 I(t))(\zeta_2(t)-\zeta_3(t)) \\ \frac{d\overline{H}(t)}{du_3(t)} &= \tau_3 u_3^*(t)+V(t)(\zeta_8(t)-\zeta_2(t)). \end{aligned}$$

Hence:

$$\begin{aligned} \frac{d\overline{H}(t)}{du_1(t)} = 0 &\Rightarrow u_1^*(t) = \tfrac{S(t)}{\tau_1}(\zeta_1(t)-\zeta_2(t)), \\ \frac{d\overline{H}(t)}{du_2(t)} = 0 &\Rightarrow u_2^*(t) = \tfrac{S(t)}{\tau_2 N}(\beta_1 E(t)+\beta_2 I(t))(\zeta_3(t)-\zeta_1(t))+\tfrac{V(t)}{\tau_2 N}(\beta_3 E(t)+\beta_4 I(t))(\zeta_3(t)-\zeta_2(t)) \\ \frac{d\overline{H}(t)}{du_3(t)} = 0 &\Rightarrow u_3^*(t) = \tfrac{V(t)}{\tau_3}(\zeta_2(t)-\zeta_8(t)). \end{aligned}$$

By applying the control space property, we obtain that:

$$\begin{cases} u_1^* = 0 & \text{if } \tfrac{S(t)}{\tau_1}(\zeta_1(t)-\zeta_2(t)) \le 0 \\ u_1^* = \tfrac{S(t)}{\tau_1}(\zeta_1(t)-\zeta_2(t)) & \text{if } 0 < \tfrac{S(t)}{\tau_1}(\zeta_1(t)-\zeta_2(t)) < 1 \\ u_1^* = 1 & \text{if } \tfrac{S(t)}{\tau_1}(\zeta_1(t)-\zeta_2(t)) \ge 1-\lambda_1. \end{cases}$$

which means that:

$$\begin{cases} u_2^* = 0 & \text{if} \quad \frac{S(t)}{\tau_2 N}(\beta_1 E(t) + \beta_2 I(t))(\zeta_3(t) - \zeta_1(t)) + \frac{V(t)}{\tau_2 N}(\beta_3 E(t) + \beta_4 I(t))(\zeta_3(t) - \zeta_2(t)) \le 0 \\ u_2^* = \omega^* & \text{if} \quad 0 < \frac{S(t)}{\tau_2 N}(\beta_1 E(t) + \beta_2 I(t))(\zeta_3(t) - \zeta_1(t)) + \frac{V(t)}{\tau_2 N}(\beta_3 E(t) + \beta_4 I(t))(\zeta_3(t) - \zeta_2(t)) < 1 \\ u_2^* = 1 & \text{if} \quad \frac{S(t)}{\tau_2 N}(\beta_1 E(t) + \beta_2 I(t))(\zeta_3(t) - \zeta_1(t)) + \frac{V(t)}{\tau_2 N}(\beta_3 E(t) + \beta_4 I(t))(\zeta_3(t) - \zeta_2(t)) \ge 1. \end{cases}$$

Since:

$$\omega^* = \frac{S(t)}{\tau_2 N}(\beta_1 E(t) + \beta_2 I(t))(\zeta_3(t) - \zeta_1(t)) + \frac{V(t)}{\tau_2 N}(\beta_3 E(t) + \beta_4 I(t))(\zeta_3(t) - \zeta_2(t)).$$

We have:

$$\begin{cases} u_3^* = 0 & \text{if} \quad \frac{V(t)}{\tau_3}(\zeta_2(t) - \zeta_8(t)) \le 0 \\ u_3^* = \frac{V(t)}{\tau_3}(\zeta_2(t) - \zeta_8(t)) & \text{if} \quad 0 < \frac{V(t)}{\tau_3}(\zeta_2(t) - \zeta_8(t)) < 1 \\ u_3^* = 1 & \text{if} \quad \frac{V(t)}{\tau_3}(\zeta_2(t) - \zeta_8(t)) \ge 1 - \lambda_2. \end{cases}$$

Thus, optimal control is defined as

$$\begin{aligned} u_1^* &= \max\{\min\{\tfrac{S(t)}{\tau_1}(\zeta_1(t) - \zeta_2(t)), 1 - \lambda_1\}, 0\}, \\ u_2^* &= \max\{\min\{\tfrac{S(t)}{\tau_2 N}(\beta_1 E(t) + \beta_2 I(t))(\zeta_3(t) - \zeta_1(t)) + \tfrac{V(t)}{\tau_2 N}(\beta_3 E(t) + \beta_4 I(t))(\zeta_3(t) - \zeta_2(t)), 1\}, 0\}. \\ u_3^* &= \max\{\min\{\tfrac{V(t)}{\tau_3}(\zeta_2(t) - \zeta_8(t)), 1 - \lambda_2\}, 0\} \end{aligned}$$

□

7. Numerical Simulation

The goal of this section is to show how the control strategies can be used to improve outcomes in vaccination campaigns in Morocco. After fitting the model to the data, the estimated parameters are taken to perform sensitivity analysis and determine the optimal control.

7.1. Parameter Estimation

We used data from a vaccination campaign in Morocco between 1 February 2021 and 25 March 2021 to validate our findings. We consider the data from the COVID-19 data [30] and the initial conditions within the values of available data on 1 February 2021 are $(S_0, V_0, E_0, I_0, Q_0, H_0, C_0, R_0, D_0)$ = (36,202,000, 200,081, 25,543, 13,099, 9824, 1572, 131, 450,052, 8287). These initial values were estimated from the data, apart from E_0 and Q_0, which were assumed. To obtain the best fitting curve for actual data, we applied the least-squares fitting technique [31]. The parameter values of the model are estimated based on this fitting and are given as follows: $\lambda_1 = 3.23 \times 10^{-3}$, $\lambda_2 = 1.5 \times 10^{-4}$, $\theta = 7.4385 \times 10^{-2}$, $\sigma_1 = 3.9601 \times 10^{-2}$, $\sigma_2 = 5.1954 \times 10^{-2}$, $\delta_1 = 2.04 \times 10^{-2}$, $\delta_2 = 1.01 \times 10^{-2}$, $\delta_3 = 0.00203$, $\beta_1 = 9.813 \times 10^{-3}$, $\beta_2 = 2.4 \times 10^{-3}$, $\beta_3 = 0.21$, $\beta_4 = 0.21$, $\gamma_1 = 9.81 \times 10^{-2}$, $\gamma_2 = 1.2 \times 10^{-2}$, $\gamma_3 = 1.02 \times 10^{-4}$ and $\mu = 1.501 \times 10^{-3}$. The value of the basic reproduction number in this case is $\mathcal{R}_0 = 0.1536999$. The fit of the number of individuals infected with COVID-19 in Morocco is described in Figure 2.

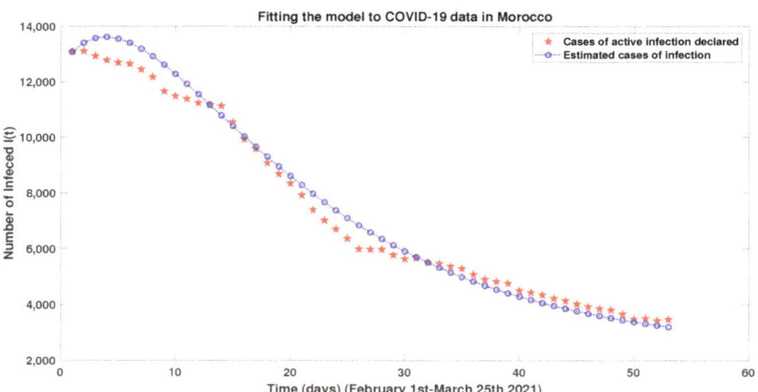

Figure 2. Fitting the infected population with real data from 1 February 2021 to 25 March 2021.

7.2. Sensitivity Analysis

In this part, we aimed to study the sensitivity of the model parameters with respect to ${\mathcal{R}_0}^{Total}$. The goal is to determine the impact of these parameters on the endemicity of the disease. The sensitivity analysis for this outbreak threshold demonstrates the importance of each parameter in the spread of COVID-19, allowing us to determine which parameter to make more sensitive on ${\mathcal{R}_0}^{Total}$ with respect to a specific parameter, ρ, via the sensitivity index define by

$$\zeta_{\rho}^{{\mathcal{R}_0}^{Total}} = \frac{\partial {\mathcal{R}_0}^{Total}}{\partial \rho} \frac{\rho}{{\mathcal{R}_0}^{Total}}. \tag{22}$$

Using the previous definition:

$$\begin{array}{llll} \zeta_{\beta_1}^{{\mathcal{R}_0}^{Total}} = \frac{1}{\theta}\frac{\beta_1}{{\mathcal{R}_0}^{Total}} & , & \zeta_{\beta_2}^{{\mathcal{R}_0}^{Total}} = \frac{1}{(\gamma_1+\gamma_2+\gamma_3)}\frac{\beta_2}{{\mathcal{R}_0}^{Total}} \\ \zeta_{\beta_3}^{{\mathcal{R}_0}^{Total}} = \frac{1}{\theta}\frac{\beta_3}{{\mathcal{R}_0}^{Total}} & , & \zeta_{\beta_4}^{{\mathcal{R}_0}^{Total}} = \frac{1}{(\gamma_1+\gamma_2+\gamma_3)}\frac{\beta_4}{{\mathcal{R}_0}^{Total}} \\ \zeta_{\gamma_1}^{{\mathcal{R}_0}^{Total}} = \frac{-(\beta_2+\beta_4)}{(\gamma_1+\gamma_2+\gamma_3)^2}\frac{\gamma_1}{{\mathcal{R}_0}^{Total}} & , & \zeta_{\gamma_2}^{{\mathcal{R}_0}^{Total}} = \frac{-(\beta_2+\beta_4)}{(\gamma_1+\gamma_2+\gamma_3)^2}\frac{\gamma_2}{{\mathcal{R}_0}^{Total}} \\ \zeta_{\gamma_3}^{{\mathcal{R}_0}^{Total}} = \frac{-(\beta_2+\beta_4)}{(\gamma_1+\gamma_2+\gamma_3)^2}\frac{\gamma_3}{{\mathcal{R}_0}^{Total}} & , & \zeta_{\theta}^{{\mathcal{R}_0}^{Total}} = \frac{-(\beta_1+\beta_3)}{\theta^2}\frac{\theta}{{\mathcal{R}_0}^{Total}}. \end{array} \tag{23}$$

Each parameter's sensitivity index, corresponding to the basic reproductive numbers ${\mathcal{R}_0}^{Total}$, was computed and displayed in Table 1, and the graphical bar-graph findings were generated in Figure 3. The sensitivity indices indicate the importance of each parameter in disease transmission and prevalence. These are some examples: if $\zeta_{\beta_1}^{{\mathcal{R}_0}^{Total}} = 0.931$, it means that if β_1 went up (or decreased) by 93.1%, ${\mathcal{R}_0}^{Total}$ is also likely to increase or decrease by 93.1%. Similarly, for $\zeta_{\theta}^{{\mathcal{R}_0}^{Total}} = -0.8202$, the decrease in the parameter θ by 82.02% will fall (or increase) ${\mathcal{R}_0}^{Total}$ by a similar proportion.

Our goal is to simulate the elasticity of ${\mathcal{R}_0}^{Total}$ with respect to model parameters in four scenarios that represent the different scenarios of the dynamics of the model as follows:

- Scenario 1: There is no disease persistence ($\mathcal{R}_0 < 1$, ${\mathcal{R}_0}^v < 1$ and ${\mathcal{R}_0}^{Total} < 1$).
- Scenario 2: Persistence of diseases with low threshold values ($\mathcal{R}_0 < 1$, ${\mathcal{R}_0}^v < 1$ and ${\mathcal{R}_0}^{Total} > 1$).
- Scenario 3: The disease persists, with a high endemicity among vaccinated people and a low endemicity among non-vaccinated people ($\mathcal{R}_0 < 1$, ${\mathcal{R}_0}^v > 1$ and ${\mathcal{R}_0}^{Total} > 1$).
- Scenario 4: The disease persists, with low endemicity among the vaccinated and high endemicity among the unvaccinated ($\mathcal{R}_0 > 1$, ${\mathcal{R}_0}^v < 1$ and ${\mathcal{R}_0}^{Total} > 1$).

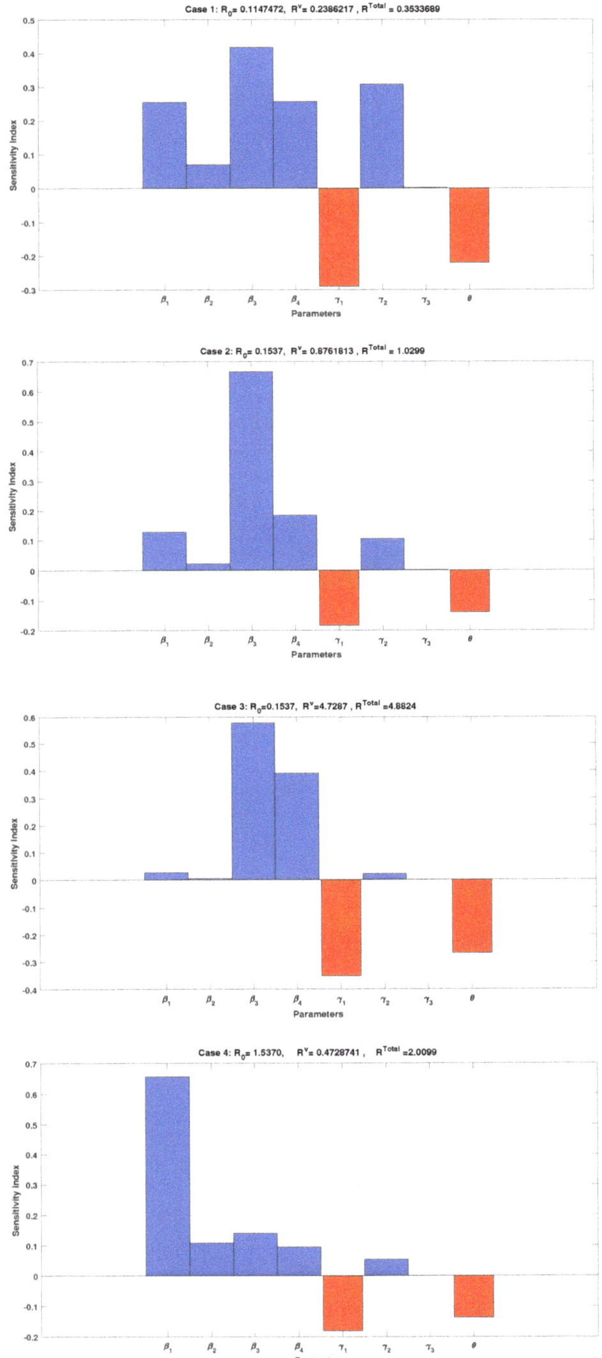

Figure 3. Scenarios of ${\mathcal{R}_0}^{Total}$ sensitivity respecting the parameters.

Table 1. Sensitivity index for each parameter that has a direct correlation to ${\mathcal{R}_0}^{Total}$.

Scenarios	Scenario 1		Scenario 2		Scenario 3		Secnario 4	
Parameters	**Values**	**S. Index**	**Values**	**S. Index**	**Values**	**S. Index**	**Values**	**S. Index**
β_1	6.71×10^{-3}	2.55×10^{-1}	9.81×10^{-3}	1.28×10^{-1}	9.81×10^{-3}	2.70×10^{-2}	9.81×10^{-2}	6.56×10^{-1}
β_2	2.70×10^{-3}	6.93×10^{-2}	2.40×10^{-3}	2.11×10^{-2}	2.40×10^{-3}	4.46×10^{-3}	2.40×10^{-2}	1.08×10^{-1}
β_3	1.10×10^{-2}	4.18×10^{-1}	5.10×10^{-2}	6.66×10^{-1}	2.10×10^{-1}	5.78×10^{-1}	2.10×10^{-2}	1.40×10^{-1}
β_4	1.00×10^{-2}	2.57×10^{-1}	2.10×10^{-2}	1.85×10^{-1}	2.10×10^{-1}	3.90×10^{-1}	2.10×10^{-2}	9.48×10^{-2}
γ_1	9.81×10^{-2}	-2.90×10^{-1}	9.81×10^{-2}	-1.84×10^{-1}	9.81×10^{-2}	-3.51×10^{-1}	9.81×10^{-2}	-1.81×10^{-1}
γ_2	1.20×10^{-2}	3.08×10^{-1}	1.20×10^{-2}	1.06×10^{-1}	1.20×10^{-2}	2.23×10^{-2}	1.20×10^{-2}	5.42×10^{-2}
γ_3	2.00×10^{-6}	2.62×10^{-3}	1.02×10^{-4}	8.99×10^{-4}	1.02×10^{-4}	1.90×10^{-4}	1.02×10^{-4}	4.61×10^{-4}
θ	7.44×10^{-2}	-2.20×10^{-1}	7.44×10^{-2}	-1.39×10^{-1}	7.44×10^{-2}	-2.66×10^{-1}	7.44×10^{-2}	-1.37×10^{-1}

All of the sensitivity scenarios described above demonstrate that the basic reproductive number ${\mathcal{R}_0}^{Total}$ is more sensitive to some parameters than others—particularly in the cases of β_1, β_3, γ_1 and θ. Scenario 1 has no persistent disease, but Scenario 3 has persistent disease with significant endemicity among the vaccinated, as can be seen from the sensitivity indices of γ_1 (rate of infected getting isolated) and θ (incubation period). The same observation applies to the scenarios of the persistence of disease with low endemicity in Scenario 2 and the persistence of disease with high endemicity among the vaccinated in Scenario 4. The level of sensitivity of these parameters is higher when endemicity is low among the non-vaccinated population. Our simulations revealed that the value of the sensitivity index changes depending on the Scenarios (1, 2, 3, 4), with β_3 being the highest value for Scenarios 1 and 2. However, in β_1, we can see more domination for the index in Scenario 4.

7.3. Simulation of Optimal Control

The time series of variables in the model without and with optimal control are shown in the figures below. The goal is to compare the effect of control on the different variables of the model.

The three variables are shown in Figure 4: susceptible, infected and recovered without and with optimal control effect. These simulations show that optimal control increases the number of susceptible and recovered people while decreasing the number of infected individuals. The effect of control on the susceptible and recovered populations is obviously more significant since the susceptible and recovered populations rose four-fold in 50 days while the infected population declined by one fold during the same period. This means that the control is very effective for all of these three compartments.

The simulation of the time series of the variables representing the vaccinated, exposed and deceased individuals is shown in Figure 5. Clearly, the vaccinated population benefits more from the optimal control than the exposed population since the control aims to improve vaccination effectiveness. The control strategy, on the other hand, has no obvious effect on the number of deaths.

Similarly, as seen in Figure 6, the control strategy reduces the number of isolated populations (at home or in hospital). However, there is more benefit to controlling the population with mild symptoms compared to people in the hospital with severe or critical symptoms. When applying the control to all three categories of quarantined, hospitalized, and critical cases of COVID-19, it is clear that the control is obvious.

The three types of optimal controls are given in Figures 7–9. Those figures show the intensity of each measure needed to be implemented in the case of imperfect vaccination. The awareness campaign should stay as long as 7 weeks with maximum intensity. At the same time, non-pharmaceutical measures must take place, including mobility restrictions or lockdown for at least 24 days. With regard to the third control measure, the improvement of the vaccine population should increase to reach possible efficacy during the first week of vaccination and stay above 30% within the first 50 days of vaccination. The outcome of this control approach shows that these three measures must be simultaneously implemented to deal with imperfect vaccination.

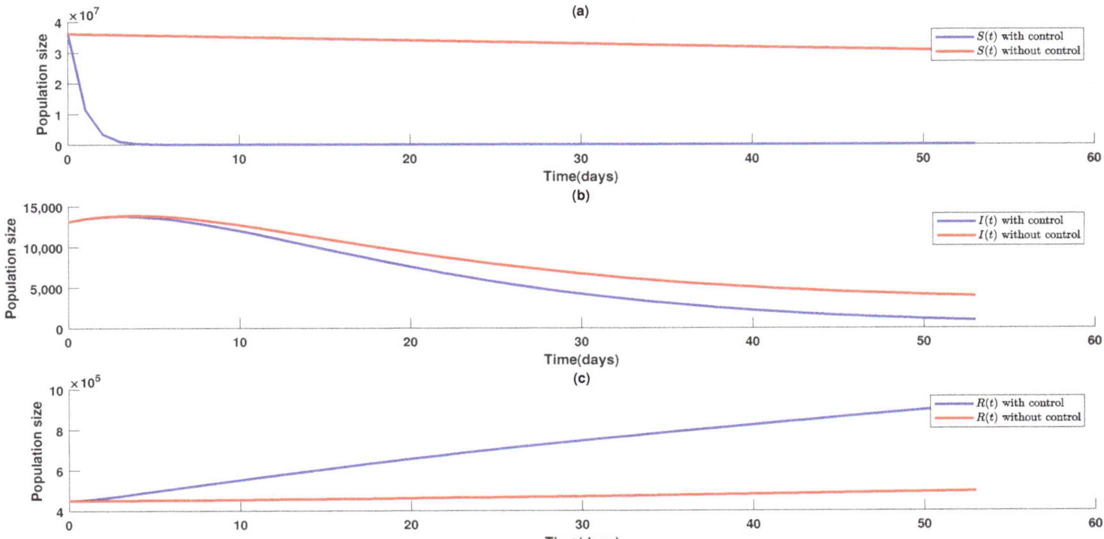

Figure 4. Time series of the states S, I and R without and with control.

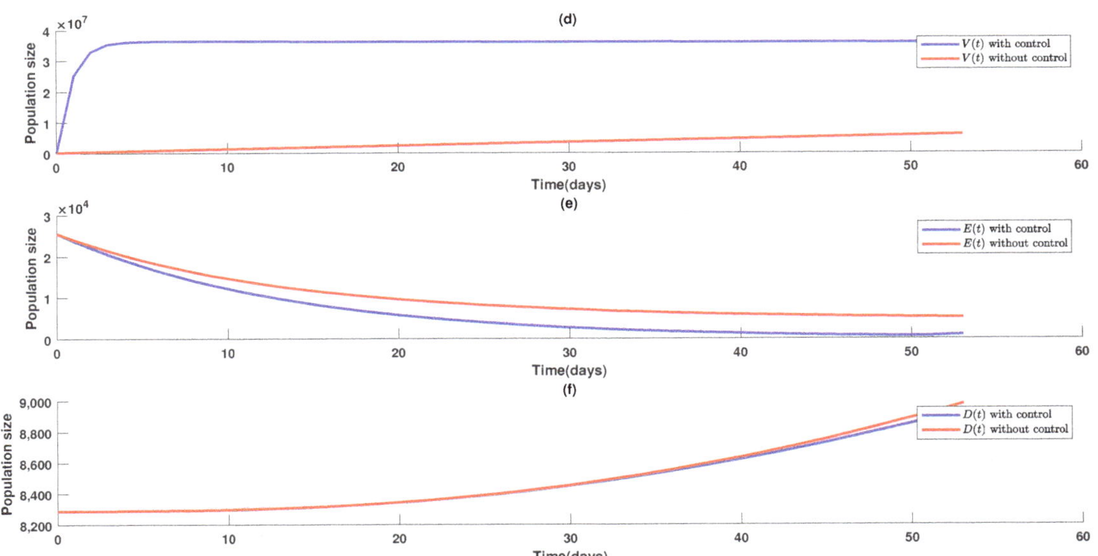

Figure 5. The time series of states V, E and D without and with control.

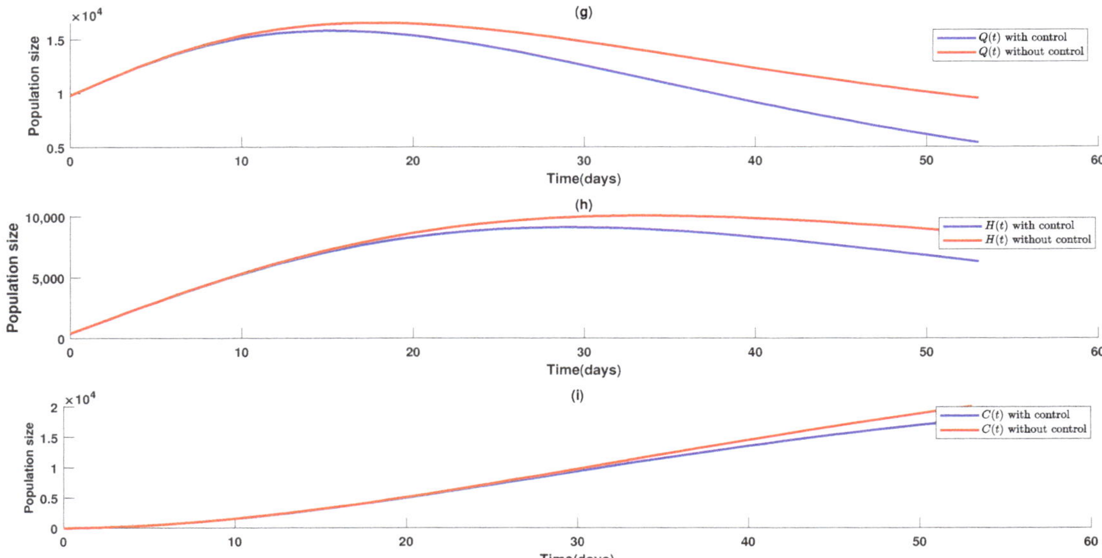

Figure 6. Time series of the states Q, H and C with and without control.

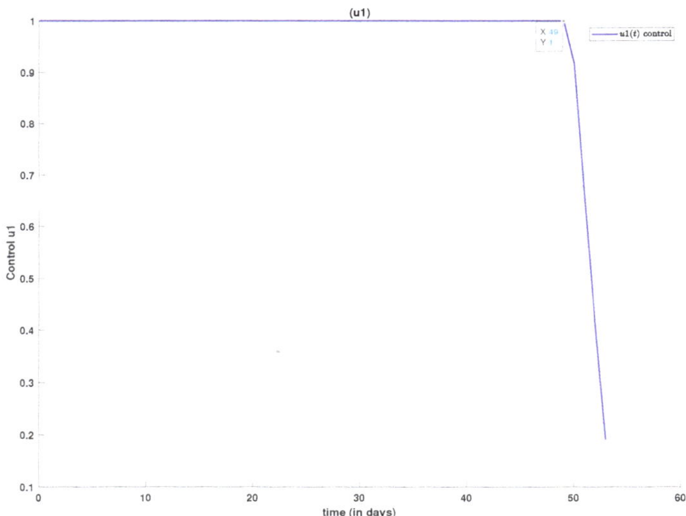

Figure 7. Evolutionary dynamics of the control $u_1(t)$.

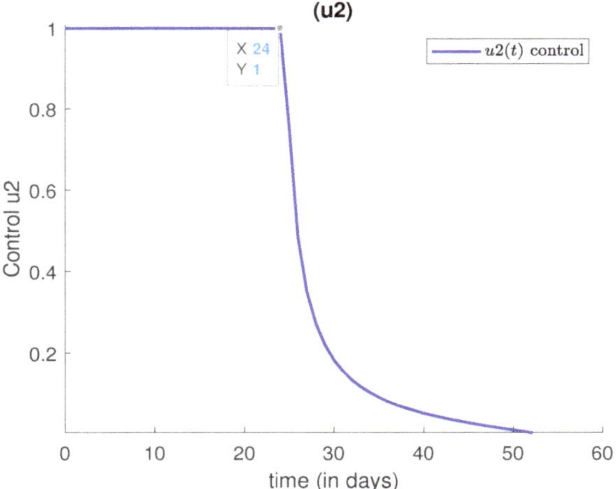

Figure 8. Evolutionary dynamics of the control $u_2(t)$.

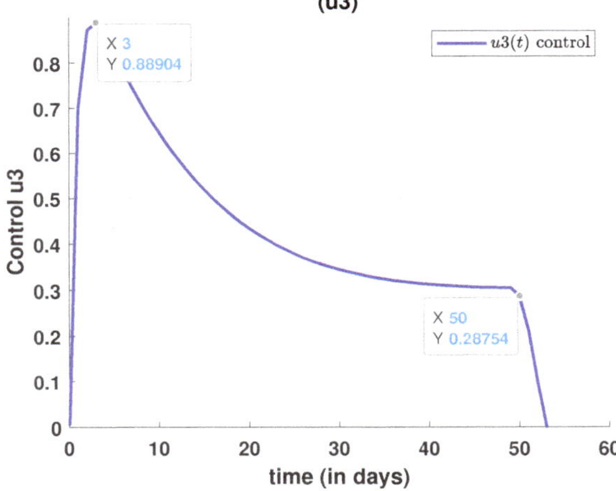

Figure 9. Evolutionary dynamics of the control $u_3(t)$.

8. Conclusions

COVID-19 is still taking a toll on people's lives all over the world. Countries are rushing to implement the vaccination to gain herd immunity, to contain the spread of the disease, and to bring the fatality rate of the disease to the lowest possible level. However, the administered vaccines have different levels of efficacy among the same population, which means the idea of relying on vaccination alone to control the pandemic would not provide total protection for the population against further waves of COVID-19. In this work, we aimed to present a mathematical model of the imperfect vaccination of COVID-19 and to study the dynamics of this model. One further element of this model is that susceptible and vaccinated people have different infection rates and the fatality rate of the disease (rate of death due to the infection) is related to the percentage of the isolated population that is in critical condition. We derive a threshold $\mathcal{R}_0$ which is the basic reproduction number of disease transmission among the population. We showed that this threshold does not give us sharp epidemiological properties of the model. In fact, we prove, via the Lyapunov method, that the disease is globally asymptotically stable if ${\mathcal{R}_0}^{Total} \leq 1$ with ${\mathcal{R}_0}^{Total}$ is the sum of $\mathcal{R}_0$ and ${\mathcal{R}_0}^{v}$, which is the threshold of transmission of the disease among the vaccinated population. This finding shows that the increase in the efficacy of the vaccination should lead to protecting the population from infection (low β_3 and β_4) which will help control the pandemic.

To make our analysis more realistic, we estimated the parameters of our model using data from Morocco between 1 February 2021 and 25 March 2021. Within the range of the estimated parameters, we performed a sensitivity analysis of $\mathcal{R}_0{}^{Total}$ with respect to the parameters of the model to find the elasticity index with respect to each parameter. Depending on the disease status, our simulation (3) produced four outcomes. Scenario 1: There is no disease persistence. Scenario 2: Persistence of diseases with low threshold values. Scenario 3: The disease persists with a high endemicity among vaccinated people and a low endemicity among non-vaccinated people. Scenario 4: The disease persists, with low endemicity among the vaccinated and high endemicity among the unvaccinated.

To further investigate the possible additional measures that help with vaccination. We introduce an optimal control problem with the goal of increasing the awareness of vaccination, limiting the probability of infection by adhering to a preventative protocol and increasing the efficacy of the vaccine. The public health authorities can easily implement these measures. In fact, as the virus mutates, many governments are pushing their populations to get vaccinated and asking people to reduce their contact and wear masks. Moreover, there is a constant effort to increase the efficacy of the vaccine by producing new ones or by boosters.

Our solution of optimal control showed that, to reduce the impact of imperfect vaccination, we needed a longer awareness campaign to engage the population in vaccination. On the other hand, the restriction on population mobility should not be long, since our simulation showed a drop of u_2 from 1 (full restriction) after 24 days. To ensure the full protection of the health population, vaccination efficacy must increase by 30% in the first 50 days.

In conclusion, our work showed that facing the imperfection of the vaccination of COVID-19, we mainly have to focus on two measures. The first one is to increase awareness of the importance of vaccination, which will increase the number of people vaccinated. The second is to work on developing vaccines with high efficacy that give more protection to the population instead of making the symptoms of viruses less severe. With these two measures, our study showed that population mobility restrictions have the lowest impact on controlling the virus spread.

Author Contributions: Conceptualization, L.B. and A.T.; methodology, M.L. and A.T.; software, L.B. and M.L.; validation, A.T., M.L. and O.Z.; formal analysis, L.B. and A.T.; investigation, L.B.; resources, M.L. and M.R.; data curation, L.B. and M.R.; writing—original draft preparation, L.B. and A.T.; writing—review and editing, L.B. and A.T.; visualization, L.B. and M.L.; supervision, M.L., A.T. and M.R. All authors have read and agreed to the published version of the manuscript.

Funding: This research received no external funding.

Acknowledgments: The authors would like to thank the reviewers for their comments and inputs that help improve the quality of our work.

Conflicts of Interest: The authors declare no conflict of interest.

References

1. Corey, L.; Mascola, J.R.; Fauci, A.S.; Collins, F.S. A strategic approach to COVID-19 vaccine R&D. *Science* **2020**, *368*, 948–950. [PubMed]
2. Arino, J.; Cooke, K.; Van Den Driessche, P.; Velasco-Hernández, J. An epidemiology model that includes a leaky vaccine with a general waning function. *Discret. Contin. Dyn. Syst.-B* **2004**, *4*, 479. [CrossRef]
3. Abu-Raddad, L.J.; Boily, M.C.; Self, S.; Longini, I.M., Jr. Analytic insights into the population level impact of imperfect prophylactic HIV vaccines. *JAIDS J. Acquir. Immune Defic. Syndr.* **2007**, *45*, 454–467. [CrossRef] [PubMed]
4. Liu, X.; Takeuchi, Y.; Iwami, S. SVIR epidemic models with vaccination strategies. *J. Theor. Biol.* **2008**, *253*, 1–11. [CrossRef] [PubMed]

5. Gulbudak, H.; Martcheva, M. A structured avian influenza model with imperfect vaccination and vaccine-induced asymptomatic infection. *Bull. Math. Biol.* **2014**, *76*, 2389–2425. [CrossRef] [PubMed]
6. Al-Darabsah, I. Threshold dynamics of a time-delayed epidemic model for continuous imperfect-vaccine with a generalized nonmonotone incidence rate. *Nonlinear Dyn.* **2020**, *101*, 1281–1300. [CrossRef] [PubMed]
7. Djilali, S.; Bentout, S. Global dynamics of SVIR epidemic model with distributed delay and imperfect vaccine. *Results Phys.* **2021**, *25*, 104245. [CrossRef]
8. Al-Darabsah, I. A time-delayed SVEIR model for imperfect vaccine with a generalized nonmonotone incidence and application to measles. *Appl. Math. Model.* **2021**, *91*, 74–92. [CrossRef] [PubMed]
9. Iboi, E.A.; Ngonghala, C.N.; Gumel, A.B. Will an imperfect vaccine curtail the COVID-19 pandemic in the US? *Infect. Dis. Model.* **2020**, *5*, 510–524. [PubMed]
10. Mancuso, M.; Eikenberry, S.E.; Gumel, A.B. Will vaccine-derived protective immunity curtail COVID-19 variants in the US? *Infect. Dis. Model.* **2021**, *6*, 1110–1134. [CrossRef]
11. Webb, G. A COVID-19 Epidemic Model Predicting the Effectiveness of Vaccination in the US. *Infect. Dis. Rep.* **2021**, *13*, 654–667. [CrossRef] [PubMed]
12. Libotte, G.B.; Lobato, F.S.; Platt, G.M.; Neto, A.J.S. Determination of an optimal control strategy for vaccine administration in COVID-19 pandemic treatment. *Comput. Methods Programs Biomed.* **2020**, *196*, 105664. [CrossRef] [PubMed]
13. Olivares, A.; Staffetti, E. Optimal control-based vaccination and testing strategies for COVID-19. *Comput. Methods Programs Biomed.* **2021**, *211*, 106411. [CrossRef] [PubMed]
14. Shim, E. Optimal allocation of the limited COVID-19 vaccine supply in South Korea. *J. Clin. Med.* **2021**, *10*, 591. [CrossRef] [PubMed]
15. Choi, W.; Shim, E. Vaccine Effects on Susceptibility and Symptomatology Can Change the Optimal Allocation of COVID-19 Vaccines: South Korea as an Example. *J. Clin. Med.* **2021**, *10*, 2813. [CrossRef] [PubMed]
16. Jordan, E.; Shin, D.E.; Leekha, S.; Azarm, S. Optimization in the Context of COVID-19 Prediction and Control: A Literature Review. *IEEE Access* **2021**, *9*, 130072–130093. [CrossRef]
17. Viana, J.; van Dorp, C.H.; Nunes, A.; Gomes, M.C.; van Boven, M.; Kretzschmar, M.E.; Veldhoen, M.; Rozhnova, G. Controlling the pandemic during the SARS-CoV-2 vaccination rollout. *Nat. Commun.* **2021**, *12*, 3674. [CrossRef] [PubMed]
18. Matrajt, L.; Eaton, J.; Leung, T.; Brown, E.R. Vaccine optimization for COVID-19: Who to vaccinate first? *Sci. Adv.* **2021**, *7*, eabf1374. [CrossRef] [PubMed]
19. Choi, Y.; Kim, J.S.; Kim, J.E.; Choi, H.; Lee, C.H. Vaccination Prioritization Strategies for COVID-19 in Korea: A Mathematical Modeling Approach. *Int. J. Environ. Res. Public Health* **2021**, *18*, 4240. [CrossRef] [PubMed]
20. Makhoul, M.; Ayoub, H.H.; Chemaitelly, H.; Seedat, S.; Mumtaz, G.R.; Al-Omari, S.; Abu-Raddad, L.J. Epidemiological impact of SARS-CoV-2 vaccination: Mathematical modeling analyses. *Vaccines* **2020**, *8*, 668. [CrossRef] [PubMed]
21. Wagner, C.E.; Saad-Roy, C.M.; Morris, S.E.; Baker, R.E.; Mina, M.J.; Farrar, J.; Holmes, E.C.; Pybus, O.G.; Graham, A.L.; Emanuel, E.J.; et al. Vaccine nationalism and the dynamics and control of SARS-CoV-2. *medRxiv* **2021**, *373*, eabj7364. [CrossRef] [PubMed]
22. Piraveenan, M.; Sawleshwarkar, S.; Walsh, M.; Zablotska, I.; Bhattacharyya, S.; Farooqui, H.H.; Bhatnagar, T.; Karan, A.; Murhekar, M.; Zodpey, S.; et al. Optimal governance and implementation of vaccination programmes to contain the COVID-19 pandemic. *R. Soc. Open Sci.* **2021**, *8*, 210429. [CrossRef] [PubMed]
23. Mohammed-Awel, J.; Numfor, E.; Zhao, R.; Lenhart, S. A new mathematical model studying imperfect vaccination: Optimal control analysis. *J. Math. Anal. Appl.* **2021**, *500*, 125132. [CrossRef]
24. Thieme, H.R. Princeton series in theoretical and computational biology. In *Mathematics in Population Biology*; Princeton University Press: Princeton, NJ, USA, 2003.
25. Birkhoff, G.; Rota, G. *Ordinary Differential Equations*; John Wiley & Sons: Hoboken, NJ, USA, 1989.
26. Van den Driessche, P.; Watmough, J. Reproduction numbers and sub-threshold endemic equilibria for compartmental models of disease transmission. *Math. Biosci.* **2002**, *180*, 29–48. [CrossRef]
27. LaSalle, J. Stability theory for ordinary differential equations. *J. Differ. Equ.* **1968**, *4*, 57–65. [CrossRef]
28. Fleming, W.; Rishel, R. Stochastic Differential Equations and Markov Diffusion Processes. In *Deterministic and Stochastic Optimal Control*; Springer: Berlin/Heidelberg, Germany, 1975; pp. 106–150.
29. Lee, D.; Milroy, I.P.; Tyler, K. Application of Pontryagin's maximum principle to the semi-automatic control of rail vehicles. In Proceedings of the Second Conference on Control Engineering, Newcastle, UK, 25–27 August 1982; pp. 233–236.
30. Roser, M.; Ritchie, H.; Ortiz-Ospina, E.; Hasell, J. Coronavirus Pandemic (COVID-19). Our World in Data. 2022. Available online: https://ourworldindata.org/coronavirus (accessed on 15 January 2022)
31. Weisstein, E.W. Least Squares Fitting. 2002. Available online: https://mathworld.wolfram.com/ (accessed on 15 January 2022).

 MDPI

Article

Determining COVID-19 Dynamics Using Physics Informed Neural Networks

Joseph Malinzi [1,2,*], Simanga Gwebu [1] and Sandile Motsa [1,3]

1 Department of Mathematics, Faculty of Science and Engineering, University of Eswatini, Kwaluseni M201, Eswatini; simangagwebu@gmail.com (S.G.); sandilemotsa@gmail.com (S.M.)
2 Institute of Systems Science, Durban University of Technology, Durban 4000, South Africa
3 School of Mathematics, Statistics and Computer Science, University of KwaZulu-Natal, Private Bag X01, Scottsville, Pietermaritzburg 3209, South Africa
* Correspondence: josephmalinzi@aims.ac.za

Abstract: The Physics Informed Neural Networks framework is applied to the understanding of the dynamics of COVID-19. To provide the governing system of equations used by the framework, the Susceptible–Infected–Recovered–Death mathematical model is used. This study focused on finding the patterns of the dynamics of the disease which involves predicting the infection rate, recovery rate and death rate; thus, predicting the active infections, total recovered, susceptible and deceased at any required time. The study used data that were collected on the dynamics of COVID-19 from the Kingdom of Eswatini between March 2020 and September 2021. The obtained results could be used for making future forecasts on COVID-19 in Eswatini.

Keywords: Physics Informed Neural Networks; mathematical modeling; data analysis; COVID-19

Citation: Malinzi, J.; Gwebu, S.; Motsa, S. Determining COVID-19 Dynamics Using Physics Informed Neural Networks. *Axioms* **2022**, *11*, 121. https://doi.org/10.3390/axioms11030121

Academic Editors: Natália Martins, Ricardo Almeida, Cristiana João Soares da Silva and Moulay Rchid Sidi Ammi

Received: 18 November 2021
Accepted: 24 January 2022
Published: 10 March 2022

Publisher's Note: MDPI stays neutral with regard to jurisdictional claims in published maps and institutional affiliations.

1. Introduction and Background

1.1. Introduction

The Severe Acute Respiratory Syndrome Coronavirus 2 (SARS-CoV-2) virus causes Coronavirus Disease of 2019 (COVID-19) [1]. On 11 March 2020, the World Health Organization (WHO) declared it a pandemic after it spread globally and inflicted havoc [2,3]. This virus is a member of the Beta coronavirus family, which are quite likely to cause severe symptoms and potentially death [4]. The massive impact of the viral disease demonstrates the need for an agent comprehension of its dynamics [3,5–8].

In order to curtail the spread of the disease, many countries have executed partial to full lockdowns, South Africa being among the first country to do so and the Kingdom of Eswatini following suit a day later. The pandemic has forced these countries to, periodically, close down most economic activities and this has had some serious ramifications on its citizenry including their currencies, the rand and emalangeni respectively, declining in value along with a rise in commodity prices [9]. It is therefore highly imperative that a solution to the pandemic is quickly found.

The spread of diseases like COVID-19 has been modelled using systems of ordinary differential equations (ODEs) amongst other types of mathematical approaches. Mathematical modeling has already been used to derive several useful insights about COVID-19 [10–22]. Several aspects have been investigated, ranging from determining the epidemic curves [11,17,20–22], investigating the role played by asymptomatic cases in the spread of the disease [15], determining the efficacy of wearing masks [12] and investigating the efficacy of the different control measures [13,16,19].

Data analysis tools such as machine learning (ML) have also been used to better understand the distribution patterns of COVID-19 [8,23]. The viability of applying artificial intelligence (AI) technologies to solve ODEs has been questioned because these equations are governed by scientific rules that are never imposed during the training process [24]. A novel framework called Physics Informed Neural Networks has been developed to address this problem (PINNs). It can also make very accurate predictions based on very small datasets [25].

Various studies conducted on COVID-19 are gradually utilizing the Physics Informed Neural Network structure. PINNs was used in one investigation to evaluate the spread of COVID-19 after quarantine restrictions were implemented. As the governing systems of equations, the researchers used the Susceptible–Exposed–Infected–Removed model. It primarily sought to comprehend the advantages of COVID-19 restrictions being implemented [6]. A time-varying set of parameters was employed in another experiment. The governing systems of equations were based on the Susceptible–Infected–Recovered–Deceased (SIRD) model. However, a recurrent neural network was used to create this model [8].

The PINNs framework is an Artificial Neural Network (ANN) architecture that exposes the generated neural network to datasets and governing laws during the training process. The governing laws are provided to the model in the form of ODEs or PDEs [24–26]. Understanding the dynamics of COVID-19 is critical for minimizing the virus's consequences. Although AI models have been built, the majority of them require a large amount of training data to obtain high accuracy. However, because COVID-19 was recently discovered it has a tiny dataset, hence these AI models are not viable. Other models just suit the facts provided, making future predictions less accurate. This necessitates the creation of models that can generate accurate predictions on the dynamics of disease spread using tiny datasets.

The goal of this research is to determine the dynamics of COVID-19 using Physics Informed Neural Networks. To achieve this, the study used the Physics Informed Neural Networks architecture. The SIRD model was employed as a PINNs mathematical model in this investigation. The dynamics that this study aims to figure out are the virus's average rates of infection, recovery, and mortality. It further uses the dynamics to predict the number of active infections, the number of people who have recovered, how many are vulnerable or susceptible, and how many are deceased at any point in time. To perform the training process of the neural network, the research utilized data collected from the Kingdom of Eswatini between March 2020 and September 2021.

The rest of the document is divided into four parts, the first of which is a literature review. This section provides a review of some previous studies which have utilized the PINNs framework. The methodology section follows, which examines the mathematical and physics informed neural network framework and its development. The results and simulations part follows, which includes the analysis as well as the results and display of inaccuracies acquired. The last section is the conclusion, which concludes the research findings and provides recommendations for further research.

1.2. Background

Artificial Neural Networks (ANNs) are the building blocks of deep learning, a branch of AI and machine learning [27]. They are computer models that try to combine the capabilities of human brains and computers [25,28]. Nodes are placed in a layer format and interlinked by connectors in the widely used ANN structure [29]. The input layer receives data in vector format and transmits a dot product of the connector weight and the received data to the next node layer. The activation function multiplies the node's dot product [30]. The activation function is a mathematical function that converts linear input values into non-linear format [31]. The feedforward process is the name given to this method. Backpropagation is the technique of taking the error and adjusting the weights during the training phase.

Definition 1. *A feedforward neural network with a total of N neurons arranged in a single layer is a function* $y : \mathbb{R}^d \rightarrow \mathbb{R}$ *of the form:*

$$y(t) = \sum_{i=1}^{N} \alpha_i \sigma(w_i^T + b_i),$$

where $t \in \mathbb{R}^d, \alpha_i, b_i \in \mathbb{R}$. σ *is the activation function,* w_i *are weights for each neuron multiplied to input value t.* α_i *are neural network weights and are applied to the output of each neuron in the layer and* b_i *is the bias of each neuron.*

There are numerous activation functions used in neural networks. This study employees the tangent hyperbolic function (tanh).

Definition 2. *A tangent hyperbolic activation function is a function* $\sigma : \mathbb{R} \rightarrow \mathbb{R}$ *such that:*

$$\sigma(t) \rightarrow \begin{cases} 1 & as \quad t \rightarrow \infty, \\ -1 & as \quad t \rightarrow -\infty. \end{cases}$$

The ability of neural networks to alter internal variables during training so that they can tackle any given problem with some degree of precision is one of its most important features. Discriminatory functions are what neural networks are by definition. As a result of this attribute, the neural network is a universal approximator.

Theorem 1. *If the* σ *in the neural network definition is continuous, then the set of all neural networks is dense in a space of continuous discriminatory functions function with domain C on* I_n *denoted by* $C(I_n)$*, where* I_n *is an n-dimensional unit cube.*

Proof. Let $\mathcal{N} \subset C(I_n)$ be the set of neural networks where $\mathcal{N}$ is a linear subspace of $C(I_n)$. To show that $\mathcal{N}$ is dense in $C(I_n)$, we show that its closure is $C(I_n)$. By contradiction, suppose $\overline{\mathcal{N}} \neq C(I_n)$. Then $\overline{\mathcal{N}}$ is a closed proper subspace of $C(I_n)$. □

The mathematical modeling approach and the data-based approach are the two primary methods for making predictions. The benefits and drawbacks of these two models are distinct. The majority of mathematical models used are generated from the underlying processes. As a result, these models follow governing laws, resulting in a directed output that, when given the correct initial values, always yields accurate results. However, one of the most significant drawbacks is that mathematical models do not account for any unanticipated changes, which is a flaw in real-time process analysis [32].

Data models that include machine learning algorithms identify patterns in incoming data and produce the desired output. Larger datasets are required to fully comprehend these trends. This means that if a limited data collection is available, other datasets that are similarly relevant can be utilised. The margin of error is increased by this compromise. The necessity for a huge amount of data and an extensive training procedure necessitates the use of a lot of computing power, which is expensive. Data can also be compressed to fit the processing power available, compromising results.

2. Review of Studies on Physics Informed Neural Networks

Multiple world problems have been analyzed and simulated using the recently established Physics Informed Neural Networks framework. This section summarizes some research on Physics Informed Neural Networks. COVID-19-related research is among the studies included.

2.1. Physics Informed Deep Learning for Traffic State Estimation

Real-time traffic states were analyzed using the Physics Informed Neural Networks framework. The method of estimating traffic variables using partial data is known as traffic state estimation. The traffic variables employed in the analysis are f, which stands for traffic flow rate, v, which stands for vehicle average speed rate, and ρ, which stands for vehicle density. The goal of traffic state analysis is to improve road planning and comprehension. This includes the early detection of traffic jams and high transit demand. A dramatic decline in average speed v, for example, could signal significant traffic congestion or an accident [24].

The methods used to conduct these traffic estimations are primarily mathematical or data-driven methods. This research takes a data-driven approach. However, data-driven technologies such as machine learning necessitate a large amount of data. This is a significant disadvantage because it necessitates the employment of a large number of sensors and other equipment to be archived, which is a very costly undertaking. This forces transportation planners to collect data primarily in cost-effective locations, resulting in noisy or error-filled data. To address these issues, the researchers used a physics informed neural network technique [24].

They set the variables based on the data acquired when developing mathematical models, q being the stated number of cars traversing a certain area at a specific time. The mean speed of vehicles is used to get the average speed v, and the vehicle density ρ is calculated as the number of vehicles in a given road distance. The cumulative traffic flow is defined as the total number of vehicles passing through a specific point x over a given time period t. The partial differential equation of cumulative flow $q(x,t)$ represents the flow with regard to time (t). The partial differential equation of cumulative flow $\rho(x,t)$ is a partial differential equation of density with regard to x. The mathematical representations of the densities is:

$$q(x,t) = \frac{\partial N(x,t)}{\partial t}, \tag{1}$$

$$\rho(x,t) = \frac{-\partial N(x,t)}{\partial x}. \tag{2}$$

The conservation law states:

$$\frac{\partial q N(x,t)}{\partial x} + \frac{\partial \rho N(x,t)}{\partial t} = 0. \tag{3}$$

The relationship between the stated variables is:

$$v(\rho) = v_f\left(1 - \frac{\rho}{\rho_m}\right), \tag{4}$$

$$q(\rho) = \rho v_f\left(1 - \frac{\rho}{\rho_m}\right), \tag{5}$$

where v_f = traffic free flow and ρ_m = maximum traffic flow.

The mean square error (MSE) of N number of outputs at point x at time t is used to construct the cost function (J_{DL}), which is utilized to increase the accuracy of the neural network. The neural network's forecast is $\rho * (x,t)$, but the genuine value is $\rho(x,t)$. The MSE J_{PHY} is found in relation to the conservation of the specified conservation laws as a result of the deployment of the physics informed neural network.

$$J_{DL} = \frac{1}{N}\sum_{i=1}^{N} |\rho(x,t) - \rho^*(x,t)|^2, \tag{6}$$

$$J_{PHY} = \frac{1}{N}\sum_{i=1}^{N} |v_f(1 - \frac{2\rho(x,t)}{\rho_m}) + \frac{\partial \rho(x,t)}{\partial x}) + \frac{\partial \rho(x,t)}{\partial t})|^2. \tag{7}$$

The neural network is then optimized using the physics informed neural network, and a parameter μ is added to give the neural network an adjustment weight. The PINNs implementation equation was then included.

$$J = \mu J_{DL} + (1 - \mu) J_{PHY}. \tag{8}$$

The Frobenius norm is then used to calculate the accuracy of the neural network. The model was put to the test with various data sizes and collection locations, with positive results [24].

This research used a physics informed neural network to analyze electricity generation. Generators are used in the power generation process, which are powered by diverse energy sources such as wind and water. The analysis and comprehension of a real-time power generation is critical in determining the amount of power generated by the generators [32]. It is not new to utilize data models to analyze power production and mathematical models to produce estimations. However, they all have disadvantages. For example, using data and machine learning models demands a large amount of data. It is also necessary to have the data analyzed by professionals before use in order to eliminate the noisy data. This data analysis paradigm also entails the creation of sophisticated neural network designs [32].

As a result, the study introduces the usage of a physics-informed neural network to construct a training process that is data and physics-based. The study employs a single machine infinite bus (SMIB) system, which is a single-generator model. The inertia constant m_1, the damping coefficient d_1, and the bus sustenance entry B_{12} are all parameters and variables in the equation. The power supplied by the generator is P_1, the voltage magnitudes of buses 1 and 2 are V_1 and V_2, and the voltage angle behind reactance is σ_1 and σ_2. The angular frequency of generators is $sigma^{\cdot}$ As a result, the final function is:

$$f_\sigma(t, P_1) = m_1 \sigma^{\cdot\cdot} + d_1 \sigma^{\cdot} + B_{12} V_1 V_2 \sin(\sigma) - P_1. \tag{9}$$

Equation (9) is used as the governing equation in the implementation of the physics informed neural network. The model adjusts σ, $\sigma^{\cdot}$ and P_1 between $[P_{min}, P_{max}]$ during the learning process. The model was simulated using datasets that were created using computer models and showed very positive results [32].

2.2. Neural Network Aided Quarantine Control Model Estimation of Global COVID-19 Spread

Two deep learning models are presented in the paper to approximate the parameters of COVID-19 spread. Forecasts are made using statistics from the United States, China (Wuhan), Italy, and South Korea. Both of the deep learning models employed in the models are physics informed neural networks. The PINNs is used to solve problems with machine learning and conventional neural network models. Overfitting of data, the necessity for high processing capacity, and the need for more data from other pandemics or diseases spread, such as MERS and SARS in this case, are all examples of these issues. Because artificial neural networks are so complicated, it is difficult to understand how the final approximation is achieved. PINNs, on the other hand, simplify the process, making it easier to comprehend and analyze COVID-19's distribution. The study's main goal is to determine the advantages of implementing COVID-19 limitations [6].

The first model employs an SEIR system of ODEs. In the model, S stands for the number of susceptible people in the population, E for the number of Exposed people individuals, I is the number of people who are infected, and R stands for the number of individuals who have been removed. The first model utilized ignores the potential consequences of COVID-19 control rules.

The common mathematical models, however, do not account for these in the predictions they make; making it hard to account for variables or elements such as over crowding, social distancing and other policies which may have been implemented by the different countries. The main policies highlighted by the authors include the use of police to enforce proper social distancing in traffic crossings, shops and other places. It also focuses on the shutdown of public transport, trains and airports. Thus, to account for these multiple policies and have a better prediction the study uses real data. This study is conducted using data and estimations. The study also estimates the effective reproduction rate. The first model:

$$\frac{dS(t)}{dt} = -\frac{\beta S(t)I(t)}{N}, \tag{10}$$

$$\frac{dE(t)}{dt} = \frac{\beta S(t)I(t)}{N} - \sigma E(t), \tag{11}$$

$$\frac{dI(t)}{dt} = \sigma E(t) - \gamma I(t), \tag{12}$$

$$\frac{dR(t)}{dt} = \gamma I(t), \tag{13}$$

subject to the initial conditions, $S = S_0$, $I = I_0$, $R = R_0$ and $E = E_0$.

The second model used in the study accounts for quarantine control. The model thus introduces a time dependent variable $T(t) = Q(t) \times I(t)$. This also changes the effective reproduction rate to $R_t = \frac{\beta}{\gamma + Q(t)}$. The parameter $Q(t)$ is also determined using a separate neural network, which takes in the data of Time, Susceptible, Exposed, Infected and Recovered as input data. The model processes the data in a 2-layer network with 10 nodes per layer and uses a ReLu activation function ($NN(W, U)$). The determined $Q(t)$ is then put in the Physics Informed Neural Network which uses the model below to make the approximations of the model.

$$\frac{dS(t)}{dt} = -\frac{\beta S(t)I(t)}{N}, \tag{14}$$

$$\frac{dI(t)}{dt} = \frac{\beta S(t)I(t)}{N} - (\gamma - Q(t))I(t), \tag{15}$$

$$\frac{dI(t)}{dt} = \frac{\beta S(t)I(t)}{N} - (\gamma - NN(W, U))I(t), \tag{16}$$

$$\frac{dR(t)}{dt} = \gamma I(t), \tag{17}$$

$$\frac{dT(t)}{dt} = Q(t)I(t) = NN(W, U)I(t). \tag{18}$$

Subjected to the initial conditions, $S = S_0$, $I = I_0$, $R = R_0$ and $T = T_0$.

The results that were attained by the study showed that the first model, which does not account for imposed restrictions, had approximations which were bigger than the real values. This means it approximated that the virus would be more catastrophic. The second model achieved a better fit showing that the imposed restrictions have had a positive impact in the spread of COVID-19. The model was also comprehensible providing parameters which can be used to make future predictions [6].

2.3. Identification and Prediction of Time-Varying Parameters of COVID-19 Model: A Data-Driven Deep Learning Approach

This study focused on finding the parameters of an SIRD model which are time based rather than to the average parameter [8]. This study also used a deep learning model and specifically a Physics Informed Neural Network. The virus spreading model employed is that study is an SIRD where S represents the number of people who are Susceptible, I represents the Infected people or active cases, R represents the number of Recovered people and D represents the number of Deaths. Where β is the spreading rate, γ is the recovery rate and δ is the death rate [8].

$$\frac{dS(t)}{dt} = -\frac{\beta S(t)I(t)}{N}, \tag{19}$$

$$\frac{dI(t)}{dt} = \frac{\beta S(t)I(t)}{N} - \gamma I(t) - \delta I(t), \tag{20}$$

$$\frac{dR(t)}{dt} = \gamma I(t), \tag{21}$$

$$\frac{dD(t)}{dt} = \delta I(t). \tag{22}$$

The model is subjected to the initial values, $S = S_0$, $I = I_0$, $R = R_0$ and $D = D_0$.

3. Problem Formulation and Methodology

The governing laws of the Physics Informed Neural Networks framework are provided as mathematical equations. This section covers the development and evaluation of the key mathematical models of focus. The mathematical model serve as the assumed physics laws that the model should adhere to.

The Susceptible–Infected–Recovered–Deceased (SIRD) model used assumes that the population can assume four states, Susceptible (S), Infected (I), Recovered (R) and Deceased (D). The susceptible population is the group which can contract the virus, this contraction occurs at the rate β. The infected population is the population group that has contacted the virus and it is still active. The infected group can be removed to either assume a recovered population at the rate γ or deceased population at the rate δ. This means δ is the death rate, β is the infection rate and γ is the recovery rate. Figure 1 shows the resulting COVID-19 transmission SIRD flow diagram.

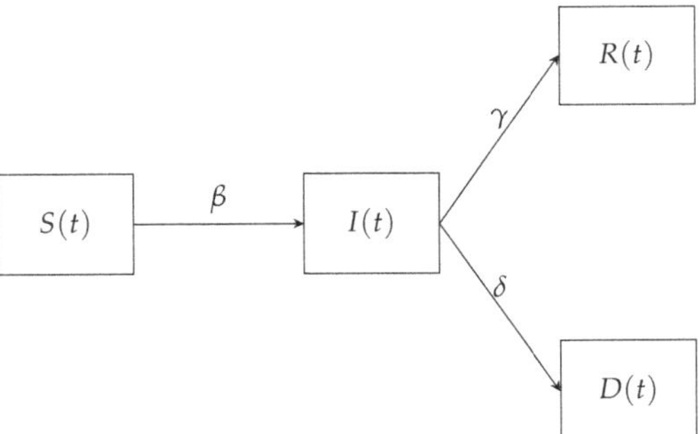

Figure 1. A schematic flow diagram representing a Susceptible-Infected-Recovered-Dead (SIRD) COVID-19 transmission.

From the flow diagram in Figure 1 we obtain the system:

$$\frac{dS(t)}{dt} = -\frac{\beta S(t)I(t)}{N}, \tag{23}$$

$$\frac{dI(t)}{dt} = \frac{\beta S(t)I(t)}{N} - \gamma I(t) - \delta I(t), \tag{24}$$

$$\frac{dR(t)}{dt} = \gamma I(t), \tag{25}$$

$$\frac{dD(t)}{dt} = \delta I(t). \tag{26}$$

The model is subjected to the initial values, $S = S_0$, $I = I_0$, $R = R_0 = 0$ and $D = D_0 = 0$.

The system in general reflects the mathematical behaviour of the virus. Equation (23) indicates the change in the number of susceptible individuals $S(t)$ with respect to time t; which is a reduction by a factor of the product of the spreading rate β, the susceptible population at the time $S(t)$ and the Infected population at time $I(t)$ divided by the total population N. Equation (24) shows that the change in the number of active infections $I(t)$ with respect to time t; which is an addition by size at which the susceptible population was reduced. This is then reduced by a factor of the product of the recovery rate γ and the Infected population at the time $I(t)$ and a product of the death rate δ and the Infected population at the time $I(t)$. Equation (25) shows that the change in the number of recoveries is an addition by a factor of the product of the recovery rate γ and the Infected population at the time $I(t)$. Equation (26) shows that the change in the number of deaths is and a product of the death rate δ and the Infected population at the time $I(t)$. The model also assumes that initially the are no recoveries or deaths and that the infected and susceptible populations are greater than zero.

Studies and implementations of neural networks have shown that using numbers less than one improves accuracy and optimisation. As a result, we must use the non-dimentionalisation technique to rescale the provided data to values between 0 and 1.

We let $w = \frac{S}{N}, \quad x = \frac{I}{N}, \quad y = \frac{R}{N}, \quad z = \frac{D}{N}, \quad t = q.$

To rescale the SIRD model, we make these assumptions, with the goal of reducing the number of variables and so obtaining new SIRD model values, thus:

$S = wN, \quad I = xN, \quad R = yN, \quad D = zN$. Substituting in the SIRD model we obtain:

$$\frac{d(wN)}{dq} = -\frac{\beta(wN)(xN)}{N}, \tag{27}$$

$$\frac{d(xN)}{dq} = \frac{\beta(wN)(xN)}{N} - \gamma xN - \delta xN, \tag{28}$$

$$\frac{d(yN)}{dq} = \gamma xN, \tag{29}$$

$$\frac{d(zN)}{dq} = \delta xN. \tag{30}$$

Hence the resulting system is:

$$\frac{dw}{dq} = -\beta wx, \tag{31}$$

$$\frac{dx}{dq} = \beta wx - \gamma x - \delta x, \tag{32}$$

$$\frac{dy}{dq} = \gamma x, \tag{33}$$

$$\frac{dz}{dq} = \delta x. \tag{34}$$

3.1. *The Neural Network*

The neural network we create, as a consequence, takes a single input value of time t. The input is processed through the layers with weights $W_{i,j}$, where i is the start node position and j is the finishing node position. The product of the weight and time is applied to an activation function *tanh* denoted by σ at every node, forming a matrix. The model's output nodes, which make up the output layer, are $S(t)$, $I(t)$, $R(t)$, and $D(t)$.

$$\sigma(x) = \frac{e^x - e^{-x}}{e^x + e^{-x}}. \tag{35}$$

The representation of a neural network matrix with m layers and n nodes per layer.

3.1.1. Residual of Model's Equations

The difference between the right and left sides of an ODE is the residual error. The residual error is utilized to determine the neural network's loss function in the construction of PINNs. We get four residual error functions from the SIRD model. We get Res_S, the residual error of the susceptible population, from Equation (23), which is the margin of how wrong the mathematically predicted susceptible population change is. We get Res_I from Equation (24), which is the residual error of the infected population, or the margin of error in the mathematically estimated active infection change. The residual error of the recovered population Res_R is given by Equation (25). We get Res_D the residual error of the deceased population from Equation (26), which is the margin of how much wrong the mathematically predicted deceased population change is.

$$Res_S = \frac{dS(t)}{dt} + \frac{\beta S(t)I(t)}{N}, \tag{36}$$

$$Res_I = \frac{dI(t)}{dt} - \frac{\beta S(t)I(t)}{N} + \gamma I(t) + \delta I(t), \tag{37}$$

$$Res_R = \frac{dR(t)}{dt} - \gamma I(t), \tag{38}$$

$$Res_D = \frac{dD(t)}{dt} - \delta I(t). \tag{39}$$

3.1.2. The Loss Function

A loss function must first be constructed before back propagation can be used to optimize a neural network. We derive the loss function $loss_T$ for the PINNs we constructed by adding the total of two loss functions $loss_1$ and $loss_2$. The total of the mean square errors of the susceptible population $MSE_{Soutput}$ is $loss_1$. The gap between the actual and forecast population sizes is the total of the sensitive population's mean square error. The total of the difference between the actual size of the infected population and the values predicted by the ANNs is the mean square errors of the infected population $MSE_{Ioutput}$. The difference between the actual size of the recovered population and the anticipated recovered population size, as well as the mean square errors of the deceased population $MSE_{Doutput}$, make up the recovered population mean square errors $MSE_{Routput}$. $MSE_{Doutput}$ is the mean square error of the difference of the predicted deceased $D^*(t_i)$ and the actual data value D_i.

The sum of the mean square errors of the susceptible population residual error MSE_{Sres}, the mean square errors of the infected population residual error MSE_{Ires}, the mean square errors of the recovered population residual error MSE_{Rres}, and the mean square errors of the deceased population residual error MSE_{Dres} is $loss_2$.

$$\begin{aligned} loss_1 &= MSE_{Soutput} + MSE_{Ioutput} + MSE_{Routput} \\ &\quad + MSE_{Doutput}, \end{aligned} \tag{40}$$

$$\begin{aligned} loss_2 &= MSE_{Sres} + MSE_{Ires} + MSE_{Rres} \\ &\quad + MSE_{Dres}, \end{aligned} \tag{41}$$

$$loss_T = loss_1 + loss_2, \tag{42}$$

$$MSE_{Sres} = \frac{1}{M}\sum_{i=1}^{M} |Res_S|^2, \tag{43}$$

$$MSE_{Ires} = \frac{1}{M}\sum_{i=1}^{M} |Res_I|^2, \tag{44}$$

$$MSE_{Rres} = \frac{1}{M}\sum_{i=1}^{M} |Res_R|^2, \tag{45}$$

$$MSE_{Dres} = \frac{1}{M}\sum_{i=1}^{M} |Res_D|^2, \tag{46}$$

$$MSE_{Soutput} = \frac{1}{M}\sum_{i=1}^{M}|S^*(t_i) - S_i|^2, \tag{47}$$

$$MSE_{Ioutput} = \frac{1}{M}\sum_{i=1}^{M}|I^*(t_i) - I_i|^2, \tag{48}$$

$$MSE_{Routput} = \frac{1}{M}\sum_{i=1}^{M}|R^*(t_i) - R_i|^2, \tag{49}$$

$$MSE_{Doutput} = \frac{1}{M}\sum_{i=1}^{M}|D^*(t_i) - D_i|^2. \tag{50}$$

We get a neural network that looks like Figure 2 using the time input, the layer matrix, the output layer, and the residual functions.

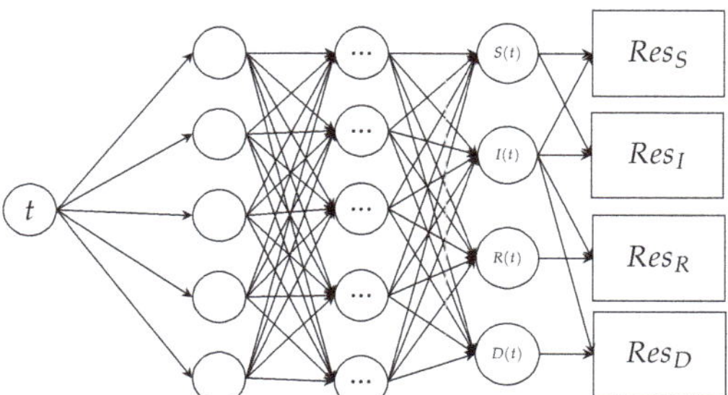

Figure 2. A schematic representation of the Physics informed neural network, which takes an input of time (t) and outputs Susceptible (S), Infected (I), Recovered (R) and Deceased D. The output is subjected to PINN.

3.2. Basic Model Properties

The analysis of the mathematical model is presented in the next part. This part examines the model's features and expected behaviour, such as determining the reproduction number, which is the minimal number of transmissions required for a pandemic to occur, and the sensitivity analysis of the ODE system.

3.2.1. Basic Reproduction Number

To comprehend COVID-19, which has now become a pandemic, we must first estimate the minimal rate of secondary infections required for a pandemic to arise. The reproduction number R_0 is also the rate at which a spread would be stopped if it fell below it. Its derivation is as follows:

$$0 < \beta S_0 I_0 - (\gamma + \delta) I_0, \tag{51}$$

$$0 < \beta S_0 - (\gamma + \delta), \tag{52}$$

$$\beta S_0 < (\gamma + \delta), \tag{53}$$

$$R_0 = \frac{\beta S_0}{\gamma + \delta}. \tag{54}$$

The change described by the left hand side, obtained after non-dimentionalization, must be strictly greater than 0 for a pandemic to occur, according to Equation (51). Equation (52) is produced by dividing both sides by I_0, and Equation (53) is obtained by moving the term containing the spreading rate to the left hand side. We can deduce from Equation (54) that the lowest needed spreading rate for a pandemic is equal to the left hand side divided by the right hand side.

3.2.2. SIRD Model Analysis

The mathematical model's sensitivity analysis also reveals some of the model's essential aspects, such as the estimated maximum number of infections I_{max}. Equations (23)–(26), which are all part of the SIRD model, are used to calculate the maximum number of infected individuals that can occur at any given time.

$$\frac{dI(t)}{dS(t)} = -1 + \frac{\gamma + \delta}{\beta S}. \tag{55}$$

To obtain this we first divide Equation (23) by Equation (24) to obtain Equation (55) and integrate it to obtain Equation (56).

$$I + S - \frac{\gamma + \delta}{\beta}(\ln S) = I_0 + S_0 - \frac{\gamma + \delta}{\beta}(\ln S_0). \tag{56}$$

To obtain the maximum possible value of infection, we find a point where the Equation (56) is equal to zero. It was determined that this occurs when $S = \frac{\beta}{\gamma+\delta}$.

$$I + S - \frac{\gamma + \delta}{\beta}(\ln S) = I_0 + S_0 - \frac{\gamma + \delta}{\beta}(\ln S_0), \tag{57}$$

$$I_{max} + S - \frac{\gamma + \delta}{\beta}(\ln S) = I_0 + S_0 - \frac{\gamma + \delta}{\beta}(\ln S_0), \tag{58}$$

$$I_{max} + \frac{\beta}{\gamma + \delta} - \frac{\gamma + \delta}{\beta}(\ln(\frac{\beta}{\gamma + \delta})) = I_0 + S_0 - \frac{\gamma + \delta}{\beta}(\ln S_0), \tag{59}$$

$$I_{max} = I_0 + S_0 - \frac{\gamma + \delta}{\beta}[1 + (\ln(\frac{\beta}{\gamma + \delta})S_0)], \tag{60}$$

$$I_{max} = I_0 + S_0 - \frac{\gamma + \delta}{\beta}[1 + (\ln(R_0))]. \tag{61}$$

In Equation (59), we substitute the value of S, to obtain the equation. In Equation (60), the model is simplified and rearranged. In Equation (61), the value R_0 is substituted where possible such that the remaining equation is attained. We now have the estimated number of persons who will become infected. Individuals infected with the virus either recover R_{end} or die D_{end}, thus we calculate the predicted number of persons who will either recover or die.

$$\overline{R}_{end} = R_{end} + D_{end}, \tag{62}$$

$$\overline{R}_{end} = S_0 + I_0 - S_{end}. \tag{63}$$

According to the created SIRD model, the whole number of people who will be infected will either recover from the disease or die from it, therefore the total number of persons affected will be equal to the sum of the recoveries R_{end} and the deceased D_{end} illustrated in Equation (62). We estimate the total infected at the conclusion of the virus spreading period to be equal to the sum of the original susceptible and infected populations minus the Susceptible population at the end of the period, as shown in Equation (63).

$$S_{end} = \frac{\gamma + \delta}{\beta} ln(S_{end}), \tag{64}$$

$$S_{end} = I_0 + S_0 - \frac{\gamma + \delta}{\beta} ln(S_0), \tag{65}$$

where S, I, R and D, respectively, represents susceptible, infectious, recovered and deceased individuals and $N = S + I + R + D$ is the total population. The parameters β, γ and δ, respectively, represent the infection, recovery rate and death rates. Since in the analysis or in the model recovered and deceased individuals have the same effects on the model, we group them as R representing removed with the removal rate of $(\gamma + \delta)$; such that we have,

$$\frac{dS(t)}{dt} = -\frac{\beta S(t)I(t)}{N}, \tag{66}$$

$$\frac{dI(t)}{dt} = \frac{\beta S(t)I(t)}{N} - (\gamma + \delta)I(t), \tag{67}$$

$$\frac{dR(t)}{dt} = (\gamma + \delta)I(t). \tag{68}$$

Which can be rewritten as:

$$\frac{dS(t)}{dt} = -\frac{\beta S(t)I(t)}{N}, \tag{69}$$

$$\frac{dI(t)}{dt} = \frac{\beta S(t)I(t)}{N} - (\gamma + \delta)I(t), \tag{70}$$

$$R = N - I - S. \tag{71}$$

The results of the simulations of the Physics Informed Neural Networks framework are presented in this section. It also includes a full analysis of the results, as well as changes in accuracy as parameters like data size change. The information was gathered through national daily updates and a Google studio analysis website created by the University of Eswatini and Wits Ithemba Labs [33]. Due to location/resource constraints, the model was constructed using Python 3 on a Spyder interface running in offline mode. Numpy, Mathplotlib, and Tensorflow were the main packages utilized.

3.3. Simulation Using Mathematica Generated Data

To test the model and validate the PINNs, we used Mathematica to create fictional data for an SIRD model. The benefit of this type of data was that it was less noisy. We started the model with a 100,000-person susceptible population, 0 recoveries and deaths, and five infections. In order to acquire the data shown in Figure 3, the average infection rate was set to be 0.14, the average healing rate was 0.037, and the average mortality rate was 0.005.

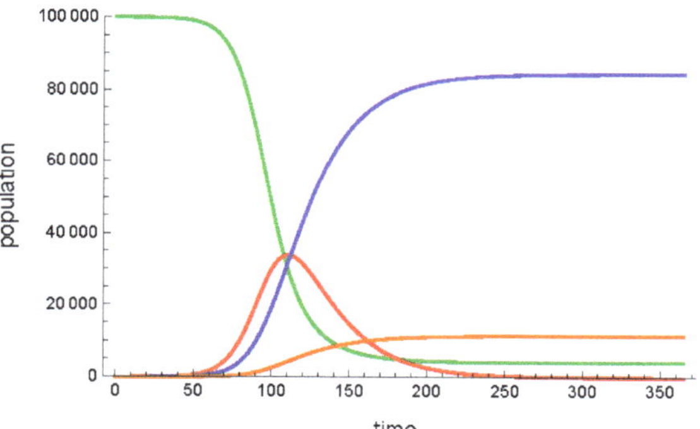

Figure 3. A Mathematica generated graph simulation of an example SIRD model. The green represents Susceptible population, blue represents the recoveries, red is the active infected population and orange is the deceased population.

The traditional behaviour of a SIR mode is seen in Figure 3, where the size of the vulnerable population decreases as the size of active illnesses increases. The size of the recoveries and deaths then increases until they reach a maximum or stabilize.

PINNs Model of Mathematica Results

The aforementioned model produced data from a PINNs of three layers, each with 30 nodes.

The findings provided after the model was trained were compared to the susceptible population, which was also utilized as the training dataset. There is only a small error between the two plots because they are so tightly aligned. Figure 4 depicts the resulting findings, which compare active infections from the dataset used to train the model to the values obtained by the trained model. This graph is well-fitting, indicating that it has a low degree of error. The graph in Figure 5 depicts the results of the recovered dataset used for training, as well as the produced results following the training process, which have a strong match and hence less errors. Figure 6 shows the outcome of comparing the actual deceased with the results acquired after the training procedure; this graph has a good match but is less accurate than the others.

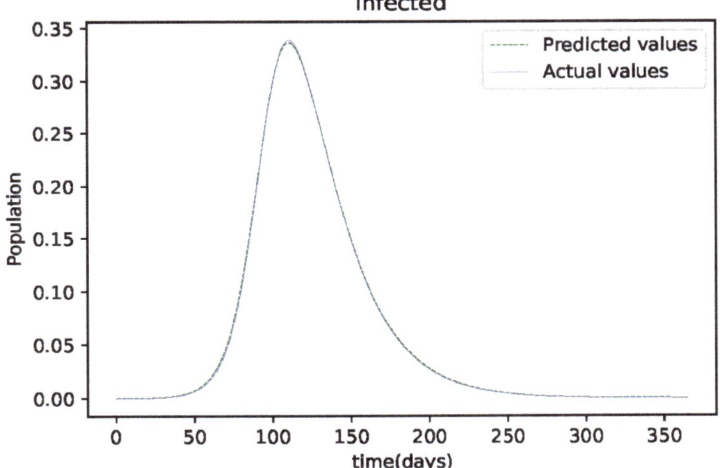

Figure 4. The resulting graph of the predicted values of the Infected population and the actual values of the infected population from the Mathematica generated data.

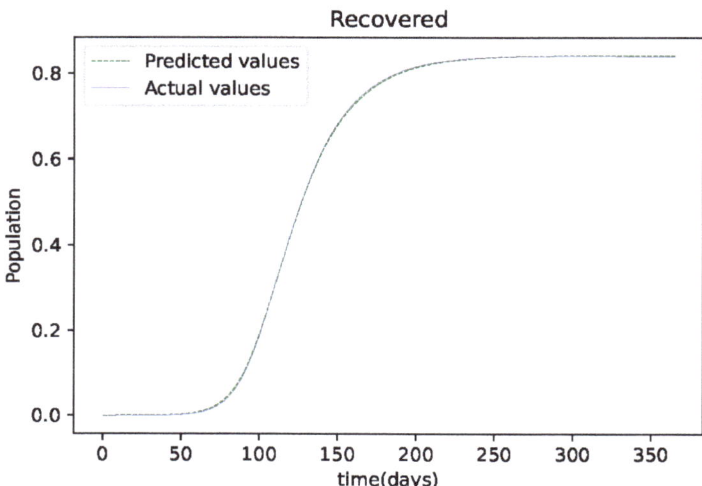

Figure 5. The graphs shows the results of the predicted values of the Recovered and the actual values of the recovered from the Mathematica generated data.

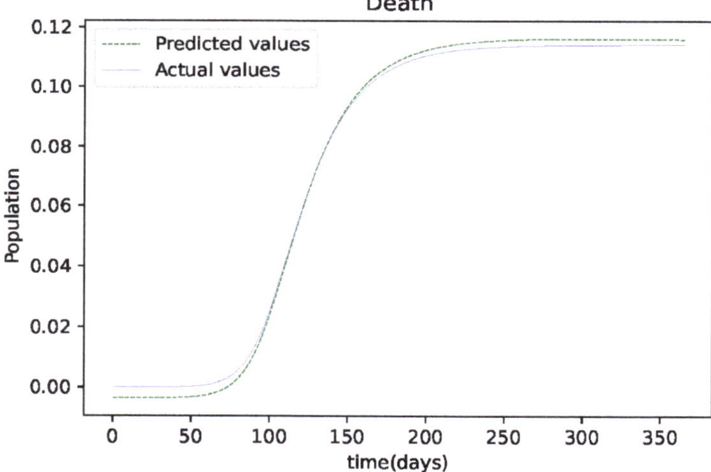

Figure 6. The resulting graph of the predicted values of the Deceased population and the actual values of the deceased population from the Mathematica generated data.

3.4. PINNs Simulations of Alabama State Data

We used a dataset from the American state of Alabama to test the SIRD model for further validation. The dataset spanned roughly 300 days, and the simulation used a three-layer neural network with 30 nodes per layer, with 1,000,000 iterations.

Figure 7 shows the outcome of data fitting, which compares the susceptible population data from the real data to that obtained by the trained PINNs model, and shows that there is a decent fit, implying that there is a little inaccuracy. The resulting graph of the data fitting is shown in Figure 8; it has a good fit, implying that there is a little error. The graph in Figure 9 compares the values acquired by the training model to the original data used in the recoveries, and it likewise has a strong match and less mistakes. Figure 10 is the last graph that compares the findings received after the model was trained to the actual data of the deceased population. This graph has a good match, but it is more erroneous than the others.

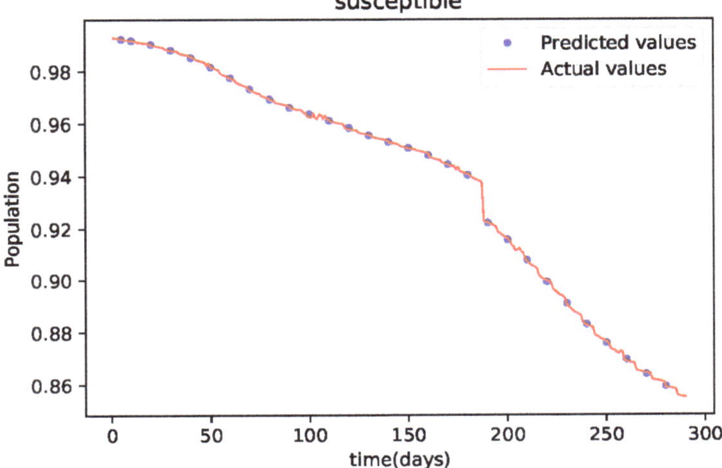

Figure 7. This graph shows a comparison of the predicted values of susceptible population and the actual data of susceptible population for the State of Alabama.

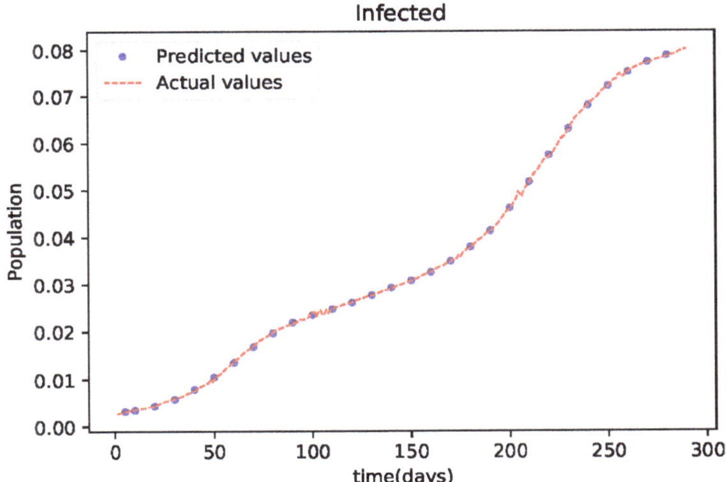

Figure 8. The graph shows a comparison of the predicted infected population values and the actual data infected population for the State of Alabama.

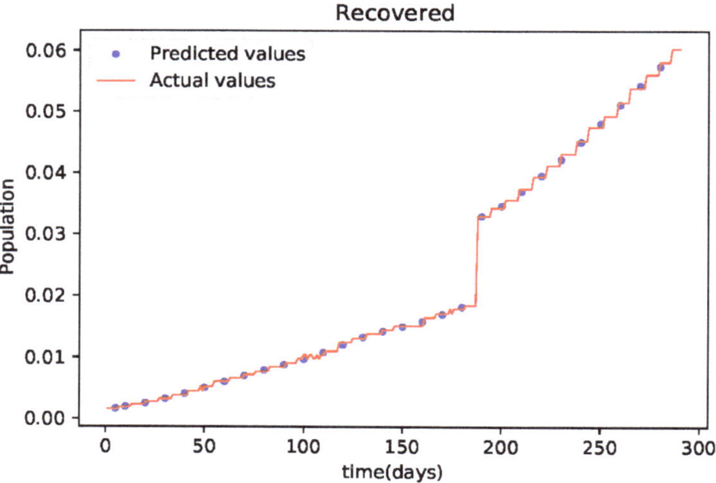

Figure 9. This graph shows a comparison of the predicted values of recovered population and the actual data of recovered population for the State of Alabama.

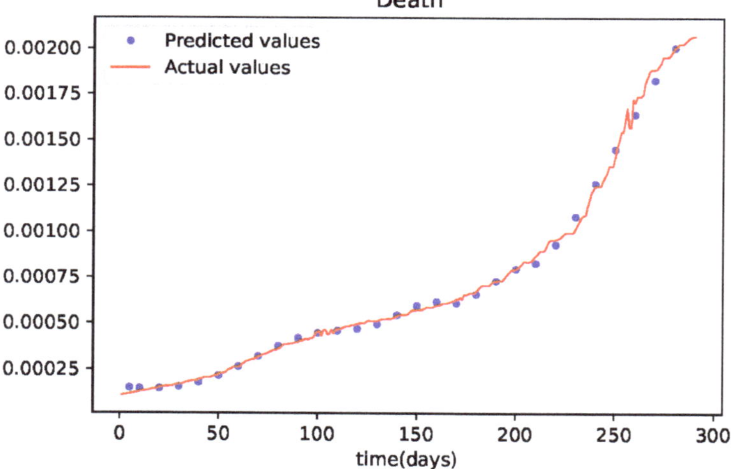

Figure 10. The graph shows a comparison of the predicted deceased population values and the actual data deceased population for the State of Alabama.

3.5. PINNs Simulation of a Model Using 170 Data Points

To put the model to the test and further test the potential of the PINNs model, a simulation using a smaller dataset was conducted. To conduct the simulation a dataset derived from the existing data to fully stretch over the period making up only 30% of the available data results were obtained.

The resulting graph, in Figure 11, compares the values acquired by the trained model to the actual data for the data fitting purposes of the vulnerable population and finds a good match, resulting in a tiny sized error. The obtained graph of data fitting is shown in Figure 12; it has a good fit, which indicates it has a minimal error. The graph in Figure 13 shows the recovered population results, which are a comparison of the trained model results and the real data, with a strong match and less mistakes. This graph has a good fit, but it has a larger error than the other graphs. Figure 14 is the resulting graph of the deceased population, which is a comparison of the trained model results and the actual data. The total result reveals that while the fitting has less mistakes, they are larger when compared to scenarios where larger data was employed.

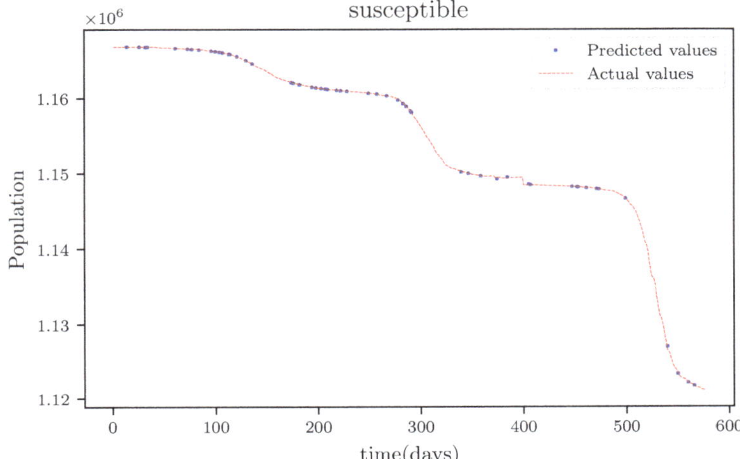

Figure 11. This graph shows a comparison of the predicted values of susceptible population and the actual data of susceptible population for a 130 data points.

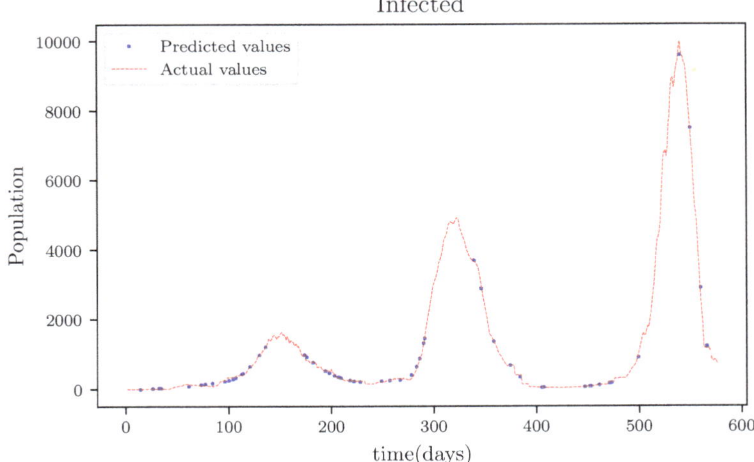

Figure 12. The graph shows a comparison of the predicted infected population values and the actual data of the infected population for a 130 data points.

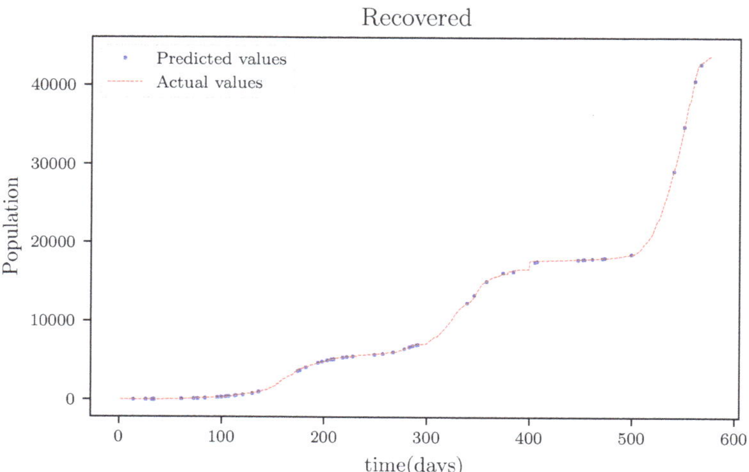

Figure 13. This graph shows a comparison of the predicted values of recovered population and the actual data of the recovered population for a 130 data points.

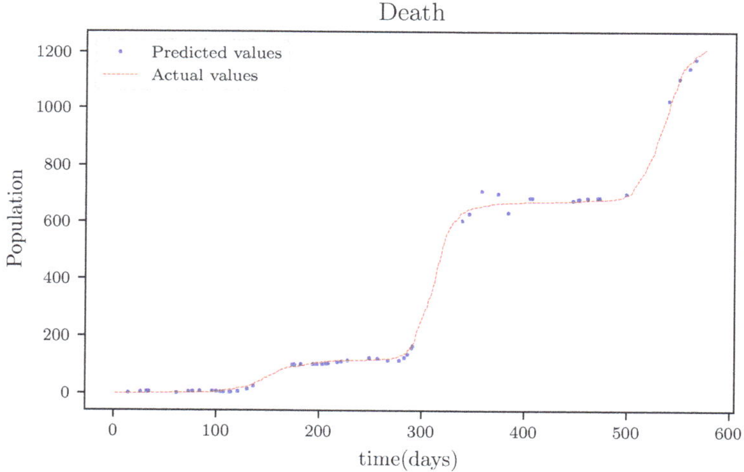

Figure 14. The graph shows a comparison of the predicted deceased population values and the actual data of the deceased population for a 130 data points.

3.6. PINNs Simulation of a Model Using All Available Data Points at the Time (576 Data Points)

The simulation was carried out with 5,000,000 iterations and four layers, each with 30 nodes. The dataset used was 576 days long, which was the maximum number of days accessible at the time.

In this situation, the best simulation is carried out utilizing all of the data available; the training data account for 70% of the data, while the testing sample accounts for just 30% of the data and is chosen at random. The resulting graph for data fitting purposes for the sensitive population is shown in Figure 15, and there is a tiny sized error and good fitting. The obtained graph of data fitting is shown in Figure 16; it has a good fit, which indicates it has a minimal error. The graph in Figure 17 depicts the recovered findings, and it has a strong fit and few mistakes. The resulting graph of deceased is shown in Figure 18; this graph has a decent fit, although it has a larger error than the other graphs. The overall result demonstrates that, while the fitting has less mistakes, they are larger than in circumstances when additional data was employed.

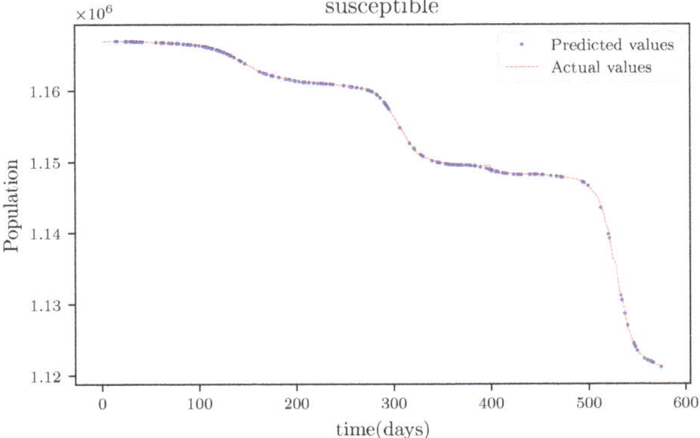

Figure 15. This graph shows a comparison of the predicted values of susceptible population and the actual data of the susceptible population for a 530 data points.

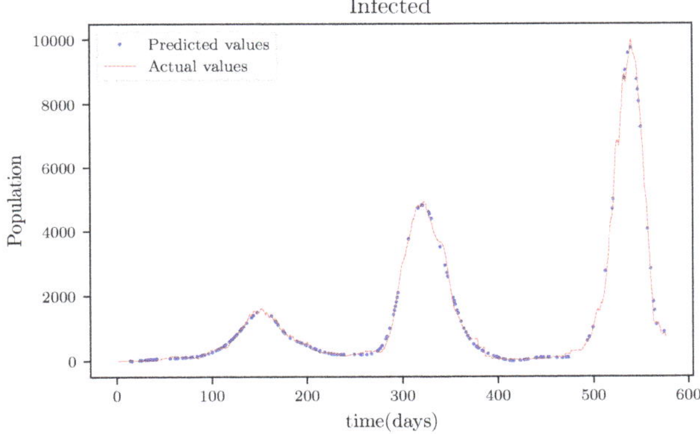

Figure 16. The graph shows a comparison of the predicted infected population values and the actual data of the infected population for a 530 data points.

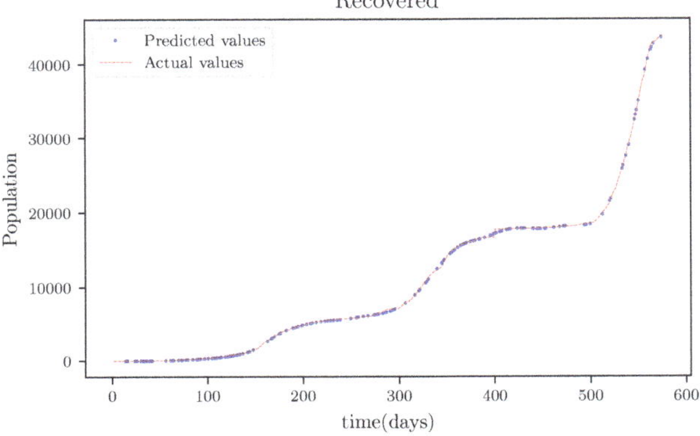

Figure 17. This graph shows a comparison of the predicted values of recovered population and the actual data of the recovered population for a 530 data points.

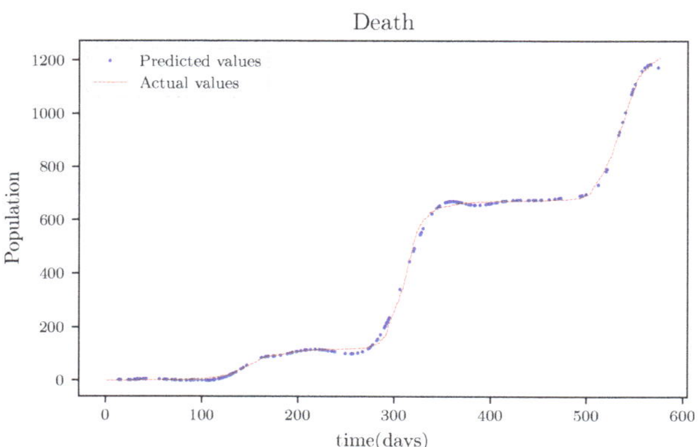

Figure 18. The graph shows a comparison of the predicted deceased population values and the actual data of the deceased population for a 530 data points.

3.7. PINNs Simulation Forecasting 30 Days

Simulations using the three layers of 30 nodes per layer was conducted and there were 5,000,000 iterations made during the training.

Figure 19 is a result graph based on the forecasting of the sensitive population for the next 30 days; the numbers appear to be declining, as expected. The resultant graph, shown in Figure 20, provides the anticipated data of active infections in a curved shape. The results of the anticipated recoveries are depicted in Figure 21. The resulting graph of the anticipated deceased population is shown in Figure 22. The prediction also determined the total predicted recoveries *Rend*, as well as the maximum expected infections *Imax*, susceptibles population expected at the end of the disease's spread *Send*, and the maximum expected infections I_{end}. Using the estimated parameter values, we obtained $I_{max} = 72{,}121$, $S_{end} = 1{,}094{,}719$, and $R_{end} = 70{,}274$.

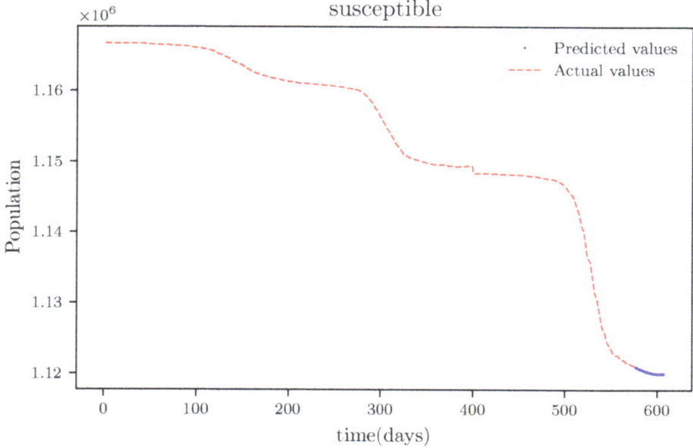

Figure 19. This graph shows a comparison of the predicted values of susceptible population and the actual data of susceptible population for a SIRD model with future predictions.

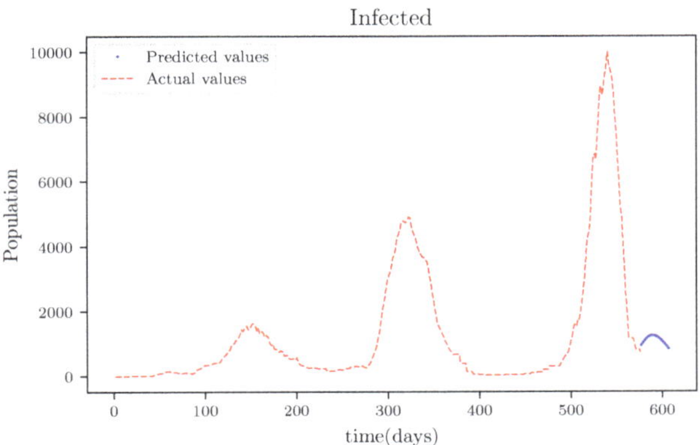

Figure 20. The graph shows a comparison of the predicted infected population values and the actual data infected population for a SIRD model with future predictions.

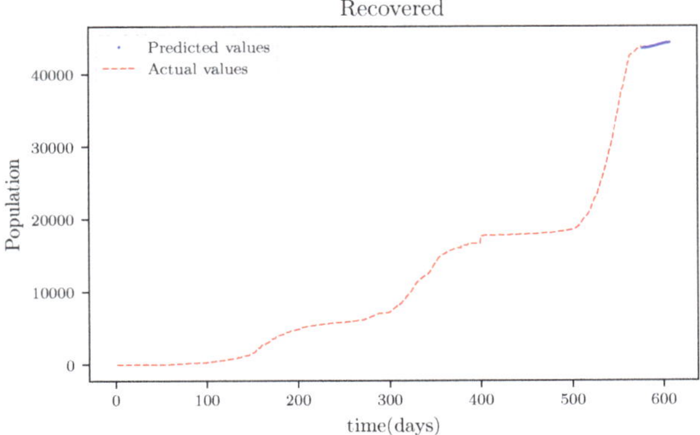

Figure 21. This graph shows a comparison of the predicted values of recovered population and the actual data of recovered population for a SIRD model with future predictions.

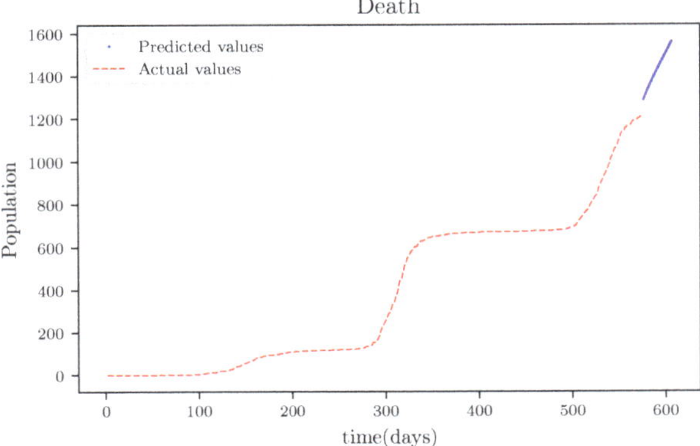

Figure 22. The graph shows a comparison of the predicted deceased population values and the actual data deceased population for a SIRD model with future predictions.

3.8. Deep Learning Sensitivity Analysis

The study's Physics Informed Neural Network framework is primarily influenced by four elements. The number of iterations conducted during training, the quantity of the training data utilized, the total number of layers in the model, and the number of nodes in each layer are all variables to consider. To begin the sensitivity analysis, the default model is set up with the following parameters: number of iterations 200,000, amount of training data 400, number of layers 3 and number of nodes 30.

Then, many simulations were run, with all model variables set to default, only two parameters changed, and all mean square errors recorded. The model setup is only subjected to a single simulation. Because the beginning values for each scenario are set at random, there are certain uncontrollable margins of error. A contingency only enables one further simulation trial in such instances.

The number of layers in the neural network and the number of performed iterations were the parameters that were varied in Table 1. The number of layers was varied between two, four and eight, while the number of iterations was varied between 100, 200, 400, and 800,000. The simulation results show that increasing the number of iterations reduces the amount of the error for the same number of layers. When the number of iterations is kept constant, the margin of error is reduced as well. This means that, as the number of iterations and layers increases, the accuracy improves.

Table 1. Results of the mean square error analyzing of varying number of iterations and number of layers in the simulations.

Iterations	Number of Layers		
	2	4	8
100,000	3.397×10^{-6}	1.996×10^{-7}	5.461×10^{-8}
200,000	2.434×10^{-6}	1.871×10^{-7}	3.866×10^{-8}
400,000	2.098×10^{-7}	2.340×10^{-8}	3.584×10^{-9}
800,000	1.454×10^{-7}	1.621×10^{-8}	3.055×10^{-9}

Table 2 compares the correlation impacts that the number of nodes per layer and the number of layers included in the neural network have. The number of nodes utilized in this test were 10, 20, 40, and 80, respectively, and the number of layers employed were two, four and eight. The data obtained, which are also displayed on the table of concern, show that increasing the number of nodes improves the margin of error given a constant number of layers. However, increasing the number of layers for a given number of layers reduces the margin of error. This means that, as the number of layers and nodes per layer are raised, the lowest margin of error is achieved for the physics informed neural network model established for this study.

Table 2. Results of the mean square error analysing of varying number of nodes in a layer and number of layers in the simulations.

Nodes	Number of Layers		
	2	4	8
10	1.870×10^{-7}	2.098×10^{-8}	4.783×10^{-9}
20	2.494×10^{-7}	2.243×10^{-8}	4.131×10^{-9}
40	2.144×10^{-7}	2.830×10^{-8}	4.723×10^{-9}
80	2.789×10^{-7}	2.941×10^{-8}	8.794×10^{-9}

The results of an analysis and simulations studying the correlation of the data size and the number of layers are shown in Table 3. The data sizes were 100, 150, 200, and 350, while the number of layers tested was two, four and eight. The results show that increasing the number of layers while keeping the data size constant reduces the margin of error. The investigation of a variety of data sizes on a fixed number of layers reveals that the values do not vary in a predictable way, but rather shift within a tiny margin. These findings reveal that data size has no effect on the number of layers, however data size has an effect on the number of layers, which reduces the margin of error.

Table 3. Results of the mean square error analysing of varying sizes of data points and number of layers in the simulations.

Data Size	Number of Layers		
	2	4	8
100	2.269×10^{-7}	2.070×10^{-8}	3.438×10^{-9}
150	2.121×10^{-7}	3.473×10^{-8}	4.123×10^{-9}
200	3.364×10^{-7}	2.904×10^{-8}	1.015×10^{-9}
350	3.214×10^{-7}	4.470×10^{-8}	3.440×10^{-9}

The results of an investigational analysis on the number of nodes per layer and the number of iterations are shown in Table 4. 10,000, 400,000, and 800,000 iterations were completed. The numbers 10, 20, 40, and 80 were used to test the nodes. The results show that increasing the number of iterations reduces the amount of the error while the number of nodes remains constant. The findings also reveal that increasing or decreasing the number of nodes has no effect on the amount of money spent. As a result, increasing the number of iterations reduces the amount of the error while having no effect on the number of nodes.

Table 4. Results of the mean square error analysing of varying number of iterations and number of nodes per layer in the simulations.

Nodes	Number of Iterations		
	100,000	400,000	800,000
10	8.991×10^{-7}	3.655×10^{-8}	5.824×10^{-8}
20	2.361×10^{-6}	2.323×10^{-8}	2.390×10^{-8}
40	5.144×10^{-7}	7.846×10^{-8}	2.632×10^{-8}
80	7.642×10^{-7}	5.921×10^{-8}	3.327×10^{-8}

The results of simulations undertaken to determine the correlation between the data size used during the training process and the number of iterations are shown in Table 5. The experiment included data sizes of 100, 150, 200, and 350, as well as iterations of 100, 400, and 800,000. The results show that increasing the number of iterations while keeping the training data size constant minimizes the margin of error. The results also show that changing the data size while maintaining a constant number of iterations has no effect on the number of iterations. As a result, increasing the number of iterations decreases the amount of the error, whereas changing the size of the training data has no effect.

Table 5. Results of the mean square error analysis of varying number of iterations and data size per layer in the simulations.

Data Size	Number of Iterations		
	100,000	400,000	800,000
100	1.292×10^{-6}	1.188×10^{-7}	2.752×10^{-8}
150	2.536×10^{-6}	3.273×10^{-8}	2.843×10^{-8}
200	1.110×10^{-6}	1.491×10^{-8}	2.064×10^{-8}
350	1.063×10^{-6}	3.399×10^{-8}	1.661×10^{-8}

The data size used during the training process and the number of nodes per layer are shown in Table 6. The experiment used 100, 150, 200, and 350 data sizes, as well as 100, 150, 200, and 350 iterations. The results suggest that increasing the number of nodes while keeping the training data size constant lowers the margin of error. The results also reveal that changing the data size while keeping the number of nodes fixed has no effect on the number of nodes. As a result, increasing the number of nodes lowers the amount of the error, however changing the size of the training data has no effect.

Table 6. Results of the mean square error analyzing of varying sizes of data points iterations and number of nodes in layers in the simulations.

Data Size	Number of Nodes		
	10	40	80
100	6.102×10^{-8}	4.882×10^{-8}	1.533×10^{-7}
150	4.574×10^{-8}	3.127×10^{-8}	1.823×10^{-7}
200	4.053×10^{-8}	3.386×10^{-8}	5.401×10^{-8}
350	1.277×10^{-7}	3.318×10^{-8}	1.231×10^{-8}

4. Discussion

The results obtained showed high accuracy especially in data fitting compared to the mathematical method's accuracy [34], particularly in data fitting. Another feature of this model is that it anticipates the wave behaviour of the active infected population as it makes forecasts. In comparison to other PINNs techniques, the model achieved roughly similar results, while a convolutional neural network time varying model [6,8] outperformed it somewhat. The method's major flaw is that it relies on previous data. This reduces its efficiency when it becomes accustomed to forecasting data, especially if the anticipated time is extended, because other previously unforeseeable factors, such as new species, are introduced.

5. Conclusions

The study's goal was to examine current data in order to determine COVID-19's behavioural dynamics. The Physics Informed Neural Networks framework was used to accomplish this analysis and to be able to estimate the dynamics of COVID-19. The focus of the research was to establish the number of patients who were susceptible, infected, recovered, and deceased in a timely manner. The study also intended to establish the disease's dissemination rate, death rate, and recovery rate based on this information. The rates were to be calculated on an average basis. The study attempted to take advantage of neural networks' ability to uncover hidden patterns in data, such as prospective increases and falls in dynamics, because they have this capability.

The Physics Informed Neural Networks framework used in the study is an ANN model that trains a neural network by exposing it to both the training data and the governing equations of the underlying problem. The Susceptible–Infected–Recovered–Deceased model was the mathematical model introduced as the governing equations of the PINNs training model. The underlying data for the simulations was collected between March 2020 and September 2021 in the kingdom of Eswatini. The main advantage of adopting the PINNs model was that it outperformed all other data analysis models even when given minimal quantities of training data, which was important given the disease's newness and the lack of data and knowledge [8].

The generated PINNs model was used to run simulations, and the results were presented in the form of tables and graphs. The study's first model was created artificially and had the advantage of being both accurate and covering a disease spread that lasted the entire life of the simulated disease spread. The acquired result had a modest margin of error, with the sole exception being the forecasts for the deceased population, which had a larger margin of error. This demonstrated that the proposed model could produce accurate findings and that the model could be used to assess the model over its entire life cycle. Another experiment was carried out with data from the state of Alabama in the United States. With a minimal margin of error, the constructed model was able to produce reliable results.This test was primarily performed to determine whether there was no overfitting; overfitting occurs when a created ANN can only handle a problem in the context of that specific data. As a result, while the study focused on using data from Eswatini, the results demonstrated that the data can be transferred and used with data from any country, region, continent, or sample of specification.

Only 170 data points were included in the first simulation results provided in the study using the main Eswatini dataset. The goal of this simulation was to see if the model could produce viable findings with limited amounts of data. This would be particularly useful in formulating predictions about the virus's behaviour in circumstances where data were sparingly collected or big gaps of missing data existed. The results collected revealed some positive outcomes with few faults. However, the results were less accurate when compared to those obtained when bigger datasets were employed. There were also substantially larger margins in the death forecasts, which was a problem that persisted across all simulations. This contradicts the conclusions of the sensitivity analysis, which found that increasing the size of the training dataset had a minimal margin of change.

To build the training and testing datasets, a bigger dataset encompassing all of the data available at the time was also employed. The results were similar to those obtained during the training of the smaller dataset, but with a far higher level of precision. This enormous dataset was also used to create forecasts for the future. These predictions were found to be quite accurate, but as the number of days forecasted increased, the accuracy declined. Even while making these future predictions, the determined dynamics demonstrated that they can both account for the creation of a wave, which was a critical and extremely unusual aspect that was found. This suggests that the model was able to adjust the dynamics both positively and negatively without any external help based on the data patterns.

The model was created over a long period of time, with early tests and simulations carried out when there were few data available. Although it is not displayed in the findings, it was discovered that the model was unable to forecast the major changes in the wave—namely the crust and thrust—during the first wave. However, as the data grew larger, the model began to recognize these patterns, and the results improved, eventually leading to those that were reached. The results produced during these simulations and tests were more accurate than those obtained during research that used the mathematical model [35,36] to conduct tests. This is because, while average rates are employed in both cases, the generated PINNs model gradually learned to modify these rates or dynamics based on hidden patterns in the data. In comparison to [6,8], the results had similar margins of error, primarily due to data fitting because no future predictions were offered in these research.

The results show that the model is well adapted to making value predictions during the training period. As a result, the model is highly adapted to data fitting in situations when data were not gathered, were incorrectly input, or were lost. Another advantage of the architecture is that it returns the spreading rate, death rate and recovery rate, which were set up to function as modifying variables for the neural network's PINNs component. However, as the number of projected days grows larger, the accuracy of the forecast decreases. This happens because the projections are based on the spread rate, mortality rate and recovery rate trends, all of which are based on outdated data. The wave format produced by active infections is also predicted by the model. The results show that the model is well equipped to make decisions. As a result, while the model is well-suited to making short-term predictions, it may also be used to make long-term predictions with a manageable margin of error.

The study had the misfortune of only lasting a few weeks. As a result, the model was unable to make a number of jumps, including the inability to forecast when a potential wave would occur. There have also been some new developments, such as the discovery that some infected individuals can become re-affected. As a result, the SIRD model would be more accurate than the SIRD model currently in use. Thus, future research into these areas, which are currently understudied, is necessary.

The lack of data and processing power was a major stumbling block during the project. As a result, we advocate developing a model that groups each of the SIRD populations by age for future research, as it has been proven that different age groups are impacted by the disease differently. Each of the metrics or rates can then be defined per age group using the well segmented data. We also suggest testing a model identical to the one employed in the study, but on a larger scale with more processing capacity.

Author Contributions: Conceptualization, J.M., S.G. and S.M.; methodology, J.M., S.G. and S.M.; software, S.G.; validation, S.G.; formal analysis, S.G.; investigation, S.G.; resources, data curation, S.G.; writing—original draft preparation, S.G.; writing—review and editing, J.M., S.G. and S.M.; visualization, J.M., S.G. and S.M.; supervision, J.M. and S.M.; project administration, J.M.; funding acquisition, S.M. All authors have read and agreed to the published version of the manuscript.

Funding: This research was funded by the National Research Foundation of South Africa, Grant number 131604 and the School of Mathematics, Statistics and Computer Science, University of KwaZulu-Natal, South Africa.

Institutional Review Board Statement: Not applicable.

Informed Consent Statement: Not applicable.

Data Availability Statement: The dataset used in this study can be obtained from: https://datastudio.google.com/reporting/b847a713-0793-40ce-8196-e37d1cc9d720/page/2a0LB (accessed on 1 October 2021).

Conflicts of Interest: The authors declare no conflict of interest.

References

1. Moriarty, L.F.; Plucinski, M.M.; Marston, B.J.; Kurbatova, E.V.; Knust, B.; Murray, E.L.; Pesik, N.; Rose, D.; Fitter, D.; Kobayashi, M.; et al. Public health responses to COVID-19 outbreaks on cruise ships worldwide, February-March 2020. *Morb. Mortal. Wkly. Rep.* **2020**, *69*, 347–352 [CrossRef] [PubMed]
2. Lu, H.; Stratton, C.W.; Tang, Y.W. Outbreak of pneumonia of unknown etiology in Wuhan, China: The mystery and the miracle. *J. Med Virol.* **2020**, *92*, 401–402. [CrossRef]
3. Chen, Y.C.; Lu, P.E.; Chang, C.S.; Liu, T.H. A time-dependent SIR model for COVID-19 with undetectable infected persons. *IEEE Trans. Netw. Sci. Eng.* **2020**, *7*, 3279–3294. [CrossRef]
4. Guo, Y.R.; Cao, Q.D.; Hong, Z.S.; Tan, Y.Y.; Chen, S.D.; Jin, H.J.; Tan, K.S.; Wang, D.Y.; Yan, Y. The origin, transmission and clinical therapies on coronavirus disease 2019 (COVID-19) outbreak-an update on the status. *Mil. Med. Res.* **2020**, *7*, 1–10. [CrossRef] [PubMed]
5. Bloomgarden, Z.T. Diabetes and COVID-19. *J. Diabetes* **2020**, *12*, 347–348. [CrossRef]
6. Dandekar, R.; Barbastathis, G. Neural Network aided quarantine control model estimation of global Covid-19 spread. *arXiv* **2020**, arXiv:2004.02752.
7. Lauer, S.A.; Grantz, K.H.; Bi, Q.; Jones, F.K.; Zheng, Q.; Meredith, H.R.; Azman, A.S.; Reich, N.G.; Lessler, J. The incubation period of coronavirus disease 2019 (COVID-19) from publicly reported confirmed cases: Estimation and application. *Ann. Intern. Med.* **2020**, 172, 577–582. [CrossRef]
8. Long, J.; Khaliq, A.Q.M.; Furati, K.M. Identification and prediction of time-varying parameters of COVID-19 model: A data-driven deep learning approach. *Int. J. Comput. Math.* **2021**, 1–19. [CrossRef]
9. Boone, L.; Haugh, D.; Pain, N.; Salins, V. 2 tackling the fallout from COVID-19. In *Economics in the Time of COVID-19*; CEPR Press: London, UK, 2020; p. 37.
10. Nyabadza, F.; Chirove, F.; Chukwu, W.; Visaya, M. Modelling the potential impact of social distancing on the covid-19 epidemic in south africa. *medRxiv* **2020**, *2020*, 5379278. [CrossRef]
11. Peng, L.; Yang, W.; Zhang, D.; Zhuge, C.; Hong, L. Epidemic analysis of covid-19 in china by dynamical modeling. *arXiv* **2020**, arXiv:2002.06563.
12. Eikenberry, S.; Mancuso, M.; Iboi, E.; Phan, T.; Eikenberry, K.; Kuang, Y.; Kostelich, E.; Gumel, A. To mask or not to mask: Modeling the potential for face mask use by the general public to curtail the covid-19 pandemic. *Infect. Dis. Model.* **2020**, *5*, 293–308. [CrossRef]

13. Tang, B.; Xia, F.; Tang, S.; Bragazzi, N.L.; Li, Q.; Sun, X.; Liang, J.; Xiao, Y.; Wu, J. The effectiveness of quarantine and isolation determine the trend of the covid-19 epidemics in the final phase of the current outbreak in china. *Int. J. Infect. Dis.* **2020**, *95*, 288–293. [CrossRef] [PubMed]
14. Ndairou, F.; Area, I.; Nieto, J.; Torres, D. Mathematical modeling of covid-19 transmission dynamics with a case study of wuhan. *Chaos Solitons Fractals* **2020**, *135*, 109846. [CrossRef]
15. Anguelov, R.; Banasiak, J.; Bright, C.; Lubuma, J.; Ouifki, R. The big unknown: The asymptomatic spread of covid-19. *BIOMATH* **2020**, *9*, 2005103. [CrossRef]
16. Jia, J.; Ding, J.; Liu, S.; Liao, G.; Li, J.; Duan, B.; Wang, G.; Zhang, R. Modeling the control of covid-19: Impact of policy interventions and meteorological factors. *arXiv* **2020**, arXiv:2003.02985.
17. Kucharski, A.; Russell, T.; Diamond, C.; Liu, Y.; Edmunds, J.; Funk, S.; Eggo, R.; Sun, F.; Jit, M.; Munday, J.; et al. Early dynamics of transmission and control of covid-19: A mathematical modelling study. *Lancet Infect. Dis.* **2020**, *20*, 553–558. [CrossRef]
18. Liu, Y.; Gayle, A.; Wilder-Smith, A.; Rocklöv, J. The reproductive number of covid-19 is higher compared to sars coronavirus. *J. Travel Med.* **2020**, *27*, aaa021. [CrossRef]
19. Shayak, B.; Sharma, M.; Rand, R.H.; Singh, A.; Misra, A. Transmission dynamics of covid-19 and impact on public health policy. *medRxiv* **2020**. [CrossRef]
20. Sun, T.; Wang, Y. Modeling covid-19 epidemic in heilongjiang province, china. *Chaos Solitons Fractals* **2020**, *138*, 109949. [CrossRef]
21. Boccaletti, S.; Ditto, W.; Mindlin, G.; Atangana, A. Modeling and forecasting of epidemic spreading: The case of COVID-19 and beyond. *Chaos Solitons Fractals* **2020**, *135*, 109794. [CrossRef]
22. Castillo, O.; Melin, P. Forecasting of covid-19 time series for countries in the world based on a hybrid approach combining the fractal dimension and fuzzy logic. *Chaos Solitons Fractals* **2020**, *140*, 110242. [CrossRef]
23. Ivorra, B.; Ferrandez, M.R.; Vela-Perez, M.; Ramosa, A.M. Mathematical modeling of the spread of the coronavirus disease 2019 (COVID-19) taking into account the undetected infections. The case of China. *Commun. Nonlinear Sci. Numer. Simul.* **2020**, *88*, 105303. [CrossRef] [PubMed]
24. Huang, J.; Agarwal, S. Physics informed deep learning for traffic state estimation. In Proceedings of the 2020 IEEE 23rd International Conference on Intelligent Transportation Systems (ITSC), Rhodes, Greece, 20–23 September 2020; pp. 1–6.
25. Raissi, M.; Perdikaris, P.; Karniadakis, G.E. Physics-informed neural networks: A deep learning framework for solving forward and inverse problems involving nonlinear partial differential equations. *J. Comput. Phys.* **2019**, *378*, 686–707. [CrossRef]
26. Jagtap, A.D.; Karniadakis, G.E. Extended Physics-Informed Neural Networks (XPINNs): A Generalized Space-Time Domain Decomposition Based Deep Learning Framework for Nonlinear Partial Differential Equations. *Commun. Comput. Phys.* **2020**, *28*, 2002–2041.
27. Goodfellow, I.; Bengio, Y.; Courville, A. *Deep Learning*; MIT Press: Cambridge, UK, 2016; Volume 1.
28. Barto, A.G.; Dietterich, T.G. Reinforcement learning and its relationship to supervised learning. *Handb. Learn. Approx. Dyn. Program.* **2004**, *10*, 9780470544785.
29. Li, S.; Ma, B.; Chang, H.; Shan, S.; Chen, X. Continuity-discrimination convolutional neural network for visual object tracking. In Proceedings of the 2018 IEEE International Conference on Multimedia and Expo (ICME), San Diego, CA, USA, 23–27 July 2018; pp. 1–6.
30. Dey, A. Machine learning algorithms: A review. International *J. Comput. Sci. Inf. Technol.* **2016**, *7*, 1174–1179.
31. Sra, S.; Nowozin, S.; Wright, S.J. (Eds.) *Optimization for Machine Learning*; MIT Press: Cambridge, MA, USA, 2012.
32. Misyris, G.S.; Venzke, A.; Chatzivasileiadis, S. Physics-informed neural networks for power systems. In Proceedings of the 2020 IEEE Power & Energy Society General Meeting (PESGM), Virtual Event, 3–6 August 2020; pp. 1–5.
33. University of Eswatini. Available online: https://datastudio.google.com/reporting/b847a713-0793-40ce-8196-e37d1cc9d720/page/2a0LB (accessed on 1 October 2021)
34. Huang, C.; Wang, Y.; Li, X.; Ren, L.; Zhao, J.; Hu, Y.; Zhang, L.; Fan, G.; Xu, J.; Gu, X.; et al. Clinical features of patients infected with 2019 novel coronavirus in Wuhan, China. *Lancet* **2020**, *395*, 497–506. [CrossRef]
35. Thabet, S.T.M.; Abdo, M.S.; Shah, K.; Abdeljawadd, T. Study of transmission dynamics of COVID-19 mathematical model under ABC fractional order derivative. *Results Phys.* **2020**, *19*, 103507. [CrossRef]
36. Inui, S.; Fujikawa, A.; Jitsu, M.; Kunishima, N.; Watanabe, S.; Suzuki, Y.; Umeda, S.; Uwabe, Y. Chest CT findings in cases from the cruise ship diamond princess with Coronavirus disease (COVID-19). *Radiol. Cardiothorac. Imaging* **2020**, *2*, e200110. [CrossRef]

MDPI

Article

Maximum Principle and Second-Order Optimality Conditions in Control Problems with Mixed Constraints

Aram V. Arutyunov [1,†], Dmitry Yu. Karamzin [2,*,†] and Fernando Lobo Pereira [3,*,†]

1 V.A. Trapeznikov Institute of Control Sciences, Russian Academy of Sciences, 117997 Moscow, Russia; arutyunov@cs.msu.ru
2 Federal Research Center "Computer Science and Control", Russian Academy of Sciences, 119333 Moscow, Russia
3 Faculty of Engineering, University of Porto, 4200-465 Porto, Portugal
* Correspondence: dmitry_karamzin@mail.ru (D.Y.K.); flp@fe.up.pt (F.L.P.)
† These authors contributed equally to this work.

Abstract: This article concerns the optimality conditions for a smooth optimal control problem with an endpoint and mixed constraints. Under the normality assumption, which corresponds to the full-rank condition of the associated controllability matrix, a simple proof of the second-order necessary optimality conditions based on the Robinson stability theorem is derived. The main novelty of this approach compared to the known results in this area is that only a local regularity with respect to the mixed constraints, that is, a regularity in an ε-tube about the minimizer, is required instead of the conventional stronger global regularity hypothesis. This affects the maximum condition. Therefore, the normal set of Lagrange multipliers in question satisfies the maximum principle, albeit along with the modified maximum condition, in which the maximum is taken over a reduced feasible set. In the second part of this work, we address the case of abnormal minimizers, that is, when the full rank of controllability matrix condition is not valid. The same type of reduced maximum condition is obtained.

Keywords: optimal control; maximum principle; mixed constraints

Citation: Arutyunov, A.V.; Karamzin, D.Y.; Pereira, F.L. Maximum Principle and Second-Order Optimality Conditions in Control Problems with Mixed Constraints. *Axioms* **2022**, *11*, 40. https://doi.org/10.3390/axioms11020040

Academic Editor: Natália Martins

Received: 28 December 2021
Accepted: 15 January 2022
Published: 20 January 2022

1. Introduction

In this article, second-order necessary optimality conditions for an optimal control problem with mixed equality and inequality constraints are investigated. Under the normality condition, which is ensured by the full rank of the controllability matrix, a rather simple proof of the optimality conditions is proposed based on Robinson's theorem on the metric regularity for set-valued mappings. For the case in which the normality condition is violated, the second-order conditions are derived based on the index approach. This means that some reduced cone of Lagrange multipliers is invoked, which is defined by using the index of the quadratic form of the Lagrange function; see, e.g., [1–3].

In work [4], the two notions of the strong and of weak regularity of an admissible trajectory with respect to mixed constraints have been considered. Strong regularity means that the constraint qualification, or the so-called Robinson condition, is satisfied for all time-points and for all admissible control values. This corresponds to the regularity condition in the classical sense. Weak regularity means that this condition is satisfied merely in some neighborhood of the optimal process. By their nature, these two concepts correspond, respectively, to global, and local regularity settings. Under weak regularity, a refined maximum condition of Pontryagin's type has been obtained, in which the maximum is taken over the closure of regular points of the feasible set, but not over the entire feasible set. In this article, the results of [4] are carried over to the second-order conditions in the case of global minimum.

The literature on optimality conditions for optimal control problems with mixed constraints is extensive. In the context of this research related to the study of mixed constraints, we note the works of [5–12]. Regarding the second-order conditions in mixed constrained problems, one may consider, e.g., [3,13,14] and the bibliography cited therein. At the same time, these selective lists of publications are far from exhaustive.

This work is organized as follows. In the next section, the problem formulation is presented, together with main definitions and notation. In Section 3, the issue of normality is discussed. In Section 4, the main result of this work—the normal maximum principle and second-order optimality conditions—is formulated and proved. In Section 5, the abnormal situation is taken into consideration, and the result of the previous section is refined. Section 6 concludes the work with a short summary.

2. Problem Formulation

Consider the following optimal control problem on the fixed time interval $[0,1]$:

$$\begin{cases} \text{Minimize} & \varphi(p) \\ \text{subject to} & \dot{x}(t) = f(x(t), u(t), t) \text{ for a.a. } t, \\ & e_1(p) \leq 0,\ e_2(p) = 0, \\ & u(t) \in U(x(t), t) \text{ for a.a. } t, \end{cases} \tag{1}$$

where $p = (x_0, x_1)$, $x_0 = x(0)$, $x_1 = x(1)$, $t \in [0,1]$, and

$$U(x,t) := \{u \in \mathbb{R}^m : r_1(x,u,t) \leq 0,\ r_2(x,u,t) = 0\}.$$

The mappings $\varphi : \mathbb{R}^{2n} \to \mathbb{R}$, $e_i : \mathbb{R}^{2n} \to \mathbb{R}^{k_i}$, $f : \mathbb{R}^n \times \mathbb{R}^m \times \mathbb{R}^1 \to \mathbb{R}^n$, $r_i : \mathbb{R}^n \times \mathbb{R}^m \times \mathbb{R}^1 \to \mathbb{R}^{q_i}$, $i = 1,2$, satisfy the following hypothesis.

Hypothesis 1 (H1). *Mappings $\varphi, e_1, e_2, f, r_1, r_2$ are twice continuously differentiable.*

The vector $p = (x_0, x_1)$ is termed the endpoint, as well as the constraints given by mappings e_1, e_2. The scalar function $\varphi(p)$ defines the minimizing functional. Mappings r_1, r_2 define the mixed constraints which are imposed on both state and control variables.

A pair of functions (x,u) is designed by the control process; if $x(\cdot)$ is absolutely continuous, $u(\cdot)$ is measurable and essentially bounded, whereas $\dot{x}(t) = f(x(t), u(t), t)$ for a.a. $t \in [0,1]$. A control process is feasible, provided that the endpoint, control and state constraints are satisfied. A feasible process $(\bar{x}, \bar{u})$ is termed optimal if, for any feasible process (x,u), $\varphi(\bar{p}) \leq \varphi(p)$, where $\bar{p} = (\bar{x}(0), \bar{x}(1))$.

This concept of the minimum is known as a *global strong minimum*. The purpose of this work is to derive the second-order necessary optimality conditions for this type of minimum under the normality assumptions. That is, to find such a set of Lagrange multipliers that simultaneously satisfies the maximum principle and Legendre's condition, and for which $\lambda^0 > 0$. Such a set of multipliers must be unique upon normalization. The abnormal situation is also examined after the normal case.

Consider the reference control process $(\bar{x}, \bar{u})$, which can be optimal, extremal, regular, or normal in what follows. Denote by $r = (r_1, r_2)$ the joint mapping acting onto $\mathbb{R}^q$, where $q = q_1 + q_2$. Let $J(x,u,t) := \{j : r^j(x,u,t) = 0\}$ be the set of active indices, where the upper index specifies the vector component. Set $J(u,t) := J(\bar{x}(t), u, t)$. Let $\mathcal{U}(\cdot)$ designate the closure of function $\bar{u}(\cdot)$ w.r.t. the Lebesgue measure; that is, for a given $t \in [0,1]$, the set $\mathcal{U}(t)$ consists of essential values of $\bar{u}(\cdot)$ at point t, [8]. Recall that the vector a is said to be the essential value of a function $u(\cdot)$ at point τ, provided that $\ell(\{t \in [\tau - \varepsilon, \tau + \varepsilon] : u(t) \in B_\varepsilon(a)\}) > 0\ \forall\, \varepsilon > 0$, where $B_\varepsilon(a)$ is the closed ball centered at a with the radius ε, and ℓ designates the Lebesgue measure on $\mathbb{R}$.

The main regularity concept is as follows.

Definition 1. *The control process $(\bar{x}, \bar{u})$ is said to be regular w.r.t. the mixed constraints, provided that, for all $t \in [0,1]$ and for all $u \in \mathcal{U}(t)$, the active gradients $(r^j)'_u(\bar{x}(t), u, t)$, $j \in J(u,t)$ are linearly independent.*

The following proposition represents an equivalent reformulation of the introduced regularity concept. For $\varepsilon \geq 0$, define the set

$$J_\varepsilon(x,u,t) := \left\{ j \in \{1,\ldots,q_1\} : r^j(x,u,t) \in [-\varepsilon, 0] \right\} \cup \{q_1+1,\ldots,q\},$$

which is subject to the same conventions as the mapping $J(x,u,t)$. It is clear that $J \subseteq J_\varepsilon$, and $J_0 = J$.

Proposition 1. *Let the control process $(\bar{x}, \bar{u})$ be regular w.r.t. the mixed constraints. Then, there exists a number $\varepsilon_0 > 0$ such that, for all $t \in [0,1]$ and for almost all $s \in [0,1]$ such that $|s-t| \leq \varepsilon_0$, the ε_0-active gradients $(r^j)'_u(\bar{x}(s), \bar{u}(s), s)$, $j \in J_{\varepsilon_0}(s)$ are linearly independent. Moreover, the number ε_0 can be chosen such that the modulus of surjectivity for this set of gradients is not lower than ε_0.*

The proof is based on a simple contradiction argument.

The point $u \in U(x,t)$ is termed regular provided that the gradients $(r^j)'_u(x,u,t)$, $j \in J(x,u,t)$ are positively linearly independent. The subset of all regular points of $U(x,t)$ is denoted as $U_R(x,t)$. Denote

$$\Theta(x,t) := \operatorname{clos} U_R(x,t).$$

It is clear that, for the regular process $(\bar{x}, \bar{u})$, one has $\mathcal{U}(t) \subseteq \Theta(\bar{x}(t),t) \neq \varnothing \; \forall\, t \in [0,1]$. Consider the Hamilton–Pontryagin function

$$\mathcal{H}(x,u,\psi,t) := \langle \psi, f(x,u,t) \rangle,$$

and the Lagrangian

$$\mathcal{L}(p,\lambda) := \lambda^0 \varphi(p) + \langle \lambda_1, e_1(p) \rangle + \langle \lambda_2, e_2(p) \rangle.$$

Here, $\psi \in (\mathbb{R}^n)^*$, $\lambda = (\lambda^0, \lambda_1, \lambda_2) \in (\mathbb{R}^{1+k_1+k_2})^*$ are the conjugate variables.

Definition 2. *The control process $(\bar{x}, \bar{u})$ is said to satisfy the maximum principle provided that there exists a vector $\lambda = (\lambda^0, \lambda_1, \lambda_2) \in (\mathbb{R}^{1+k_1+k_2})^*$, where $\lambda^0 \geq 0$ and $\lambda_1 \geq 0$, an absolutely continuous vector-valued function $\psi \in \mathbb{W}_{1,\infty}([0,1]; (\mathbb{R}^n)^*)$, and a measurable, essentially bounded, vector-valued function $\nu \in \mathbb{L}_\infty([0,1]; (\mathbb{R}^q)^*)$ of which the j-th component is nonnegative for $j = 1,\ldots,q_1$, such that $\lambda \neq 0$, and, on $[0,1]$, it holds that:*

$$\dot{\psi}(t) = -\mathcal{H}'_x(\bar{x}(t), \bar{u}(t), \psi(t), t) + \nu(t) r'_x(\bar{x}(t), \bar{u}(t), t) \text{ for a.a. } t, \tag{2}$$

$$\psi(\alpha) = (-1)^\alpha \mathcal{L}'_{x_\alpha}(\bar{p}, \lambda) \text{ for } \alpha = 0,1, \tag{3}$$

$$\max_{u \in \Theta(\bar{x}(t),t)} \mathcal{H}(\bar{x}(t), u, \psi(t), t) = \mathcal{H}(\bar{x}(t), \bar{u}(t), \psi(t), t) \text{ for a.a. } t, \tag{4}$$

$$\mathcal{H}'_u(\bar{x}(t), \bar{u}(t), \psi(t), t) - \nu(t) r'_u(\bar{x}(t), \bar{u}(t), t) = 0 \text{ for a.a. } t, \tag{5}$$

$$\langle \lambda_1, e_1(\bar{p}) \rangle = 0, \text{ and } \int_0^1 \langle \nu(t), r(\bar{x}(t), \bar{u}(t), t) \rangle dt = 0. \tag{6}$$

Here, Condition (2) is the co-state equation, that is, the differential equation for the conjugate variable ψ. Equalities (3) are the transversality conditions. Equality (4) is the maximum condition. Equality (5) is the so-called Euler–Lagrange equation. Equalities (6) are known as the complementary slackness condition. Furthermore, (λ, ψ, ν) are known as the Lagrange multipliers.

Under the regularity condition given in Definition 3, the multipliers ψ, and ν are uniquely defined by the vector λ, where (λ, ψ, ν) is the set of Lagrange multipliers corresponding to $(\bar{x}, \bar{u})$ in view of the maximum principle. This assertion simply follows from the Euler–Lagrange equation. Then, denote by $\Lambda = \Lambda(\bar{x}, \bar{u})$ the set of vectors $\lambda \in (\mathbb{R}^{1+k_1+k_2})^*$ for which there exist (ψ, ν) such that the corresponding set of Lagrange multipliers (λ, ψ, ν) generated by λ satisfies the maximum principle.

3. Normality Condition

Let us introduce the notion of normality. This notion is based on the concept of linearization of the control problem and the corresponding variational differential system. Consider the reference control process $(\bar{x}, \bar{u})$, and a pair $(\delta x_0, \delta u) \in \mathcal{X} := \mathbb{R}^n \times \mathbb{L}_2([0,1];\mathbb{R}^m)$. Denote by $\delta x(\cdot)$ the solution to the variational differential equation on the time interval $[0,1]$, which corresponds to $(\delta x_0, \delta u)$, that is,

$$\dot{\delta x}(t) = f'_x(\bar{x}(t), \bar{u}(t), t)\delta x(t) + f'_u(\bar{x}(t), \bar{u}(t), t)\delta u(t), \tag{7}$$

where $\delta x(0) = \delta x_0$. Such a solution exists on the entire time interval $[0,1]$ and, as soon as δu is an L_2-function, one finds that $\delta x \in \mathbb{W}_{1,2}([0,1];\mathbb{R}^n)$.

In what follows, it is not restrictive to set $e_1(\bar{p}) = 0$. Thus, all the endpoint constraints of the inequality type are assumed to be active. Consider the two following subspaces in $\mathcal{X}$:

$$\mathcal{N}_e := \Big\{(\delta x_0, \delta u) \in \mathcal{X} : e'(\bar{p})\delta p = 0\Big\},$$
$$\mathcal{N}_r := \Big\{(\delta x_0, \delta u) \in \mathcal{X} : D(t)\Big[r'_x(\bar{x}(t), \bar{u}(t), t)\delta x(t) + r'_u(\bar{x}(t), \bar{u}(t), t)\delta u(t)\Big] = 0\Big\}.$$

Here, e is the joint mapping of e_1, e_2; $\delta p = (\delta x_0, \delta x_1)$, where $\delta x_1 = \delta x(1)$, and $D(t)$ is the diagonal $q \times q$-matrix which has 1 in the position (j,j) iff $j \in J(t)$ and 0 otherwise.

Consider the matrix

$$R(t) = r'_u(\bar{x}(t), \bar{u}(t), t)^* D(t) r'_u(\bar{x}(t), \bar{u}(t), t).$$

Set

$$M(t) := R(t)^+ r'_u(\bar{x}(t), \bar{u}(t), t)^* D(t) r'_x(\bar{x}(t), \bar{u}(t), t).$$

Here, A^+ stands for the generalized inverse [15]. Here, the generalized inverse $R(t)^+$ can be computed as follows. Let $T(t)$ be a non-singular orthogonal linear transform which maps the subspace $\ker R(t)^\perp = \operatorname{im} R(t)$ onto the subspace of $\mathbb{R}^m$ with the first $m - q(t)$ coordinates vanished, where $q(t) = |J(t)|$ is the number of active indices. Then, $R(t)^+ = T^{-1}(t)[T(t)R(t)T^{-1}(t)]^{-1}T(t)$, where the pseudo-inverse of the block $\begin{pmatrix} 0 & 0 \\ 0 & A \end{pmatrix}^{-1}$ is understood as $\begin{pmatrix} 0 & 0 \\ 0 & A^{-1} \end{pmatrix}$.

Consider the matrix differential system

$$\dot{\Phi}(t) = f'_x(\bar{x}(t), \bar{u}(t), t)\Phi(t) - f'_u(\bar{x}(t), \bar{u}(t), t)M(t)\Phi(t), \tag{8}$$

where $\Phi(0) = I$. Let $\Phi(t)$ be the solution to (8), and $P(t)$ be the matrix of orthogonal projection onto $\ker R(t)$.

It is clear that by virtue of the construction, any element $(\delta x_0, \delta u) \in \mathcal{N}_r$ can be represented as

$$(\delta x_0, \delta u) = (\delta x_0, P\delta u + \mathcal{V}[\delta x_0, \delta u]), \tag{9}$$

where $\mathcal{V}[\delta x_0, \delta u] = M(t)\delta x(t)$, whereas

$$\delta x(t) = \Phi(t)\Big(\delta x_0 + \int_0^t \Phi^{-1}(s) f'_u(\bar{x}(s), \bar{u}(s), s)P(s)\delta u(s)ds\Big). \tag{10}$$

Conversely, any $\delta x_0 \in \mathbb{R}^n$ and $\delta v \in \mathbb{L}_2([0,1];\mathbb{R}^m)$: $\delta v(t) \in \ker R(t)$ a.e. yields an element of $\mathcal{N}_r$ as $(\delta x_0, \delta v + \mathcal{V}[\delta x_0, \delta v]) \in \mathcal{N}_r$. Therefore, there is a one-to-one correspondence between $\mathcal{N}_r$ and the space of the above-specified elements $(\delta x_0, \delta v)$. At the same time, the formula for the solution δx in $\mathcal{N}_r$ is given by (10).

Let us proceed with the construction of the controllability matrix. Define the $\mathbb{R}^{n\times k}$-matrix A as

$$A = e'_{x_0}(\bar{p}) + e'_{x_1}(\bar{p})\Phi(1),$$

with the $\mathbb{R}^{m\times k}$-matrix $B(t)$ given as

$$B(t) = e'_{x_1}(\bar{p})\Phi(1)\Phi^{-1}(t)f'_u(\bar{x}(t),\bar{u}(t),t).$$

Now, the controllability matrix Q is introduced as the $\mathbb{R}^{k\times k}$-matrix:

$$Q = AA^* + \int_0^1 B(t)P(t)B^*(t)dt.$$

Definition 3. *The regular control process $(\bar{x},\bar{u})$ is said to be normal, provided that $Q > 0$, or equivalently,* $\operatorname{rank} Q = k$.

4. Main Result

In this section, the second-order necessary optimality conditions are addressed. Consider the two cones

$$C_e := \Big\{y \in \mathbb{R}^k : y^j \le 0 \text{ for } j = 1,\dots,k_1, \text{ and } y^j = 0 \text{ for } j = k_1+1,\dots,k\Big\},$$

$$C_r := \Big\{y \in \mathbb{R}^q : y^j \le 0 \text{ for } j = 1,\dots,q_1, \text{ and } y^j = 0 \text{ for } j = q_1+1,\dots,q\Big\}.$$

Define the cone

$$\mathcal{K} := \Big\{(\delta x_0, \delta u) \in \mathcal{X} : e'(\bar{p})\delta p \in C_e,\\ D(t)\big[r'_x(\bar{x}(t),\bar{u}(t),t)\delta x(t) + r'_u(\bar{x}(t),\bar{u}(t),t)\delta u(t)\big] \in C_r\Big\}.$$

On the space $\mathcal{X}$, consider the quadratic form

$$\Omega_\lambda[(\delta x_0,\delta u)]^2 = \mathcal{L}''_{pp}(\bar{p},\lambda)[\delta p]^2 - \int_0^1 \mathcal{H}''_{ww}(\bar{x}(t),\bar{u}(t),\psi(t),t)[\delta w(t)]^2 dt \\ + \int_0^1 \Big\langle \nu(t), r''_{ww}(\bar{x}(t),\bar{u}(t),t)[\delta w(t)]^2 \Big\rangle dt.$$

Here and further, for convenience of notation: $w = (x,u)$, $\delta w(t) = (\delta x(t), \delta u(t))$.

The main result of this section consists in the following theorem.

Theorem 1. *Let $(\bar{x},\bar{u})$ be an optimal control process in Problem (1). Suppose that this process is normal.*

Then, $\Lambda \neq \varnothing$. Moreover, $\dim\operatorname{span}(\Lambda) = 1$*, and, for $\lambda = (\lambda^0, \lambda_1, \lambda_2) \in \Lambda$, it holds that $\lambda^0 > 0$, and*

$$\Omega_\lambda[(\delta x_0,\delta u)]^2 \ge 0 \;\; \forall\, (\delta x_0,\delta u) \in \mathcal{K}. \tag{11}$$

The proof is preceded with the following auxiliary assertion.

Lemma 1. *Consider linear bounded operators A and A_i, $i = 1,2,\dots$, acting in a given Hilbert space X, such that $A_i \to A$ pointwise. Assume that the spaces $\operatorname{im} A_i$ and $\operatorname{im} A_i^*$ are closed and that $\operatorname{im} A \subseteq \operatorname{im} A_i$ for all i. Assume also that the sequence of norms $\|(A_iA_i^*)^{-1}\|_{\operatorname{im} A_i}$ is uniformly bounded. Let $C \subseteq X$ be a closed and convex set.*

Then,

$$A^{-1}(C) = \operatorname*{Limsup}_{i\to\infty} A_i^{-1}(C). \tag{12}$$

Proof. Let $\xi_i \in A_i^{-1}(C)$ and $\xi_i \to \xi_0$ as $i \to \infty$. By virtue of the uniform boundedness principle, $A_i\xi_i \to A\xi_0$. Thus, $A\xi_0 \in C$ and the embedding '$\supseteq$' is proven.

Let us confirm the inverse embedding. Given $\xi_0 \in A^{-1}(C)$, it is necessary to indicate a sequence of elements $\xi_i \in A_i^{-1}(C)$, such that $\xi_i \to \xi_0$.

Consider the extremal problem

$$\|\xi - \xi_0\|^2 \to \min, \quad A_i\xi = A\xi_0.$$

Denote the solution to this problem as ξ_i. The solution exists since $\operatorname{im} A \subseteq \operatorname{im} A_i$ and since the quadratic functional is weakly lower semi-continuous, whereas the closed convex set $A_i^{-1}(A\xi_0)$ is weakly closed. Since the image of A_i is closed, one can apply the Lagrange multiplier rule as follows. There exists a non-zero vector $\lambda_i \in \operatorname{im} A_i$ such that

$$\xi_i - \xi_0 + A_i^*\lambda_i = 0.$$

Applying A_i, the multiplier is expressed as follows

$$\lambda_i = (A_iA_i^*)^{-1}(A_i\xi_0 - A_i\xi_i).$$

Therefore,

$$\xi_i = \Big(I - A_i^*(A_iA_i^*)^{-1}(A_i - A)\Big)\xi_0.$$

Note that $\|A_i^*(A_iA_i^*)^{-1}\| \le \|A_i\| \cdot \|(A_iA_i^*)^{-1}\| \le \text{const}$ by the assumption of the lemma. However, $A_i\xi_0 \to A\xi_0$, and thus, $\xi_i \to \xi_0$. □

Proof to Theorem 1. By virtue of Theorem 3.5 in [4] and the regularity of the process $(\bar{x}, \bar{u})$, there exists a set of multipliers (λ, ψ, ν) satisfying the maximum principle, such that $\lambda \neq 0$.

Firstly, we prove that the given λ satisfies the following Lagrange multipliers rule:

$$\lambda^0\big(\langle \varphi'_{x_0}(\bar{p}), \delta x_0\rangle + \langle \varphi'_{x_1}(\bar{p}), \delta x_1\rangle\big) + (\lambda_1, \lambda_2)\big(e'_{x_0}(\bar{p})\delta x_0 + e'_{x_1}(\bar{p})\delta x_1\big) = 0 \tag{13}$$

for all $(\delta x_0, \delta u) \in \mathcal{N}_r$.

Due to the maximum principle, one has

$$\begin{aligned}
\langle \psi(1), \delta x(1)\rangle &= \langle \psi(0), \delta x(0)\rangle + \int_0^1 \big(\langle \dot\psi(t), \delta x(t)\rangle + \langle \psi(t), \dot{\delta x}(t)\rangle\big)dt \\
&= \langle \psi(0), \delta x(0)\rangle + \int_0^1 \big(\langle -\psi(t) f'_x(\bar{x}(t), \bar{u}(t), t) + \nu(t) r'_x(\bar{x}(t), \bar{u}(t), t), \delta x(t)\rangle \\
&\qquad \langle \psi(t), f'_x(\bar{x}(t), \bar{u}(t), t)\delta x(t) + f'_u(\bar{x}(t), \bar{u}(t), t)\delta u(t)\rangle\big)dt \\
&= \langle \mathcal{L}'_{x_0}(\bar{p}, \lambda), \delta x_0\rangle \\
&\quad + \int_0^1 \big(\nu(t) r'_x(\bar{x}(t), \bar{u}(t), t)\delta x(t) + \psi(t) f'_u(\bar{x}(t), \bar{u}(t), t)\delta u(t)\big)dt.
\end{aligned}$$

Let us add and subtract the term $\nu(t) r'_u(\bar{x}(t), \bar{u}(t), t)\delta u(t)$ under the integral. Then, by virtue of (5), and also by taking into account that

$$\nu(t)\big(r'_x(t)\delta x(t) + r'_u(t)\delta u(t)\big) = 0 \ \text{ for a.a. } t$$

when $(\delta x_0, \delta u) \in \mathcal{N}_r$, we derive $\langle \psi(1), \delta x(1)\rangle = \langle \mathcal{L}'_{x_0}(\bar{p}, \lambda), \delta x_0\rangle$. Then, from (3),

$$\langle -\mathcal{L}'_{x_1}(\bar{p}, \lambda), \delta x_1\rangle = \langle \psi(1), \delta x(1)\rangle = \langle \mathcal{L}'_{x_0}(\bar{p}, \lambda), \delta x_0\rangle.$$

Hence,

$$\langle \mathcal{L}'_{x_0}(\bar{p},\lambda),\delta x_0\rangle + \langle \mathcal{L}'_{x_1}(\bar{p},\lambda),\delta x_1\rangle = 0,$$

and therefore, Condition (13) is proven.

Consider the endpoint constraint operator

$$\mathcal{E}(\delta x_0,\delta u) := e'(\bar{p})\delta p$$

acting from $\mathcal{X}$ to $\mathbb{R}^k$. It is a straightforward task to derive that condition $Q > 0$ implies that $\mathcal{E}(\mathcal{N}_r) = \mathbb{R}^k$. Then, using $\lambda \neq 0$, Equation (13) yields that $\lambda^0 > 0$. Moreover, the multiplier λ is unique, upon normalization.

Let us proceed to the proof of the second-order condition (11). Take a number $\varepsilon > 0$. Let $D_\varepsilon(t)$ designate the diagonal $q \times q$-matrix defined as $D(t)$ but, now, with the set $J(t)$ replaced by $J_\varepsilon(t)$. Define the cone $\mathcal{K}_\varepsilon$ in the same way as $\mathcal{K}$, but with the matrix $D(t)$ replaced by $D_\varepsilon(t)$. It is clear that $\mathcal{K}_\varepsilon \subseteq \mathcal{K}$ for all $\varepsilon > 0$. Firstly, we prove (11) for the reduced cone $\mathcal{K}_\varepsilon$. Consider the space $\mathcal{X}_\infty = \mathbb{R}^n \times \mathbb{L}_\infty([0,1];\mathbb{R}^n)$ as the L_∞-analogue of $\mathcal{X}$, and the following image-space $\mathcal{Y}_\varepsilon := \prod_{j=1}^{q} \mathbb{L}_\infty(T_j^\varepsilon;\mathbb{R}) \times \mathbb{R}^k$, where $T_j^\varepsilon = \{t \in [0,1] : j \in J_\varepsilon(t)\}$, $j = 1,\ldots,q$.

Define the mapping $\mathcal{F}_\varepsilon : \mathcal{X}_\infty \to \mathcal{Y}_\varepsilon$ as follows

$$\mathcal{F}_\varepsilon(x_0,u(\cdot)) = \Big(r^1(x(\cdot),u(\cdot),\cdot)|_{T_1^\varepsilon},\ldots,r^q(x(\cdot),u(\cdot),\cdot)|_{T_q^\varepsilon}, e^1(p), e^2(p),\ldots,e^k(p)\Big).$$

Set $\bar{y}_\varepsilon = \mathcal{F}_\varepsilon(\bar{x}_0,\bar{u}(\cdot))$. Let $\mathcal{C}_\varepsilon \subset \mathcal{Y}_\varepsilon$ be the closed cone such that, for all pairs $(\xi(\cdot),\gamma) \in \mathcal{C}_\varepsilon$, one has $\xi(t) \in C_r$ for a.a. $t \in [0,1]$, and $\gamma \in C_e$. Here, $\xi^j(t) = 0$ when $t \notin T_j^\varepsilon$.

Consider the inclusion

$$\mathcal{F}_\varepsilon(x_0,u(\cdot)) \in \bar{y}_\varepsilon + \mathcal{C}_\varepsilon,\quad (x_0,u(\cdot)) \in \mathcal{X}_\infty. \tag{14}$$

The Fréchet derivative $\mathcal{F}'_\varepsilon(\bar{x}_0,\bar{u}(\cdot))$ is the linear mapping $(\mathcal{A}_\varepsilon,\mathcal{B}) : \mathcal{X}_\infty \to \mathcal{Y}_\varepsilon$, where $\mathcal{A}_\varepsilon = (\mathcal{A}_\varepsilon^1,\ldots,\mathcal{A}_\varepsilon^q)$,

$$\mathcal{A}^j(\delta x_0,\delta u) := (r^j)'_x(\bar{x}(t),\bar{u}(t),t)\delta x(t) + (r^j)'_u(\bar{x}(t),\bar{u}(t),t)\delta u(t),\quad t \in T_j^\varepsilon,$$

and $\mathcal{B}(\delta x_0,\delta u) := e'(\bar{p})\delta p$. The proof of this fact involves a standard argument.

Firstly, consider this derivative as the extended linear mapping acting from $\mathcal{X}$ to $\prod_{j=1}^{q} \mathbb{L}_2(T_j^\varepsilon;\mathbb{R}) \times \mathbb{R}^k$, that is, in Hilbert spaces. Let us prove its surjection. Since the linear mapping $\mathcal{A}_\varepsilon$ is surjective due to regularity w.r.t. the mixed and state constraints (this is a simple task to ensure by solving the corresponding Volterra equation and using Proposition 1 in this way), it is sufficient to show that the linear mapping $\mathcal{B}$ is surjective on $\ker \mathcal{A}_\varepsilon$. Let Q_ε be the matrix constructed as Q; however, with the matrix $D(t)$ replaced by $D_\varepsilon(t)$. It is clear that $Q_\varepsilon \to Q$ as $\varepsilon \to 0$. Therefore, one has that $Q_\varepsilon > 0$ for all sufficiently small ε. At the same time, this condition implies that $\mathcal{E}(\ker \mathcal{A}_\varepsilon) = \mathbb{R}^k$. Therefore, it is simple to conclude that $(\mathcal{A}_\varepsilon,\mathcal{B})$ is a surjective linear mapping for all sufficiently small ε.

The surjection of $(\mathcal{A}_\varepsilon,\mathcal{B})$ as the linear mapping from $\mathcal{X}_\infty$ to $\mathcal{Y}_\varepsilon$ results from the following simple argument. Firstly, notice that, in space $\mathcal{X}$, one has the relation

$$\mathrm{clos}(\ker \mathcal{A}_\varepsilon \cap \mathcal{X}_\infty) = \ker \mathcal{A}_\varepsilon, \tag{15}$$

which is clear due to Formulas (9) and (10), as these still hold when $D(t)$ is replaced by $D_\varepsilon(t)$ for a sufficiently small ε. Then, simply, $\mathcal{N}_r = \mathcal{N}_r(\varepsilon) = \ker \mathcal{A}_\varepsilon$. At the same time, the linear operator $\mathcal{A}_\varepsilon$ is surjective as the mapping from $\mathcal{X}_\infty$ to $\prod_{j=1}^{q} \mathbb{L}_\infty(T_j^\varepsilon;\mathbb{R})$ by virtue of the same arguments involving the solution to a Volterra equation. However, the image of $\mathcal{B}$ is finite-dimensional, whereas, as has already been confirmed, $\mathcal{B}$ is surjective on the space $\ker \mathcal{A}_\varepsilon$. Therefore, by virtue of (15), one finds that $\mathcal{B}$ is surjective on the subspace $\ker \mathcal{A}_\varepsilon \cap \mathcal{X}_\infty$. Thus, the derivative $\mathcal{F}'_\varepsilon(\bar{x}_0,\bar{u}(\cdot))$ is surjective and, thereby, $(\bar{x}_0,\bar{u}(\cdot))$ is a normal point for the mapping $\mathcal{F}_\varepsilon$.

The Robinson theorem (see Theorem 1 in [16]) asserts the existence of a neighbourhood O_ε of point $((\bar{x}_0, \bar{u}(\cdot)), \bar{y}_\varepsilon) \in \mathcal{X}_\infty \times \mathcal{Y}_\varepsilon$ such that

$$\operatorname{dist}\Big((x_0, u(\cdot)), \mathcal{F}_\varepsilon^{-1}(y + \mathcal{C}_\varepsilon)\Big) \leq c \operatorname{dist}(\mathcal{F}_\varepsilon(x_0, u(\cdot)), y + \mathcal{C}_\varepsilon) \quad \forall ((x_0, u(\cdot)), y) \in O_\varepsilon. \quad (16)$$

Consider an arbitrary element $h = (\delta x_0, \delta u) \in \mathcal{K}_\varepsilon \cap \mathcal{X}_\infty$, and a number $\tau > 0$. By putting in (16) the value $\xi(\tau) = (\bar{x}_0, \bar{u}(\cdot)) + \tau h$ for $(x_0, u(\cdot))$, and $\bar{y}_\varepsilon$ for y, one obtains

$$\operatorname{dist}\Big(\xi(\tau), \mathcal{F}_\varepsilon^{-1}(\bar{y}_\varepsilon + \mathcal{C}_\varepsilon)\Big) \leq o(\tau),$$

and, thus, the set $\mathcal{K}_\varepsilon \cap \mathcal{X}_\infty$ is tangent to the solution set $\mathcal{F}_\varepsilon^{-1}(\bar{y}_\varepsilon + \mathcal{C}_\varepsilon)$; see Corollary 2 in [16]. This means that, for every small τ, there exists a vector $\omega(\tau) = (a(\tau), v(\cdot;\tau)) \in \mathcal{X}_\infty$ such that $\dfrac{\|\omega(\tau)\|}{\tau} \to 0$ as $\tau \to 0$, whereas $\mathcal{F}_\varepsilon(\xi(\tau) + \omega(\tau)) \in \bar{y}_\varepsilon + \mathcal{C}_\varepsilon$.

Let us set $x_0(\tau) := \bar{x}_0 + \tau\delta x_0 + a(\tau)$, and $u(t;\tau) := \bar{u}(t) + \tau\delta u(t) + v(t;\tau)$. Then, one may verify that the control pair $(x_0(\tau), u(\cdot;\tau)) \in \mathcal{X}_\infty$ is admissible to Problem (1) for all sufficiently small τ due to the construction of the mapping $\mathcal{F}_\varepsilon$. At this point, it is essential that $\varepsilon > 0$. Let $x(\cdot;\tau)$ be the trajectory corresponding to the pair $(x_0(\tau), u(\cdot;\tau))$, and $p(\tau) = (x(0;\tau), x(1;\tau))$. Take the multiplier $\lambda = (1, \lambda_1, \lambda_2) \in \Lambda(\bar{x}, \bar{u})$, and the corresponding multiplier ν entailed by λ. Let $\varphi(\bar{p}) = 0$.

Consider the inequality

$$\mathcal{L}(p(\tau), \lambda) + \int_0^1 \langle \nu(t), r(x(t;\tau), u(t;\tau), t)\rangle dt \geq 0, \quad (17)$$

which results from the condition of minimum and from the fact that the control process $(x(\cdot;\tau), u(\cdot;\tau))$ is admissible.

Consider the second-order variational system

$$\begin{cases} \dot{\delta x}_{(1)}(t;\tau) = f_x'(\bar{x}(t), \bar{u}(t), t)\delta x_{(1)}(t;\tau) + f_u'(\bar{x}(t), \bar{u}(t), t)\delta u(t;\tau), \\ \dot{\delta x}_{(2)}(t;\tau) = f_x'(\bar{x}(t), \bar{u}(t), t)\delta x_{(2)}(t;\tau) + \frac{1}{2} f_{ww}''(\bar{x}(t), \bar{u}(t), t)[(\delta x_{(1)}(t;\tau), \delta u(t;\tau))]^2, \\ \delta x_{(1)}(0;\tau) = \delta x_0(\tau),\ \delta x_{(2)}(0;\tau) = 0. \end{cases}$$

Here, $\delta x_0(\tau) = \delta x_0 + \dfrac{a(\tau)}{\tau}$, and $\delta u(t;\tau) = \delta u(t) + \dfrac{v(t;\tau)}{\tau}$. It is clear that

$$x(t;\tau) = \bar{x}(t) + \tau\delta x_{(1)}(t;\tau) + \tau^2 \delta x_{(2)}(t;\tau) + \mathbf{o}(\tau^2).$$

Therefore, by expanding in the Taylor series in (17), one has

$$\begin{aligned} o(\tau^2) \leq\ & \langle \mathcal{L}_{x_0}'(\bar{p}, \lambda), \tau\delta x_0(\tau)\rangle + \Big\langle \mathcal{L}_{x_1}'(\bar{p}, \lambda), \tau\delta x_{(1)}(1;\tau) + \tau^2\delta x_{(2)}(1;\tau)\Big\rangle \\ & + \frac{\tau^2}{2}\mathcal{L}_{pp}''(\bar{p}, \lambda)\Big[(\delta x_0(\tau), \delta x_{(1)}(1;\tau))\Big]^2 + \int_0^1 \Big\langle \nu(t), r_x'(\bar{x}(t), \bar{u}(t), t) \\ & [\tau\delta x_{(1)}(t;\tau) + \tau^2\delta x_{(2)}(t;\tau)] + \tau r_u'(\bar{x}(t), \bar{u}(t), t)\delta u(t;\tau) \\ & + \frac{\tau^2}{2} r_{ww}''(\bar{x}(t), \bar{u}(t), t)[(\delta x_{(1)}(t;\tau), \delta u(t;\tau))]^2 \Big\rangle dt. \end{aligned}$$

Using the adjoint equation, one has

$$\frac{d}{dt}\Big\langle \psi(t), \delta x_{(1)}(t;\tau)\Big\rangle = \nu(t) r_x'(t)\delta x_{(1)}(t;\tau) + \psi(t) f_u'(t)\delta u(t;\tau),$$

$$\frac{d}{dt}\Big\langle \psi(t), \delta x_{(2)}(t;\tau)\Big\rangle = \nu(t) r_x'(t)\delta x_{(2)}(t;\tau) + \frac{1}{2}\psi(t) f_{ww}''(t)[(\delta x_{(1)}(t;\tau), \delta u(t;\tau))]^2.$$

Here, and from now on, the dependence on the optimal process is, for simplicity, omitted. Therefore, using these relations and the transversality conditions (3), and by gathering the terms with τ, and τ^2 in two different groups, we obtain

$$o(\tau^2) \leq -\tau \int_0^1 [\mathcal{H}'_u(t) - \nu(t) r'_u(t)] \delta u(t;\tau) dt + \frac{\tau^2}{2} \Big(\mathcal{L}''_{pp}(\bar{p},\lambda)[(\delta x_0(\tau), \delta x_{(1)}(1;\tau))]^2 - \int_0^1 (\psi(t) f''_{ww}(t) + \nu(t) r''_{ww}(t)) [(\delta x_{(1)}(t;\tau), \delta u(t;\tau))]^2 dt \Big).$$

Now, as the implication of (5), we obtain (11) for the given $(\delta x_0, \delta u) \in \mathcal{K}_\varepsilon \cap \mathcal{X}_\infty$. Then, Estimate (11) is proven on the cone $\mathcal{K}_\varepsilon$ by a simple passage to the limit in $\mathcal{X}$.

Let us pass to the limit as $\varepsilon \to 0$, and prove (11) on the entire cone $\mathcal{K}$. Take $h \in \mathcal{K}$. One needs to justify that, for all $\varepsilon > 0$, there exists $h_\varepsilon \in \mathcal{K}_\varepsilon$ such that $h_\varepsilon \to h$ in $\mathcal{X}$. Indeed, then, (11) is proven due to a simple passage to the limit. However, the existence of such h_ε is yielded by Lemma 1. Indeed, it is a straightforward task to verify that the derivative operator $\mathcal{F}'_\varepsilon$ satisfies all the assumptions of this assertion; it obviously converges pointwise to $\mathcal{F}'_0$ as $\varepsilon \to 0$, while its image is closed, as was confirmed above. The image of the conjugate operator is closed due to the regularity of the reference control process with respect to the mixed constraints which is merely a technical step to ensure. Another technical step is to assert that $\mathcal{F}'_\varepsilon[\mathcal{F}'_\varepsilon]^*$ is positive due to normality. Moreover, the constant of covering does not depend on ε. It is also a straightforward task to verify that the rest of the assumptions hold if we consider $\mathcal{C}_0$ as C.

The proof is complete. □

5. Abnormal Case

In this section, we consider the case when rank $Q < k$. This case, when the normality condition is not satisfied, is called abnormal. Then, as a simple example can show that Theorem 1 fails to hold. Firstly, the normalized multiplier λ is no longer unique, and moreover, there may not exist such a multiplier from the cone Λ for which Estimate (11) is still valid everywhere on $\mathcal{K}$. Therefore, Theorem 1 requires a certain refinement in the abnormal case. Let us formulate the "abnormal" version of this statement. In this enterprise, we follow the method based on the so-called index approach.

Consider the reduced cone of Lagrange multipliers $\Lambda_a = \Lambda_a(\bar{x}, \bar{u})$, which contains multipliers $\lambda \in \Lambda$ such that

$$\operatorname{ind}_{\mathcal{N}} \Omega_\lambda = k - \operatorname{rank} Q.$$

Here, the notation ind_X stands for the index of a quadratic form over the space X, and $\mathcal{N} = \mathcal{N}_e \cap \mathcal{N}_r$. Consider also the following extra hypotheses.

Hypothesis 2 (H2). *Mappings f, r_2 are affine w.r.t. u, while r_1 is convex w.r.t. u.*

Hypothesis 3 (H3). *Mixed constraints are globally regular, that is, $U(x,t) = U_R(x,t)$ for all x and t. Moreover, the set-valued mapping $U(x,t)$ is uniformly bounded.*

The main result of this section is as follows.

Theorem 2. *Let $(\bar{x}, \bar{u})$ be an optimal control process in Problem (1). Suppose that this process is regular with respect to the mixed constraints.*

Then, $\Lambda_a \neq \varnothing$. Moreover, under (H2) and (H3), one has

$$\max_{\lambda \in \Lambda_a} \Omega_\lambda[(\delta x_0, \delta u)]^2 \geq 0 \quad \forall\, (\delta x_0, \delta u) \in \mathcal{K}. \tag{18}$$

In the case of local weak minimum, Estimate (18) has been proven in [14]. Here, our task is to prove it in the case of global strong, or Pontryagin's type of the minimum. In [3], the condition that the cone Λ_a is non-empty has been proven in the class of generalized controls. Note that, under the normality condition of the optimal control process, Estimate (18) implies (11) since the normalized multiplier is unique. Thus, Theorem 2, in essence, represents a stronger assertion than Theorem 1, albeit under some extra assumptions such as (H2) and (H3). These two assumptions are meant to simplify the presentation. Note that (H3) is sufficient to suppose on the optimal trajectory only.

Proof. The proof of theorem is divided into the two stages.

STAGE 1. In this stage, we prove that $\Lambda_a \neq \varnothing$. In the beginning, suppose that Hypothesis (H3) is valid. Under (H1) and (H3), it is convenient to assume that $f(x,u,t)$, and $r(x,u,t)$ are constant with respect to (x,u), and t, outside of some sufficiently large ball. This can be obtained due to a simple problem reduction. In what follows, it will not also be restrictive to consider that $\varphi(p^*) = 0$, and, for the simplicity of exposition, to consider that all the constraints are scalar-valued, i.e., $k_1 = k_2 = q_1 = 1$, while $q_2 = 0$.

Let a, b be non-negative numbers. Consider the mapping

$$\Delta(a,b) := \begin{cases} ab^{-4} & \text{if } b > 0, \\ 1 & \text{if } a > 0,\ b = 0, \\ 0 & \text{if } a = b = 0. \end{cases}$$

This function is lower semi-continuous. It will serve as a penalty function in the applied method below.

Take a pair $(x_0, u) \in \mathcal{X}$, and consider the unique solution to the Cauchy problem $\dot{x}(t) = f(x(t), u(t), t)$, $x(0) = x_0$, which exists on the entire time interval $[0,1]$ due to the above assumptions. Set $p = (x_0, x_1)$, where $x_1 = x(1)$. Note that p depends on (x_0, u). Let $\{\varepsilon_i\}$ be an arbitrary sequence of positive numbers converging to zero. Consider the mapping $\varphi_i^+(p) = (\varphi(p) + \varepsilon_i)^+$, where $a^+ = \max\{a, 0\}$ for $a \in \mathbb{R}$. Thus, the following functional over the space $\mathcal{X}$ is well-defined:

$$F_i(x_0,u) := \varphi_i^+(p) + \Delta\left((e_1^+(p))^2 + |e_2(p)|^2 + \int_0^1 (r(x,u,t)^+)^2 dt, \varphi_i^+(p) \right).$$

Functional F_i is lower semi-continuous which is a straightforward exercise to verify due to the assumptions made above regarding the mappings f and r. At the same time, this functional is positive everywhere: $F_i > 0$.

Consider the following problem

$$\text{Minimize } F_i(x_0,u), \ (x_0,u) \in \mathcal{X}.$$

Note that $F_i(\bar{x}_0, \bar{u}) = \varepsilon_i$. By applying the smooth variational principle, see, e.g., in [17], for each i, there exists an element $(x_{0,i}, u_i) \in \mathcal{X}$ and a sequence of elements $(\tilde{x}_j, \tilde{u}_j) \in \mathcal{X}$, $j = 1, 2, \ldots$, converging to $(x_{0,i}, u_i)$ such that

$$F_i(x_{0,i}, u_i) \le F_i(\bar{x}_0, \bar{u}) = \varepsilon_i, \tag{19}$$

$$|x_{0,i} - \bar{x}_0|^2 + \int_0^1 |u_i(t) - \bar{u}(t)|^2 dt \le \sqrt[3]{\varepsilon_i^2}, \tag{20}$$

and the pair $(x_{0,i}, u_i)$ is the unique solution to the following problem:

$$\text{Minimize } F_i(x_0,u) + \sqrt[3]{\varepsilon_i} \sum_{j=1}^{\infty} 2^{-j} \left(|x_0 - \tilde{x}_j|^2 + \int_0^1 |u - \tilde{u}_j(t)|^2 dt \right), \ (x_0,u) \in \mathcal{X}.$$

Suppose that $\varphi_i^+(p_i) = 0$. Then, $\varphi(p_i) < 0$ and, in view of optimality, taking into account that $\varphi(\bar{p}) = 0$, it follows that some of constraints in (1): e_1, or e_2, or r, are violated. Therefore, by definition of Δ, one has $F_i(x_{0,i}, u_i) \ge 1$. However, this contradicts (19) for

$i > 1$. Thus, $\varphi_i^+(p_i) > 0$. Consider a number $\delta_i > 0$ such that $\varphi_i^+(p) > 0 \; \forall \, p$: $|p - p_i| \le \delta_i$. Then, by virtue of, again, the definition of Δ, the pair $(x_{0,i}, u_i)$ is the unique global minimum to the following control problem:

$$\begin{aligned} \text{Minimize} \quad & z_0 + z_0^{-4}\Big((e_1(p)^+)^2 + |e_2(p)|^2\Big) + \int_0^1 z^{-4}(r(x,u,t)^+)^2 dt \\ & \qquad + \sqrt[3]{\varepsilon_i}\sum_{j=1}^{\infty} 2^{-j}\Big(|x_0 - \tilde{x}_j|^2 + \int_0^1 |u - \tilde{u}_j(t)|^2 dt\Big), \\ \text{subject to} \quad & \dot{x} = f(x,u,t), \\ & \dot{z} = 0, \quad \text{for a.a.}\ t \in [0,1], \\ & |p - p_i| \le \delta_i, \; z_0 = \varphi_i^+(p). \end{aligned} \tag{21}$$

Denote by x_i, z_i the solution to (21), that is, the trajectory corresponding to the pair $(x_{0,i}, u_i(\cdot))$. Note that function $z_i(\cdot)$ is constant, and thus it can be treated simply as number $z_i \in \mathbb{R}$.

Problem (21) is, as a matter of fact, unconstrained. Consider the first and second-order necessary optimality conditions for this problem.

The first-order conditions are stated as follows. There exist a number $\lambda_i^0 > 0$, and absolutely continuous conjugate functions ψ_i and σ_i which correspond to x_i, and z_i, respectively, such that, for a.a. $t \in [0,1]$,

$$\begin{aligned} \dot{\psi}_i(t) &= -\mathcal{H}_x'(x_i(t), u_i(t), \psi_i(t), t) + 2\lambda_i^0 z_i^{-4} r^+(x_i(t), u_i(t), t) r_x'(x_i(t), u_i(t), t), \\ \dot{\sigma}_i(t) &= -4\lambda_i^0 z_i^{-5}(r(x_i(t), u_i(t), t)^+)^2, \end{aligned} \tag{22}$$

$$\begin{aligned} \psi_i(s) &= (-1)^s \lambda_i^0 \Big(2z_i^{-4}\Big(e_1^+(p_i)\frac{\partial e_1}{\partial x_s}(p_i) + e_2(p_i)\frac{\partial e_2}{\partial x_s}(p_i)\Big) + (1-s)\sqrt[3]{\varepsilon_i}\,\omega_{1,i}'(x_{0,i})\Big) \\ &\qquad - (-1)^s \rho_i \frac{\partial \varphi}{\partial x_s}(p_i), \quad s = 0,1, \\ \sigma_i(0) &= \lambda_i^0\Big(1 - 4z_i^{-5}(e_1^+(p_i))^2 - 4z_i^{-5}|e_2(p_i)|^2\Big) + \rho_i, \\ \sigma_i(1) &= 0, \end{aligned} \tag{23}$$

$$\begin{aligned} &\max_{u \in \mathbb{R}^m}\Big(\mathcal{H}(x_i(t), u, \psi_i(t), t) - \lambda_i^0 z_i^{-4}(r(x_i(t), u, t)^+)^2 - \lambda_i^0 \sqrt[3]{\varepsilon_i}\cdot\omega_{2,i}(u,t)\Big) \\ &\quad = \mathcal{H}(x_i(t), u_i(t), \psi_i(t), t) - \lambda_i^0 z_i^{-4}(r(x_i(t), u_i(t), t)^+)^2 - \lambda_i^0 \sqrt[3]{\varepsilon_i}\cdot\omega_{2,i}(u_i(t), t), \end{aligned} \tag{24}$$

Here, $\rho_i \in \mathbb{R}$ is the multiplier corresponding to the constraint $z_0 = \varphi_i^+(p)$,

$$\omega_{1,i}(x) := \sum_{j=1}^{\infty} 2^{-j}|x - \tilde{x}_j|^2, \text{ and } \omega_{2,i}(u,t) := \sum_{j=1}^{\infty} 2^{-j}|u - \tilde{u}_j(t)|^2.$$

Conditions (22)–(24) are the first-order optimality conditions in the form of the maximum principle. Consider the second-order optimality conditions for Problem (21).

Take an element $(\delta x_0, \delta u) \in \mathcal{X}$. Consider the variational differential equation related to (21), that is,

$$\begin{aligned} \dot{\delta x}_i(t) &= \frac{\partial f}{\partial x}(x_i(t), u_i(t), t)\delta x_i(t) + \frac{\partial f}{\partial u}(x_i(t), u_i(t), t)\delta u(t), \\ \dot{\delta z}_i(t) &= 0, \end{aligned} \tag{25}$$

for a.a. $t \in [0,1]$, where

$$\begin{aligned} \delta x_i(0) &= \delta x_0, \\ \delta z_i(0) &= \varphi'(p_i)\delta p_i. \end{aligned}$$

Here, $\delta p_i := (\delta x_0, \delta x_i(1))$.

The solution to (25) exists, and it is unique on the entire time interval $[0,1]$ due to the assumptions made above. The function $\delta z_i(\cdot)$ is obviously constant, thus, it is treated just as number δz_i in what follows.

On the space $\mathcal{X}$, consider the quadratic form

$$\begin{aligned}\Omega_i[(\delta x_0,\delta u)]^2 = {} & \lambda_i^0 2z_i^{-4}e_1(p_i)^+e_1''(p_i)[\delta p_i]^2+2z_i^{-4}e_2(p_i)e_2''(p_i)[\delta p_i]^2\\ & +20z_i^{-6}\Big((e_1(p_i)^+)^2+|e_2(p_i)|^2\Big)\delta z_i^2-\Big(\rho_i\varphi''(p_i)+\lambda_i^0\sqrt[3]{\varepsilon_i}\omega_{1,i}''(p_i)\Big)[\delta p_i]^2\\ & -\int_0^1 \mathcal{H}_{ww}''(x_i(t),u_i(t),\psi_i(t),t)[(\delta x_i(t),\delta u(t))]^2dt\\ & +\lambda_i^0\int_0^1 2z_i^{-4}r(x_i(t),u_i(t),t)^+r_{ww}''(x_i(t),u_i(t),t)[(\delta x_i(t),\delta u(t))]^2dt\\ & +\int_0^1 20z_i^{-6}(r(x_i(t),u_i(t),t)^+)^2\delta z_i^2dt+\lambda_i^0\sqrt[3]{\varepsilon_i}\int_0^1\omega_{2,i}''(u_i(t),t)[\delta u(t)]^2dt.\end{aligned}$$

Consider the closed subspace $\mathcal{N}_i\subseteq\mathcal{X}$ of pairs $(\delta x_0,\delta u)$ such that

(a) $e'(p_i)\delta p_i=0$;

(b) $r_x'(x_i(t),u_i(t),t)\delta x_i(t)+r_u'(x_i(t),u_i(t),t)\delta u(t)=0$ for a.a. $t\in T_i$.

Here, $T_i:=\{t\in[0,1]:r(x_i(t),u_i(t),t)\geq 0\}$.

Then, the second-order necessary optimality condition is given by the inequality

$$\Omega_i[(\delta x_0,\delta u)]^2\geq 0\ \ \forall\,(\delta x_0,\delta u)\in\mathcal{N}_i. \tag{26}$$

(Note that functional $F(x_0,u)$ is not twice continuously differentiable. At the same time, the scalar function $F(x_{0,i}+\tau\delta x_0,u_i+\tau\delta u)$ of τ possesses the second derivative w.r.t. τ at $\tau=0$, provided that $(\delta x_0,\delta u)\in\mathcal{N}_i$. Using this fact, and the fact that Problem (21) is unconstrained, it is simple to derive (26) by applying direct variations arguments.)

The next step is to pass to the limit as $i\to\infty$ in the obtained optimality conditions. Firstly, it follows from (20) that $x_{0,i}\to\bar{x}_0$, and $u_i(t)\to\bar{u}(t)$ strongly in L_2, and, thereby, $x_i(t)\rightrightarrows\bar{x}(t)$ uniformly on $[0,1]$. Then, $z_i\to 0$. Define

$$\begin{aligned}\lambda_i^1 &:= 2\lambda_i^0 z_i^{-4}e_1(p_i)^+;\\ \lambda_i^2 &:= 2\lambda_i^0 z_i^{-4}e_2(p_i);\\ \nu_i(t) &:= 2\lambda_i^0 z_i^{-4}r(x_i(t),u_i(t),t)^+,\end{aligned}$$

and consider the following normalization for the multipliers

$$|\lambda_i|+|\psi_i(0)|+\|\nu_i\|_{L_2}=1, \tag{27}$$

where $\lambda_i=(\lambda_i^0,\lambda_i^1,\lambda_i^2)$.

Let us show that $\sigma_i(0)\to 0$. Indeed, one has

$$\sigma_i(0)=4\int_0^1\lambda_i^0 z_i^{-5}r(x_i(t),u_i(t),t)^+)^2dt=2\int_0^1\nu_i(t)z_i^{-1}r(x_i(t),u_i(t),t)^+dt.$$

However, due to (19), one has $\|z_i^{-1}r(x_i(t),u_i(t),t)^+\|_{L_2}\to 0$. This, together with (27), implies that $\sigma_i(0)\to 0$. Then, the transversality condition and, again, (19) and (27) simply yield that $\lambda_i^0-\rho_i\to 0$.

By passing to a subsequence, in view of the compactness argument, one may assume from (27) that $\lambda_i\to\lambda$, $\psi_i(t)\rightrightarrows\psi(t)$ and $\nu_i\overset{w}{\to}\nu$ weakly in L_2 as $i\to\infty$ for some multipliers $\lambda=(\lambda^0,\lambda_1,\lambda_2)$, ψ and ν. Then, $\rho_i\to\lambda^0$. It is also clear that by passing to a subsequence, one can assert that $\lambda_i^0z_i^{-4}\to\infty$. Indeed, otherwise all the multipliers converge to zero, contradicting (27). By virtue of the regularity of the optimal control process with respect to mixed constraints, for each i, there exists a control function ζ_i such that $\zeta_i(t) \in U(x_i(t), t)$ a.e., and $\zeta_i \to \bar{u}$ in L_∞. Thus, from the maximum condition (24), it follows that $r(x_i(t), u_i(t), t)^+ \to 0$ uniformly. Since the set $U(\bar{x}(t), t)$ is uniformly bounded, this implies, again due to regularity, that the control function u_i is essentially bounded uniformly with respect to i, that is $\|u_i\|_{L_\infty} \le \text{const}$.

From (24), one derives that

$$\nu_i(t) \cdot r'_u(x_i(t), u_i(t), t) = \mathcal{H}'_u(x_i(t), u_i(t), \psi_i(t), t) - \lambda_i^0 \sqrt[3]{\varepsilon_i} \cdot \frac{\partial \omega_{2,i}}{\partial u}(u_i(t), t).$$

Whence, using regularity and the above obtained facts, one has

$$|\nu_i(t)| \le \text{const}\Big(|\psi_i(t)| + \lambda_i^0\Big) \ \ \forall\, i, \ \ t \in [0,1]. \tag{28}$$

Using the facts and estimates obtained above, one can simply pass to the limit in (22)–(24) and prove that the set of multipliers (λ, ψ, ν) satisfies the maximum principle. At the same time, the fact that $\lambda \neq 0$ follows from (28).

Now, let us pass to the limit in (26). Take numbers $\varepsilon > 0$ and $\sigma > 0$. Restricting to a subsequence, one can state that $u_i(t) \to \bar{u}(t)$ for a.a. $t \in [0,1]$. Due to Egorov's theorem, there is a subset E_σ, the measure of which equals $1 - \sigma$, such that $u_i(t) \rightrightarrows \bar{u}(t)$ uniformly on E_σ. Denote $T_\varepsilon := \{t \in [0,1] : r(\bar{x}(t), \bar{u}(t), t) \ge -\varepsilon\}$, $T_\varepsilon(\sigma) := T_\varepsilon \cap E_\sigma$.

Consider the bounded linear operator $\mathcal{A}_i : \mathcal{X}(E_\sigma) \to \mathbb{L}_2(T_\varepsilon(\sigma); \mathbb{R})$ such that

$$\mathcal{A}_i(\delta x_0, \delta u) = r'_x(x_i(t), u_i(t), t)\delta x_i(t) + r'_u(x_i(t), u_i(t), t)\delta u(t)|_{T_\varepsilon(\sigma)},$$

where $\mathcal{X}(E_\sigma)$ is defined as $\mathcal{X}$, but $\delta u(t) = 0$ for a.a. $t \notin E_\sigma$. It is a simple matter to show that, due to the uniform convergence, one has $\mathcal{A}_i \to \mathcal{A}$ strongly, where

$$\mathcal{A}(\delta x_0, \delta u) = r'_x(\bar{x}(t), \bar{u}(t), t)\delta x(t) + r'_u(\bar{x}(t), \bar{u}(t), t)\delta u(t)|_{T_\varepsilon(\sigma)}.$$

Then, as is known, $\ker \mathcal{A}_i \to \mathcal{X}_\varepsilon(\sigma) := \ker \mathcal{A} \subseteq \mathcal{X}(E_\sigma)$. It is clear that $\mathcal{X}_\varepsilon(\sigma) \to \mathcal{X}_\varepsilon := \mathcal{X}_\varepsilon(0)$ as $\sigma \to 0$ by virtue of its definition and the regularity condition. One needs to use the solution to a corresponding Volterra equation to prove this simple fact. Then, Lemma 1 yields that $\mathcal{X}_\varepsilon \to \mathcal{N}_r$ as $\varepsilon \to 0$ if we consider $C = \{0\}$. Here, when treating the convergence of spaces, the symbol '$\to$' stands for Limsup.

Let $\Pi_i \subseteq \mathcal{X}$ denote the kernel of the endpoint operator $e'(p_i)\delta p_i$. It is clear that $\operatorname{codim} \Pi_i \le k$. Then, this is a simple exercise to ensure the existence of a subspace $\Pi \subseteq \mathcal{N}_e$ such that $\operatorname{codim} \Pi \le k$ and

$$\Pi \cap \mathcal{N}_r \subseteq \operatorname{Limsup} \Pi_i \cap \ker \mathcal{A}_i,$$

where Limsup is total: firstly as $i \to \infty$, then, as $\sigma \to 0$ and finally, as $\varepsilon \to 0$. At the same time, note that $T_i \cap E_\sigma \subseteq T_\varepsilon(\sigma)$ for all large i. Therefore, one has the embedding $\ker \mathcal{A}_i \cap \Pi_i \subseteq \mathcal{N}_i \cap \mathcal{X}(E_\sigma)$, and then, the passage to the limit in (26) gives the condition $\Lambda_a \neq \varnothing$. In the latter deduction, Proposition 1 of [14] has been used and also the fact that the terms with δz_i^2 in Ω_i converge to zero in view of (19) and (27).

Now, it is necessary to remove the extra assumptions imposed in (H3) regarding the boundedness and global regularity. However, this can be done following precisely the same method as presented in [4]. Take $c > 0$, and consider the additional control constraint $|u| \le c$. For each $\varepsilon > 0$, there will be N specifically constructed regular selectors of $U(x,t)$ which are surrounded with N ε-tubes as in the above-cited source. Then, the passage to the limit, firstly as $\varepsilon \to 0$, then, as $N \to \infty$, and, at the end, as $c \to \infty$ will complete the proof for Stage 1.

The full proof for the next stage is rather lengthy. Therefore, let us present it schematically, in a sketch-form, exposing the main idea.

STAGE 2. Here, under (H2) and (H3), we prove Estimate (18). For this purpose, the notion of χ-problem is used. Take any $\varepsilon, \delta > 0$ and $(\delta x_0, \delta u) \in \mathcal{K}$. It is not restrictive to assume that the minimum in (1) is absolute.

Consider the problem

$$\begin{cases} \text{Minimize} & \varphi(p) - \chi\varphi(\bar{p}_\varepsilon) + (\max\{0, \chi - 1\})^4 + \delta|x_0 - \bar{x}_0 - \varepsilon\delta x_0|^2 \\ & \qquad + \delta \int_0^1 |u(t) - \bar{u}(t) - \varepsilon\delta u(t)|^2 dt, \\ \text{subject to} & \dot{x}(t) = f(x(t), u(t), t) \text{ for a.a. } t, \\ & e(p) - \chi e(\bar{p}_\varepsilon) \in C_e, \ |x_0 - \bar{x}_0| \leq \delta, \\ & r(x(t), u(t), t) - \chi r(\bar{x}_\varepsilon(t), \bar{u}(t) + \varepsilon\delta u(t), t) \in C_r \text{ for a.a. } t, \\ & \chi \geq 0. \end{cases} \tag{29}$$

Here, $\bar{x}_\varepsilon(\cdot)$ is the trajectory corresponding to the perturbed pair $(\bar{x}_0 + \varepsilon\delta x_0, \bar{u}(t) + \varepsilon\delta u(t))$, whereas $\bar{p}_\varepsilon$ is the corresponding endpoint vector.

Note that the infimum in Problem (29) is finite due to the imposed assumptions. Moreover, it is not greater than zero, since the process $(\bar{x}_\varepsilon(t), \bar{u}(t) + \varepsilon\delta u(t))$, $\chi = 1$ is feasible, whereas the value of the cost equals zero. At the same time, when $\chi = 0$, the infimum over $\mathcal{X}$ is positive due to the absolute optimality in (1) and since $\varphi(\bar{p}) = 0$. This suggests the application of the smooth variational principle, albeit in the version from Ref. [18], so the finite-dimensional variable χ is not subject to perturbation. That is, the adding to the cost due to the variational principle is within the space $\mathcal{X}$ only. Then, for any sufficiently small $\alpha > 0$, one can assert the solution $(x_{0,\alpha}, u_\alpha(\cdot))$, χ_α to the perturbed problem such that $\chi_\alpha > 0$. Indeed, otherwise there is a contradiction with the range of the infimum.

The remaining arguments are somewhat standard: the results of Stage 1 are applied to the α-solutions which are regular due to (H3). By taking $\alpha = \alpha(\varepsilon)$ appropriately small according to the given ε, one can prove the convergence of this solution to the optimal solution $(\bar{x}_0, \bar{u}(t))$ as $\varepsilon \to 0$. (In this enterprise, Hypothesis (H2) is essentially used together with the weak sequential compactness of controls $u_{\alpha(\varepsilon)}$ implemented by virtue of a standard technique. It is also needed to use the form of the minimizing functional in (29) and the above obtained fact that the infimum is not greater than zero, in order to prove the strong convergence of these controls.) Then, it is necessary to pass to the limit in the obtained conditions, firstly as $\alpha \to 0$, then, as $\varepsilon \to 0$ and finally, as $\delta \to 0$. At the same time, the transversality condition with respect to the χ-variable will yield the desired Estimate (18) by virtue of the expansion in Taylor series.

The proof is complete. □

6. Conclusions

In this article, second-order necessary conditions in the form of Estimate (18) have been derived for both normal and abnormal cases. The notion of normality is defined as the condition of full rank for the corresponding controllability matrix. In the normal case, the set of Lagrange multipliers is unique, upon normalization, while the multiplier λ^0 is positive. In the abnormal case, it is essential that the reduced cone of Lagrange multipliers Λ_a is considered and that it has been proven non-empty. Along with the second-order necessary conditions, a refined version of the maximum condition in the form (4) has been obtained. The principal feature of the obtained result is that the maximum is taken over the reduced feasible set $\Theta(x, t)$, which is the closure of the set of regular points.

Author Contributions: Conceptualization and methodology, A.V.A.; writing—original draft preparation, D.Y.K.; writing—review and editing, D.Y.K. and F.L.P.; supervision and project administration, F.L.P. All authors have read and agreed to the published version of the manuscript.

Funding: This research was funded by the Ministry of Science and Higher Education of the Russian Federation, project no 075-15-2020-799.

Data Availability Statement: Not applicable.

Acknowledgments: The support of SYSTEC UID/EEA/00147, Ref. POCI-01-0145-FEDER-006933; SNAP project, Ref. NORTE-01-0145-FEDER-000085; and MAGIC project, Ref. POCI-01-0145-FEDER-032485; all funded by ERDF | NORTE 2020, PT2020, FEDER | COMPETE2020 | POCI | PIDDAC | FCT/MCTES, are acknowledged. Theorem 1 is obtained by A.V. Arutyunov under financial support of Russian Science Foundation (Project No 22-21-00863). Theorem 2 was obtained by A.V. Arutyunov under financial support of Russian Science Foundation (Project No 20-11-20131). The work of D.Yu. Karamzin was carried out according to the State Assignment of Russian Federation (State registration number AAAA-A19-119092390082-8, Topic 0063-2019-0010). The useful remarks of the anonymous reviewers are also acknowledged.

Conflicts of Interest: The authors declare no conflict of interest.

References

1. Agrachev, A.A.; Gamkrelidze, R.V. Index of Extremality and Quasiextremality. *Russ. Math. Dokl.* **1985**, *284*, 11–14.
2. Agrachev, A.A.; Gamkrelidze, R.V. Quasi-extremality for control systems. *J. Sov. Math.* **1991**, *55*, 1849–1864. [CrossRef]
3. Arutyunov, A.V. Perturbations of extremal problems with constraints and necessary optimality conditions. *J. Sov. Math.* **1991**, *54*, 1342–1400. [CrossRef]
4. Arutyunov, A.V.; Karamzin, D.Y.; Pereira, F.L.; Silva, G.N. Investigation of regularity conditions in optimal control problems with geometric mixed constraints. *Optimization* **2016**, *65*, 185–206. [CrossRef]
5. Pontryagin, L.S.; Boltyanskii, V.G.; Gamkrelidze, R.V.; Mishchenko, E.F. *Mathematical Theory of Optimal Processes*; Nauka: Moscow, Russia, 1983.
6. Hestenes, M.R. *Calculus of Variations and Optimal Control Theory*; Wiley: New York, NY, USA, 1966.
7. Neustadt, L.W. *Optimization*; Princeton University Press: Princeton, NJ, USA, 1976.
8. Dubovitskii, A.Y.; Milyutin, A.A. Necessary conditions for a weak extremum in optimal control problems with mixed constraints of the inequality type. *USSR Comput. Math. Math. Phys.* **1968**, *8*, 24–98. [CrossRef]
9. Devdariani, E.N.; Ledyaev, Y.S. Maximum Principle for Implicit Control Systems. *Appl. Math. Optim.* **1999**, *40*, 79–103. [CrossRef]
10. Milyutin, A.A. *Maximum Principle in a General Optimal Control Problem*; Fizmatlit: Moscow, Russia, 2001. (In Russian)
11. Pinho, M.R.; Vinter, R.B.; Zheng, H. A maximum principle for optimal control problems with mixed constraints. *IMA J. Math. Control Inform.* **2001**, *18*, 189–205. [CrossRef]
12. Clarke, F.; Pinho, M.R. Optimal control problems with mixed constraints. *SIAM J. Control Optim.* **2010**, *48*, 4500–4524. [CrossRef]
13. Milyutin, A.A.; Osmolovskii, N.P. *Calculus of Variations and Optimal Control*; American Mathematical Society: Providence, RI, USA, 1998.
14. Arutyunov, A.V.; Karamzin, D.Y. Necessary conditions for a weak minimum in an optimal control problem with mixed constraints. *Differ. Equ.* **2005**, *41*, 1532–1543. [CrossRef]
15. Penrose, R. A generalized inverse for matrices. *Math. Proc. Camb. Philos. Soc.* **1955**, *51*, 406–413. [CrossRef]
16. Robinson, S.M. Regularity and stability for convex multivalued functions. *Math. Oper. Res.* **1976**, *1*, 130–143. [CrossRef]
17. Ioffe, A.D.; Tikhomirov, V.M. Some remarks on variational principles. *Math. Notes* **1997**, *61*, 48–253. [CrossRef]
18. Arutyunov, A.V.; Karamzin, D. Square-root metric regularity and related stability theorems for smooth mappings. *SIAM J. Optim.* **2021**, *31*, 1380–1409. [CrossRef]

MDPI

Article

Riemann–Liouville Fractional Sobolev and Bounded Variation Spaces †

Antonio Leaci [1,*] and Franco Tomarelli [2]

1 Dipartimento di Matematica e Fisica "Ennio De Giorgi", Università del Salento, 73100 Lecce, Italy
2 Politecnico di Milano, Dipartimento di Matematica, 20133 Milan, Italy; franco.tomarelli@polimi.it
* Correspondence: antonio.leaci@unisalento.it
† Dedicated to Delfim F. M. Torres on the Occasion of His 50th Birthday.

Abstract: We establish some properties of the bilateral Riemann–Liouville fractional derivative D^s. We set the notation, and study the associated Sobolev spaces of fractional order s, denoted by $W^{s,1}(a,b)$, and the fractional bounded variation spaces of fractional order s, denoted by $BV^s(a,b)$. Examples, embeddings and compactness properties related to these spaces are addressed, aiming to set a functional framework suitable for fractional variational models for image analysis.

Keywords: fractional derivatives; distributional derivatives; Sobolev spaces; bounded variation functions; embeddings; compactness; calculus of variations; Abel equation

Citation: Leaci, A.; Tomarelli, F. Riemann–Liouville Fractional Sobolev and Bounded Variation Spaces. *Axioms* **2022**, *11*, 30. https://doi.org/10.3390/axioms11010030

Academic Editors: Natália Martins, Ricardo Almeida, Moulay Rchid Sidi Ammi and Cristiana João Soares da Silva

Received: 27 November 2021
Accepted: 11 January 2022
Published: 14 January 2022

1. Introduction

Among several different available definitions for fractional derivatives and corresponding functional spaces, this paper focuses the analysis on some classical pointwise defined or distributional fractional derivatives connected to integral-convolution operators. Precisely, we refer to bilateral definitions of Riemann–Liouville fractional derivatives and related Sobolev and bounded variation spaces that we introduced in [1]: here, we show some compactness and embedding properties of these spaces.

First, we recall the classical Riemann–Liouville left and right fractional derivatives $(d/dx)^s_+$ and $(d/dx)^s_-$ and introduce the distributional Riemann–Liouville left and right fractional derivatives D^s_+, D^s_- together with their bilateral even and odd versions, respectively D^s_e, D^s_o, all of them defined for non-integer orders s, $0<s<1$ (see Definition 4).

Second, we provide the definitions of the fractional Sobolev spaces $W^{s,1}$ and fractional bounded variation spaces BV^s, associated to these bilateral derivatives (see Definitions 9 and 10). These function spaces are studied here (see Theorem 6, Examples 2–5, 6 and 8) in comparison with their non-bilateral counterpart ([2–8]).

The spaces $W^{s,1}$ and BV^s turn out to be the natural setting for data of Abel integral equations in order to make them well-posed problems in the distributional framework too: see Propositions 2 and 3 showing that if $f \in BV^s(a,b)$ with $-\infty < a \le b \le +\infty$, then the distributional Abel integral equation $I^s_{a+}[u] = f$ admits a unique solution and provides an explicit resolvent formula. Corollaries 1 and 2 state analogous results for backward equations. This approach provides an alternative formulation of classical L^1 representability (see [9]); precisely, this approach leads to a straightforward extension of solvability for the Abel integral equation under conditions weaker than L^1 representability, namely with data possibly belonging to $BV^s(a,b)$.

Basic properties of the functional spaces introduced in present article (weak compactness property stated by Theorems 3 and 11 together with comparison embeddings and strict embeddings stated in Theorems 6 and 8 and by (92) and (93)), namely

$$BV(a,b) \underset{\neq}{\subset} \bigcap_{\sigma\in(0,1)} W^{\sigma,1}(a,b) \underset{\neq}{\subset} W^{s,1}(a,b) \underset{\neq}{\subset} BV^s_+(a,b) \qquad \forall s \in (0,1),$$

$$W^{s,1}(a,b) \subsetneq BV^s_+(a,b), \qquad W^{s,1}(a,b) \subsetneq BV^s_-(a,b), \qquad \forall s \in (0,1).$$

are studied with the aim of providing a functional framework suitable to fractional variational models for image analysis ([10–17]), which are the object of a forthcoming paper [18]. The present preliminary study deals with the one-dimensional case only.

We thank an anonymous referee for useful remarks and pointing us to the recent article [19] containing a different approach to the Sonin–Abel equation in weighted Lebesgue spaces.

2. Bilateral Fractional Integral and Derivative

In this paper $(a,b) \subset \mathbb{R}$ is a nonempty (possibly unbounded) open interval, u is a real function of one variable and $0 < s < 1$. The support of a function u is denoted by $\mathrm{spt}\, u$. The notation d/dx stands for the classical pointwise derivative; D_x, or shortly D, denotes the distributional derivative with respect to the variable x. For every open interval $A \subset \mathbb{R}$, we denote by $AC(A)$ the set of absolutely continuous functions with the domain in the interval A, which coincides ([20]) with the space the Gagliardo–Sobolev space $W^{1,1}_G(A) = \{u \in L^1(A) \mid Du \in L^1(A)\}$ when they are both endowed with the standard norm $\|u\|_{L^1(A)} + \|Du\|_{L^1(A)}$. Moreover, we set $L^1_{loc}(A)$ as the set of measurable functions which are Lebesgue integrable on every compact subset of A, and $AC_{loc}(A) = W^{1,1}_{G,loc}(A) = \{u \in L^1_{loc}(A) \mid Du \in L^1_{loc}(A)\}$ and $BV(A) = \{u \in L^1(A) \mid Du \in \mathcal{M}(A)\}$, where $\mathcal{M}(A)$ denotes the measures whose total variation on A is bounded. We denote by $\mathcal{D}'(A)$ and $\mathcal{S}'(A)$ respectively the space of distributions and the space of tempered distributions on the open set A. We denote by $C^{0,\alpha}(K)$ the space of of Hölder continuous functions on the set K.

For the reader's convenience, we recall the definition of Gagliardo's fractional Sobolev spaces $W^{s,1}_G$ ([21,22]). For any $s \in (0,1)$, we set

$$W^{s,1}_G = \left\{ u \in L^1(a,b) \ : \ \frac{|u(x) - u(y)|}{|x-y|^{1+s}} \in L^1([a,b] \times [a,b]) \right\}, \tag{1}$$

which is a Banach space endowed with the norm

$$\|u\|_{W^{s,1}_G} = \left[\int_{[a,b]} |u(x)| dx + \int_{[a,b]} \int_{[a,b]} \frac{|u(x) - u(y)|}{|x-y|^{1+s}} dx\, dy \right],$$

and we recall also the definition of the Riemann–Liouville fractional integral and derivative of order s for L^1-functions, whose standard references can be found in the book by Samko et al. [9].

In the sequel, H denotes the Heaviside function $H(x)=1$ if $x \geq 0$, $H(x)=0$ if $x<0$, while sign denotes the sign function $\mathrm{sign}(x)=1$ if $x>0$, $\mathrm{sign}(x)=-1$ if $x<0$, $\mathrm{sign}(0)=0$.

Definition 1. ***(Riemann–Liouville fractional integral)***
Assume $u \in L^1(a,b)$ and $s > 0$.

The left-side and right-side Riemann–Liouville fractional integrals ${}_{RL}I^s_{a+}$ and ${}_{RL}I^s_{b-}$ are defined by setting, respectively,

$$_{RL}I^s_{a+}[u](x) = \frac{1}{\Gamma(s)} \int_a^x \frac{u(t)}{(x-t)^{1-s}} dt, \qquad x \in [a,b], \tag{2}$$

$$_{RL}I^s_{b-}[u](x) = \frac{1}{\Gamma(s)} \int_x^b \frac{u(t)}{(t-x)^{1-s}} dt, \qquad x \in [a,b], \tag{3}$$

Here, Γ denotes the Euler gamma function [23].

Notice that ${}_{RL}I^1_{a+}[u](x) = \int_a^x u(t)\, dt$ and in general, for every strictly positive integer value, $s = n \in \mathbb{N}$, ${}_{RL}I^n_{a+}[u]$ coincides with the n-th order primitive, vanishing at $x = a$ together with all derivatives up to order $n-1$.

Both ${}_{RL}I^s_{a+}[u]$ and ${}_{RL}I^s_{b-}[u]$ are absolutely continuous functions if $s \geq 1$ since they are primitives of L^1 functions, whereas we can only say that they are L^q functions if $0<s<1$,

$1 \le q < 1/(1-s)$ (see [9]): indeed, jump discontinuities are allowed if $0 < s < 1$, as shown by the next example.

Example 1. *Set $(a,b) = (-1,1)$. Then for every $0 < s < 1$ there is $u \in L^p(-1,1)$, $1 \le p < 1/s$, s.t. $I^s_{(-1)+}[u]$ is discontinuous. For instance, consider $u(x) = H(x)\,x^{-s}$, thus, exploiting the Euler beta function $B(\nu,\mu) = \int_0^1 y^{\nu-1}(1-y)^{\mu-1}dy = \frac{\Gamma(\nu)\Gamma(\mu)}{\Gamma(\nu+\mu)}$, one gets*

$$ {}_{RL}I^s_{(-1)+}[u](x) = \frac{1}{\Gamma(s)}\int_{-1}^{x} \frac{H(t)}{t^s\,(x-t)^{1-s}}\,dt = H(x)\frac{B(s,1-s)}{\Gamma(s)} = \begin{cases} 0 & \text{if } -1 < x \le 0 \\ \Gamma(1-s) & \text{if } 0 < x < 1 . \end{cases} $$

Thus, $I^s_{(-1)+}[u]$ is a piecewise constant function on $(-1,+1)$ with a jump at $x = 0$.

Next, we recall the classical definition of left and right Riemann–Liouville fractional derivatives as in [9,24–27].

Definition 2. ***(Classical Riemann–Liouville fractional derivative)***
Assume $u \in L^1(a,b)$ and $0 < s < 1$.
The left Riemann–Liouville derivative of u at $x \in [a,b]$ is defined by

$$ {}_{RL}\left(\frac{d}{dx}\right)^s_{a+}[u](x) = \frac{d}{dx}\,{}_{RL}I^{1-s}_{a+}[u](x) = \frac{1}{\Gamma(1-s)}\frac{d}{dx}\int_a^x \frac{u(t)}{(x-t)^s}dt \tag{4} $$

for every value of x such that this derivative exists.
Similarly, we may define the right Riemann–Liouville derivative of u at $x \in [a,b]$ as

$$ {}_{RL}\left(\frac{d}{dx}\right)^s_{b-}[u](x) = -\frac{d}{dx}\,{}_{RL}I^{1-s}_{b-}[u](x) = \frac{-1}{\Gamma(1-s)}\frac{d}{dx}\int_x^b \frac{u(t)}{(t-x)^s}dt \tag{5} $$

for every value of x such that this derivative exists.

Then we introduce the distributional Riemann–Liouville fractional derivative as in [25]: a refinement of the Riemann–Liouville fractional derivative, obtained by the plain substitution of the pointwise classical derivative with the distributional derivative

Definition 3. ***(Distributional Riemann–Liouville fractional derivative)***
Assume $u \in L^1(a,b)$ and $0 < s < 1$. The distributional left Riemann–Liouville derivative of u, ${}_{RL}D^s_{a+}[u] \in \mathcal{D}'(a,b)$, is defined by

$$ {}_{RL}D^s_{a+}[u](x) = D_x\,{}_{RL}I^{1-s}_{a+}[u](x) = \frac{1}{\Gamma(1-s)}\,D_x\int_a^x \frac{u(t)}{(x-t)^s}dt\,. \tag{6} $$

Similarly, we may define the distributional right Riemann–Liouville derivative of u, ${}_{RL}D^s_{b-}[u] \in D'(a,b)$, as

$$ {}_{RL}D^s_{b-}[u](x) = -D_x\,{}_{RL}I^{1-s}_{b-}[u](x) = \frac{-1}{\Gamma(1-s)}\,D_x\int_x^b \frac{u(t)}{(t-x)^s}dt\,. \tag{7} $$

Remark 1. *The distributional Riemann–Liouville fractional derivatives $D^s_\pm$ provide a suitable refinement of the classical ones $(d/dx)^s_\pm$ for the purposes of the present paper. However, we emphasize that they coincide on every L^1 function u such that $I^{1-s}[u]$ is absolutely continuous, as it was always the case in the classical applications of fractional derivatives ([28–30]).*

In Lemma 5 below, we examine the case when the above pointwise-defined derivative d/dx exists a.e. and defines an L^1 function coincident with the distributional derivative D, respectively of ${}_{RL}I^{1-s}_{a+}[u]$ and ${}_{RL}I^{1-s}_{b-}[u]$.

In the sequel, we omit the suffix RL of the interval without loss of information, since in this paper, we do not consider any other fractional derivative than the Riemann–Liouville one; we omit also the endpoints a_+ and b_- suffix whenever they are clearly established.

Therefore, we will write shortly $I^s_+[u]$, $I^s_-[u]$, $D^s_+[u]$, $D^s_-[u]$, $(d/dx)^s_+[u]$ and $(d/dx)^s_-[u]$, respectively, in place of $_{RL}I^s_{a+}[u]$, $_{RL}I^s_{b-}[u]$, $_{RL}D^s_{a+}[u]$, $_{RL}D^s_{b-}[u]$, $_{RL}(d/dx)^s_{a+}[u]$ and $_{RL}(d/dx)^s_{b-}[u]$.

One of the disadvantages of the one-side Riemann–Liouville derivative and integral, as defined above, is the fact that only one endpoint of the interval plays a role (see (56) and (57)) since they are "anisotropic" definitions (see [31] and Lemma 6). On the other hand, if we aim to exploit such definitions in a variational context, we have to deal with boundary conditions so that both interval endpoints must play a role ([32]). Therefore, we introduced the bilateral fractional integral and derivative, by keeping separate their "even" and "odd" parts:

Definition 4. *For every $u \in L^1(a,b)$ we set the even and odd versions of bilateral fractional integrals and derivatives:*

$$\begin{aligned} I^s_e[u](x) &:= \frac{1}{2}(I^s_+[u](x) + I^s_-[u](x)) \\ &= \frac{1}{2\,\Gamma(s)} \int_a^b \frac{u(t)}{|x-t|^{1-s}} dt \;=\; \frac{(u * 1/|t|^{1-s})(x)}{2\,\Gamma(s)}, \end{aligned} \tag{8}$$

$$\begin{aligned} D^s_e[u](x) &:= D_x\, I^{1-s}_e[u](x) \\ &= \frac{1}{2}(D^s_+[u](x) - D^s_-[u](x)) \;=\; D_x \frac{(u * 1/|t|^s)(x)}{2\,\Gamma(1-s)}, \end{aligned} \tag{9}$$

$$\begin{aligned} I^s_o[u](x) &:= \frac{1}{2}(I^s_+[u](x) - I^s_-[u](x)) \\ &= \frac{1}{2\,\Gamma(s)} \int_a^b u(t)\, \frac{\operatorname{sign}(x-t)}{|x-t|^{1-s}}\, dt = \frac{(u * \frac{\operatorname{sign}(t)}{|t|^{1-s}})(x)}{2\,\Gamma(s)}, \end{aligned} \tag{10}$$

$$\begin{aligned} D^s_o[u](x) &:= D_x\, I^{1-s}_o[u](x) \\ &= \frac{1}{2}(D^s_+[u](x) + D^s_-[u](x)) \;=\; D_x \frac{(u * \operatorname{sign}(t)/|t|^s)(x)}{2\,\Gamma(1-s)}. \end{aligned} \tag{11}$$

So that

$$I^s_+[u] = I^s_e[u] + I^s_o[u], \qquad D^s_+[u] = D^s_e[u] + D^s_o[u], \tag{12}$$

$$I^s_-[u] = I^s_e[u] - I^s_o[u], \qquad D^s_-[u] = D^s_o[u] - D^s_e[u]. \tag{13}$$

Whenever $(a,b) \neq \mathbb{R}$, the convolution in (8)–(11) has to be understood, without relabeling, as the convolution of the trivial extension of u (still an $L^1(\mathbb{R})$ function with support on $[a,b]$) with either $1/|t|^s$ or $\operatorname{sign}(t)/|t|^s$ (both belonging to $L^1_{loc}(\mathbb{R})$). Also $I^s_\pm[u](x)$, $I^s_e[u](x)$, $I^s_o[u](x)$ have to be understood, without relabeling, as the natural extension for $x \in \mathbb{R}\setminus[a,b]$, provided by the convolution of the trivial extension of u with the corresponding kernels (here, H denotes the Heaviside function):

$$I^s_+[u] = u * \frac{H(x)}{\Gamma(s)|x|^{1-s}}, \qquad I^s_-[u] = u * \frac{H(-x)}{\Gamma(s)|x|^{1-s}}, \qquad \text{for every } x \in \mathbb{R}, \tag{14}$$

$$I^s_e[u] = u * \frac{1}{2\Gamma(s)|x|^{1-s}}, \qquad I^s_o[u] = u * \frac{\operatorname{sign}(x)}{2\Gamma(s)|x|^{1-s}}, \qquad \text{for every } x \in \mathbb{R}, \tag{15}$$

namely

$$
\begin{aligned}
I_+^s[u](x) &= \frac{1}{\Gamma(s)} \int_a^b \frac{u(t)\,H(x-t)}{|x-t|^{1-s}}\,dt && \text{for every } x \in \mathbb{R}\,, && (16)\\
I_-^s[u](x) &= \frac{1}{\Gamma(s)} \int_a^b \frac{u(t)\,H(t-x)}{|x-t|^{1-s}}\,dt && \text{for every } x \in \mathbb{R}\,, && (17)\\
I_e^s[u](x) &= \frac{1}{2\,\Gamma(s)} \int_a^b \frac{u(t)}{|x-t|^{1-s}}\,dt && \text{for every } x \in \mathbb{R}\,, && (18)\\
I_o^s[u](x) &= \frac{1}{2\,\Gamma(s)} \int_a^b \frac{u(t)\ \mathrm{sign}(x-t)}{|x-t|^{1-s}}\,dt && \text{for every } x \in \mathbb{R}\,. && (19)
\end{aligned}
$$

Moreover

$$
\mathrm{spt}\, I_+^s[u] \subset [a, +\infty) \qquad \mathrm{spt}\, I_-^s[u] \subset (-\infty, b]\,. \tag{20}
$$

Remark 2. *In (9) and (11), D_x denotes the distributional derivative in $\mathcal{D}'(\mathbb{R})$, but obviously its restriction as a distribution on the open set (a,b) is understood whenever one works in the bounded interval (a,b).*

Up to a normalization constant (see (25)), $I_e^s[u]$ is called the Riesz potential of u ([1,9]). These fractional integrals $I_+^s[u]$, $I_-^s[u]$, $I_e^s[u]$, $I_o^s[u]$ turn out to be in $L_{loc}^p(\mathbb{R})$ (thus $L^p(I)$ on every bounded interval I) for every $1 \le p < 1/(1-s)$, since they are convolutions of $u \in L^1(\mathbb{R})$ with an $L_{loc}^p(\mathbb{R})$ kernel. Moreover, we have the next result.

Lemma 1. *If $-\infty < a < b < +\infty$, $u \in L^\infty(\mathbb{R})$, $\mathrm{spt}(u) \subset [a,b]$ and $0 < s < 1$ then $I_+^s[u]$, $I_-^s[u]$, $I_e^s[u]$, $I_o^s[u]$ belong to $L^\infty(\mathbb{R}) \cap C^{0,s}$.*

Proof. See Lemmas 2.5 and 3.6 (iii) in [1]. □

The behavior of all the above operators, as $s \to 0_+$ or $s \to 1_-$, is clarified by subsequent Lemmas of the present section, whose proof can be found in [1].

Notice that both $1/|x|^s$ and $\mathrm{sign}(x)/|x|^s$ belong to $L_{loc}^1(\mathbb{R})$, for $0 < s < 1$; hence, the convolution with any L^1 function is well defined and belongs to $L_{loc}^1(\mathbb{R})$; moreover $\mathrm{sign}(x)/|x|^s \to \text{p.v.}\,\frac{1}{x}$ in $\mathcal{S}'$ as $s \to 1_-$, while $1/|x|^s$ has no limit in $\mathcal{S}'$ as $s \to 1_-$, where $\mathcal{S}'$ denotes the space of tempered distributions.

Fractional derivatives degenerate developing singularities as $s \to 1_-$; nevertheless, they can be made convergent to meaningful limits by suitable normalization.

Lemma 2. *Assume $0 < s < 1$, $u \in W_G^{1,2}(\mathbb{R})$ and choose the constants in the Fourier transform such that $\widehat{u}(\xi) = \int_{\mathbb{R}} \exp(-i\xi x)\, u(x)\, dx$. Then*

$$
\frac{D_e^s[u]}{\sin(s\,\pi/2)} \longrightarrow \mathcal{F}^{-1}\{\, i\,\xi\,\widehat{u}(\xi)\,\} = Du \quad \text{in } L^2(\mathbb{R}) \quad \text{as } s \to 1_-\,, \tag{21}
$$

$$
\frac{D_o^s[u]}{\cos(s\,\pi/2)} \longrightarrow \mathcal{F}^{-1}\{\, |\xi|\,\widehat{u}(\xi)\,\} \quad \text{in } L^2(\mathbb{R}) \quad \text{as } s \to 1_-\,, \tag{22}
$$

$$
D_+^s[u] \longrightarrow Du \quad \text{in } L^2(\mathbb{R}) \quad \text{as } s \to 1_-\,, \tag{23}
$$

$$
D_-^s[u] \longrightarrow -Du \quad \text{in } L^2(\mathbb{R}) \quad \text{as } s \to 1_-\,. \tag{24}
$$

Remark 3. *Notice that relations (21), (23) and (24) tell that, as $s \to 1_-$, both $D_+^s[u]$ (left Riemann–Liouville fractional derivative of order s of u) and $D_e^s[u]$ (even Riemann–Liouville fractional derivative of order s of u) converge in L^2 to the distributional derivative Du, while $D_-^s[u]$ converges in L^2 to $-Du$.*

On the other hand relationship (22) means that $D_o^s[u]$ (odd Riemann–Liouville fractional derivative of order s of u) fades as $s \to 1^-$ but, when suitably normalized as $D_o^s[u]/\cos(s\,\pi/2)$, it con-

verges in L^2 *to the Gagliardo fractional derivative of order 1 of* u, *say* $(-\Delta)^{1/2}u := \mathcal{F}^{-1}\{\,|\xi|\,\widehat{u}(\xi)\,\}$.

Fractional integrals degenerate developing singularities as $s \to 0_+$; indeed the convolution term fulfills $|x|^{s-1}/(2\Gamma(s)\cos(s\pi/2)) \to \delta$ in $\mathcal{S}'$ as $s \to 0_+$; nevertheless, fractional integrals are convergent to meaningful limits by suitable normalization.

Lemma 3. *Assume* $0 < s < 1$, $u \in L^1(\mathbb{R})$ *with* $\widehat{u} \in L^1(\mathbb{R})$ *and set the constants in the Fourier transform such that* $\widehat{u}(\xi) = \int_{\mathbb{R}} \exp(-i\xi x)\,u(x)\,dx$. *Then*

$$\frac{1}{\cos(s\,\pi/2)}\, I_e^s[u](x) \longrightarrow u(x) \qquad \textit{uniformly in } \mathbb{R} \quad \textit{as } s \to 0_+\,, \tag{25}$$

$$\frac{\pi}{\sin(s\,\pi/2)}\, I_o^s[u](x) \longrightarrow (\text{p.v.}\,1/x) * u \quad \textit{in } \mathcal{S}'(\mathbb{R}) \quad \textit{as } s \to 0_+\,. \tag{26}$$

Lemma 4. *Assume* $0 < s < 1$, $u \in L^1(\mathbb{R})$.
If $I_o^{1-s}[u] \in AC_{loc}(\mathbb{R})$, *then*

$$\frac{1}{(\cos(s\,\pi/2))^2}\, I_e^s[\,D_o^s[u]\,] = u\,. \tag{27}$$

If $I_e^{1-s}[u] \in AC_{loc}(\mathbb{R})$, *then*

$$\frac{1}{(\sin(s\,\pi/2))^2}\, I_o^s[\,D_e^s[u]\,] = u\,. \tag{28}$$

If $I_o^{1-s}[I_e^s[u]] \in AC_{loc}(\mathbb{R})$, *then*

$$\frac{1}{(\cos(s\,\pi/2))^2}\, D_o^s[\,I_e^s[u]\,] = u\,. \tag{29}$$

If $I_e^{1-s}[I_o^s[u]] \in AC_{loc}(\mathbb{R})$, *then*

$$\frac{1}{(\sin(s\,\pi/2))^2}\, D_e^s[\,I_o^s[u]\,] = u\,. \tag{30}$$

Lemma 5. *Assume* $0 < s < 1$, $u \in L^1(\mathbb{R})$. *Then*

$$D_+^s[\,I_+^s[u]\,] = u \tag{31}$$

$$D_-^s[\,I_-^s[u]\,] = u\,. \tag{32}$$

If in addition $I_+^{1-s}[u] \in AC_{loc}(\mathbb{R})$, *then*

$$I_+^s[\,D_+^s[u]\,] = u\,. \tag{33}$$

If in addition $I_-^{1-s}[u] \in AC_{loc}(\mathbb{R})$, *then*

$$I_-^s[\,D_-^s[u]\,] = u\,. \tag{34}$$

Remark 4. *Every distributional fractional derivative (left, right, even, and odd) appearing in the statements of Lemmas 3–5, which are proved in [1] with fractional classical derivatives* $(d/dx)_\pm^s$, *still hold true in the present formulation with corresponding distributional derivatives* $(D_x)_\pm^s$ *by exactly the same proof, since the assumptions ensure that all derivatives are evaluated on local absolute continuous functions.*

Remark 5. *Notice that, when $\mathbb{R}$ is replaced by a bounded interval, the identities (27), (28), (33) and (34) require an additional correction term, taking into account of boundary values (see (52) and (53) in Theorem 1), whereas (31) and (32) remain true (see (136), (137)).*

Symmetries of even or odd functions are inherited neither by fractional integrals, nor by fractional derivatives. Nevertheless, the next lemma holds true.

Lemma 6. *For every $s \in (0,1)$, $0 < a \leq +\infty$ and $v \in L^1(-a,a)$, by setting*

$$\check{v}(x) = v(-x)\,, \tag{35}$$

we obtain

$$I^s_+[\check{v}](x) = I^s_-[v](-x) \qquad \text{on } (-a,a)\,, \tag{36}$$

$$D^s_+[\check{v}](x) = -\,D^s_-[v](-x) \qquad \text{on } (-a,a)\,. \tag{37}$$

For every $s \in (0,1)$, $0 < a \leq +\infty$ and every even function $v \in L^1(-a,a)$, we get

$$I^s_+[v](x) = I^s_+[\check{v}](x) = I^s_-[v](-x) \qquad \text{on } (-a,a)\,, \tag{38}$$

$$D^s_+[v](x) = D^s_+[\check{v}](x) = -\,D^s_-[v](-x) \qquad \text{on } (-a,a)\,, \tag{39}$$

For every $s \in (0,1)$ and every odd function $v \in L^1(-a,a)$, we obtain

$$I^s_+[v](x) = -\,I^s_+[\check{v}](x) = -\,I^s_-[v](-x) \qquad \text{on } (-a,a)\,, \tag{40}$$

$$D^s_+[v](x) = -D^s_+[\check{v}](x) = D^s_-[v](-x) \qquad \text{on } (-a,a)\,, \tag{41}$$

Proof.

$$\begin{aligned} I^s_+[\check{v}](x) &= \frac{1}{\Gamma(s)}\int_{-a}^{x}\frac{v(-t)}{(x-t)^{1-s}}\,dt = \frac{1}{\Gamma(s)}\int_{-a}^{x}\frac{v(-t)}{\big(-t-(-x)\big)^{1-s}}\,dt \overset{s=-t}{=} \\ &= -\frac{1}{\Gamma(s)}\int_{a}^{-x}\frac{v(s)}{\big(s-(-x)\big)^{1-s}}\,ds = \frac{1}{\Gamma(s)}\int_{-x}^{a}\frac{v(s)}{\big(s-(-x)\big)^{1-s}}\,ds = I^s_-[v](-x). \end{aligned}$$

By inserting $1-s$ in place of s in (36), if v is even we obtain (37) via

$$D^s_+[\check{v}](x) = D_x\, I^{1-s}_+[\check{v}](x) = D_x I^{1-s}_-[v](-x) = -D^s_-[v](-x)\,.$$

Even v entails $v(x) = v(-x)$, $v = \check{v}$, $I^s_+[v](x) = I^s_+[\check{v}](x)$ and $D^s_+[v](x) = D^s_+[\check{v}](x)$; hence, (36) and (37) entail, respectively, (38) and (39).

Odd v entails $v(x) = -v(-x)$, $v = -\check{v}$, $I^s_+[v](x) = -I^s_+[\check{v}](x)$ and $D^s_+[v](x) = -D^s_+[\check{v}](x)$; hence, (36), (37) entail, respectively, (40), (41). □

Results listed above (mainly Lemmas 3 and 4 proved in [1]) lead to the natural definition of the operators representing the bilateral version of Riemann–Liouville fractional derivatives and integrals, as stated below. Results similar to the ones in Lemma 6 can be found also in [33].

Definition 5. *(Bilateral Riemann–Liouville fractional integral of order s)*

$$I^s[u] = \frac{1}{\cos(s\,\pi/2)}\, I^s_e[u] = \frac{1}{2\,\Gamma(s)\,\cos(s\pi/2)}\,\big(I^s_+[u] + I^s_-[u]\big)\,.$$

Definition 6. *(Bilateral Riemann–Liouville fractional derivative of order s)*

$$D^s[u] = \frac{1}{\cos(s\,\pi/2)}\, D^s_o[u] = \frac{1}{2\,\Gamma(s)\,\cos(s\pi/2)}\,\big(D^s_+[u] - D^s_-[u]\big)\,.$$

3. The Bilateral Fractional Sobolev Space

From now on, we consider only functions defined on a bounded interval (a,b).

As already mentioned in [25], possible naïve definitions of bilateral fractional Sobolev spaces could be set by $U^{s,1} = U_+^{s,1} \cap U_-^{s,1}$, where $s \in (0,1)$ and

$$U_+^{s,1} = \{u \in L^1(a,b) \mid (d/dx)_+^s \, u \in L^1(a,b)\},$$

$$U_-^{s,1} = \{u \in L^1(a,b) \mid (d/dx)_-^s \, u \in L^1(a,b)\},$$

for example, a definition which refers to L^1-functions whose classical Riemann–Liouville fractional derivative of prescribed order $s \in (0,1)$ exists finitely almost everywhere and belongs to $L^1(a,b)$.

Actually, if the classical Riemann–Liouville fractional derivative $(d/dx)_+^s\,[u](x)$ of u exists a.e. for x for some $s \in (0,1)$, then $I_{a+}^{1-s}[u]$ is differentiable almost everywhere, referring to the same s; nevertheless, such an a.e. derivative does not provide complete information about the distributional derivative of the fractional integral $I_{a+}^{1-s}[u]$, when $I_{a+}^{1-s}[u]$ is not an absolutely continuous function. Thus, the differential properties are not completely described by the pointwise fractional derivative, though existing almost everywhere in (a,b). This shows that the previous definitions $U_+^{s,1}$ and $U_-^{s,1}$ are not suitable to obtain an integration by parts formula, whereas the appropriate ones refer to distributional Riemann–Liouville fractional derivative $D_+^s\,[u](x)$ in Definition 3, namely, they are given by

$$\mathfrak{U}_+^{s,1} = \{u \in L^1(a,b) \mid D_+^s u \in L^1(a,b)\}, \tag{42}$$

$$\mathfrak{U}_-^{s,1} = \{u \in L^1(a,b) \mid D_-^s u \in L^1(a,b)\}, \tag{43}$$

Therefore, to develop a satisfactory theory of fractional Sobolev spaces, we introduced a more effective function space in [25], by defining the fractional Sobolev spaces related to one-sided fractional derivatives, which are recalled in subsequent Definition 7, where we confine to the case $p = 1$.

Definition 7. *We recall the definitions of Riemann–Liouville fractional Sobolev spaces related to one-sided fractional derivatives, as introduced in [25]:*

$$W_+^{s,1}(a,b) := \{u \in L^1(a,b) \mid I_+^{1-s}[u] \in W_G^{1,1}(a,b)\} = \mathfrak{U}_+^{s,1}, \tag{44}$$

$$W_-^{s,1}(a,b) := \{u \in L^1(a,b) \mid I_-^{1-s}[u] \in W_G^{1,1}(a,b)\} = \mathfrak{U}_-^{s,1}. \tag{45}$$

Explicitly, the properties $u \in W_\pm^{s,1}(a,b)$ entail, respectively, that the distributional derivatives $D\left[I_\pm^{1-s}[u]\right]$ belong to $L^1(a,b)$, thus $W_+^{s,1}(a,b) = \mathfrak{U}_+^{s,1}$ and $W_-^{s,1}(a,b) = \mathfrak{U}_-^{s,1}$.

Here, we introduce also the "even" and "odd" fractional Sobolev spaces.

Definition 8. *The even/odd Riemann–Liouville fractional Sobolev spaces are*

$$W_e^{s,1}(a,b) := \{u \in L^1(a,b) \mid I_e^{1-s}[u] \in W_G^{1,1}(a,b)\}, \tag{46}$$

$$W_o^{s,1}(a,b) := \{u \in L^1(a,b) \mid I_o^{1-s}[u] \in W_G^{1,1}(a,b)\}. \tag{47}$$

Eventually, we define the bilateral Riemann–Liouville fractional Sobolev spaces, with the aim to achieve a symmetric framework.

Definition 9. ***The (Bilateral) Riemann–Liouville Fractional Sobolev spaces.***
For every $s \in (0,1)$, we set $W^{s,1}(a,b) = W_+^{s,1}(a,b) \cap W_-^{s,1}(a,b)$, that is,

$$W^{s,1}(a,b) := \{u \in L^1(a,b) \mid I_+^{1-s}[u] \in W_G^{1,1}(a,b) \text{ and } I_-^{1-s}[u] \in W_G^{1,1}(a,b)\}. \tag{48}$$

Notice that, concerning Definition 9, by exploiting (12) and (13), we get also

$$W^{s,1}(a,b) \;=\; W^{s,1}_{+}(a,b)\cap W^{s,1}_{-}(a,b) \;=\; W^{s,1}_{o}(a,b)\cap W^{s,1}_{e}(a,b)\,. \tag{49}$$

Theorem 1. *Assume $0<s<1$ and (a,b) is bounded. Then, the (bilateral) Riemann–Liouville fractional Sobolev space $W^{s,1}(a,b)$ (Definition 9) is a normed space when endowed with the natural norm*

$$\|u\|_{W^{s,1}} := \|u\|_{L^1(a,b)} + \|\, D^s_{a+}[u]\,\|_{L^1(a,b)} + \|\, D^s_{b-}[u]\,\|_{L^1(a,b)}\,. \tag{50}$$

The set $W^{s,1}(a,b)$ is a Banach space and, for every $q\in[1,1/(1-s))$ there is $C=C(s,q,a,b)$ such that

$$\|u\|_{L^q(a,b)} \;\le\; C(s,q,a,b)\,\|u\|_{W^{s,1}(a,b)}\,. \tag{51}$$

Every $u\in W^{s,1}(a,b)$ can be represented by both

$$u(x) \;=\; I^s_{a+}\big[D^s_{a+}[u]\big](x) + \frac{I^{1-s}_{a+}[u](a)}{\Gamma(s)}\,(x-a)^{s-1} \qquad a.e\ x\in(a,b), \tag{52}$$

and

$$u(x) \;=\; I^s_{b-}\big[D^s_{b-}[u]\big](x) + \frac{I^{1-s}_{b-}[u](b)}{\Gamma(s)}\,(b-x)^{s-1} \qquad a.e\ x\in(a,b). \tag{53}$$

Proof. The map $u\mapsto\|u\|_{W^{s,1}}$ is a norm on $W^{s,1}(a,b)$, indeed, $\|u\|_{L^1(a,b)}+\|D^s_{a+}[u]\|_{L^1(a,b)}$ is equivalent to the norm $\|I^{1-s}_{a+}[u]\|_{W^{1,1}_G(a,b)}$ since $I^{1-s}_{a+}[u]$ belongs to $W^{1,1}_G$, $D^s_{a+}[u] = \frac{d}{dx}I^{1-s}_{a+}[u] = DI^{1-s}_{a+}[u]$ and $\|I^{1-s}_{a+}[u]\|_{L^1(a,b)} \le C\|u\|_{L^1(a,b)}$; analogously $\|u\|_{L^1(a,b)}+\|D^s_{b-}[u]\|_{L^1(a,b)}$ is a norm for $I^{1-s}_{b-}[u]$, due to $I^{1-s}_{b-}[u]\in W^{1,1}_G$, $D^s_{b-}[u]=\frac{d}{dx}I^{1-s}_{b-}[u]=DI^{1-s}_{b-}[u]$ and $\|I^{1-s}_{a+}[u]\|_{L^1(a,b)}\le C\|u\|_{L^1(a,b)}$.

The completeness of $W^{s,1}(a,b)$ with respect to such a norm when (a,b) is bounded, follows by the completeness of L^1 and $W^{1,1}_G$ together with the fact that u_k is a Cauchy sequence in the norm $W^{s,1}(a,b)$ if and only if u_k is a Cauchy sequence in $L^1(a,b)$ and $I^{1-s}_{\pm}[u_k]$ are Cauchy sequences in $W^{1,1}_G$.

Estimate (51) and representations (52) and (53) follow by (129), (130) and (135) of Proposition 12, which is shown in Section 5. □

Remark 6. *Thanks to $\|I^{1-s}_{a+}[u]\|_{L^1(a,b)}\le\|u\|_{L^1(a,b)}$, $\|I^{1-s}_{b-}[u]\|_{L^1(a,b)}\le\|u\|_{L^1(a,b)}$, we have replaced the terms $\|I^{1-s}_{\pm}[u]\|_{W^{1,1}(a,b)}$ in the norm (50) by $\|D\left[I^{1-s}_{\pm}[u]\right]\|_{L^1(a,b)}$, where D denotes the distributional derivative.*

Obviously, the terms $\|D^s_{+}[u]\|_{L^1(a,b)}$ and $\|D^s_{-}[u]\|_{L^1(a,b)}$ in the norm (50) could be alternatively replaced by $\|D^s_e[u]\|_{L^1(\mathbb{R})}$ and $\|D^s_o[u]\|_{L^1(\mathbb{R})}$, referring to (8), (16) and (17), still achieving an equivalent norm on $W^{s,1}(a,b)$.

Example 2. *For every $s\in(0,1)$, the constant functions and $v(x)=x(1-x)$ both belong to the space $W^{s,1}(a,b)$. Spaces $C^\infty_0(a,b)$, test functions on (a,b), and $C^1([a,b])$, continuously differentiable functions, are contained in $W^{s,1}(a,b)$.*

Example 3. *For every $0<s<1$, the discontinuous piecewise constant $H(x)$ belongs to $W^{s,1}(-1,1)\backslash W^{1,1}_G(-1,1)$.*

Indeed, both $I^{1-s}_{+}[H](x) = H(x)\frac{|x|^{1-s}}{\Gamma(2-s)}$ and $I^{1-s}_{-}[H](x) = \frac{H(x)(1-x)^{1-s}+H(-x)(|1-x|^{1-s}-|x|^{1-s})}{\Gamma(2-s)}$ belong to $W^{1,1}_G(-1,1)$.

Example 4. *For every $s\in(0,1)$ and $\beta\ge 0$, the function x^β belongs to $W^{s,1}(0,1)$. This claim is straightforward for $\beta=0$ and $\beta\ge 1$; we refer to (97) and (98) for $0<\beta<1$.*

Example 5. *Function $x^{-\alpha}$ with $\alpha > 0$ belongs to $W^{s,1}(0,1)$ if and only if $0 < \alpha < 1 - s$.*

Indeed, $I_+^{1-s}[x^{-\alpha}](x) = \frac{\Gamma(1-\alpha)}{\Gamma(2-s-\alpha)} x^{1-s-\alpha} \in W_G^{1,1}(0,1)$ if $-s-\alpha > -1$, while $I_+^{1-s}[x^{s-1}]$ $(x) = \frac{1}{\Gamma(1-s)} \in W_G^{1,1}(0,1)$, summarizing $x^{-\alpha}$ belongs to $W_+^{s,1}(0,1)$ for $\alpha \le 1-s$; on the other hand, $I_-^{1-s}[x^{-\alpha}]$ belongs to $L^1(a,b)$ for $\alpha < 1$ and is bounded on $(0,1)$ if and only if $\alpha < 1-s$, due to

$$\begin{aligned} I_-^{1-s}[\,x^{-\alpha}\,](x) &= \frac{1}{\Gamma(1-s)} \int_x^1 t^{-\alpha}(t-x)^{-s}\,dt \overset{t=xy}{=} \\ &= \frac{x^{1-s-\alpha}}{\Gamma(1-s)} \int_1^{1/x} y^{-\alpha}(y-1)^{-s}\,dy\,. \end{aligned}$$

Summarizing, and taking into account Example 4 for $\alpha \le 0$,

$$x^{-\alpha} \in W_+^{s,1}(0,1) \cap W_-^{s,1}(0,1) \quad \text{if and only if} \quad \alpha < 1-s\,. \tag{54}$$

In the particular case $\alpha = 1-s$, we recover

$$x^{s-1} \in W_+^{s,1}(0,1) \setminus W_-^{s,1}(0,1) \tag{55}$$

with $I_+^{1-s}[x^{s-1}](x) = \Gamma(s)$ and $I_-^{1-s}[x^{s-1}](x)$ unbounded in a right neighborhood of $x = 0$.

Theorem 2. ***(Integration by parts in $W^{s,1}(a,b)$)***
Next, identities hold true for $0 < s < 1$, $-\infty < a < b < +\infty$:

$$\begin{cases} \displaystyle\int_a^b u(x)\, D_+^s[v](x)\,dx = -\int_a^b \frac{d}{dx}u(x)\, I_+^{1-s}[v](x)\,dx + u(b)\, I_+^{1-s}[v](b) \\ \hfill \forall\, v \in W_+^{s,1}(a,b),\ \forall\, u \in W_G^{1,1,}(a,b)\,, \end{cases} \tag{56}$$

$$\begin{cases} \displaystyle\int_a^b u(x)\, D_-^s[v](x)\,dx = +\int_a^b \frac{d}{dx}u(x)\, I_-^{1-s}[v](x)\,dx + u(a)\, I_-^{1-s}[v](a) \\ \hfill \forall\, v \in W_-^{s,1}(a,b),\ \forall\, u \in W_G^{1,1,}(a,b)\,, \end{cases} \tag{57}$$

$$\begin{cases} \displaystyle\int_a^b u(x)\, D_e^s[v](x)\,dx = \\ \displaystyle\quad -\int_a^b \frac{d}{dx}u(x)\, I_e^{1-s}[v](x)\,dx + \frac{1}{2}\Big(u(b)\, I_+^{1-s}[v](b) - u(a)\, I_-^{1-s}[v](a)\Big) \\ \hfill \forall\, v \in W^{s,1}(a,b),\ \forall\, u \in W_G^{1,1,}(a,b)\,, \end{cases} \tag{58}$$

$$\begin{cases} \displaystyle\int_a^b u(x)\, D_o^s[v](x)\,dx = \\ \displaystyle\quad -\int_a^b \frac{d}{dx}u(x)\, I_0^{1-s}[v](x)\,dx + \frac{1}{2}\Big(u(b)\, I_+^{1-s}[v](b) + u(a)\, I_-^{1-s}[v](a)\Big) \\ \hfill \forall\, v \in W^{s,1}(a,b),\ \forall\, u \in W_G^{1,1,}(a,b)\,, \end{cases} \tag{59}$$

Proof. Identity (56) follows by (2), (5) and

$$\begin{aligned} \int_a^b u(x)\, D_+^s[v](x)\,dx &= \frac{1}{\Gamma(1-s)} \int_a^b u(x) \Big(\frac{d}{dx} \int_a^x \frac{v(t)}{(x-t)^s}\,dt\Big)\,dx = \\ = -\frac{1}{\Gamma(1-s)} \int_a^b \frac{d}{dx}u(x) &\Big(\int_a^x \frac{v(t)}{(x-t)^s}\,dt\Big)\,dx + u(b)\, \frac{1}{\Gamma(1-s)} \int_a^b \frac{v(t)}{(x-t)^s}\,dt\,. \end{aligned}$$

Identity (57) follows by (3), (4) and similar computations.
Identities (58) and (59) follow by the subtraction and sum of (56) and (57). □

Remark 7. *Notice that when u is representable, e.g., under a slightly stronger condition, then we find a more symmetric formulation. For instance,* (59) *translates into*

$$\begin{cases} \int_a^b I_o^{1-s}[w](x)\, D_o^s[v](x)\, dx = \\ \quad - \int_a^b D_o^s[w](x)\, I_o^{1-s}[v](x)\, dx + \frac{1}{2}\left(I_o^{1-s}[w](b)\, I_+^{1-s}[v](b) + I_o^{1-s}[w](a)\, I_-^{1-s}[v](a) \right) \\ \qquad \forall\, v,\, w \in W^{s,1}(a,b)\,. \end{cases} \tag{60}$$

Lemma 7. *The strict embedding*

$$W_G^{1,1}(a,b) \underset{\neq}{\subset} W^{s,1}(a,b) \qquad 0 < s < 1 \tag{61}$$

holds true with the related uniform estimate: there is a constant $K = K(s,a,b)$ such that

$$\|v\|_{W^{s,1}(a,b)} \,\leq\, K\, \|v\|_{W_G^{1,1}(a,b)} \tag{62}$$

Proof. By computations in Example 3, we know that the Heaviside function belongs to $W^{s,1}(-1,1)\backslash W_G^{1,1}(-1,1)$; thus if the embedding holds true, then it is strict.

Recalling the definition ([34]) of right Caputo fractional derivative ${}_C D_+^s[u]$ and its relationship with the right Riemann–Liouville fractional derivative

$${}_C D_+^s[u](x) \,:=\, I_+^{1-s}[u'](x) \,=\, \frac{1}{\Gamma(1-s)} \int_a^x \frac{u'(t)}{(x-t)^s}\, dt \qquad \forall u \in W_G^{1,1}(a,b)\,,$$

$$\begin{aligned} {}_{RL} D_+^s[u](x) \;&=\; {}_C D^s s_+[u](x) + \frac{u(a)}{\Gamma(1-s)}\,(x-a)^{-s} = \\ &=\; I_+^{1-s}[u'](x) + \frac{u(a)}{\Gamma(1-s)}\,(x-a)^{-s} \qquad \forall u \in W_G^{1,1}(a,b)\,, \end{aligned}$$

and taking into account

$$\|I_+^{1-s}[u']\|_{L^1(a,b)} \,\leq\, K_1\, \|u'\|_{L^1(a,b)} \,\leq\, K_1\, \|u\|_{W_G^{1,1}(a,b)}\,, \quad |u(a)| \,\leq\, K_2\, \|u\|_{W_G^{1,1}(a,b)}\,,$$

we get (7) and (62). □

Theorem 3. [*Compactness in $W^{s,1}(a,b)$*]
Assume that the interval (a,b) is bounded, the parameter s fulfills $0<s<1$, and

$$\|u_n\|_{W^{s,1}(a,b)} \,\leq\, C\,. \tag{63}$$

Then there exist $u \in L^q(a,b)$, $\forall q \in [1, 1/(1-s))$, and a subsequence such that, without relabeling,

$$\begin{cases} (i) & u_n \rightharpoonup u & \text{weakly in } L^q(a,b)) \quad \forall q \in [1, 1/(1-s))\,, \\ (ii) & I_+^{1-s}[u_n] \to I_+^{1-s}[u] & \text{strongly in } L^p(a,b)\,,\ \forall p \in [1,+\infty)\,, \\ (iii) & I_-^{1-s}[u_n] \to I_-^{1-s}[u] & \text{strongly in } L^p(a,b)\,,\ \forall p \in [1,+\infty)\,, \end{cases}$$

$$I_+^{1-s}[u_n] \rightharpoonup I_+^{1-s}[u]\,, \qquad I_-^{1-s}[u_n] \rightharpoonup I_-^{1-s}[u] \qquad \text{weakly in } BV(a,b). \tag{64}$$

Proof. Claim (i) follows by (51) and (63) and reflexivity of $L^q(a,b)$ for any fixed $1 < q_k < 1/(1-s)$; thus, by choosing a sequence $q_k \to 1/(1-s)$ and extracting a diagonal sequence, we get the claim for a unique subsequence and unique u valid for every q fulfilling $1 < q < 1/(1-s)$. Moreover, such u belongs to $L^1(a,b)$. Eventually, for $q = 1$

there is a measure μ such that $u_n \rightharpoonup \mu$ in $\mathcal{M}(a,b)$, but such μ must be equal to u, then $u_n \rightharpoonup u$ in $L^1(a,b)$.

The compact embedding $W_G^{1,1}(a,b) \hookrightarrow L^p(a,b)$ valid for any $p \in [1,+\infty)$ (Rellich Theorem) entails the existence of z_+ and z_- in $L^p(a,b)$ fulfilling, up to subsequences,

$$I_+^{1-s}[u_n] \to z_+ \quad \text{strongly in } L^p(a,b)\,,\ \forall p \in [\,1,+\infty\,)\,, \tag{65}$$

$$I_-^{1-s}[u_n] \to z_- \quad \text{strongly in } L^p(a,b)\,,\ \forall p \in [\,1,+\infty\,)\,, \tag{66}$$

$$I_e^{1-s}[u_n] \to \frac{1}{2}(z_+ + z_-) \quad \text{strongly in } L^p(a,b)\,,\ \forall p \in [\,1,+\infty\,)\,, \tag{67}$$

$$I_o^{1-s}[u_n] \to \frac{1}{2}(z_+ - z_-) \quad \text{strongly in } L^p(a,b)\,,\ \forall p \in [\,1,+\infty\,)\,. \tag{68}$$

By (i) and the Mazur Theorem, there is a sequence of convex combinations y_n, which is strongly converging: precisely, $y_n \to u$ strongly in $L^q(a,b)$ for every $q \in [1, 1/(1-s))$ with $y_n = \sum_{j=1}^n c_{n,j} u_j$, $c_{n,j} \geq 0$, $\sum_{j=1}^n c_{n,j} = 1$. Hence, by (63),

$$\|y_n\|_{W^{s,1}} \leq \sum_{j=1}^{n} c_{n,j} \|u_j\|_{W^{s,1}} \leq C\,. \tag{69}$$

I^{1-s} is a continuous map from L^q to L^r, $q \in [1, 1/(1-s))$ and $r \in [1, q/(1-(1-s)sq))$, hence, we obtain

$$I_\pm^{1-s}[y_n] \to I_\pm^{1-s}[u] \quad \text{strongly in } L^r(a,b),\ r \in [1, q/(1-(1-s)sq)) \tag{70}$$

and hence, in $\mathcal{D}'(a,b)$. Moreover, by (69), $I^{1-s}[y_n]$ is bounded in $W_G^{1,1}(a,b)$; then, there exists $w_\pm \in BV(a,b)$ such that, possibly up to subsequences,

$$I_+^{1-s}[y_n] \rightharpoonup w_+, \quad I_-^{1-s}[y_n] \rightharpoonup w_- \qquad \text{weakly in } BV(a,b). \tag{71}$$

Taking into account (70), (71) and the uniqueness of limit in $\mathcal{D}'(a,b)$, we obtain

$$w_+ = I_+^{1-s}[u] \in BV(a,b),\ w_- = I_-^{1-s}[u] \in BV(a,b), \quad I_\pm^{1-s}[y_n] \rightharpoonup I_\pm^{1-s}[u] \ \text{ in } BV(a,b). \tag{72}$$

Taking into account $u_n \in W^{s,1}(a,b)$, we set $f_n = I^{1-s}[u_n]$, so u_n solves Abel integral equation $I^{1-s}[u_n] = f_n$. By the semigroup property, $I_+^s[f_n](x) = I^s\big[I_+^{1-s}[u_n]\big](x) = I_+^1[u_n] = \int_a^x u_n(t)\,dt$; hence, $I_+^s[f_n](a) = 0$. Therefore (by Proposition 2), the Abel integral equations have a unique solution in $L^1(a,b)$, given by $u_n = D_+^{1-s}[f_n] = D_+^{1-s}\big[I_+^{1-s}[u_n]\big]$, $n \in \mathbb{N}$.

Set $f = I^{1-s}[u] \in BV(a,b)$. $u \in L^1(a,b)$. So u solves the Abel equation $I^{1-s}[u] = f$. Moreover, by the semigroup property, $I_+^s[f](x) = I^s\big[I_+^{1-s}[u]\big](x) = I_+^1[u] = \int_a^x u(t)\,dt$; hence, $I_+^s[f](a_+) = 0$. Therefore (Proposition 3), the Abel equation has a unique solution in $L^1(a,b)$, given by $u = D_+^{1-s}[f] = D_+^{1-s}\big[I_+^{1-s}[u]\big]$.

By (i), $u_n \to u$ strongly in L^q, hence,

$$f_n = I_+^{1-s}[u_n] \to I_+^{1-s}[u] = f \quad \text{strongly in } L^q,\ q \in [1, 1/(1-s)).$$

Then, by (65), $I^{1-s}[u] = f = z_+$. Hence we have shown claims (ii) and (iii).

Moreover, the convergence is also in the sense of distributions and the sequence is bounded in $W_G^{1,1}$; therefore, $I_+^{1-s}[u]$ belongs to $BV(a,b)$ and, again up to subsequences,

$$f_n = I_+^{1-s}[u_n] \rightharpoonup I_+^{1-s}[u] = f \quad \text{weakly in } BV(a,b).$$

We can deal with z_- by the same argument, exploiting Corollary 1 for the backward Abel integral equation $I_-^{1-s}[u_n] = g_n$, leading to $I^{1-s}[u] = g = z_-$. $\square$

Remark 8. *The boundedness of (a,b) is an essential assumption in the previous compactness theorem, not only to exploit the Rellich theorem, but also to avoid slow non-integrable decay at infinity of the fractional integral: indeed, even for an integrable compactly supported u, we may have $I_+^{1-s}[u](x) \sim (x-a)^{-s}$ at $+\infty$, e.g., if $u = \chi_{[a,b]}$.*

Remark 9. *We emphasize that in Theorem 3, we cannot improve (64), since $I_+^{1-s}[u]$ may belong to $BV(a,b) \setminus W_G^{1,1}(a,b)$.*

Indeed, we can choose $f(x) = \mathrm{sign}(x)$ if $-1 < x < 1$, $f_n(x) = -1$ if $-1 < x < -1/n$, $f_n(x) = n\,x$ if $-1/n < x < 1/n$ and $f_n(x) = 1$ if $1/n < x < 1$. Thus, f belongs to $BV(-1,1)\setminus W^{1,1}(-1,1)$ and $\|f_n\|_{W_G^{1,1}(-1,1)}$ is uniformly bounded. Solving the Abel equations $I_+^{1-s}[u_n] = f_n$ and $I_+^{1-s}[u] = f$ with Propositions 2 and 3 provides $u = D_+^s[f] \in \mathcal{M}(-1,1)\setminus L^1(-1,1)$, whereas $u_n = D_+^s[f_n]$ is uniformly bounded in L^1; hence, u_n is uniformly bounded in $W^{s,1}(-1,1)$, due to Lemma 7.

We recall a well-known result [9] (Theorem 2.1) concerning the L^1-representability of functions.

Theorem 4. **[*L^1-representability*]** *Given $f \in L^1(a,b)$, then*

$f \in I_+^s(L^1(a,b))$ for some $s \in (0,1)$ if and only if

$$I_+^{1-s}[f] \in W^{1,1}(a,b) \quad \text{and} \quad I_+^{1-s}[f](a) = 0\,.$$

$f \in I_-^s(L^1(a,b))$ for some $s \in (0,1)$ if and only if

$$I_-^{1-s}[f] \in W^{1,1}(a,b) \quad \text{and} \quad I_-^{1-s}[f](b) = 0\,.$$

Moreover, in the affirmative case, say, when there exists $u \in L^1(a,b)$ such that $f = I_+^s[u]$ (resp. $f = I_-^s[u]$), we obtain

$$u = D_+^s f \quad (\text{respectively } u = D_-^s f)\,. \tag{73}$$

In Section 5, we provide a self-contained proof of the above result together with a discussion of the related forward and backward Abel equation in the distributional framework, even in the cases when $I_+^{1-s}[f](a) \neq 0$ or $I_-^{1-s}[f](b) \neq 0$ (see Propositions 2 and 3 and Corollaries 1 and 2).

Here, we show that the representability result has a natural extension to the bilateral case.

Theorem 5. *Assume $0 < s < 1$. Then*

$$f \in L^1(a,b) \quad \text{and} \quad f \in I_+^s(L^1(a,b)) \cap I_-^s(L^1(a,b))$$

if and only if

$$f \in W^{s,1}(a,b), \qquad 2I^{1-s}[f](a) - I_-^{1-s}[f](a) = 0 = 2I^{1-s}[f](b) - I_+^{1-s}[f](b)\,,$$

if and only if

$$f \in W^{s,1}(a,b), \qquad 2I_+^{1-s}[f](b) - I_-^{1-s}[f](a) = 0 = 2I_-^{1-s}[f](a) - I_+^{1-s}[f](b)\,.$$

Proof. Since

$$I^{1-s}[f](a) = \frac{1}{2\Gamma(1-s)} \int_a^b \frac{f(t)}{(t-a)^s}dt = I_+^{1-s}[f](b)\,,$$

$$I^{1-s}[f](b) = \frac{1}{2\Gamma(1-s)} \int_a^b \frac{f(t)}{(b-t)^s}dt = I_-^{1-s}[f](a)\,,$$

$$I_+^{1-s}[f](a) = 2\,I^{1-s}[f](a) - I_-^{1-s}[f](a) = 2\,I_+^{1-s}[f](b) - I_-^{1-s}[f](a)\,,$$

$$I_-^{1-s}[f](b) = 2\,I^{1-s}[f](b) \;-\; I_+^{1-s}[f](b) = 2\,I_-^{1-s}[f](a) \;-\; I_+^{1-s}[f](b)\,,$$

the claim follows by Definition 9, Theorem 4, Proposition 1 (semigroup property of fractional integral), Propositions 2 and 3 and Corollaries 1 and 2. □

Next, we make explicit some embedding relationship between $W_\pm^{s,1}$ and $U_\pm^{s,1}$.

Theorem 6. *The following strict embeddings hold true:*

$$W^{s,1} \underset{\neq}{\subset} W_+^{s,1} \underset{\neq}{\subset} U_+^{s,1}\,, \qquad W^{s,1} \underset{\neq}{\subset} W_-^{s,1} \underset{\neq}{\subset} U_-^{s,1} \qquad \forall s \in (0,1)\,, \tag{74}$$

where we refer to Definitions 7–9 about $W^{s,1}(a,b)$, shortly denoted $W^{s,1}$ here, versus the naïve definition $U_\pm^{s,1}$ at the beginning of the present Section 3.

Proof. Without loss of generality, we assume $(a,b) = (0,1)$.

Strict embeddings of $W^{s,1}$ in $W_+^{s,1}$ and of $W^{s,1}$ in $W_-^{s,1}$ are shown respectively by x^{s-1} and $(1-x)^{s-1}$: see (55) in Example 5.

Therefore, in order to show (74), it is sufficient to show an example for the strict embedding $W_+^{s,1} \underset{\neq}{\subset} U_+^{s,1}$: indeed, the proof of $W_-^{s,1} \underset{\neq}{\subset} U_-^{s,1}$ is achieved by replacing the variable t with $(1-t)$ in the counterexample showing the other strict embedding by exploiting the symmetry with respect to $x = 1/2$, analogous to the one with respect to $x = 0$ in (36) and (37).

We first note that $W_+^{s,1} \subset U_+^{s,1}$ follows by definition (4): the existence of a weak derivative in L^1 of $I_\pm^{1-s}[u]$ entails the existence of the fractional derivative $D_\pm^s[u]$, coincident with the almost everywhere defined fractional derivative $(d/dx)_\pm^s[u](x)$.

The strict embeddings $W_+^{s,1} \underset{\neq}{\subset} U_+^{s,1}$, and $W_-^{s,1} \underset{\neq}{\subset} U_-^{s,1}$ follows by the subsequent argument, which, for any fixed $s \in (0,1)$, provides the existence of a function in $U_+^{s,1} \backslash W_+^{s,1}$ and a function in $U_-^{s,1} \backslash W_-^{s,1}$.

Given $s \in (0, \ln 2/\ln 3)$, we show a function z in $U_+^{s,1}$ such that $z \notin W_+^{s,1}$.

Precisely, by denoting V the Cantor–Vitali function on $[0,1]$ ([21]), we claim that

$$z := D_+^{1-s}[V] \;\in\; U_+^{s,1} \backslash W_+^{s,1} \qquad s \in (\alpha, 1)\,.$$

Indeed V is α-Hölder continuous with $\alpha := \ln 2/\ln 3$. So $I_+^s[V]$ belongs to $C^1(a,b)$ for every $s \in (1-\alpha, 1)$ by Theorem 3.1 in [9] and the fact that $V(0) = 0$. Therefore $I_+^s[V] \in W_G^{1,1}$ (hence $V \in W_+^{1-s,1}(0,1)$) for $s \in (1-\alpha, 1)$. Moreover, $I_+^s[V](0) = 0$ for $s \in (0,1)$: indeed, due to continuity of V in $[0,1]$, $I_+^s[V]$ is continuous in $[0,1]$ and we obtain

$$\Gamma(s) I_+^s[V](0) = \lim_{x \to 0} \int_0^x \frac{V(t)}{(x-t)^{1-s}}\,dt = \lim_{x \to 0} \left(\int_0^x \frac{V(t) - V(x)}{(x-t)^{1-s}}\,dt + \frac{x^s V(x)}{s} \right).$$

By Hölder continuity ($V \in C^{0,\alpha}(a,b)$), we obtain $|V(t) - V(x)| \le C|x-t|^\alpha$. Then

$$\left| \int_0^x \frac{V(t)}{(x-t)^{1-s}}\,dt \right| \le C \int_0^x (x-t)^{\alpha+s-1}\,dt + \frac{x^s V(x)}{s} = \frac{C x^{s+\alpha}}{s+\alpha} + \frac{x^s V(x)}{s}.$$

Therefore, the limit above is equal to 0, as $x \to 0^+$, thus proving the claim $I_+^s[V](0) = 0$.

Summarizing, $V \in BV(0,1)$, $I_+^s[V](0) = 0$ and $I_+^s[V] \in W^{1,1}(0,1)$, for $s \in (1-\alpha, 1)$.

Therefore we can consider the Abel integral equation in the distributional setting

$$\text{find } z \in L^1(0,1): \qquad I_+^{1-s}[z] \;=\; V \qquad \text{on } (0,1), \tag{75}$$

and solve it; by Proposition 3, the unique solution is given by $z = D_+^{1-s}[V] = D\,I_+^s[V]$, and fulfils $V = I_+^{1-s}[z]$. Moreover $(d/dx)_+^s[z] = (d/dx) I_+^{1-s}[z] = (d/dx)V = 0$ a.e. on

$(0,1)$, whereas $D^s_+[z] = D\, I^{1-s}_+[z] = DV$, which is a nontrivial bounded measure. Explicitly $z = D^{1-s}_+[V]$ fulfills $z \in \mathcal{U}^{s,1}_+ \setminus W^{s,1}_+$. So far, we have proved the first embedding chain in (74) for $s \in (1-\alpha, 1) = (1 - \ln 2/\ln 3,\ 1)$.

In the sequel, we show that, given any $\sigma \in (0,1)$, we can adapt the Cantor–Vitali function in such a way that it is s-Hölder continuous for any $s \in (0,\sigma]$; hence, we recover the strict embedding for any s in $(1-\sigma, 1)$, and hence, for any s in $(0,1)$, due to the generic choice of σ.

Indeed, given $\tau \in (0,1)$, we can replace the construction of Cantor $1/3$ - middle set $\mathcal{C}$ (say, a set whose Hausdorff dimension is $\ln 2/\ln 3$, which leads to the $\alpha = \alpha(1/3) = \ln 2/\ln 3$ Hölder continuous Cantor–Vitali function $V_{1/3} := V$) with the Cantor-like τ - middle set $\mathcal{C}_\tau$, with Hausdorff dimension $\dim(\mathcal{C}_\tau) = \ln 2/\big(\ln 2 - \ln(1-\tau)\big)$, which leads to the $\alpha(\tau)$ Hölder continuous Cantor–Vitali generalized function $V = V_\tau$, where

$$\alpha(\tau) = \dim(\mathcal{C}_\tau) = \ln 2/\big(\ln 2 - \ln(1-\tau)\big)\,.$$

Notice that $\alpha(\tau) \to 1_-$ as $\tau \to 0_+$ and $\alpha(\tau) \to 0_+$ as $\tau \to 1_-$, so that $\alpha(\tau)$ spans the interval $(0,1)$ as τ runs over $(0,1)$. Moreover, $V_\tau \in (BV \cap C^0)\setminus W^{1,1}$ for $\tau \in (0,1)$.

Again by Proposition 3, we get that V_τ is representable, say there exists (unique) $z_\tau \in L^1(0,1)$ s.t. $z_\tau = D^{1-s}_+[V_\tau]$ s.t. $V_\tau = I^{1-s}_+[z_\tau]$ for $s \in (1-\alpha(\tau), 1)$, and we claim that $V_\tau(0) = 0$, $I^s_+[V_\tau](0) = 0$, $z_\tau \in \mathcal{U}^{s,1}_+ \setminus W^{s,1}_+$ for $s \in (1-\alpha(\tau), 1)$: indeed these claims about the generalized Cantor–Vitali function V_τ can be proved by the same procedure dealing with the definition of $V = V_{1/3}$, as it is sketched below.

The function V_τ is of bounded variations since is monotone, as it is the uniform limit of a sequence of monotone nondecreasing functions.

Continuity of V_τ follows from uniform convergence of standard iterative approximations by piecewise linear functions. The absolutely continuous part of the distributional derivative $D\,V_\tau$ is identically 0 since V_τ is locally constant on an open set of Lebesgue measure 1: indeed, it is a union of open intervals, which is iteratively obtained by approximation with finite unions A_n whose measure ℓ_n fulfills the recursive scheme: $\ell_1 = \tau$, $\ell_{n+1} = \ell_n + 2^n \tau \frac{1-\ell_n}{2^n}$, so that $\ell_n = 1 - (1-\tau)^n \ \to\ 1$ as $n \to \infty$.

The worst case for differential quotients of n-th approximations of V_τ is provided by $(1/2)^n/\big((1-\ell_n)/2^n\big) = 1/(1-\tau)^n$, so that $\alpha(\tau)$ is the biggest real α s.t.

$$(1/2)^n/\big((1-\ell_n)/2^n\big)^\alpha = 2^{n(\alpha-1)}/(1-\tau)^{n\alpha}$$

is uniformly bounded for $n \in \mathbb{N}$, say

$$0 < \alpha \le \alpha(\tau) = \ln 2/\big(\ln 2 - \ln(1-\tau)\big)\,.$$

So $I^s_+[V_\tau]$ belongs to $\mathcal{C}^1(a,b)$ for every $s \in (1-\alpha(\tau), 1)$ by Theorem 3.1 of [9] and taking into account that $V_\tau(0) = 0$. Therefore, $I^{1-s}_+[V_\tau] \in W^{1,1}$ that is $V_\tau \in W^{s,1}_+(0,1)$ for $s \in (0, \alpha(\tau))$.

Moreover, $I^{1-s}_+[V_\tau](0) = 0$. Indeed, by continuity of V_τ in $[0,1]$, we obtain

$$\Gamma(s)\, I^s_+[V_\tau](0) = \lim_{x\to 0}\int_0^x \frac{V_\tau(t)}{(x-t)^{1-s}}dt = \lim_{x\to 0}\left(\int_0^x \frac{V_\tau(t)-V_\tau(x)}{(x-t)^{1-s}}dt + \frac{x^s V_\tau(x)}{s}\right).$$

Since $V_\tau \in \mathcal{C}^{0,\alpha(\tau)}(a,b)$, we get $|V_\tau(t) - V_\tau(x)| \le C|x-t|^{\alpha(\tau)}$. Thus

$$\left|\int_0^x \frac{V_\tau(t)}{(x-t)^{1-s}}dt\right| \le C\int_0^x (x-t)^{\alpha(\tau)+s-1}dt + \frac{x^s V_\tau(x)}{s} = \frac{C x^{s+\alpha(\tau)}}{s+\alpha(\tau)} + \frac{x^s V_\tau(x)}{s}\,.$$

Summarizing, if V_τ is the generalized Cantor–Vitali function and $\mathcal{U}_\tau(x) = V_\tau(1-x)$,

$$z_\tau := D^{1-s}_+[V_\tau] \in \mathcal{U}^{s,1}_+ \setminus W^{s,1}_+ \qquad s \in (1-\alpha(\tau), 1)\,, \tag{76}$$

since $D^s_+[z_\tau] = D\,\mathcal{V}_\tau$ is a nontrivial Cantor measure with no atomic part, whereas $(d/dx)^s_+[z_\tau] = 0$ a.e.; moreover,

$$u_\tau := D^{1-s}_-[\mathcal{U}_\tau] \in U^{s,1}_- \setminus W^{s,1}_- \qquad s \in (1-\alpha(\tau), 1), \tag{77}$$

where, to achieve (77), we exploit Proposition 2 to solve the backward Abel integral equation in the distributional framework $I^{1-s}[u_\tau] = \mathcal{U}_\tau$; indeed, $\mathcal{U}_\tau \in BV(0,1)$, $I^s_-[\mathcal{U}_\tau](1) = 0$, then the unique solution $v \in L^1(0,1)$ of $I^{1-s}_-[v] = \mathcal{U}_\tau$ is $v = u_\tau = D^{1-s}_-[\mathcal{U}_\tau]$, which fulfills $I^{1-s}_-[u_\tau] = \mathcal{U}_\tau$. Hence, by evaluating the distributional derivative D, we get $D^s_-[u_\tau] = -D\,\mathcal{U}_\tau$ which is a nontrivial Cantor measure with no atomic part, whereas $(d/dx)^s_-[u_\tau] = 0$ a.e. □

We list some properties concerning the comparison of bilateral Riemann–Liouville fractional Sobolev spaces $W^{s,1}$ with classical spaces: Gagliardo fractional Sobolev spaces $W^{s,1}_G$, functions of bounded variation $BV(0,1)$ and $SBV(0,1)$, De Giorgi's space of special bounded variation functions, whose derivatives have no Cantor part ([21,35] for example).

Theorem 7. *Let be $s, r \in (0,1)$ such that $r > s$. Then*

$$W^{r,1}_G(0,1) \cap I^s_+(L^1(a,b)) \cap I^s_-(L^1(a,b)) \subset W^{s,1}(0,1)$$

with continuous injection, say $\forall u \in W^{r,1}_G(0,1) \cap I^s_+(L^1(a,b)) \cap I^s_-(L^1(a,b))$

$$\|u\|_{W^{s,1}} := \|u\|_{L^1(a,b)} + \|I^{1-s}_+ u\|_{W^{1,1}(0,1)} + \|I^{1-s}_- u\|_{W^{1,1}(0,1)} \leq C\|u\|_{W^{r,1}_G(0,1)},$$

Proof. Straightforward consequence of Theorem 3.2 of [25] and Definition 9. □

In [25], we have compared $W^{s,1}_+(0,1)$ and $W^{s,1}_-(0,1)$ with $SBV(0,1)$, and proved

$$SBV(0,1) \subset \bigcap_{s\in(0,1)} W^{s,1}(0,1).$$

This inclusion was refined by a recent result (Theorem 3.4 in [2]) showing

$$BV(a,b) \subset \bigcap_{s\in(0,1)} W^{s,1}(a,b). \tag{78}$$

On the other hand, for every $s \in (0,1)$, $W^{s,1}(a,b)$ is contained neither in $W^{1,1}_G(a,b)$ nor in $BV(a,b)$, due to remarkable examples of Weierstrass-type functions. Indeed a Weierstrass function w can be defined ([36]) so that w belongs to $W^{s,1}$, but w does not belong to $BV(0,1)$ since it is nowhere differentiable. Fix $q > 1$ and set

$$w(x) = \sum_{n=0}^{\infty} q^{-n}\left(\exp(iq^n x) - \exp(iq^n a)\right). \tag{79}$$

Notice that the the constant subtraction entails $w(a) = 0$, thus preventing a singularity of $D^s_{a+}[w](x)$ at $x = a$.

Theorem 8. *Let $s, r \in (0,1)$ be such that $r > s$. Then*

$$W^{r,1}_G(0,1) \cap I^s_+(L^1(a,b)) \cap I^s_-(L^1(a,b)) \subset W^{s,1}(0,1)$$

with continuous injection. Precisely,

$$\|u\|_{W^{s,1}} := \|u\|_{L^1(a,b)} + \|I^{1-s}_+ u\|_{W^{1,1}(0,1)} + \|I^{1-s}_- u\|_{W^{1,1}(0,1)} \leq C\|u\|_{W^{r,1}_G(0,1)},$$

$\forall u \in W^{r,1}_G(0,1) \cap I^s_+(L^1(a,b)) \cap I^s_-(L^1(a,b))$.

We emphasize that in the case of an unbounded interval (a,b), there is no chance for a compactness statement analogous to Theorem 3 in $W^{s,1}$(a,b), since the Rellich theorem cannot be applied.

On the other hand, the property $u \in W^{s,1}(\mathbb{R})$ entails a stronger qualitative condition on u than in the case of $u \in W^{s,1}(a,b)$ with a boundedness of (a,b), as clarified by the next remark.

Remark 10. *If $u \in W^{s,1}(\mathbb{R})$, $0 < s < 1$, then $\int_{\mathbb{R}} u(t)\,dt = 0$ and $|\xi|^{s-1}\widehat{u}(\xi)$ is bounded in a neighborhood of $\xi=0$. Property $\int_a^b u(t)\,dt=0$ may fail for $u \in W^{s,1}(a,b)$ if $(a,b) \neq \mathbb{R}$.*

Indeed, $(a,b)=\mathbb{R}$ entails $\widehat{u} \in C^0 \cap L^\infty(\mathbb{R})$, hence $u \in L^1(\mathbb{R})$, $u \in W^{s,1}(\mathbb{R}) \subset W_e^{s,1}(\mathbb{R})$, hence

$$I_e^{1-s}[u] \;=\; u * \frac{1}{2\,\Gamma(1-s)\,|t|^s} \;\in\; W_G^{1,1}(\mathbb{R}) \;\subset\; L^1(\mathbb{R})\,,$$

then, exploiting the Fourier transform $\mathcal{F}$, $|\xi|^{s-1}\widehat{u}(\xi) \in C^0 \cap L^\infty(\mathbb{R})$, hence $|\xi|^{s-1}\widehat{u}(\xi)$ is bounded and $\int_{\mathbb{R}} u(t)dt = \widehat{u}(0) = 0$.

If $(a,b) \neq \mathbb{R}$, then, referring to (8) and (10), neither $I_e^{1-s}[u]$ nor $I_o^{1-s}[u]$ belong to $L^1(\mathbb{R})$ (or even to $L^p(\mathbb{R})$ with $1 < p \leq 2$, when $s > 1/2$), moreover the summability of I_o^{1-s} may fail at infinity due to a decay of order $|x|^{-s}$, therefore $\widehat{u}$ may be unbounded around $\xi = 0$.

Remark 11. *Notwithstanding Remark 10 (excluding nontrivial constant functions from the space $W^{s,1}(\mathbb{R})$), if we restrict to bounded intervals, a constant function $u \equiv K$ belongs to $W^{s,1}(a,b)$, for every bounded interval (a,b) and every value of K. Indeed,*

$$I_+^{1-s}[K](x) = \frac{K}{\Gamma(2-s)}(x-a)^{1-s} \;\in\; L^1(a,b)\,,$$

$$D_+^s[K](x) = D_x\left[I_+^{1-s}[K]\right](x) = \frac{d}{dx}I_+^{1-s}[K](x) \;=\; \frac{K}{\Gamma(1-s)}\,(x-a)^{-s} \in L^1(a,b)\,.$$

4. Bilateral Fractional Bounded Variation Space

Possible naïve definitions could be provided, for $s \in (0,1)$, by

$$A_+^s \;=\; \left\{ u \in L^1(a,b) \mid \left(\frac{d}{dx}\right)_+^s u \text{ is a bounded measure} \right\},$$

$$A_-^s \;=\; \left\{ u \in L^1(a,b) \mid \left(\frac{d}{dx}\right)_-^s u \text{ is a bounded measure} \right\},$$

$$A^s \;=\; A_+^s \cap A_-^s,$$

which refer to L^1-functions whose classical pointwise-defined Riemann–Liouville fractional derivative of prescribed order $s \in (0,1)$ is a bounded measure.

Actually, if the Riemann–Liouville fractional derivative $D^s[u](x)$ of u exists for a.e. x for some $s \in (0,1)$, then $I_{a+}^s[u]$ is differentiable almost everywhere, referring to the same s; nevertheless, we have no information on the distributional derivative of the fractional integral $I_{a+}^s[u]$.

These differential properties are not completely described by the point-wise derivative, though it exists almost everywhere. This shows that the previous definitions of A^s, A_+^s and A_-^s are not suitable to obtain an integration by the parts formula. Therefore, to develop a satisfactory theory of fractional bounded variation spaces, as we did for fractional Sobolev spaces in [25], we introduce a more suitable function space: the *bilateral fractional bounded variation space* BV^s, as defined in the sequel.

Remark 12. *We recall that, as long as these classical fractional derivatives are evaluated on absolutely continuous functions, as it was done in all previous section, using the operators of the classical Definition 2 provides the same results as the distributional Definition 3: for this reason,* *we keep the usual classical notations ($_{RL}D^s_{a+}$, $_{RL}D^s_{b-}$ and the corresponding short forms D^s_+, D^s_-). However, in the present section, we evaluate fractional derivatives on functions of bounded variations, a setting where the two definitions provide different evaluations.*

Next, inspired by [2], where the nonsymmetric spaces are studied also in the case of higher order derivatives, we introduce the bilateral Riemann–Liouville bounded variation space, with the aim to achieve a symmetric framework.

Definition 10. *The (bilateral) Riemann–Liouville fractional bounded variation spaces. For every $s \in (0,1)$, we set*

$$BV^s \;=\; BV^s_+ \cap BV^s_- \tag{80}$$

where, referring to Definition 3,

$$BV^s_+ = \{u \in L^1(a,b) \mid I^{1-s}_+[u] \in BV(a,b)\} \;=\; \{u \in L^1(a,b) \mid D^s_+[u] \in \mathcal{M}(a,b)\}\,,$$

$$BV^s_- = \{u \in L^1(a,b) \mid I^{1-s}_-[u] \in BV(a,b)\} \;=\; \{u \in L^1(a,b) \mid D^s_-[u] \in \mathcal{M}(a,b)\}\,.$$

Theorem 9. *Assume that the interval (a,b) is bounded and the parameter s fulfills $0<s<1$. Then, the space $BV^s(a,b)$ is a normed space endowed with the norm*

$$\|u\|_{BV^s} := \|u\|_{L^1(a,b)} + \|\, D^s_{a+}[u]\,\|_{\mathcal{M}(a,b)} + \|\, D^s_{b-}[u]\,\|_{\mathcal{M}(a,b)}\,. \tag{81}$$

Contribution $\|D^s_+[u]\|_{\mathcal{M}(a,b)} + \|D^s_-[u]\|_{\mathcal{M}(a,b)}$ in the norm (81) can be replaced by $\|D^s_e[u]\|_{\mathcal{M}(a,b)} + \|D^s_o[u]\|_{\mathcal{M}(a,b)}$.

Moreover, $BV^s(a,b)$ is a Banach space and for every $q \in [1, 1/(1-s))$, there is $C = C(s,q,a,b)$, such that

$$\|u\|_{L^q(a,b)} \;\le\; C(s,q,a,b)\,\|u\|_{BV^s(a,b)}, \tag{82}$$

Every $u \in BV^s(a,b)$ can be represented by both

$$u(x) \;=\; I^s_{a+}\big[D^s_{a+}[u]\big](x) + \frac{I^{1-s}_{a+}[u](a_+)}{\Gamma(s)}\,(x-a)^{s-1} \qquad a.e\; x \in (a,b), \tag{83}$$

and

$$u(x) \;=\; I^s_{b-}\big[D^s_{b-}[u]\big](x) + \frac{I^{1-s}_{b-}[u](b_-)}{\Gamma(s)}\,(b-x)^{s-1} \qquad a.e\; x \in (a,b). \tag{84}$$

Proof. We emphasize that here $I^{1-s}_{a+}[u](a_+)$ and $I^{1-s}_{b-}[u](b_-)$ replace, respectively, $I^{1-s}_{a+}[u](a)$ and $I^{1-s}_{b-}[u](b)$ which were in representations (52) and (53) of $W^{s,1}$ functions, since in the present BV setting, there are not pointwise defined values, though there are well-defined finite right and left limits at every point in (a,b).

The map $u \mapsto \|u\|_{BV^s}$ is a norm on $W^{s,1}(a,b)$, indeed,

$\|u\|_{L^1(a,b)} + \|D^s_{a+}[u]\|_{\mathcal{M}(a,b)}$ is equivalent to the norm $\|I^{1-s}_{a+}[u]\|_{BV^{s,1}(a,b)}$, since $I^{1-s}_{a+}[u]$ belongs to BV, $D^s_{a+}[u] = DI^{1-s}_{a+}[u]$ and $\|I^{1-s}_{a+}[u]\|_{L^1(a,b)} \le C\|u\|_{L^1(a,b)}$; analogously $\|u\|_{L^1(a,b)}$ $+\|D^s_{b-}[u]\|_{\mathcal{M}(a,b)}$ is a norm for $I^{1-s}_{b-}[u]$, due to $I^{1-s}_{b-}[u] \in BV$, $D^s_{b-}[u] = DI^{1-s}_{b-}[u]$ and $\|I^{1-s}_{a+}[u]\|_{L^1(a,b)} \le C\|u\|_{L^1(a,b)}$. Therefore, terms $\|I^{1-s}_{\pm}[u]\|_{BV^s(a,b)}$ can be replaced, respectively, by $\|D\left[I^{1-s}_{\pm}[u]\right]\|_{\mathcal{M}(a,b)}$ in the natural norm

$$\|u\|_{BV^s} := \|u\|_{L^1(a,b)} + \|I^{1-s}_+[u]\|_{BV^s(a,b)} + \|I^{1-s}_-[u]\|_{BV^s(a,b)}\,.$$

The other claims follow by the same proof of Theorem 1 for the fractional Sobolev setting, where actually only the Proposition 2 and Corollary 1 about Abel forward and backward integral equation must be suitably tuned as stated in Remark 17. □

Example 6. *The constant functions and* $v(x) = x(1-x)$ *belong to the space* $BV^s(0,1)$*. In general, the space* $C_0^\infty(a,b)$ *of test function on* (a,b) *is contained in* $BV^s(a,b)$*.*

Example 7. *Heaviside function H belongs to* $BV^s(-1,1)$*, thanks to Example 3.*

Example 8. *Function* $H(x)|x|^{s-1}$ *belongs to* $BV_+^s(-1,1)\setminus W_+^{s,1}(-1,1)$ *if* $0 < s < 1$*, since* $I_+^{1-s}[H(x)\,|x|^{s-1}] = \Gamma(s)H(x) \in BV_+^s(-1,1)$ *due to Example 1.*

Due to the unboundedness of $I_-^{1-s}[\,H(x)\,|x|^{s-1}\,](x)$ *in a right neighborhood of* $x=0$ *(due to Example 5), we obtain that* $H(x)\,|x|^{s-1}$ *does not belong to* $BV_-^s(-1,1)$*.*

In general, for $0<s<1$, $H(x)\,|x|^{-\alpha}$ *belongs to* $BV_+^s(-1,1)\setminus BV_-^s(-1,1)$ *if* $0<\alpha<1-s$*.*

Theorem 10. ***(integration by parts in*** $\boldsymbol{BV^s(a,b)}$***)***

Next, identities hold true for $0<s<1$, $-\infty<a<b<+\infty$*:*

$$\begin{cases} \displaystyle\int_a^b u(x)\,d\,D_+^s[v](x) \;=\; -\int_a^b D_x u(x)\;I_+^{1-s}[v](x)\,dx \;+\; u(b)\,I_+^{1-s}[v](b) \\ \qquad\qquad\qquad\qquad\qquad\qquad \forall\, v\in BV_+^s(a,b),\ \forall\, u\in W_G^{1,1,}(a,b)\,, \end{cases} \tag{85}$$

$$\begin{cases} \displaystyle\int_a^b u(x)\,d\,D_-^s[v](x) \;=\; +\int_a^b D_x u(x)\;I_-^{1-s}[v](x)\,dx \;+\; u(a)\,I_-^{1-s}[v](a) \\ \qquad\qquad\qquad\qquad\qquad\qquad \forall\, v\in BV_-^s(a,b),\ \forall\, u\in W_G^{1,1,}(a,b)\,, \end{cases} \tag{86}$$

$$\begin{cases} \displaystyle\int_a^b u(x)\,d\,D_e^s[v](x) \;= \\ \displaystyle\quad -\int_a^b D_x u(x)\;I_e^{1-s}[v](x)\,dx + \frac{1}{2}\left(u(b)\,I_+^{1-s}[v](b) - u(a)\,I_-^{1-s}[v](a)\right) \\ \qquad\qquad\qquad\qquad\qquad\qquad \forall\, v\in BV^s(a,b),\ \forall\, u\in W_G^{1,1,}(a,b)\,, \end{cases} \tag{87}$$

$$\begin{cases} \displaystyle\int_a^b u(x)\,d\,D_o^s[v](x) \;= \\ \displaystyle\quad -\int_a^b D_x u(x)\;I_0^{1-s}[v](x)\,dx + \frac{1}{2}\left(u(b)\,I_+^{1-s}[v](b) + u(a)\,I_-^{1-s}[v](a)\right) \\ \qquad\qquad\qquad\qquad\qquad\qquad \forall\, v\in BV^s(a,b),\ \forall\, u\in W_G^{1,1,}(a,b)\,, \end{cases} \tag{88}$$

Proof. Exactly the same proof of Theorem 2, but the facts that, here, the distributional derivatives D_x in BV replaces the almost everywhere pointwise derivative d/dx in $W_G^{1,1}$ and the integrals at the left-hand side are evaluated with respect to the measures $D_+^s[v]$, $D_-^s[v]$, $D_e^s[v]$ and $D_o^s[v]$, in place of Lebesgue measure. □

Theorem 11. **[*Compactness in*** $\boldsymbol{BV^s(a,b)}$**]**

Assume that $0<s<1$*, the interval* (a,b) *is bounded and*

$$\|u_n\|_{BV^s(a,b)} \;\le\; C\,. \tag{89}$$

Then, there exist $u \in L^1(a,b)$ *and a subsequence such that, without relabeling,*

$$\begin{cases} (i) & u_n \rightharpoonup u & \text{weakly in } L^q(a,b)) \quad \forall q\in[1,1/(1-s))\,, \\ (ii) & I_+^{1-s}[u_n] \to I_+^{1-s}[u] & \text{strongly in } L^p(a,b)\,,\ \forall p<+\infty\,, \\ (iii) & I_-^{1-s}[u_n] \to I_-^{1-s}[u] & \text{strongly in } L^p(a,b)\,,\ \forall p<+\infty\,. \end{cases}$$

$$I_+^{1-s}[u_n] \rightharpoonup I_+^{1-s}[u]\,, \qquad I_-^{1-s}[u_n] \rightharpoonup I_-^{1-s}[u] \qquad \text{weakly in } BV(a,b). \tag{90}$$

Proof. The proof can be achieved by exactly the same argument used in the proof of compactness in $W^{s,1}(a,b)$ (Theorem 3). □

Remark 13. *We emphasize that*

$$BV(a,b) \subsetneq BV^s(a,b) \qquad \forall s \in (0,1), \tag{91}$$

since $BV \subsetneq W^{s,1}$ *and* $W^{s,1} \subset BV^s$.

Moreover,

$$BV(a,b) \subsetneq \bigcap_{\sigma\in(0,1)} W^{\sigma,1}(a,b) \subsetneq W^{s,1}(a,b) \subsetneq BV^s_+(a,b) \qquad \forall s \in (0,1). \tag{92}$$

Indeed, the first embedding follows by (78) and is strict due to (79); the second embedding is obviously strict, about the third embedding notice that $H(x)|x|^{s-1} \in BV^s_+(-1,1)\setminus W^{s,1}_+(-1,1)$ *(see Example 8).*

In addition, we can rewrite (74) as follows

$$W^{s,1}(a,b) \subsetneq BV^s_+(a,b)\,, \qquad W^{s,1}(a,b) \subsetneq BV^s_-(a,b)\,, \qquad \forall s \in (0,1). \tag{93}$$

since, referring to notations (76) and (77) in the proof of Theorem 6,

$$\exists\, z_\tau \in BV^s_+(-1,1)\setminus W^{s,1}(-1,1)\,, \qquad s \in \Big(1-\ln 2/\big(\ln 2-\ln(1-\tau)\big)\,,\,1\Big)\,, \tag{94}$$

$$\exists\, u_\tau \in BV^s_-(-1,1)\setminus W^{s,1}(-1,1)\,, \qquad s \in \Big(1-\ln 2/\big(\ln 2-\ln(1-\tau)\big)\,,\,1\Big) \tag{95}$$

5. Abel Equation in $\mathcal{D}'(\mathbb{R})$ and Some Useful Relationships

Here, for the reader's convenience, first, we recall some basic algebra of fractional differential calculus, then we extend to the distributional setting some classical results about Abel integral equations: these suitably tuned claims are exploited in Sections 3 and 4 to prove the main properties of $W^{\alpha,1}(a,b)$ and $BV^\alpha(a,b)$, with $\alpha \in (0,1)$: Theorems 1, 3, 5, 6, 9 and 11.

All the results stated in this section are independent of the ones of previous sections.

To avoid confusion with the standard notation of variable s in the Laplace transform, here, we label by α, instead of s, the index of fractional integral, fractional derivative and fractional Sobolev space.

All along this section: the *Laplace transformable function* refers to a measurable function v on $\mathbb{R}$ with support contained in $[0,+\infty)$ such that there exists $\lambda \in \mathbb{R}$ for which $e^{-\lambda x}v(x)$ is a Lebesgue-integrable function; the *Laplace transformable distribution* refers to a distribution v on $\mathbb{R}$ with support contained in $[0,+\infty)$ such that there exists $\lambda \in \mathbb{R}$ for which $e^{-\lambda x}v(x)$ is a tempered distribution; and in all cases, $V = \mathcal{L}\{v\}$ denotes the Laplace transform of v.

First we recall some relationships concerning fractional integral of powers of x in $(0,1)$:

$$I^{1-\alpha}_{0+}[x^\beta] = \frac{\Gamma(1+\beta)}{\Gamma(2+\beta-\alpha)}\, x^{1+\beta-\alpha} \qquad \alpha \in (0,1)\,,\ \beta > -1\,, \tag{96}$$

$$I^{\alpha}_{0+}[x^\beta] = \frac{\Gamma(1+\beta)}{\Gamma(1+\beta+\alpha)}\, x^{\beta+\alpha} \qquad \alpha \in (0,1)\,,\ \beta > -1\,, \tag{97}$$

$$I^{1-\alpha}_{1-}[x^\beta] = \left(\frac{\Gamma(\alpha-\beta-1)}{\Gamma(-\beta)} - \frac{B(x,\alpha-\beta-1,1-\alpha)}{\Gamma(1-\alpha)}\right) x^{1+\beta-\alpha} \qquad \alpha,\beta \in (0,1), \tag{98}$$

where $B = B(x,\mu,\nu)$ is the incomplete Beta function: $B(x,\nu,\mu) = \int_0^x y^{\nu-1}(1-y)^{\mu-1}dy$.

Hence, since both conditions $I_{0+}^{1-\alpha}[x^\beta] \in W_G^{1,1}(0,1)$ and $I_{0+}^{1-\alpha}[x^\beta](0) = 0$ hold true when $\beta > \alpha - 1$, one obtains the fractional derivative of power functions of x in $(0,1)$:

$$\begin{aligned} D_{0_+}^{\alpha}[x^\beta] &= D_x\, I_{0_+}^{1-\alpha}[x^\beta] = \frac{d}{dx}\, I_{0_+}^{1-\alpha}[x^\beta] \\ &= \frac{\Gamma(1+\beta)}{(1+\beta-\alpha)\Gamma(1+\beta-\alpha)}\,\frac{d}{dx}x^{1+\beta-\alpha} \\ &= \frac{\Gamma(1+\beta)}{\Gamma(1+\beta-\alpha)}\,x^{\beta-\alpha} \qquad \alpha\in(0,1)\,,\ \beta>\alpha-1\,. \end{aligned} \tag{99}$$

Moreover,

$$I_{0+}^{1-\alpha}[x^{\alpha-1}] = \frac{1}{\Gamma(1-\alpha)}\int_0^x \frac{t^{\alpha-1}dt}{(x-t)^\alpha} = \frac{1}{\Gamma(1-\alpha)}\int_0^1 \frac{dy}{y^{1-\alpha}(1-y)^\alpha} = \frac{B(\alpha,1-\alpha)}{\Gamma(1-\alpha)} = \Gamma(\alpha)$$

entails

$$D_{0_+}^{\alpha}[x^{\alpha-1}] \equiv 0\,, \qquad \alpha\in(0,1)\,. \tag{100}$$

In the particular case $\alpha = 1/2$ we obtain

$$I_{0+}^{1/2}[x^{-1/2}](x) = \sqrt{\pi} \quad \text{and} \quad D_{0+}^{1/2}[x^{-1/2}] \equiv 0\,. \tag{101}$$

Thus $D^{1/2}$ has a nontrivial kernel, as it is the case of the linear operators D^α. More in general, by (100), we know that

$$D_{a_+}^{\alpha}[\,H(x-a)(x-a)^{\alpha-1}\,] \equiv 0 \qquad \forall\,\alpha\in(0,1),\ \forall\,a\in\mathbb{R},\ x\in\mathbb{R}. \tag{102}$$

The converse holds too (see Proposition 8).

Proposition 1. *(Semigroup property of fractional integral* $\mathbf{I}_{a+}^{\alpha}$ *)*
For every $0<\alpha<1$, $v\in L^1(\mathbb{R})$, $\operatorname{spt} v\subset[a,+\infty)$ *with* $a\in\mathbb{R}$, *we have*

$$I_{a+}^{1-\alpha}\left[I_{a+}^{\alpha}[v]\right](x) = I_{a+}^{1}[v] = \int_a^x v(t)\,dt \qquad x\in\mathbb{R}\,, \tag{103}$$

$$D_{a+}^{1-\alpha}\left[D_{a+}^{\alpha}[v]\right] = D_x[v] \qquad in\ \mathcal{D}'(\mathbb{R})\,. \tag{104}$$

In general, if $-\infty<a<b\le+\infty$, $\alpha,\beta\in(0,1)$, $\alpha+\beta<1$, $v\in L^1(a,b)$, *then*

$$I_{a+}^{\alpha}\left[I_{a+}^{\beta}[v]\right](x) = I_{a+}^{\alpha+\beta}[v](x) \qquad x\in\mathbb{R}\,, \tag{105}$$

$$D_{a+}^{\alpha}\left[D_{a+}^{\beta}[v]\right](x) = D_{a+}^{\alpha+\beta}[v] \qquad \mathcal{D}'(\mathbb{R})\,. \tag{106}$$

if $-\infty\le a<b<+\infty$, $\alpha,\beta\in(0,1)$, $\alpha+\beta<1$, $v\in L^1(a,b)$, *then*

$$I_{b-}^{\alpha}\left[I_{b-}^{\beta}[v]\right](x) = I_{b-}^{\alpha+\beta}[v](x) \qquad x\in\mathbb{R}\,, \tag{107}$$

$$D_{b-}^{\alpha}\left[D_{b-}^{\beta}[v]\right](x) = D_{b-}^{\alpha+\beta}[v] \qquad \mathcal{D}'(\mathbb{R})\,. \tag{108}$$

Proof. Consider the trivial extension of v and the standard extension of related subsequent fractional integrals as defined by

$$I_{a+}^{\alpha}[v] = v * \frac{1}{\Gamma(\alpha)}\,\frac{H(x)}{|x|^{1-\alpha}} \qquad x\in\mathbb{R}\,.$$

We assume first $a = 0$. By denoting V, the Laplace transform of v and taking into account of $\mathcal{L}\{H(x)\,x^{\beta}\} = \Gamma(\beta+1)/s^{\beta+1}$ and (97), we obtain, for $\mathrm{Re}\, s > 0$,

$$\begin{aligned}\mathcal{L}\Big\{I_{0+}^{1-\alpha}\big[I_{0+}^{\alpha}[v]\big](x)\Big\} &= \frac{1}{\Gamma(1-\alpha)}\,\frac{1}{\Gamma(\alpha)}\,\mathcal{L}\Big\{v * \frac{H(x)}{|x|^{\alpha}} * \frac{H(x)}{|x|^{1-\alpha}}\Big\} \\ &= \frac{1}{s}\,V(s) = \mathcal{L}\Big\{\int_{-\infty}^{x} v(t)dt\Big\} = \mathcal{L}\Big\{\int_{0}^{x} v(t)dt\Big\},\end{aligned}$$

hence, claim (103) follows by the injectivity of the Laplace transform.

$$\begin{aligned}\mathcal{L}\Big\{D_{0+}^{1-\alpha}\big[D_{0+}^{\alpha}[v]\big](x)\Big\} &= \frac{1}{\Gamma(\alpha)}\,\frac{1}{\Gamma(1-\alpha)}\,\mathcal{L}\Big\{D_x\Big(\frac{H(x)}{|x|^{1-\alpha}} * D_x\Big(\frac{H(x)}{|x|^{\alpha}} * v\Big)\Big)\Big\} \\ &= \frac{1}{\Gamma(\alpha)}\,\frac{1}{\Gamma(1-\alpha)}\,s\,\frac{\Gamma(\alpha)}{s^{\alpha}}\,s\,\frac{\Gamma(1-\alpha)}{s^{1-\alpha}}\,V(s) = s\,V(s) = \mathcal{L}\{D_x v\}\,,\end{aligned}$$

hence, claim (104) follows by the injectivity of the Laplace transform.

In general, we obtain, for $\mathrm{Re}\, s > 0$,

$$\begin{aligned}\mathcal{L}\Big\{I_{0+}^{\alpha}\big[I_{0+}^{\beta}[v]\big](x)\Big\} &= \frac{1}{\Gamma(\alpha)}\,\frac{1}{\Gamma(\beta)}\,\mathcal{L}\Big\{v * \frac{H(x)}{|x|^{1-\alpha}} * \frac{H(x)}{|x|^{1-\beta}}\Big\} \\ &= \frac{1}{s^{\alpha+\beta}}\,V(s) = \frac{1}{\Gamma(\alpha+\beta)}\mathcal{L}\Big\{v * \frac{H(x)}{|x|^{1-(\alpha+\beta)}}\Big\} = \mathcal{L}\Big\{I_{0+}^{\alpha+\beta}[v](x)\Big\},\end{aligned}$$

which proves (105). Identity (106) is achieved in the same way. The case of general $a \in \mathbb{R}$ is achieved by translation. Moreover, given $v \in L^1(a,b)$ with $-\infty < a < b \le +\infty$, by considering the trivial extension of v on $\mathbb{R}$, we obtain

$$\begin{aligned}I_{b-}^{\alpha}\big[I_{b-}^{\beta}[v]\big] &\overset{(35)}{=} \Big(I_{a+}^{\alpha}\Big[\big(I_{b-}^{\beta}[v]\big)^{\check{}}\Big]\Big)^{\check{}} = \Big(I_{a+}^{\alpha}\Big[\Big(\big(I_{a+}^{\beta}[\check{v}]\big)^{\check{}}\Big)^{\check{}}\Big]\Big)^{\check{}} = \Big(I_{a+}^{\alpha}\big[I_{a+}^{\beta}[\check{v}]\big]\Big)^{\check{}} = \\ &\overset{(105)}{=} \Big(I_{a+}^{\alpha+\beta}[\check{v}]\Big)^{\check{}} \overset{(35)}{=} \Big(\Big(I_{b-}^{\alpha+\beta}[v]\Big)^{\check{}}\Big)^{\check{}} = I_{b-}^{\alpha+\beta}[v]\,.\end{aligned}$$

hence, (107) is proved. Identity (108) is achieved in the same way. □

Proposition 2. *Assume $\alpha \in (0,1)$, $-\infty < a < b \le +\infty$, $I_{a+}^{1-\alpha}[f](a) = 0$ and f belongs to $W_{+}^{\alpha,1}(a,b) := \Big\{v \in L^1(a,b) \,|\, I_{a+}^{1-\alpha}[v] \in W_{G}^{1,1}(a,b)\Big\}$.*

*Then the **Abel integral equation***

$$I_{a+}^{\alpha}[u](x) = f(x) \qquad \text{for a.e. } x \text{ in the interval } (a,b) \tag{109}$$

admits the solution u given by

$$u(x) = D_{a+}^{\alpha}[f](x) \quad \text{for a.e. } x \text{ in the interval } (a,b). \tag{110}$$

which is unique among Laplace-transformable functions evaluated with translated variable $x - a$.

Proof. Whenever necessary, we consider the trivial extension (namely, 0 valued) on $(-\infty, a)$ of every function and if necessary on $(b, +\infty)$, without relabeling the function name. Thus, every related fractional integral set as a function defined over (a,b) has a trivial extension, which coincides on (a,b) with the same fractional integral of the trivial extension, namely, it has support contained on $[a, +\infty)$.

First, we assume $a = 0$. In such a case, f is a Laplace-transformable function: we denote by $F(s) = \mathcal{L}\{f\}(s)$ and $\mathrm{U}(s) = \mathcal{L}\{\mathrm{u}\}(s)$ their Laplace transform evaluated at the variable s. If a Laplace transformable solution u exists, then its Laplace transform U must fulfill the transformed equation. We have

$$I^{\alpha}_{0+}[u](x) = \frac{1}{\Gamma(\alpha)} \int_0^x \frac{u(t)\,dt}{(x-t)^{1-\alpha}} = \frac{1}{\Gamma(\alpha)} \left(H(x)\,x^{\alpha-1}\right) * u$$

$$I^{\alpha}_{0+}[u](x) = f(x)$$

$$\frac{1}{\Gamma(\alpha)} \frac{\Gamma(\alpha)}{s^{\alpha}} \mathrm{U(s)} = \mathrm{F(s)}$$

$$\mathrm{U}(s) = s^{\alpha} F(s) = s\,\frac{F(s)}{s^{1-\alpha}}$$

We evaluate $u = \mathcal{L}^{-1}\mathrm{U}$: reminding that $\mathcal{L}\{D_x w(x)\} = s\,\mathcal{L}\{w\}$, where w is any $\mathcal{L}$-transformable distribution, and here, D_x and d/dx denote respectively the distributional derivative on the open set $\mathbb{R}$ and on $(a,+\infty)$. By taking into account that $I^{1-\alpha}_{+}[f]$ belongs to $W^{1,1}_G(0,b) \subset L^{\infty}(0,b) \cap C^0[0,b]$, we know that $I^{1-\alpha}_{+}[f](0_+)$ is a well-defined real value. Thus, by formula $s\,G(s) = \mathcal{L}\{D_x g\} = \mathcal{L}\{(d/dx)g\} + g(0_+)$ applied to $g = I^{1-\alpha}_{a+}[f]$ under the assumption $I^{1-\alpha}_{a+}[f](0) = 0$, we obtain

$$\begin{aligned} u(x) &= \mathcal{L}^{-1}\{U(s)\} = \mathcal{L}^{-1}\left\{s\,\frac{F(s)}{s^{1-\alpha}}\right\} \\ &= D_x\left(f * \frac{1}{\Gamma(1-\alpha)\,x^{\alpha}}\right) = D_x\left(I^{1-\alpha}_{+}[f](x)\right) \\ &= \frac{d}{dx}\left(I^{1-\alpha}_{+}[f](x)\right) = \frac{1}{\Gamma(1-\alpha)}\frac{d}{dx}\int_0^x \frac{f(t)\,dt}{(x-t)^{\alpha}} = D^{\alpha}_{+}[f](x), \end{aligned} \tag{111}$$

where the four last equalities are understood in the sense a.e. on $(-\infty,b)$, coherently with the fact that $u \in L^1(-\infty,b)$ because it coincides with the derivative of the function $I^{1-\alpha}_{+}[f] \in W^{1,1}_G(0,b)$ and vanishes on $(-\infty,0)$. Moreover u is unique due to the injectivity of the Laplace transform. Then (110) is proved when $a=0$.

If $a \neq 0$, we can exploit the solution Formula (111) proved in case $a=0$: assume $I^{\alpha}_{a+}[u] = f$ on (a,b) and set $u(t) = v(t-a)$; then $t \mapsto v(t) = u(t+a)$ and $t \mapsto f(t+a)$ have support on $[0,+\infty)$ and, hence, are Laplace transformable functions.

$$\begin{aligned} I^{\alpha}_{a+}[u](x) &= \frac{1}{\Gamma(\alpha)}\int_a^x \frac{u(t)dt}{(x-t)^{1-\alpha}} = \frac{1}{\Gamma(\alpha)}\int_a^x \frac{v(t-a)\,dt}{\left((x-a)-(a+t)\right)^{1-\alpha}} \\ &= \frac{1}{\Gamma(\alpha)}\int_0^{x-a} \frac{v(\tau)\,d\tau}{\left((x-a)-\tau\right)^{1-\alpha}} = I^{\alpha}_{0+}[v](x-a) = I^{\alpha}_{0+}[u(t+a)](x-a) \end{aligned}$$

Thus we have the Abel equation $I^{\alpha}_{0+}[u(t+a)](x-a) = f(x)$, that is

$$I^{\alpha}_{0+}[u(t+a)](x) = f(x+a).$$

By (111), we get $u(x+a) = v(x) = D^{\alpha}_{0+}[f(x+a)](x+a)$, that is

$$u(x) = D^{\alpha}_{a+}[f](x). \tag{112}$$

□

Remark 14. *At a first glance, both technical assumptions in Proposition 2, namely $f \in W^{\alpha,1}_{+}(a,b)$ and $I^{1-\alpha}_{a+}[f](a) = 0$, may look strange or unnatural.*

However, they cannot be circumvented: actually, they are both necessary conditions for the existence of a solution $u \in L^1(a,b)$ of Equation (109).

Let us check this claim: if such a solution as $u \in L^1(a,b)$ exists, then $f = I^{\alpha}_{a+}[u]$ belongs to $L^1(a,b)$; moreover, due to the semigroup property of fractional integrals (see Proposition 1),

$$I^{1-\alpha}_{a+}[f](x) = I^{1-\alpha}_{a+}[I^{\alpha}_{+}[u]](x) = I^{1}_{a+}[[u]](x) = \int_a^x u(t)\,dt\,, \tag{113}$$

hence, $I^{1-\alpha}_{a+}[f]$ is the primitive of an $L^1(a,b)$ function; thus $I^{1-\alpha}_{a+}[f]$ belongs to $W^{1,1}_G(a,b)$ and $I^{1-\alpha}_{a+}[f](a) = 0$.

Remark 15. *Condition $I^{1-\alpha}_{+}[f](a) = 0$ may be not easy to check. However, it can be replaced by stronger conditions, which are much easier to check. Indeed, if either there exists a finite value $f(a_+) := \lim_{x \to a_+} f(x)$ or f is bounded in a neighborhood of 0, then $I^{1-\alpha}_{+}[f](a) = 0$.*

Remark 16. *For the unnormalized Abel equation , $\Gamma(\alpha)\, I^{\alpha}_{a+}[u] = f$, namely*

$$\int_a^x \frac{u(t)}{(x-t)^{1-\alpha}}\,dt \;=\; f(x) \qquad \text{for } x \text{ in the interval } (a,b), \tag{114}$$

as a straightforward consequence of Proposition 2 and Euler reflection formula, *$\Gamma(z)\Gamma(1-z) = \pi/\sin(\pi z)\ \forall z \in \mathbb{C}\backslash\mathbb{Z}$, under the assumption $f \in W^{s,1}(a,b)$, we recover the next formula for the unique solution u in $L^1(a,b)$:*

$$u(x) = \frac{1}{\Gamma(\alpha)} D^{\alpha}_{a+}[f](x) = \frac{1}{\Gamma(\alpha)\Gamma(1-\alpha)}\frac{d}{dx}\int_a^x \frac{f(t)\,dt}{(x-t)^{1-\alpha}} = \frac{\sin(\alpha\pi)}{\pi}\frac{d}{dx}\int_a^x \frac{f(t)\,dt}{(x-t)^{1-\alpha}} \tag{115}$$

still under the requirement that necessary conditions $f \in W^{\alpha,1}_{+}(a,b)$ and $I^{1-\alpha}_{+}[f](a) = 0$ hold true.

Now, we remove the assumption $I^{1-\alpha}_{a+}[v](a) = 0$ and look for solutions in $\mathcal{D}'(\mathbb{R})$.

Proposition 3. *Assume that $\alpha \in (0,1)$, $-\infty < a < b \le +\infty$ and f belongs to the space $BV^{\alpha}_{+}(a,b) := \left\{ v \in L^1(a,b) \,|\, I^{1-\alpha}_{a+}[v] \in BV(a,b) \right\}$.*

Then, the ***Abel integral equation in the distributional framework***

$$I^{\alpha}_{a+}[u] \;=\; f \qquad \text{in } \mathcal{D}'(\mathbb{R}), \tag{116}$$

admits a unique solution u among Laplace transformable distributions evaluated at $x - a$ (variable translation), which is the bounded measure on $\mathbb{R}$ with support contained in $[a, +\infty)$ given by

$$u(x) \;=\; D^{\alpha}_{a+}[f](x) \;+\; I^{1-\alpha}_{a+}[f](a_+)\,\delta(x-a) \quad \text{in } \mathcal{D}'(\mathbb{R}). \tag{117}$$

In (116), actually u denotes the trivial extension outside (a,b), and

$$I^{\alpha}_{a+}[u] \;=\; u * \frac{1}{\Gamma(\alpha)}\frac{H(x)}{|x|^{1-\alpha}}$$

represents the distributional convolution whose evaluation, namely f, is identically 0 on $(-\infty, a)$ and possibly non-zero on $[b, +\infty)$.

Proof. Same proof of Proposition 2. Only the step in (111) with $a = 0$ has to be slightly modified: denoting by $\widetilde{\widetilde{D}}_x$ and $\widetilde{D}_x$ the distributional derivative respectively in $\mathcal{D}'(\mathbb{R}\backslash\{0\})$ and $\mathcal{D}'(0,+\infty)$, setting $F(s) = \mathcal{L}\{f\}$, $g(x) = I^{1-\alpha}_{a+}[f](x)$, $G(s) = \mathcal{L}\{g\} = F(s)/s^{1-\alpha}$, $\mathcal{L}\{D_x g\} = sG(s)$ and

$$D_x\, g = \widetilde{\widetilde{D}}_x\, g + g(0_+)\,\delta(x) = \widetilde{D}_x\, g + g(0_+)\,\delta(x)\,,$$

we exploit the fact that $I_+^{1-\alpha}[f](0_+)$ is a finite well-defined value (since $f \in BV_+^\alpha(0,b)$ entails $I_+^{1-\alpha}[f] \in BV(0,b)$), and we replace (111) by

$$\begin{aligned} u(x) &= \mathcal{L}^{-1}\{U(s)\} = \mathcal{L}^{-1}\left\{ s\,\frac{F(s)}{s^{1-\alpha}} \right\} = \mathcal{L}^{-1}\{\, s\, G(s)\,\} \\ &= D_x\, g = \frac{d}{dx}\, g + g(0_+)\,\delta(x) \\ &= \frac{d}{dx}\left(f * \frac{1}{\Gamma(1-\alpha)\, x^\alpha} \right) + I_{a+}^{1-\alpha}[f](0_+)\,\delta(x) \\ &= \frac{d}{dx}\left(I_+^{1-\alpha}[f](x) \right) + I_{a+}^{1-\alpha}[f](0_+)\,\delta(x) \\ &= D_+^\alpha[f](x) + I_{a+}^{1-\alpha}[f](0_+)\,\delta(x)\,. \end{aligned}$$

□

Corollary 1. *Assume $\alpha \in (0,1)$, $-\infty \le a < b < +\infty$, the value $f(b_-) := \lim_{x\to b_-} f(x)$ exists and is finite (possibly substituted by weaker condition $I_{b-}^{1-\alpha}[f](b) = 0$) and f belongs to $W_-^{\alpha,1}(a,b) := \left\{ v \in L^1(a,b) |\; I_{b-}^{1-\alpha}[v] \in W_G^{1,1}(a,b) \right\}$.*

Then, the ***backward Abel integral equation***

$$I_{b-}^\alpha[u](x) = f(x) \qquad \text{for a.e. } x \text{ in the interval } (a,b) \tag{118}$$

admits a solution u, unique among Laplace transformable functions evaluated at $b-x$ (sign change and translation), which is given by

$$u(x) = D_{b-}^\alpha[f](x) \qquad \text{for a.e. } x \text{ in the interval } (a,b). \tag{119}$$

Proof. Taking into account that $b \in \mathbb{R}$, set $v(t) = \check{u}(t) := u(-t)$, and hence, $u : (a,b) \to \mathbb{R}$, $v : (-b,-a) \to \mathbb{R}$, and choose $x \in (a,b)$. Then

$$\begin{aligned} f(x) &= I_{b-}^\alpha[u](x) = \frac{1}{\Gamma(\alpha)} \int_x^b \frac{u(\tau)\, d\tau}{(\tau - x)^{1-\alpha}} \\ &= \frac{1}{\Gamma(\alpha)} \int_{x-b}^0 \frac{u(t+b)\, dt}{\big((b+t)-x\big)^{1-\alpha}} = \frac{1}{\Gamma(\alpha)} \int_{x-b}^0 \frac{v(-(b+t))\, dt}{\big(-x+(b+t)\big)^{1-\alpha}} \\ &= \frac{-1}{\Gamma(\alpha)} \int_0^{x-b} \frac{v(-(b+t))\, dt}{\big(-x+(b+t)\big)^{1-\alpha}} = \frac{1}{\Gamma(\alpha)} \int_{-b}^{-x} \frac{v(y)\, dy}{\big(-x-y)\big)^{1-\alpha}} = I_{(-b)+}^\alpha[v](-x) \end{aligned}$$

Therefore, we can apply Proposition 2 to an Abel equation on $(-b,-a)$:

$$I_{(-b)+}^\alpha[v](x) = f(-x)$$

$$I_{(-b)+}^\alpha[\check{u}](x) = \check{f}(x)$$

$$\check{u}(x) = D_{(-b)+}^\alpha[\check{f}](x)$$

$$\begin{aligned} u(-x) &= \frac{1}{\Gamma(1-\alpha)} \frac{d}{dx} \int_{-b}^x \frac{f(-t)\, dt}{(x-t)^\alpha} = \frac{-1}{\Gamma(1-\alpha)} \frac{d}{dx} \int_b^{-x} \frac{f(\tau)\, d\tau}{(x+\tau)^\alpha} \\ &= \frac{1}{\Gamma(1-\alpha)} \frac{d}{dx} \int_{-x}^b \frac{f(\tau)\, d\tau}{(x+\tau)^\alpha} = \frac{d}{dx}\left(I_{b-}^{1-\alpha}[f](-x) \right) \overset{\boxed{\text{chain-rule}}}{=} \\ &= -\frac{d}{dx}\left(I_{b-}^{1-\alpha}[f] \right)(-x) = D_{b-}^\alpha[f](-x) \qquad x \in (-b,-a)\,. \end{aligned}$$

$$u(x) = D^{\alpha}_{b-}[f](x) \qquad x \in (a,b) \quad \square$$

Corollary 2. *Assume that* $\alpha \in (0,1)$, $-\infty < a < b \leq +\infty$ *and* f *belongs to the space* $BV^{\alpha}_{-}(a,b) := \left\{ v \in L^1(a,b) | \ I^{1-\alpha}_{b-}[v] \in BV(a,b) \right\}$.

Then the ***backward Abel integral equation in the distributional framework***

$$I^{\alpha}_{b-}[u](x) = f(x) \qquad in\ \mathcal{D}'(\mathbb{R}), \tag{120}$$

admits a unique solution u among Laplace transformable distributions evaluated at $b - x$ *(say with sign change and translation), which is the bounded measure with support contained in* $(-\infty, b]$ *given by*

$$u(x) = D^{\alpha}_{b-}[f](x) + I^{1-\alpha}_{b-}[f](b_-)\,\delta(x-b) \qquad in\ \mathcal{D}'(\mathbb{R}). \tag{121}$$

In (120), actually u denotes the trivial extension outside (a,b), *and*

$$I^{\alpha}_{b-}[u] = u * \frac{1}{\Gamma(\alpha)} \frac{H(-x)}{|x|^{1-\alpha}}\,.$$

represents the distributional convolution whose evaluation, namely f, *is identically 0 on* $[b,+\infty)$ *and possibly non-zero on* $(-\infty, a)$.

Proof. Same proof of Corollary 1, but exploiting Proposition 3 instead of Proposition 2. Notice that the trivial extension of a function in $L^1(a,b)$ has compact support and can be dealt with as a Laplace transformable distribution evaluated at the variable $(b-x)$. $\square$

Example 9. *We mention some basic examples of solution u for Abel integral equation* $I^{\alpha}_{0_+}[u] = f$ *on* $(0,b)$ *with* $0 < b \leq +\infty$ *and, more in general for distributional Abel integral equation* $I^{\alpha}_{a_+}[u] = f$ *and* $I^{\alpha}_{b_-}[u] = f$ *with support condition.*

1. *If* $f = x^{\alpha}$, *then* $u = D^{\alpha}_{a+}[t^{\alpha}](x) = \Gamma(\alpha+1)\,H(x)$, *for* $\alpha \in (0,1)$, *due to Proposition 2.*
2. *If* $f = H(x)$, *then* $u = D^{\alpha}_{a+}[H](x) = \dfrac{x^{-\alpha}}{\Gamma(1-\alpha)}$, *for* $\alpha \in (0,1)$, *due to Proposition 2.*
3. *If* $f = x^{\beta}$, *then* $u = \dfrac{\Gamma(\beta+1)}{\Gamma(1+\beta-\alpha)}\, x^{\beta-\alpha}$, *for* $\alpha \in (0,1)$, $\beta > \alpha - 1$, *due to Proposition 2.*

   These relationships are deduced by Proposition 2: in the first and second item, notice that $f \in L^{\infty}(a,b)$ *entails* $I^{1-\alpha}_{0+}[u](0) = 0$ *(see Remark 15), while in third item* $\beta > \alpha - 1$ *entails both* $I^{1-\alpha}_{+}[x^{\beta}](x) = \frac{\Gamma(1+\beta)}{\Gamma(2+\beta-\alpha)} x^{1+\beta-\alpha} \in W^{1,1}_{G}(0,1)$ *and* $I^{1-\alpha}_{0+}[x^{\beta}](0) = 0$. *Thus, we get the three claims above by applying the relationships*

$$D^{\alpha}_{0_+}[x^{\beta}] = \frac{\Gamma(1+\beta)}{\Gamma(1+\beta-\alpha)}\, x^{\beta-\alpha}\,, \qquad (1-\alpha) \neq \alpha \in (0,1),\ \beta > \alpha - 1\,, \tag{122}$$

$$D^{\alpha}_{0_+}[x^{\alpha-1}] = 0\,, \qquad \alpha \in (0,1)\,. \tag{123}$$

4. *If* $f(x) = (x-a)^{\alpha-1}$, $x \in (a,b)$, $\alpha \in (0,1)$, *then the solution u with support on* $[a,+\infty)$ *to distributional backward Abel equation* $I^{\alpha}_{a+}[u] = (x-a)^{\alpha-1}H(x-a)$ *is given by* $u(x) = \Gamma(\alpha)\,\delta(x-a)$.

   Indeed $I^{1-\alpha}_{a+}[(t-a)^{\alpha-1}](x) = \Gamma(\alpha)$, $D^{\alpha}_{a+}[t^{\alpha-1}](x) = D_x I^{1-\alpha}_{a_+}[t^{\alpha-1}](x) \equiv 0$ *thus, by Proposition 3,* $u = D^{\alpha}_{a+}[(t-a)^{\alpha-1}](x) + I^{1-\alpha}_{a+}[(t-a)^{1-\alpha}](a_+)\delta(x-a) = \Gamma(\alpha)\delta(x-a)$.

   Then, u solves the Abel equation since, by representation (14), we obtain

$$I^{\alpha}_{a+}[\Gamma(\alpha)\,\delta(x-a)] = \Gamma(\alpha)\,\delta(x-a) * \frac{H(x)\,(x)^{\alpha-1}}{\Gamma(\alpha)} = H(x-a)\,(x-a)^{\alpha-1} \quad in\ \mathcal{D}'(\mathbb{R}).$$

5. If $f(x) = x^2(1-x)$, $x \in (0,1)$, $\alpha \in (0,1)$, then the solution u to backward Abel equation $I^{\alpha}_{1-}[u] = f(x)$ is given by $u(x) = D^{\alpha}_{1-}f(x) = \frac{(1-x)^{1-\alpha}(6x^2-4\alpha x+(\alpha-1)\alpha)}{\Gamma(4-\alpha)}$, due to Corollary 1 since $I^{1-\alpha}_{b-}[f](b_-) = 0$.
6. If $f(x) = (b-x)^{\alpha-1}$, $x \in (a,b)$, $\alpha \in (0,1)$, then the solution u with support on $(-\infty, b]$ to backward distributional Abel equation $I^{\alpha}_{b-}[u] = (b-x)^{\alpha-1}H(b-x)$ is given by $u(x) = \Gamma(\alpha)\,\delta(x-b)$. Indeed

$$\begin{aligned} I^{1-\alpha}_{b-}[(b-t)^{\alpha-1}](x) &= \frac{1}{\Gamma(1-\alpha)}\int_x^b \frac{1}{(b-t)^{1-\alpha}(t-x)^{\alpha}}\,dt \overset{\boxed{y=(b-t)x/(b-x)}}{=} \\ &= \frac{1}{\Gamma(1-\alpha)}\int_0^x \frac{1}{y^{1-\alpha}(1-y)^{\alpha}}\,dy = \frac{B(\alpha,1-\alpha)}{\Gamma(1-\alpha)} = \Gamma(\alpha), \end{aligned}$$

so $I^{1-\alpha}_{b-}[(b-t)^{\alpha-1}](x) = \Gamma(\alpha)$, $D^{\alpha}_{b-}[t^{\alpha-1}](x) = -D_x I^{1-\alpha}_{b-}[t^{\alpha-1}](x) \equiv 0$ so, by Corollary 2, $u = D^{\alpha}_{b-}[(b-t)^{\alpha-1}](x) + I^{1-\alpha}_{b-}[(b-t)^{\alpha-1}](b_-)\,\delta(b-x) = \Gamma(\alpha)\,\delta(b-x)$.
Then, such u solves the Abel equation since, by representation (14),

$$I^{\alpha}_{-}[\Gamma(\alpha)\,\delta(b-x)] = \Gamma(\alpha)\,\delta(b-x) * \frac{H(-x)\,(x)^{\alpha-1}}{\Gamma(\alpha)} = H(b-x)\,(b-x)^{\alpha-1} \quad \text{in } \mathcal{D}'(\mathbb{R}).$$

Lemma 8. *Fix a value $\alpha \in (0,1)$.*

If a Laplace transformable function u fulfils $D^{\alpha}_{0+}[u] \equiv 0$ on the half-line $(0,+\infty)$, then $u(x) = C\,x^{\alpha-1}$, for a suitable constant C.

If a function $u \in L^1(a,b)$, with $-\infty < a < b < +\infty$, fulfils $D^{\alpha}_{a+}[u] \equiv 0$ on (a,b), then $u(x) = C\,(x-a)^{\alpha-1}$, for a suitable constant K.

If a function $u \in L^1(a,b)$, with $-\infty < a < b < +\infty$, fulfils $D^{\alpha}_{b-}[u] \equiv 0$ on (a,b), then $u(x) = C\,(b-x)^{\alpha-1}$, for a suitable constant C.

Proof. The property

$$D_x\, I^{1-\alpha}_{a+}[u] = D^{\alpha}_{a+}[u] \equiv 0$$

entails $I^{1-\alpha}_{a+}[u]$ is constant. Thus, for a suitable constant function K, we have that u fulfills the Abel integral equation: $I^{1-\alpha}_{a+}[u] = K$, moreover $I^{\alpha}_{a+}[K](a) = 0$ since $K \in L^{\infty}(a,b)$, and due to (96) and the boundedness of (a,b)

$$I^{\alpha}_{a+}[K] = K\frac{(x-a)^{\alpha}}{\alpha\,\Gamma(\alpha)} \in W^{1,1}_G(a,b) \qquad \forall\,\alpha \in (0,1).$$

Then, by Proposition 2, the solution u of the Abel equation $I^{1-\alpha}_{a+}[u] = K$ is

$$u(x) = D^{1-\alpha}_{+}[K] = D_x\, I^{\alpha}_{a+}[K] = \frac{K}{\Gamma(\alpha)}\frac{d}{dx}\frac{(x-a)^{\alpha}}{\alpha} = \frac{K}{\Gamma(\alpha)}(x-a)^{\alpha-1}.$$

This proves the first and second claim, since an $\mathcal{L}$-transformable function is an L^1 function on every bounded interval. The third one follows in the same way, by applying Corollary 1 to the backward Abel equation $I^{1-\alpha}_{b-}[u] = K$. □

Lemma 8 provides the inverse of (100). Hence, summarizing

$$u \in L^1(a,b) \text{ fulfils } D^{\alpha}_{a+}[u] \equiv 0 \text{ on } (a,b) \qquad \text{iff} \qquad u(x) = C(x-a)^{\alpha-1}. \tag{124}$$

Lemma 9. *Assume that the interval (a,b) is bounded, $0 < \alpha < 1$, $v \in L^1(a,b)$, $I^{1-\alpha}_{a+}[v]$ belongs to $W^{1,1}_G(a,b)$ and $I^{1-\alpha}_{a+}[v](a_+) = 0$.*

Then

$$\exists \text{ unique } \mathcal{V} \in L^1(a,b): \quad v = I^{\alpha}_{a+}[\mathcal{V}]; \text{ and } \mathcal{V}(x) = D^{\alpha}_{a+}[v] = D_x I^{1-\alpha}_{a+}[v], \tag{125}$$

and $v \in L^q(a,b)$ *for every* $q \in [0, 1/(1-\alpha))$*; moreover, there is* $C = C(q)$ *such that*

$$\|v\|_{L^q(a,b)} \leq C(q)\left(\|v\|_{L^1(a,b)} + \|I^{1-\alpha}_{a+}[v]\|_{W^{1,1,}_G(a,b)}\right). \tag{126}$$

The same claims hold true when $I^\alpha_{a+}[v]$*,* $I^{1-\alpha}_{a+}[v]$ *and* $D^\alpha_{a+}[v]$ *are replaced, respectively, by* $I^\alpha_{b-}[v]$*,* $I^{1-\alpha}_{b-}[v]$ *and* $D^\alpha_{b-}[v]$ *in the assumptions and the claims.*

Proof. By considering $\mathcal{V}$ as the unknown in the Abel integral equation

$$I^\alpha_{a+}[\mathcal{V}] \; = \; v \qquad \text{on } (a,b) \tag{127}$$

we know by Proposition 2 that there is a solution $\mathcal{V} \in L^1(a,b)$ fulfilling the integral equation: such $\mathcal{V}$ is the unique solution in $L^1(a,b)$ and fulfills

$$\mathcal{V}(x) = D^\alpha_{a+}[v] = D_x\left(I^{1-\alpha}_{a+}[v]\right). \tag{128}$$

Thus $\mathcal{V}(x) \in L^1(a,b)$. Moreover, by (127), $I^{1-\alpha}_{a+}[v](a) = 0$ and the semigroup property of $s \mapsto I^s_{a+}$ (Proposition 1),

$$I^{1-\alpha}_{a+}[v](x) = I^{1-\alpha}_{a+}[I^\alpha_{a+}[\mathcal{V}]] = I^1_{a+}[\mathcal{V}] = \int_a^x \mathcal{V}(t)\,dt\,.$$

Summing up $\mathcal{V} \in L^1(a,b)$, $I^{1-\alpha}_{a+}[\mathcal{V}](a_+) = 0$, $I^{1-\alpha}_{a+}[v](a_+) = 0$, $I^{1-\alpha}_{a+}[v - I^\alpha_{a+}[\mathcal{V}]](a_+) \equiv 0$; then $v = I^\alpha_{a+}[\mathcal{V}]$, $v \in I^\alpha_{a+}(L^1(a,b))$ and (125), (126) follow by standard embedding of fractional integrals. □

If we remove the assumption $I^{1-\alpha}_{a+}[v](a_+) = 0$ in Lemma 9, then we must add suitable corrections to both v and $\mathcal{V}$, as stated by the next theorem.

Theorem 12. *Assume that* (a,b) *bounded,* $0 < \alpha < 1$*,* $v \in L^1(a,b)$*,* $I^{1-\alpha}_{a+}[v]$ *belongs to* $BV(a,b)$*. Then*

$$\exists\, \mathcal{V} \in \mathcal{M}(\mathbb{R}),\ \mathrm{spt}\,\mathcal{V} \subset [a,+\infty),\ \exists\, \mathfrak{K} \in \mathbb{R}:\ v = I^\alpha_{a+}[\mathcal{V}] + \frac{\mathfrak{K}}{\Gamma(\alpha)}(x-a)^{\alpha-1};\ D^\alpha_{a+}[v] = \mathcal{V}(x), \tag{129}$$

and $v \in L^q(a,b)$ *for every* $q \in [1, 1/(1-\alpha))$*; moreover, there is* $B = B(q, \mathfrak{K}, \alpha)$ *such that*

$$\|v\|_{L^q(a,b)} \leq B(q,\mathfrak{K},\alpha)\left(\|v\|_{L^1(a,b)} + \|I^{1-\alpha}_{a+}[v]\|_{W^{1,1,}_G(a,b)}\right). \tag{130}$$

Explicitly, for every given $\alpha \in (0,1)$*, we have*

$$v = I^\alpha_{a+}[D^\alpha_{a+}[v]] + \frac{I^{1-\alpha}_{a+}[v](a_+)}{\Gamma(\alpha)}(x-a)^{\alpha-1}, \quad \text{a.e. } x \in (a,b), \forall v \in L^1(a,b): I^{1-\alpha}_{a+}[v] \in BV(a,b). \tag{131}$$

The same claims hold true when $I^\alpha_{a+}[v]$*,* $I^{1-\alpha}_{a+}[v]$ *and* $D^\alpha_{a+}[v]$ *are replaced respectively by* $I^\alpha_{b-}[v]$*,* $I^{1-\alpha}_{b-}[v]$ *and* $D^\alpha_{b-}[v]$ *in the assumptions and the claims.*

Proof. Since $I^{1-\alpha}_{a+}[v]$ belongs to $BV(a,b)$, it has a finite right value $I^{1-\alpha}_{a+}[v](a_+)$ at $x = a$, labeled by $\mathfrak{K}$, say $\mathfrak{K} := I^{1-\alpha}_{a+}[v](a_+)$. By (97), $I^{1-\alpha}_{a+}[(x-a)]^{\alpha-1} \equiv \Gamma(\alpha)$, $0 < \alpha < 1$. We set

$$w(x) = v(x) - \frac{\mathfrak{K}}{\Gamma(\alpha)}(x-a)^{\alpha-1}$$

then $w \in L^1(a,b)$, $I^{1-\alpha}_{a+}[w] \in BV(a,b)$ and $I^{1-\alpha}_{a+}[w](a_+) = 0$. We know by Proposition 3 that there is a solution $\mathcal{W} \in \mathcal{M}(\mathbb{R})$ with $\mathrm{spt}\,\mathcal{W} \subset [a,+\infty)$ fulfilling the integral equation

$$v = I^\alpha_{a+}[\mathcal{V}] \qquad \text{in } \mathcal{D}'(\mathbb{R}), \tag{132}$$

such $\mathcal{W}$ is the unique solution with support on $[a,+\infty)$ and fulfills

$$\mathcal{W}(x) \; = \; D^\alpha_{a+}[w](x) \tag{133}$$

Thus $\mathcal{W}(x) \in \mathcal{M}$. By (132), $I_{a+}^{1-\alpha}[w](a_+) = 0$ and the semigroup property of $s \mapsto I_{a+}^{s}$

$$I_{a+}^{1-\alpha}[w](x) = I_{a+}^{1-\alpha}[I_{a+}^{\alpha}[\mathcal{W}]] = I_{a+}^{1}[\mathcal{W}] = \int_a^x \mathcal{W}(t)\,dt\,.$$

Hence, by setting $\mathcal{V} = \mathcal{W} + I_{a+}^{1-\alpha}[v](a_+)\,\delta(x-a)$ and taking into account (99), we obtain

$$v(x) = w(x) + \frac{\mathfrak{K}}{\Gamma(\alpha)}(x-a)^{\alpha-1} = I_{a+}^{\alpha}[\mathcal{V}](x) + \frac{\mathfrak{K}}{\Gamma(\alpha)}(x-a)^{\alpha-1}\,. \tag{134}$$

$$D_{a+}^{\alpha}[v] = D_{a+}^{\alpha}[w] + D_{a+}^{\alpha}\left[\frac{\mathfrak{K}}{\Gamma(\alpha)}(x-a)^{\alpha-1}\right] = D_x I_{a+}^{1-\alpha}[v] = \mathcal{V}(x) = D_{a+}^{\alpha}[w]. \tag{135}$$

Thus $\mathcal{V} \in L^1(a,b)$, $I_{a+}^{1-\alpha}[\mathcal{V}](a_+) = 0$, $I_{a+}^{1-\alpha}[v](a_+) = 0$, $I_{a+}^{1-\alpha}[v - I_{a+}^{\alpha}[\mathcal{V}]](a_+) \equiv 0$; then $v = I_{a+}^{\alpha}[\mathcal{V}]$, $v \in I_{a+}^{\alpha}(L^1(a,b))$. The function $(x-a)^{\alpha-1}$ belongs to $L^q(a,b)$ for every $q \in [1, 1/(1-\alpha))$, due to the boundedness of the interval. By standard embedding of fractional integrals, the function $w = I_{a+}^{\alpha}(\mathcal{V})$ belongs to $L^q(a,b)$ for every $q \in [1, 1/(1-\alpha))$.

Summarizing, $v \in L^q(a,b)$ for every $q \in [1, 1/(1-\alpha))$ and (129) and (130) hold true. □

Remark 17. *We emphasize that in Theorem 12 the fractional integrals and derivatives $I_{a+}^{1-\alpha}$, $I_{b-}^{1-\alpha}$, D_{a+}^{α} and D_{b-}^{α} are understood in the distributional sense provided by Definitions 1 and 3. Referring to Definition 10, with (a,b) bounded, (131) reads as follows*

$$v(x) = I_{a+}^{\alpha}[D_{a+}^{\alpha}[v]](x) + \frac{I_{a+}^{1-\alpha}[v](a_+)}{\Gamma(\alpha)}(x-a)^{\alpha-1} \quad \text{a.e. } x \in (a,b), \forall v \in BV_+^s(a,b), \alpha \in (0,1). \tag{136}$$

Moreover, in a bounded interval (a,b) we have

$$v = D_{a+}^{\alpha}[I_{a+}^{\alpha}[v]] \quad \text{a.e. on } \in (a,b),\ \forall v \in L^1(a,b),\ \forall \alpha \in (0,1), \tag{137}$$

since $D_{a+}^{\alpha}[I_{a+}^{\alpha}[v]] = D_x[I_{a+}^{1-\alpha}[I_{a+}^{\alpha}[v]]] = D_x[[I_{a+}^{1}[v]] = D_x[\int_a^x v] = v$; whereas

$$v = D_{a+}^{\alpha}[I_{a+}^{\alpha}[v]] + C\,\delta(x-a) \quad \text{on } \mathcal{D}'(\mathbb{R}),\ \forall v \in \mathcal{M}(\mathbb{R}), \operatorname{spt} v \subset [a,b],\ \forall \alpha \in (0,1), \tag{138}$$

where $C = \lim_{x \to a_+} I^1[v](x)$, indeed by Lemma 8 the kernel of D_{a+}^{α} is made by functions of the kind $K(x-a)^{\alpha-1}$, which all belong to $BV_+^s(a,b)$ and fulfill on $\mathbb{R}$

$$C\,\Gamma(\alpha) H(x-a)(x-a)^{\alpha-1} = C\,\Gamma(\alpha)\,\delta(x-a) * \frac{1}{\Gamma(\alpha)}\frac{H(x)}{x^{\alpha-1}} = I_{a+}^{\alpha}[C\,\Gamma(\alpha)\,\delta(x-a)]\,.$$

6. Conclusions

We establish some properties of the bilateral Riemann–Liouville fractional derivative D^s.

We set the notation and study the associated Sobolev spaces of fractional order s, denoted by $W^{s,1}(a,b)$, and the fractional bounded variation spaces of fractional order s, denoted by $BV^s(a,b)$. The basic properties of these spaces are proved: weak compactness properties, and comparison embeddings and strict embeddings with several related spaces, namely,

$$BV(a,b) \underset{\neq}{\subset} \bigcap_{\sigma \in (0,1)} W^{\sigma,1}(a,b) \underset{\neq}{\subset} W^{s,1}(a,b) \underset{\neq}{\subset} BV_+^s(a,b) \qquad \forall s \in (0,1),$$

$$W^{s,1}(a,b) \underset{\neq}{\subset} BV_+^s(a,b)\,, \qquad W^{s,1}(a,b) \underset{\neq}{\subset} BV_-^s(a,b)\,, \qquad \forall s \in (0,1).$$

Spaces $W^{s,1}$ and BV^s are the natural setting for data of Abel integral equations in order to make them well-posed problems in the distributional framework too.

Author Contributions: Conceptualization, A.L. and F.T.; methodology, A.L. and F.T.; formal analysis, A.L. and F.T.; writing—original draft preparation, A.L. and F.T.; writing—review and editing, A.L. and F.T. All authors have read and agreed to the published version of the manuscript.

Funding: The authors are members of the Gruppo Nazionale per l'Analisi Matematica, la Probabilità e le loro Applicazioni (GNAMPA) of the Istituto Nazionale di Alta Matematica (INdAM). This research was partially funded by Italian M.U.R. PRIN: grant number 2017BTM7SN "Variational Methods for stationary and evolution problems with singularities and interfaces".

Institutional Review Board Statement: Not applicable.

Informed Consent Statement: Not applicable.

Data Availability Statement: Not applicable.

Acknowledgments: We thank Maïtine Bergounioux for many helpful discussions.

Conflicts of Interest: The authors declare no conflict of interest.

References

1. Leaci, A.; Tomarelli, F. Bilateral Riemann-Liouville Fractional Sobolev spaces. *Note Mat.* **2021**, *41*, 61–84.
2. Carbotti, A.; Comi, G.E. A note on Riemann-Liouville fractional Sobolev spaces. *Commun. Pure Appl. Anal.* **2021**, *20*, 17–54. [CrossRef]
3. Comi, G.E.; Spector, D.; Stefani, G. The fractional variation and the precise representative of $BV^{\alpha,p}$ functions. *arXiv* **2021**, arXiv:2109.15263.
4. Comi, G.E.; Stefani, G. A distributional approach to fractional Sobolev spaces and fractional variation: Existence of blow-up. *J. Funct. Anal.* **2019**, *277*, 3373–3435. [CrossRef]
5. Comi, G.E.; Stefani, G. A distributional approach to fractional Sobolev spaces and fractional variation: Asymptotics I. *arXiv* **2019**, arXiv:1910.13419.
6. Shieh, T.-T.; Spector, D. On a new class of fractional partial differential equations. *Adv. Calc. Var.* **2015**, *8*, 321–336. [CrossRef]
7. Shieh, T.-T.; Spector, D. On a new class of fractional partial differential equations II. *Adv. Calc. Var.* **2018**, *11*, 289–307. [CrossRef]
8. Spector, D. A Noninequality for the Fractional Gradient. *Port. Math.* **2019**, *76*, 153–168. [CrossRef]
9. Samko, S.G.; Kilbas, A.; Marichev, O. *Fractional Integrals and Derivatives-Theory and Applications, Gordon and Breach*; CRC Press: Boca Raton, FL, USA, 1993.
10. Aubert, G.; Kornprobst, P. Mathematical problems in image processing, Partial Differential Equations and the Calculus of Variations. In *Applied Mathematical Sciences*, 2nd ed.; Springer: New York, NY, USA, 2006.
11. Bergounioux, M.; Trélat, E. A variational method using fractional order Hilbert spaces for tomographic reconstruction of blurred and noised binary images. *J. Funct. Anal.* **2010**, *259*, 2296–2332. [CrossRef]
12. Carriero, M.; Leaci, A.; Tomarelli, F. A candidate local minimizer of Blake & Zisserman functional. *J. Math. Pures Appl.* **2011**, *96*, 58–87. [CrossRef]
13. Carriero, M.; Leaci, A.; Tomarelli, F. Image inpainting via variational approximation of a Dirichlet problem with free discontinuity. *Adv. Calc.Var.* **2014**, *7*, 267–295. [CrossRef]
14. Carriero, M.; Leaci, A.; Tomarelli, F. A Survey on the Blake–Zisserman Functional. *Milan J. Math.* **2015**, 83, 397–420. [CrossRef]
15. Carriero, M.; Leaci, A.; Tomarelli, F. Euler equations for Blake & Zisserman functional. *Calc. Var. Partial. Differ. Equations* **2008**, *32*, 81–110. [CrossRef]
16. Carriero, M.; Leaci, A.; Tomarelli, F. Segmentation and Inpainting of Color Images. *J. Convex Anal.* **2018**, *25*, 435–458.
17. Valdinoci, E. A Fractional Framework for Perimeters and Phase Transitions. *Milan J. Math.* **2013**, *81*, 1–23. [CrossRef]
18. Leaci, A.; Tomarelli, F. Symmetrized fractional total variation models for image analysis. *article in preparation*.
19. Kukushkin, M.V. On Solvability of the Sonin-Abel Equation in the Weighted Lebesgue Space. *Fractal Fract.* **2021**, *5*, 77. [CrossRef]
20. Royden, H.L. *Real Analysis*, 2nd ed.; Macmillan: New York, NY, USA, 1968.
21. Ambrosio, L.; Fusco, N.; Pallara, D. *Functions of Bounded Variation and Free Discontinuity Problems;* Oxford Mathematical Monographs; Oxford University Press: Oxford, UK, 2000.
22. Di Nezza, E.; Palatucci, G.; Valdinoci, E. Hitchhiker's guide to the fractional Sobolev spaces. *Bull. Sci. Math.* **2012**, *136*, 521–573. [CrossRef]
23. Oldham, K.; Myl, J.; Spanier, J. *An Atlas of Functions*, 2nd ed.; Springer: Berlin/Heidelberg, Germany, 2009.
24. Anastassiou, G. *Fractional Differentiation Inequalities;* Springer: New York, NY, USA, 2009.
25. Bergounioux, M.; Leaci, A.; Nardi, G.; Tomarelli, F. Fractional Sobolev spaces and functions of bounded variation of one variable. *Fract. Calc. Appl. Anal.* **2017**, *20*, 936–962. [CrossRef]
26. Herrmann, R. *Fractional Calculus: An Introduction for Physicists*, 2nd ed.; World Scientific: Singapore, 2014.
27. Malinowska, A.B.; Torres, D.F.M. *Introduction to the Fractional Calculus of Variations;* World Scientific: Singapore, 2012.
28. Kilbas, A.A.; Srivastava, H.M.; Trujillo, J.J. *Theory and Applications of Fractional Differential Equations;* Elsevier: Amsterdam, The Netherlands, 2006.

29. Povstenko, Y. *Linear Fractional Diffusion-Wave Equation for Scientists and Engineers;* Birkhäuser: New York, NY, USA, 2015.
30. Uchaikin, V.V. *Fractional Derivatives for Physicists and Engineers;* Springer: Berlin, Germany, 2013.
31. Bourdin, L.; Idczak, D. A fractional fundamental lemma and a fractional integration by parts formula- Application to critical points of Bolza functionals and to linear boundary value problems. *Adv. Diff. Eq.* **2015**, *20*, 213–232.
32. Almeida, R.; Martins, N. A Generalization of a Fractional Variational Problem with Dependence on the Boundaries and a Real Parameter. *Fractal Fract.* **2021**, *5*, 24. [CrossRef]
33. Kukushkin, M.V. Riemann-Liouville operator in weighted L_p spaces via the Jacoby series expansion. *Axioms* **2019**, *8*, 75. [CrossRef]
34. Caputo, M. Linear Models of Dissipation Whose Q is Almost Frequency Independent. *Geophys. J. Int.* **1967**, *13*, 529–539. [CrossRef]
35. De Giorgi, E.; Ambrosio, L. Un nuovo tipo di funzionale del calcolo delle variazioni. *Atti Accad. Naz. Lincei Rend. Cl. Sci. Fis. Mat. Natur.* **1988**, *82*, 199–210.
36. Ross, B.; Samko, S.G.; Love, R.E. Functions that have no first order derivative might have fractional derivatives of all orders less that one. *Real Anal. Exch.* **1994**, *2*, 140–157. [CrossRef]

 MDPI

Article

Hybrid Method for Simulation of a Fractional COVID-19 Model with Real Case Application

Anwarud Din [1], Amir Khan [2], Anwar Zeb [3], Moulay Rchid Sidi Ammi [4,*], Mouhcine Tilioua [4] and Delfim F. M. Torres [5]

1 Department of Mathematics, Sun Yat-sen University, Guangzhou 510275, China; anwarud@mail.sysu.edu.cn
2 Department of Mathematics and Statistics, University of Swat, Mingora 19130, Khyber Pakhtunkhwa, Pakistan; amirkhan@uswat.edu.pk
3 Department of Mathematics, Abbottabad Campus, COMSATS University Islamabad, Abbottabad 22060, Khyber Pakhtunkhwa, Pakistan; anwar@cuiatd.edu.pk
4 MAIS Laboratory, AMNEA Group, FST Errachidia, Moulay Ismaïl University of Meknès, P.O. Box 509 Boutalamine, Errachidia 52000, Morocco; m.tilioua@umi.ac.ma
5 Center for Research and Development in Mathematics and Applications (CIDMA), Department of Mathematics, University of Aveiro, 3810-193 Aveiro, Portugal; delfim@ua.pt
* Correspondence: sidiammi@ua.pt

Abstract: In this research, we provide a mathematical analysis for the novel coronavirus responsible for COVID-19, which continues to be a big source of threat for humanity. Our fractional-order analysis is carried out using a non-singular kernel type operator known as the Atangana-Baleanu-Caputo (ABC) derivative. We parametrize the model adopting available information of the disease from Pakistan in the period 9 April to 2 June 2020. We obtain the required solution with the help of a hybrid method, which is a combination of the decomposition method and the Laplace transform. Furthermore, a sensitivity analysis is carried out to evaluate the parameters that are more sensitive to the basic reproduction number of the model. Our results are compared with the real data of Pakistan and numerical plots are presented at various fractional orders.

Keywords: coronavirus disease 2019 (COVID-19); ABC derivative; hybrid method; existence analysis; semi-analytical solution

MSC: 34C60; 26A33; 92D30

Citation: Din, A.; Khan, A.; Zeb, A.; Sidi Ammi, M.R.; Tilioua, M.; Torres, D.F.M. Hybrid Method for Simulation of a Fractional COVID-19 Model with Real Case Application. *Axioms* **2021**, *10*, 290. https://doi.org/10.3390/axioms10040290

Academic Editors: Stevan Pilipović and Chris Goodrich

Received: 1 August 2021
Accepted: 28 October 2021
Published: 1 November 2021

1. Introduction

The novel coronavirus SARS-CoV-2, responsible for COVID-19, which is member of the family of Severe Acute Respiratory Syndrome (SARS) viruses, has been recognized as the most dangerous virus of this decade [1]. This virus has become the new novel strain of the SARS family, which was not recognized in humans before [2]. COVID-19 has not just affected humans, but also a number of animals have been infected by the virus. The SARS-CoV-2 virus has been transmitted from human to human and similarly in animals, but its origin is still a controversy [3]. Infected humans and different species of various animals are recognized as active causes of spreading of the virus [1]. In the past, some similar viruses, like the Middle East Respiratory Syndrome Coronavirus (MERS-CoV), were spread out from camels to human population, and for SARS-CoV-1 the civet was recognized as the source of spreading into humans. For COVID-19, the main source or the major reason of spreading is human-to-human interaction, where the virus transmission is easily made by an infected person to a susceptible one. Currently, thousands of research studies have been proposed and many predictions have been given on COVID-19 dynamics, see in [4–8] and references therein. Our paper is, however, different from those in the literature. In [4], special focus is given to the transmissibility of the so-called superspreaders, with numerical simulations being given for data of Galicia, Spain, and Portugal. It turns out that, for each region, the order of the Caputo derivative takes a different value, always different from one, showing the relevance of considering fractional-order models to investigate COVID-19. The work in [5] studies the COVID-19 pandemic in Portugal until the end of the three states of emergency, describing well what has happen in Portugal with respect to the evolution of active infected and hospitalized individuals. In [6,7], a non-fractional but stochastic time-delayed model for COVID-19 is given, with the aim to study the situation of Morocco. In [8], the authors provide a S-E-I-P-A-H-R-F model while here we propose a much simpler P-I-Q model (our model has only three state variables while the model in [8] is much more complex, with eight state variables). In [8], the authors use the classical operator of Caputo; differently, here we use the more recent ABC derivatives that, in contrast, use non-singular kernels that allow us to consider a much simpler model. While the main result in [8] is the proof of the global stability of the disease free equilibrium; in contrast, here we prove Ulam–Hyers stability. We also construct a practical algorithm to compute numerically the solution of the model (see Section 5), while such algorithmic approach is not addressed in [8]. Moreover, we do a sensitivity analysis to the parameters of the model. Such sensitivity analysis is also not addressed in [8]. In contrast with the work in [8], that investigates the realities of Wuhan, Spain, and Portugal, we study the case of Pakistan.

COVID-19 generally transfers by interaction with humans in close contact for a particular time period, with most common symptoms of sneezing and coughing. The virus droplets stay on the layer of matters and when they come to the contact with any susceptible human, then the virus symptoms easily transfer to the individuals. Such infected humans can pass the infection to others by touching their mouth, eyes, or nose. This virus has the strength to be alive on different surfaces, like cardboard and copper, for many hours up to some days. As the time passes, the amount of the virus symptoms decreases over a time span and might not be alive in sufficient amount for spreading the infection. It has been recorded that the symptoms appearance and COVID-19 infection initial stage lies between 1 to 14 days [1]. Several countries have prepared and implemented a COVID-19 vaccine program and are trying to protect their populations. However, to date, there is yet no treatment available. At present, the most effect way to protect ourselves from the virus remains the quarantine or isolation, effective use of mask, following the guidelines that have been passed by governments of all countries along with the World Health Organization (WHO).

Modeling of infectious diseases has a rich literature and a number of research articles have been developed, both using classical dynamical systems as well as fractional models [4,8]. Fractional-order derivatives can be useful and helpful as compared to classical derivatives, because the dynamics of real phenomena can be comprehensively understood by fractional-order derivatives due to its special properties, i.e., hereditary and memory [9–16]. For a comparison between classical (integer-order) and fractional-order models, see in [4,8]. Roughly speaking, ordinary derivatives cannot distinguish the phenomenon at two distinct closed points. To sort out this problem of ordinary derivatives, generalized derivatives have been introduced in the framework of fractional calculus [17]. The first concept of fractional-order derivative was given by Leibniz and L'Hôpital in 1695. Aiming quantitative analysis, optimization, and numerical estimations, many number of attempts have been made by employing fractional differential equations (FDEs) [1,2,18–38]. The growing interest in the modeling of complex real-world issues with the use of FDEs is due to its numerous properties that can not be found in the ordinary sense. These characteristics allow FDEs to model effectively not only non-Markovian processes but also non-Gaussian phenomena [39]. Different non-classical fractional-order derivatives and different kinds of FDEs were proposed [40–42]. Among them, one has the Atangana-Baleanu-Caputo (ABC) derivative, which is a nonlocal fractional derivative with a non-singular kernel, connected with various applications. For a discussion of the ABC and related operators see in [43,44], and for their use on contemporary modeling we refer the interested reader to the works in [45–48].

The famous method of decomposition was developed from the 1970s to the 1990s by George Adomian, to analytically handle nonlinear problems. After that, the Adomian decomposition method became a powerful tool to simulate analytical or approximate solutions for various problems of an applied nature. Many mathematical models have been studied by the applications of homotopy, Laplace Adomian Decomposition Method (LADM), and variational methods [49–51]. To the best of our knowledge, no one has studied a variable order epidemic model with ABC derivatives by the LADM. Motivated by this fact, here we study a fractional-order COVID-19 epidemic model with ABC derivatives by the Laplace Adomian decomposition algorithm. In particular, we use Banach and Krassnoselskii fixed point theorems to define some sufficient conditions to prove existence and uniqueness of solution. As stability is important for the estimated solution, we consider Ulam type stability through nonlinear functional analysis. The aforementioned stability is investigated for ordinary fractional derivatives in many research papers, see, e.g., in [52–54], but research on Ulam type stability regarding ABC derivatives is a rarity. At the end of the paper, our results are illustrated with real data based on Pakistan COVID-19 cases in March 2020.

The paper is organized as follows. Section 2 is devoted to the model formulation. Section 3 is concerned with some preliminary results on fractional differential equations. Existence and uniqueness are carried out in Section 4. Section 5 deals with the solution of the COVID-19 model using the LADM. Some plots are given in Section 6, showing the simplicity and reliability of the proposed algorithm. In Section 7, a sensitivity analysis is given to find the most sensitive parameter with respect to the basic reproduction number. We end with Section 8 of conclusions, including some possible future directions of research.

2. Model Formulation

Mathematical modeling plays a major role in investigating and thus controlling the dynamics of a disease, particularly in the vaccination privation or at the initial phases of the epidemic. Several mathematical models can be found in [12–15]. We formulated a fractional COVID-19 epidemic model, similar to other diseases [49,50], and predict its future behavior. Inspired by FDEs using the ABC derivative, we aim to simulate the COVID-19 transmission in the form of

$$\begin{cases} {}^{ABC}\mathbf{D}_t^{\theta} P(t) = \lambda - \gamma P(t)I(t) - d_0 P(t), \\ {}^{ABC}\mathbf{D}_t^{\theta} I(t) = \gamma P(t)I(t) - (d_0 + h + \eta)I(t) + \sigma Q(t), \\ {}^{ABC}\mathbf{D}_t^{\theta} Q(t) = \eta I(t) - (d_0 + \mu + \sigma)Q(t), \end{cases} \tag{1}$$

along with initial conditions

$$P(0) = P_0,\ I(0) = I_0,\ Q(0) = Q_0, \tag{2}$$

where ${}^{ABC}\mathbf{D}_t^{\theta}$ is the ABC fractional derivative of order $0 < \theta \leq 1$ (see Definition 1 in Section 3). In this model, $P(t)$ represents the amount of susceptible humans, $I(t)$ stands for the population of infected humans, and $Q(t)$ represents the population of quarantined humans at time t. The meaning of the parameters of Model (1) are given in Table 1. We take the below assumptions to the given system:

A_1. All the variables and parameters of the system are non-negative.

A_2. The susceptible people transfer to the infectious compartment with a constant susceptible inflow into population.

A_3. Originally infectious or susceptible persons transfer to the quarantined class while reported cases return to the infected class from quarantined classes.

The basic reproduction number R_0, which represents the secondary cases for the Model (1), is easily demonstrated to be given by

$$R_0 = \frac{\gamma\lambda(d_C + \mu + \sigma)}{d_0(d_0 + \mu + \sigma)(d_0 + h) + \eta(d_0 + \mu)}. \tag{3}$$

Table 1. Parameters description defined in the given Model (1).

Notation	Description
λ	Rate of recruitment
γ	Transmission rate of disease
d_0	Natural death rate
η	Transmission rate of infected to quarantine
μ	Deaths in quarantined zone
σ	Transmission flow of quarantined to become infectious
h	Rate of deaths in infected zone

In addition, $I(t) + P(t) + Q(t) = N(t)$, where N represents the total population.

3. Preliminary Results

For completeness, here we recall necessary definitions and results from the literature.

Definition 1 (See [11,48]). *If x is an absolutely continuous function and $0 < \theta \leq 1$, then the ABC derivative is given by*

$$^{\mathcal{ABC}}\mathbf{D}_t^\theta \phi(t) = \frac{\mathcal{ABC}(\theta)}{1-\theta} \int_0^t \frac{d}{dy} x(\omega) \mathcal{M}_\theta \left[\frac{-\theta}{1-\theta} (t-\omega)^\theta \right], \tag{4}$$

where $\mathcal{ABC}(\theta)$ is a normalization function such that $1 = \mathcal{ABC}(1) = \mathcal{ABC}(0)$ and $\mathcal{M}_\theta$ is a special Mittag–Leffler function.

Remark 1. *By replacing $\mathcal{M}_\theta \left[\frac{-\theta}{1-\theta} \left(t - \omega \right)^\theta \right]$ with $\mathcal{M}_1 = \exp \left[\frac{-\theta}{1-\theta} \left(t - \omega \right) \right]$ one obtains the so-called Caputo–Fabrizio derivative. Additionally, we have*

$$^{\mathcal{ABC}}\mathbf{D}_0^\theta [constant] = 0.$$

Remark 2. *Let $x(t)$ be a function having fractional ABC derivative. Then, the Laplace transform of $^{\mathcal{ABC}}\mathbf{D}_0^\theta x(t)$ is given by*

$$\mathscr{L}\left[^{\mathcal{ABC}}\mathbf{D}_0^\theta x(t) \right] = \frac{\mathcal{ABC}(\theta)}{[s^\theta (1-\theta) + \theta]} \left[s^\theta \mathscr{L}[x(t)] - s^{\theta-1} x(0) \right].$$

Lemma 1 (See [52]). *The solution to*

$$^{\mathcal{ABC}}\mathbf{D}_0^\theta x(t) = z(t), \quad t \in [0, T],$$
$$x(0) = x_0,$$

$1 > \theta > 0$, is given by

$$x(t) = x_0 + \frac{(1-\theta)}{\mathcal{ABC}(\theta)} z(t) + \frac{\theta}{\mathcal{ABC}(\theta)\Gamma(\theta)} \int_0^t (t-\omega)^{\theta-1} z(\omega) d\omega.$$

Theorem 1 (See [54]). *Let $\mathbf{X} = C[0, T]$ and consider the Banach space defined by $\mathbf{Z} = \mathbf{X} \times \mathbf{X} \times \mathbf{X}$ with the norm-function $\|A\| = \|(P, I, Q)\| = \max_{t \in [0,T]} [|P(t) + |I(t)| + |Q(t)|]$. Consider $\mathbf{B}$ to be a convex subset of $\mathbf{Z}$ and $\mathbf{F}$, and $\mathbf{G}$ be operators such that*

1. $\mathbf{F}u + \mathbf{G}u \in \mathbf{B} \; \forall \; u \in \mathbf{B}$,
2. *$\mathbf{F}$ is a contraction, and*
3. *$\mathbf{G}$ is compact and continuous.*

Then, $\mathbf{F}u + \mathbf{G}u = u$ possesses at least one solution.

4. Qualitative Analysis of the Proposed Model

Here, we rewrite the right-hand sides of (1) as

$$\begin{aligned} f_1(t,P(t),I(t),Q(t)) &= -\gamma I(t)P(t) + \lambda - d_0P(t), \\ f_2(t,P(t),I(t),Q(t)) &= \gamma I(t)P(t) - (d_0 + h + \eta)I(t) + \sigma Q(t), \\ f_3(t,P(t),I(t),Q(t)) &= \eta I(t) - (d_0 + \sigma + \mu)Q(t). \end{aligned} \tag{5}$$

By using (5), we have

$$\begin{aligned} {}^{ABC}\mathbf{D}^{\theta}_{+0}\mathcal{A}(t) &= \Phi(t,\mathcal{A}(t)), \quad t \in [0,\tau], \quad 0 < \theta \leq 1, \\ \mathcal{A}(0) &= \mathcal{A}_0. \end{aligned} \tag{6}$$

In view of Lemma 1, (6) yields

$$\begin{aligned} \mathcal{A}(t) = \mathcal{A}_0(t) + [\Phi(t,\mathcal{A}(t)) - \Phi_0(t)]\frac{(1-\theta)}{\mathcal{ABC}(\theta)} \\ + \frac{\theta}{\Gamma(\theta)\mathcal{ABC}(\theta)}\int_0^t (t-\omega)^{\theta-1}\Phi(\omega,\mathcal{A}(\omega))d\omega, \end{aligned} \tag{7}$$

where

$$\begin{aligned} \mathcal{A}(t) = \begin{cases} P(t) \\ I(t) \\ Q(t) \end{cases}, \quad \mathcal{A}_0(t) = \begin{cases} P_0 \\ I_0 \\ Q_0 \end{cases}, \\ \Phi(t,\mathcal{A}(t)) = \begin{cases} f_1(t,P,I,Q) \\ f_2(t,P,I,Q) \\ f_3(t,P,I,Q). \end{cases}, \quad \Phi_0(t) = \begin{cases} f_1(0,P_0,I_0,Q_0) \\ f_2(0,P_0,I_0,Q_0) \\ f_3(0,P_0,I_0,Q_0). \end{cases} \end{aligned} \tag{8}$$

Using (7) and (8), we define the two operators $\mathbf{F}$ and $\mathbf{G}$ as follows:

$$\begin{aligned} \mathbf{F}(\mathcal{A}) &= \mathcal{A}_0(t) + [\Phi(t,\mathcal{A}(t)) - \Phi_0(t)]\frac{(1-\theta)}{\mathcal{ABC}(\theta)}, \\ \mathbf{G}(\mathcal{A}) &= \frac{\theta}{\Gamma(\theta)\mathcal{ABC}(\theta)}\int_0^t (t-\omega)^{\theta-1}\Phi(\omega,\mathcal{A}(\omega))d\omega. \end{aligned} \tag{9}$$

For existence and uniqueness, we assume some basic axioms and a Lipschitz hypothesis:

(H1) there is C_Φ and D_Φ such that

$$|\Phi(t,\mathcal{A}(t))| \leq C_\Phi\|\mathcal{A}\| + D_\Phi;$$

(H2) there is $L_\Phi > 0$ such that $\forall\ \mathcal{A},\ \bar{\mathcal{A}} \in \mathbf{Z}$ one has

$$|\Phi(t,\mathcal{A}) - \Phi(t,\bar{\mathcal{A}})| \leq L_\Phi[\|\mathcal{A}\| - \|\bar{\mathcal{A}}\|].$$

Theorem 2. *Under hypotheses* $(H1)$ *and* $(H2)$, *Equation* (7) *possesses at least one solution, which implies that* (1) *possesses an equal number of solutions if* $\frac{(1-\theta)}{\mathcal{ABC}(\theta)}L_\Phi < 1$.

Proof. The theorem is proved in two steps, with the help of Theorem 1. (i) Consider $\bar{\mathcal{A}} \in \mathbf{B}$, where $\mathbf{B} = \{\mathcal{A} \in \mathbf{Z} : \|\mathcal{A}\| \leq \rho, \rho > 0\}$ is a closed and convex set. Then, for $\mathbf{F}$ in (9), we have

$$\begin{aligned}\|\mathbf{F}(\mathcal{A})-\mathbf{F}(\bar{\mathcal{A}})\| &= \frac{(1-\theta)}{\mathcal{ABC}(\theta)}\max_{t\in[0,\tau]}|\Phi(t,\mathcal{A}(t))-\Phi(t,\bar{\mathcal{A}}(t))| \\ &\leq \frac{(1-\theta)}{\mathcal{ABC}(\theta)}L_{\Phi}\|\mathcal{A}-\bar{\mathcal{A}}\|.\end{aligned} \tag{10}$$

Therefore, **F** is a contraction. (ii) We want **G** to be relatively compact. For that it suffices that **G** is equicontinuous and bounded. Obviously, **G** is continuous as Φ is continuous and for all $\mathcal{A}\in\mathbf{B}$ one has

$$\begin{aligned}\|\mathbf{G}(\mathcal{A})\| &= \max_{t\in[0,\tau]}\left|\frac{\theta}{\Gamma(\theta)\mathcal{ABC}(\theta)}\int_0^t(t-\omega)^{\theta-1}\Phi(\omega,\mathcal{A}(\omega))d\omega\right| \\ &\leq \frac{\theta}{\Gamma(\theta)\mathcal{ABC}(\theta)}\int_0^{\tau}(\tau-\omega)^{\theta-1}|\Phi(\omega,\mathcal{A}(\omega))|d\omega \\ &\leq \frac{\tau^{\theta}}{\mathcal{ABC}(\theta)\Gamma(\theta)}[C_{\Phi}\rho+D_{\Phi}].\end{aligned} \tag{11}$$

Thus, (11) shows the boundedness of **G**. For equi-continuity, we assume $t_1 > t_2 \in [0,\tau]$, so that

$$\begin{aligned}&|\mathbf{G}(\mathcal{A}(t_1))-\mathbf{G}(\mathcal{A}(t_2))| \\ &= \frac{\theta}{\mathcal{ABC}(\theta)\Gamma(\theta)}\left|\int_0^{t_1}(t_1-\omega)^{\theta-1}\Phi(\omega,\mathcal{A}(\omega))d\omega-\int_0^{t_2}(t_2-\omega)^{\theta-1}\Phi(\omega,\mathcal{A}(\omega))d\omega\right| \\ &\leq \frac{[C_{\Phi}\rho+D_{\Phi}]}{\mathcal{ABC}(\theta)\Gamma(\theta)}|t_1^{\theta}-t_2^{\theta}|.\end{aligned} \tag{12}$$

The right-hand side in (12) goes to zero at $t_1\to t_2$. Remembering that **G** is continuous,

$$|\mathbf{G}(\mathcal{A}(t_1))-\mathbf{G}(\mathcal{A}(t_2))|\to 0 \text{ as } t_1\to t_2.$$

Having the boundedness and continuity of **G**, we conclude that **G** is uniformly continuous and bounded. According to the theorem of Arzelá–Ascoli, **G** is relatively compact and therefore entirely continuous. It follows from Theorem 1 that the integral Equation (7) has at least one solution. □

Now, we show uniqueness.

Theorem 3. *Under hypotheses* $(H1)$ *and* $(H2)$*, Equation* (7) *possesses a unique solution and this implies that* (1) *possesses also a unique solution if* $\frac{(1-\theta)L_{\Phi}}{\mathcal{ABC}(\theta)}+\frac{\tau^{\theta}L_{\Phi}}{\mathcal{ABC}(\theta)\Gamma(\theta)}<1$.

Proof. Let the operator $\mathbf{T}:\mathbf{Z}\to\mathbf{Z}$ be defined by

$$\begin{aligned}\mathbf{T}\mathcal{A}(t) = \mathcal{A}_0(t)+\Big[\Phi(t,\mathcal{A}(t))-\Phi_0(t)\Big]\frac{(1-\theta)}{\mathcal{ABC}(\theta)} \\ +\frac{\theta}{\mathcal{ABC}(\theta)\Gamma(\theta)}\int_0^t(t-\omega)^{\theta-1}\Phi(\omega,\mathcal{A}(\omega))d\omega, \quad t\in[0,\tau].\end{aligned} \tag{13}$$

Let $\mathcal{A}, \bar{\mathcal{A}} \in \mathbf{Z}$. Then, one can take

$$\begin{aligned} \|\mathbf{T}\mathcal{A} - \mathbf{T}\bar{\mathcal{A}}\| \leq & \frac{(1-\theta)}{\mathcal{ABC}(\theta)} \max_{t\in[0,\tau]} \left|\Phi(t,\mathcal{A}(t)) - \Phi(t,\bar{\mathcal{A}}(t))\right| \\ & + \frac{\theta}{\Gamma(\theta)\mathcal{ABC}(\theta)} \max_{t\in[0,\tau]} \left| \int_0^t (t-\omega)^{\theta-1}\Phi(\omega,\mathcal{A}(\omega))d\omega \right. \\ & \left. - \int_0^t (t-\omega)^{\theta-1}\Phi(\omega,\bar{\mathcal{A}}(\omega))d\omega \right| \\ \leq & \Xi\|\mathcal{A} - \bar{\mathcal{A}}\|, \end{aligned} \tag{14}$$

where

$$\Xi = \frac{(1-\theta)L_\Phi}{\mathcal{ABC}(\theta)} + \frac{\tau^\theta L_\Phi}{\Gamma(\theta)\mathcal{ABC}(\theta)}. \tag{15}$$

Thus, $\mathbf{T}$ is a contraction from (14). Therefore, (7) possesses a unique solution. □

Next, in order to investigate the stability of our problem, we consider a small disturbance $\phi \in C[0,T]$, with $\phi(0) = 0$, that depends only on the solution.

Lemma 2. *Let $\phi \in C[0,T]$ with $\phi(0) = 0$ such that $|\phi(t)| \leq \varepsilon$ for $\varepsilon > 0$ and consider the problem*

$$\begin{aligned} {}^{\mathcal{ABC}}\mathbf{D}^{\theta}_{+0}\mathcal{A}(t) &= \Phi(t,\mathcal{A}(t)) + \phi(t), \\ \mathcal{A}(0) &= \mathcal{A}_0. \end{aligned} \tag{16}$$

The solution of (16) satisfies the following relation:

$$\begin{aligned} & \left|\mathcal{A}(t) - \left(\mathcal{A}_0(t) + \left[\Phi(t,\mathcal{A}(t)) - \Phi_0(t)\right]\frac{(1-\theta)}{\mathcal{ABC}(\theta)} \right.\right. \\ & \left.\left. + \frac{\theta}{\mathcal{ABC}(\theta)\Gamma(\theta)} \int_0^t (t-\omega)^{\theta-1}\Phi(\omega,\mathcal{A}(\omega))d\omega\right)\right| \\ & \leq \frac{\Gamma(\theta) + \tau^\theta}{\Gamma(\theta)\mathcal{ABC}(\theta)}\varepsilon = \Omega_{\tau,\theta}. \end{aligned} \tag{17}$$

Proof. The proof is standard and is omitted here. □

Theorem 4. *Consider hypotheses (H1) and (H2) along with (17) of Lemma 2. Then, the solution to Equation (7) is Ulam–Hyers stable if $\Xi < 1$, where Ξ is defined by (15).*

Proof. Assume $\mathcal{A} \in \mathbf{Z}$ and let $\bar{\mathcal{A}} \in \mathbf{Z}$ be the unique solution of (7). Then,

$$\begin{aligned} \|\mathcal{A} - \bar{\mathcal{A}}\| &= \max_{t\in[0,T]} \left|\mathcal{A}(t) - \left(\mathcal{A}_0(t) + \left[\Phi(t,\bar{\mathcal{A}}(t)) - \Phi_0(t)\right]\frac{(1-\theta)}{\mathcal{ABC}(\theta)} \right.\right. \\ & \quad \left.\left. + \frac{\theta}{\Gamma(\theta)\mathcal{ABC}(\theta)} \int_0^t (t-\omega)^{\theta-1}\Phi(\omega,\bar{\mathcal{A}}(\omega))d\omega\right)\right| \\ &\leq \max_{t\in[0,T]} \left|\mathcal{A}(t) - \left(\mathcal{A}_0(t) + \left[\Phi(t,\mathcal{A}(t)) - \Phi_0(t)\right]\frac{(1-\theta)}{\mathcal{ABC}(\theta)} \right.\right. \\ & \quad \left.\left. + \frac{\theta}{\Gamma(\theta)\mathcal{ABC}(\theta)} \int_0^t (t-\omega)^{\theta-1}\Phi(\omega,\mathcal{A}(\omega))d\omega\right)\right| \\ & \quad + \left|\left(\mathcal{A}_0(t) + \left[\Phi(t,\mathcal{A}(t)) - \Phi_0(t)\right]\frac{(1-\theta)}{\mathcal{ABC}(\theta)} \right.\right. \\ & \quad \left.\left. + \frac{\theta}{\Gamma(\theta)\mathcal{ABC}(\theta)} \int_0^t (t-\omega)^{\theta-1}\Phi(\omega,\mathcal{A}(\omega))d\omega\right)\right. \end{aligned}$$

$$-\left(\mathcal{A}_0(t)+\left[\Phi(t,\bar{\mathcal{A}}(t))-\Phi_0(t)\right]\frac{(1-\theta)}{\mathcal{ABC}(\theta)}\right.$$
$$\left.+\frac{\theta}{\Gamma(\theta)\mathcal{ABC}(\theta)}\int_0^t(t-\omega)^{\theta-1}\Phi(\omega,\bar{\mathcal{A}}(\omega))d\omega\right)\Bigg|$$
$$\leq \Omega_{\tau,\theta}+\frac{(1-\theta)L_{\Phi}}{\mathcal{ABC}(\theta)}\|\mathcal{A}-\bar{\mathcal{A}}\|+\frac{\tau^{\theta}L_{\Phi}}{\Gamma(\theta)\mathcal{ABC}(\theta)}\|\mathcal{A}-\bar{\mathcal{A}}\|$$
$$\leq \Omega_{\tau,\theta}+\Xi\|\mathcal{A}-\bar{\mathcal{A}}\|. \tag{18}$$

From (18), we can write that

$$\|\mathcal{A}-\bar{\mathcal{A}}\|\leq\frac{\Omega_{\tau,\theta}}{1-\Xi}\|\mathcal{A}-\bar{\mathcal{A}}\|. \tag{19}$$

The proof is complete. □

5. Construction of an Algorithm for Deriving the Solution of the Model

Herein, we derive a general series-type solution for the proposed system with ABC derivatives. Taking the Laplace transform in Model (1), we transform both sides of each equation and we use the initial conditions to obtain that

$$\begin{cases}\mathscr{L}[P(t)]=\dfrac{P_0}{s}+\dfrac{[s^{\theta}(1-\theta)+\theta]}{s^{\theta}\mathcal{ABC}(\theta)}\mathscr{L}[\lambda-\gamma P(t)I(t)-d_0P(t)],\\ \mathscr{L}[I(t)]=\dfrac{I_0}{s}+\dfrac{[s^{\theta}(1-\theta)+\theta]}{s^{\theta}\mathcal{ABC}(\theta)}\mathscr{L}[\gamma I(t)P(t)-I(t)(d_0+h+\eta)+\sigma Q(t)],\\ \mathscr{L}[Q(t)]=\dfrac{Q_0}{s}+\dfrac{[s^{\theta}(1-\theta)+\theta]}{s^{\theta}\mathcal{ABC}(\theta)}\mathscr{L}[\eta I(t)-(d_0+\mu+\sigma)Q(t)].\end{cases} \tag{20}$$

Now, considering each solution in the form of series,

$$P(t)=\sum_{n=0}^{\infty}P_n(t),\quad I(t)=\sum_{n=0}^{\infty}I_n(t),\quad Q(t)=\sum_{n=0}^{\infty}Q_n(t), \tag{21}$$

we separate the nonlinear term $P(t)I(t)$ in terms of Adomian polynomials as

$$P(t)I(t)=\sum_{n=0}^{\infty}H_n(t),\ \text{where}\ H_n(t)=\frac{1}{n!}\frac{d^n}{d\lambda^n}\left[\sum_{k=0}^{n}\lambda^kP_k(t)\sum_{k=0}^{n}\lambda^kI_k(t)\right]\Bigg|_{\lambda=0}. \tag{22}$$

Therefore, from (21) and (22), we obtain from (20) that

$$\begin{cases}\mathscr{L}\left[\sum\limits_{n=0}^{\infty}P_n(t)\right]=\dfrac{P_0}{s}+\dfrac{[s^{\theta}(1-\theta)+\theta]}{s^{\theta}\mathcal{ABC}(\theta)}\mathscr{L}\left[\lambda-\gamma\sum\limits_{n=0}^{\infty}H_n(t)-d_0\sum\limits_{n=0}^{\infty}P_n(t)\right],\\ \mathscr{L}\left[\sum\limits_{n=0}^{\infty}I_n(t)\right]=\dfrac{I_0}{s}\\ \quad+\dfrac{[s^{\theta}(1-\theta)+\theta]}{s^{\theta}\mathcal{ABC}(\theta)}\mathscr{L}\left[\gamma\sum\limits_{n=0}^{\infty}H_n(t)-(d_0+h+\eta)\sum\limits_{n=0}^{\infty}I_n(t)+\sigma\sum\limits_{n=0}^{\infty}Q_n(t)\right],\\ \mathscr{L}\left[\sum\limits_{n=0}^{\infty}Q_n(t)\right]=\dfrac{Q_0}{s}+\dfrac{[s^{\theta}(1-\theta)+\theta]}{s^{\theta}\mathcal{ABC}(\theta)}\mathscr{L}\left[\eta\sum\limits_{n=0}^{\infty}I_n(t)-(d_0+\mu+\sigma)\sum\limits_{n=0}^{\infty}Q_n(t)\right].\end{cases} \tag{23}$$

Now, comparing the terms on both sides of (23), one has

$$\begin{cases} \mathscr{L}[P_0(t)] = \frac{P_0}{s}, \quad \mathscr{L}[I_0(t)] = \frac{I_0}{s}, \quad \mathscr{L}[Q_0(t)] = \frac{Q_0}{s}, \\ \mathscr{L}[P_1(t)] = \frac{[s^\theta(1-\theta)+\theta]}{s^\theta ABC(\theta)}\mathscr{L}[\lambda - \gamma H_0(t) - d_0 P_0(t)], \\ \mathscr{L}[I_1(t)] = \frac{[s^\theta(1-\theta)+\theta]}{s^\theta ABC(\theta)}\mathscr{L}[\gamma H_0(t) - (d_0+h+\eta)I_0(t) + \sigma Q_0(t)], \\ \mathscr{L}[Q_1(t)] = \frac{[s^\theta(1-\theta)+\theta]}{s^\theta ABC(\theta)}\mathscr{L}[\eta I_0(t) - (d_0+\mu+\sigma)Q_0(t)], \\ \mathscr{L}[P_2(t)] = \frac{[s^\theta(1-\theta)+\theta]}{s^\theta ABC(\theta)}\mathscr{L}\Big[\lambda - \gamma H_1(t) - d_0 P_1(t)\Big], \\ \mathscr{L}[I_2(t)] = \frac{[s^\theta(1-\theta)+\theta]}{s^\theta ABC(\theta)}\mathscr{L}[\gamma H_1(t) - (d_0+h+\eta)I_1(t) + \sigma Q_1(t)], \\ \mathscr{L}[Q_2(t)] = \frac{[s^\theta(1-\theta)+\theta]}{s^\theta ABC(\theta)}\mathscr{L}[\eta I_1(t) - (d_0+\mu+\sigma)Q_1(t)], \\ \qquad \vdots \\ \mathscr{L}[P_{n+1}(t)] = \frac{[s^\theta(1-\theta)+\theta]}{s^\theta ABC(\theta)}\mathscr{L}[\lambda - \gamma H_n(t) - d_0 P_n(t)], \ n \geq 0, \\ \mathscr{L}[I_{n+1}(t)] = \frac{[s^\theta(1-\theta)+\theta]}{s^\theta ABC(\theta)}\mathscr{L}[\gamma H_n(t) - (d_0+h+\eta)I_n(t) + \sigma Q_n(t)], \ n \geq 0, \\ \mathscr{L}[Q_{n+1}(t)] = \frac{[s^\theta(1-\theta)+\theta]}{s^\theta ABC(\theta)}\mathscr{L}[\eta I_n(t) - (d_0+\mu+\sigma)Q_n(t)], \quad n \geq 0. \end{cases} \tag{24}$$

Applying the inverse Laplace transform to (24), we obtain that

$$\begin{cases} P_0(t) = P_0, \quad I_0(t) = I_0, \quad Q_0(t) = Q_0, \\ P_1(t) = \mathscr{L}^{-1}\left[\frac{[s^\theta(1-\theta)+\theta]}{s^\theta ABC(\theta)}\mathscr{L}[\lambda - \gamma H_0(t) - d_0 P_0(t)]\right], \\ I_1(t) = \mathscr{L}^{-1}\left[\frac{[s^\theta(1-\theta)+\theta]}{s^\theta ABC(\theta)}\mathscr{L}[\gamma H_0(t) - (d_0+h+\eta)I_0(t) + \sigma Q_0(t)]\right], \\ Q_1(t) = \mathscr{L}^{-1}\left[\frac{[s^\theta(1-\theta)+\theta]}{s^\theta ABC(\theta)}\mathscr{L}[\eta I_0(t) - (d_0+\mu+\sigma)Q_0(t)]\right], \\ P_2(t) = \mathscr{L}^{-1}\left[\frac{[s^\theta(1-\theta)+\theta]}{s^\theta ABC(\theta)}\mathscr{L}[\lambda - \gamma H_1(t) - d_0 P_1(t)]\right], \\ I_2(t) = \mathscr{L}^{-1}\left[\frac{[s^\theta(1-\theta)+\theta]}{s^\theta ABC(\theta)}\mathscr{L}[\gamma H_1(t) - (d_0+h+\eta)I_1(t) + \sigma Q_1(t)]\right], \\ Q_2(t) = \mathscr{L}^{-1}\left[\frac{[s^\theta(1-\theta)+\theta]}{s^r ABC(\theta)}\mathscr{L}[\eta I_1(t) - (d_0+\mu+\sigma)Q_1(t)]\right], \\ \qquad \vdots \\ P_{n+1}(t) = \mathscr{L}^{-1}\left[\frac{[s^\theta(1-\theta)+\theta]}{s^\theta ABC(\theta)}\mathscr{L}[\lambda - \gamma H_n(t) - d_0 P_n(t)]\right], \quad n \geq 0, \\ I_{n+1}(t) = \mathscr{L}^{-1}\left[\frac{[s^\theta(1-\theta)+\theta]}{s^\theta ABC(\theta)}\mathscr{L}[\gamma H_n(t) - (d_0+h+\eta)I_n(t) + \sigma Q_n(t)]\right], \quad n \geq 0, \\ Q_{n+1}(t) = \mathscr{L}^{-1}\left[\frac{[s^\theta(1-\theta)+\theta]}{s^\theta ABC(\theta)}\mathscr{L}[\eta I_n(t) - (d_0+\mu+\sigma)Q_n(t)]\right], \quad n \geq 0. \end{cases} \tag{25}$$

6. Numerical Interpretation and Discussion

To illustrate the dynamical structure of our infectious disease model, we now consider a practical case study under various numerical observations and given parameter values. The concrete parameter values we have used are shown in Table 2.

Table 2. Numerical values for the parameters of Model (1).

Notation	Parameters Description	Numerical Value
λ	Rate of recruitment	0.003
γ	Transmission rate of disease	0.009
d_0	Natural death rate	0.009
η	Transmission rate of infected to quarantine	0.004
μ	Death rate in quarantine	0.004
σ	Transmission flow of quarantined to infectious	0.003
h	Rate of death for infected	0.007
P_0	Initial population of susceptible	10 millions
I_0	Initially infected population	0.01 millions
Q_0	Quarantined population at $t = 0$	0.0011 millions

We assume that the initial susceptible, infected, and isolated populations are 10, 0.01, and 0.0011 million, respectively. Among the 21,000 selected population, the density of susceptible population is about 0.6 percent, the infected population is 0.2 percent, and the isolated population is 0.2 percent.

By using the parameter values in Table 2, we computed the first three terms of the general series solution (25) with $\mathcal{ABC}(\theta) = 1$ as

$$\begin{aligned} P(t) &= 0.6 + 1.99712\left[1 - \theta + \frac{\theta t^{\theta}}{\Gamma(\theta)}\right] \\ &\quad - 0.00681\left[(1-\theta)^2 t + \frac{\theta^2 t^{2\theta}}{\Gamma(2\theta+1)} + \frac{2\theta(1-\theta)t^{\theta+1}}{\Gamma(\theta+2)}\right] + \cdots, \\ I(t) &= 0.2 - 0.5976\left[1 - \theta + \frac{\theta t^{\theta}}{\Gamma(\theta)}\right] \\ &\quad - 0.003096\left[(1-\theta)^2 t + \frac{2\theta(1-\theta)t^{\theta}}{\Gamma(\theta)} + \frac{\theta^2 t^{2\theta}}{\Gamma(2\theta+1)}\right] + \cdots, \\ Q(t) &= 0.2 + 0.14\left[1 - \theta + \frac{\theta t^{\theta}}{\Gamma(\theta)}\right] \\ &\quad - 0.004058\left[(1-\theta)^2 t + \frac{2\theta(1-\theta)t^{\theta}}{\Gamma(\theta)} + \frac{\theta^2 t^{2\theta}}{\Gamma(2\theta+1)}\right] + \cdots. \end{aligned} \tag{26}$$

We have utilized the numeric computing environment MATLAB, version 2016, and plotted the solution (25) in Figure 1 by considering the first fifteen terms of the series (21).

Figure 1 shows the dynamics of each one of the state variables in the classical sense when $\theta = 1$ (green curves). Similarly, by considering the model in ABC sense, that is, for $\theta \in (0, 1)$, we plot in Figure 1 each of the state variables to analyze the changes in comparison with the classical case. From Figure 1a, we see that as we increase the order θ of the fractional ABC derivative, the susceptibility increases. Further, all fractional order derivatives shows no effect after about 60 days, i.e., the susceptible population stabilizes. Figure 1b shows that infected individuals tend to increase, with different rates, when we decrease the fractional-order: the smaller the fractional order θ, faster the increase rate, and vice versa. All obtained curves for infected individuals, for different values of the fractional order derivatives, approach towards a non-zero steady state, which shows that the disease will persist in the community if not properly managed. On the other hand, Figure 1c shows that during the first month the disease progress with more and more people getting quarantined, irrespective of the order of the derivative. However, after that, the quarantined population tends to decline and, at the end, there will be no quarantined individuals in the community.

In Section 6.1, we show that the fractional Model (1) with ABC derivatives has the ability to describe effectively the dynamics of transmission of the current COVID-19 outbreak.

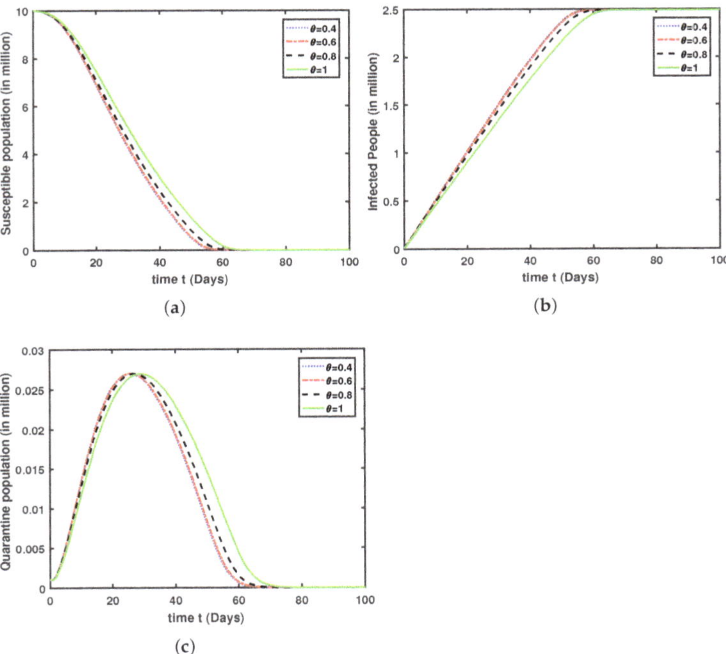

Figure 1. Dynamical nature of susceptible, infected and quarantined individuals of the fractional ABC Model (1) for different values of the fractional-order θ. (**a**) $P(t)$—susceptible individuals along time t; (**b**) $I(t)$—infected individuals along time t; (**c**) $Q(t)$—quarantined individuals along time t.

6.1. Case Study with Real Data: Khyber Pakhtunkhawa (Pakistan)

The Khyber Pakhtunkhawa Province, like other provinces of Pakistan and the rest of the world, is also being affected by COVID-19. We decided to calibrate our model with real data of COVID-19 from Khyber Pakhtunkhawa, Pakistan, from 9 April to 2 June 2020. For that, we have used the minimization method of MATLAB taking the initial weights

$$P(0) = 35{,}525{,}047, \quad I(0) = 10{,}485, \quad Q(0) = 18{,}000,$$

determined from the work in [55], and $\theta = 1$, from which we arrived to the values of the parameters shown in Table 3.

Figure 2 shows the total number of individuals infected by COVID-19 as registered from 9 April to the 2nd in June 2020, which corresponds to the period of one month and 24 days used to calibrate our model.

Figure 3 compares the actual/real data of COVID-19 with the curve of infected given by Model (1), clearly showing the appropriateness of our model to describe the COVID-19 outbreak.

Table 3. Parameter values for the case of Khyber Pakhtunkhawa, Pakistan.

Notation	Value	Reference
λ	0.028	[55]
γ	0.2	Estimated
d_0	0.011	[55]
μ	0.2	Estimated
h	0.06	[55]
σ	0.04	Estimated
η	0.3	Estimated

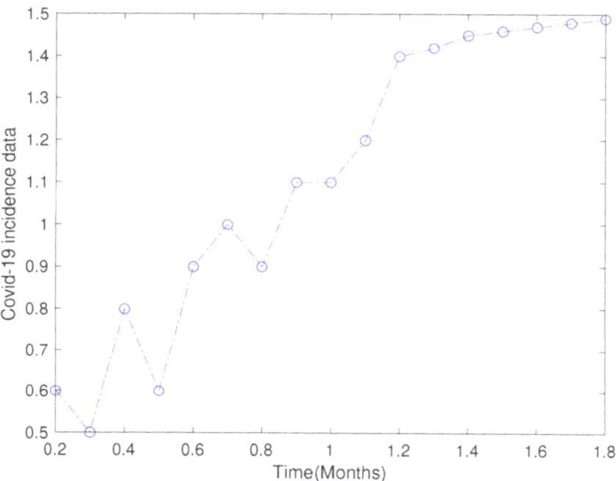

Figure 2. Real data of infected individuals by COVID-19 from Khyber Pakhtunkhwa, Pakistan, from 9 April to 2 June 2020.

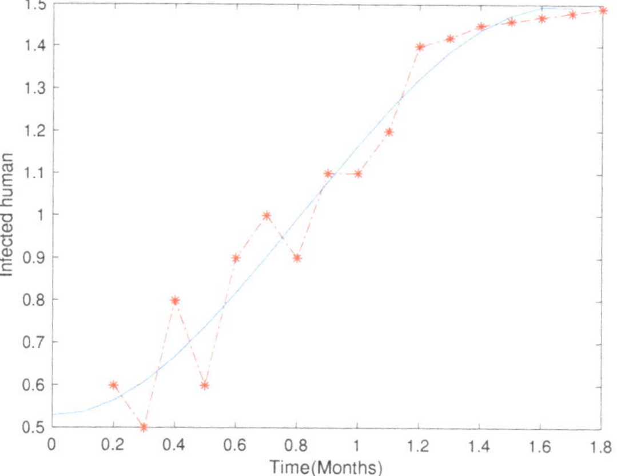

Figure 3. Comparison of infected individuals by COVID-19: Model (1) output (in blue) versus real data of Khyber Pakhtunkhawa, Pakistan, from 9 April to 2 June 2020 (in red).

Figure 4 projects the long-term behavior of the COVID-19 outbreak during a period of eight months. We can see the data matches during the first 1.8 months and, additionally, we observe that the long-term behavior consists on a rise of infected individuals with time. This means that if the government did not apply proper strategies, the incidence could increase drastically in the coming months.

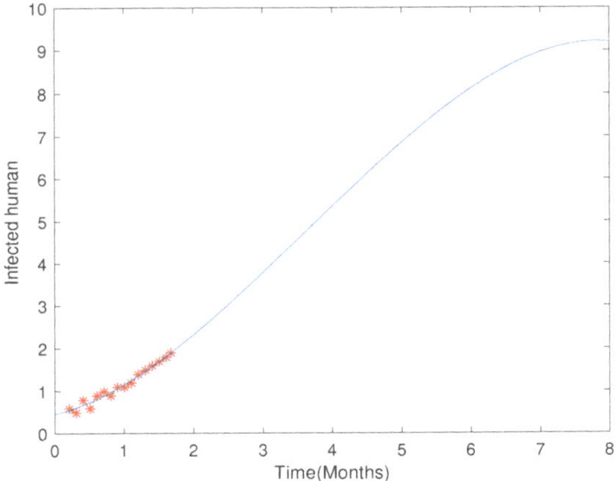

Figure 4. Real data of infected individuals by COVID-19 in Khyber Pakhtunkhwa, Pakistan (first 1.8 months, in red) and prediction from Model (1) during a period of 8 months (in blue).

7. Sensitivity Analysis

Here, we conduct a sensitivity analysis to evaluate the parameters that are sensitive in minimizing the propagation of the ailment. Although its computation is tedious for complex biological models, forward sensitivity analysis is recorded as an important component of epidemic modeling: the ecologist and epidemiologist gain a lot of insight from the sensitivity study of the basic reproduction number R_0 [56]. In Definition 2, we assume that the basic reproduction number R_0 is differentiable with respect to parameter ω. Given (3), this means that Definition 2 makes sense for $\omega \in \{\gamma, \lambda, d_0, \mu, \sigma, h, \eta\}$.

Definition 2. *The normalized forward sensitivity index of R_0 with respect to parameter ω is defined by*

$$S_\omega = \frac{\omega}{R_0}\frac{\partial R_0}{\partial \omega}. \tag{27}$$

As we have an analytical form for the basic reproduction number, recall (3), we apply the direct differentiation process given in (27). Not only do the sensitivity indexes show us the impact of various factors associated with the spread of the infectious disease, but they also provide us with valuable details on the comparative change between R_0 and the parameters. Moreover, they also assist in the production of control strategies [57].

Table 4 demonstrates that γ, h, and σ parameters have a positive effect on the basic reproduction number R_0, which means that the growth or decay of these parameters by 10% would increase or decrease the reproduction number by 10%, 6.36%, and 0.31%, respectively. On the other hand, d_0-, μ-, and η-sensitive indexes indicate that increasing their values by 10% would decrease the basic reproduction number R_0 by 14.89%, 0.09%, and 1.68%, respectively.

Table 4. Sensitivity indexes of the basic reproduction number R_0 (3) (see Definition 2) for relevant parameters of Model (1).

Parameters	Sensitivity	Value	Parameters	Sensitivity	Value
γ	S_γ	1.00000000	h	S_h	0.63636363
d_0	S_{d_0}	−1.48944805	μ	S_μ	−0.00974026
σ	S_σ	0.03165584	η	S_η	−0.16883117

The sensitivity of the basic reproduction number R_0 is also seen graphically in Figure 5.

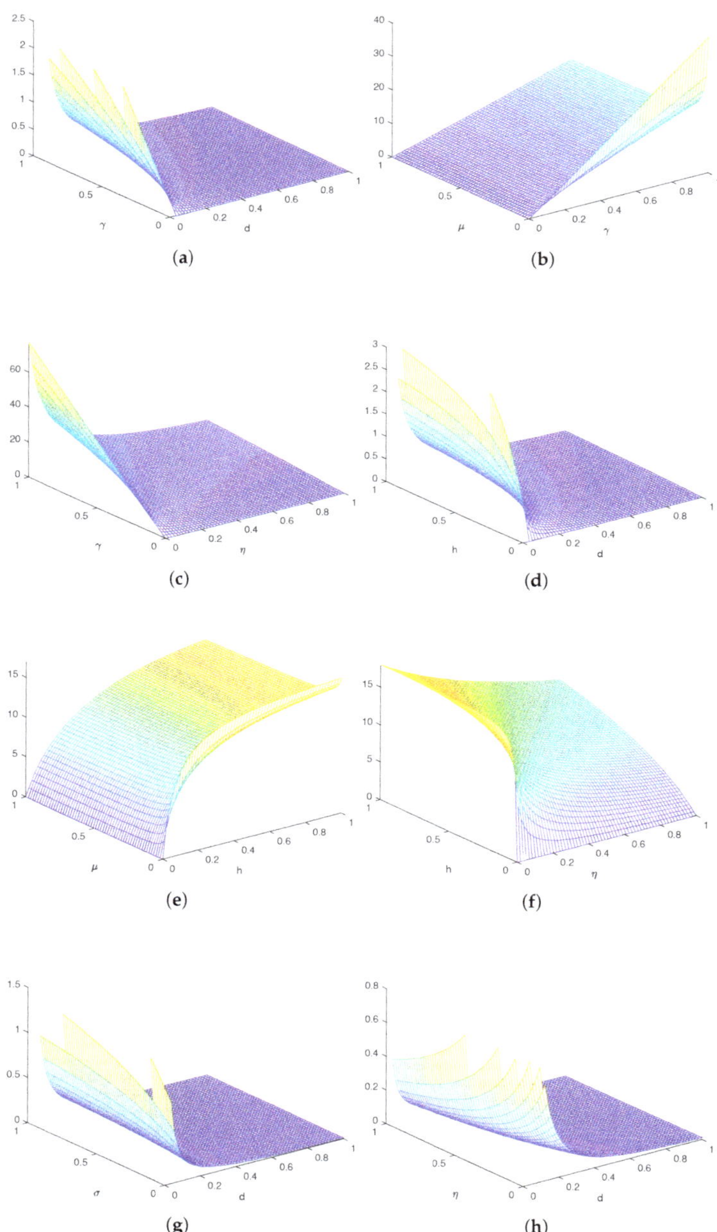

Figure 5. Sensitivity of the basic reproduction number R_0 (3) for relevant parameters of Model (1). (**a**) R_0 versus γ and d; (**b**) R_0 versus γ and μ; (**c**) R_0 versus γ and η; (**d**) R_0 versus h and d; (**e**) R_0 versus h and μ; (**f**) R_0 versus h and η; (**g**) R_0 versus d and σ; (**h**) R_0 versus d and η.

8. Conclusions and Future Work

In this manuscript, we studied a COVID-19 disease model providing a detailed qualitative analysis and showed its usefulness with a case study of Khyber Pakhtunkhawa, Pakistan. Our sensitivity analysis shows that the transmission rate γ has a huge effect on the model as compared to other parameters: the basic reproduction number varies directly with the transmission rate γ. The sensitivity analysis also showed that the death rate parameter μ has no effect on spreading the infection, which seems biologically correct. The transmission rate will be small by keeping a social distancing and self-quarantine situation that causes a decrease in the infection. In this way, one can control COVID-19 infection from rapid spreading in the community. In the future, we plan to analyze optimal control techniques to reduce the population of infected individuals by adopting a number of control measures. A modification of the given model is also possible by introducing more parameters for analyzing the early outbreaks of COVID-19 and then transmission and treatment aspects can be recalled. The given system can be also simulated by adding exposed and hospitalized classes and taking a stochastic fractional derivative. Here, we have provided a case study with real data from Pakistan, but other case studies can also be done.

Author Contributions: Conceptualization, M.R.S.A., A.K., A.Z., and D.F.M.T.; Formal analysis, M.R.S.A., A.D., A.K., and D.F.M.T.; Investigation, A.D. and A.Z.; Methodology, M.T. and A.Z.; Software, A.D. and A.K.; Supervision, D.F.M.T.; Validation, M.R.S.A. and D.F.M.T.; Writing—original draft, A.D., A.K., and A.Z.; Writing—review and editing, M.T., A.K., and D.F.M.T. All authors have read and agreed to the published version of the manuscript.

Funding: This research was partially funded by Fundação para a Ciência e a Tecnologia (FCT) grant number UIDB/04106/2020 (CIDMA).

Data Availability Statement: Not applicable.

Acknowledgments: The authors are grateful to reviewers for their comments, questions, and suggestions, which helped them to improve the manuscript.

Conflicts of Interest: The authors declare no conflict of interest. The funders had no role in the design of the study; in the collection, analyses, or interpretation of data; in the writing of the manuscript; or in the decision to publish the results.

References

1. Din, A.; Li, Y.; Khan, T.; Zaman, G. Mathematical analysis of spread and control of the novel corona virus (COVID-19) in China. *Chaos Solitons Fractals* **2020**, *141*, 110286. [CrossRef] [PubMed]
2. Din, A.; Khan, A.; Baleanu, D. Stationary distribution and extinction of stochastic coronavirus (COVID-19) epidemic model. *Chaos Solitons Fractals* **2020**, *139*, 110036. [CrossRef] [PubMed]
3. Koopmans, M.; Daszak, P.; Dedkov, V.G.; Dwyer, D.E.; Farag, E.; Fischer, T.K.; Hayman, D.T.S.; Leendertz, F.; Maeda, K.; Nguyen-Viet, H.; et al. Origins of SARS-CoV-2: Window is closing for key scientific studies. *Nature* **2021**, *596*, 482–485. [CrossRef] [PubMed]
4. Ndaïrou, F.; Area, I.; Nieto, J.J.; Silva, C.J.; Torres, D.F.M. Fractional model of COVID-19 applied to Galicia, Spain and Portugal. *Chaos Solitons Fractals* **2021**, *144*, 110652. [CrossRef] [PubMed]
5. Lemos-Paião, A.P.; Silva, C.J.; Torres, D.F.M. A New Compartmental Epidemiological Model for COVID-19 with a Case Study of Portugal. *Ecol. Complex.* **2020**, *44*, 100885. [CrossRef]
6. Zine, H.; Boukhouima, A.; Lotfi, E.M.; Mahrouf, M.; Torres, D.F.M.; Yousfi, N. A stochastic time-delayed model for the effectiveness of Moroccan COVID-19 deconfinement strategy. *Math. Model. Nat. Phenom.* **2020**, *15*, 50. [CrossRef]
7. Mahrouf, M.; Boukhouima, A.; Zine, H.; Lotfi, E.M.; Torres, D.F.M.; Yousfi, N. Modeling and Forecasting of COVID-19 Spreading by Delayed Stochastic Differential Equations. *Axioms* **2021**, *10*, 18. [CrossRef]
8. Ndaïrou, F.; Torres, D.F.M. Mathematical Analysis of a Fractional COVID-19 Model Applied to Wuhan, Spain and Portugal. *Axioms* **2021**, *10*, 135. [CrossRef]
9. Murray, J.D. *Mathematical Biology I. An Introduction*; Springer Science & Business Media: Berlin/Heidelberg, Germany, 2007.
10. Stewart, I.W. *The Static and Dynamic Continuum Theory of Liquid Crystals*; CRC Press: Boca Raton, FL, USA, 2019.
11. Samko, S.G.; Kilbas, A.A.; Marichev, O.I. *Fractional Integrals and Derivatives*; Gordon and Breach Science Publishers: Yverdon, Switzerland, 1993.

12. Toledo-Hernandez, R.; Rico-Ramirez, V.; Iglesias-Silva, G.A.; Diwekar, U.M. A Fractional calculus approach to the dynamic optimization of biological reactive systems. Part I: Fractional models for biological reactions. *Chem. Eng. Sci.* **2014**, *117*, 217–228. [CrossRef]
13. Miller, K.S.; Bertram, R. *An Introduction to the Fractional Calculus and Fractional Differential Equations*; Wiley-Interscience: New York, NY, USA, 1993.
14. Kilbas, A.A.; Srivastava, H.M.; Trujillo, J.J. *Theory and Applications of Fractional Differential Equations*; North-Holland Mathematics Studies; Elsevier Science B.V.: Amsterdam, The Netherlands, 2006; Volume 204.
15. Rahimy, M. Applications of fractional differential equations. *Appl. Math. Sci.* **2010**, *4*, 2453–2461. [CrossRef]
16. Rossikhin, Y.A.; Shitikova, M.V. Applications of fractional calculus to dynamic problems of linear and nonlinear hereditary mechanics of solids. *Appl. Mech. Rev.* **1997**, *50*, 15–67. [CrossRef]
17. Magin, R.L. *Fractional Calculus in Bioengineering*; Begell House: Redding, CA, USA, 2006.
18. Kamal, S.; Jarad, F.; Abdeljawad, T. On a nonlinear fractional order model of dengue fever disease under Caputo-Fabrizio derivative. *Alex. Eng. J.* **2020**, *59*, 2305–2313.
19. Biazar, J. Solution of the epidemic model by Adomian decomposition method. *Appl. Math. Comput.* **2006**, *173*, 1101–1106. [CrossRef]
20. Rafei, M.; Ganji, D.D.; Daniali, H. Solution of the epidemic model by homotopy perturbation method. *Appl. Math. Comput.* **2007**, *187*, 1056–1062. [CrossRef]
21. Abdelrazec, A. Adomian Decomposition Method: Convergence Analysis and Numerical Approximations. Master's Thesis, McMaster University, Hamilton, ON, Canada, 2008.
22. Rezapour, S.; Mohammadi, H.; Jajarmi, A. A new mathematical model for Zika virus transmission. *Adv. Differ. Equ.* **2020**, *2020*, 1–15. [CrossRef]
23. Baleanu, D.; Ghanbari, B.; Asad, J.H.; Jajarmi, A.; Pirouz, H.M. Planar System-Masses in an Equilateral Triangle: Numerical Study within Fractional Calculus. *Comput. Model. Eng. Sci.* **2020**, *124*, 953–968. [CrossRef]
24. Jajarmi, A.; Baleanu, D. A New Iterative Method for the Numerical Solution of High-Order Non-linear Fractional Boundary Value Problems. *Front. Phys.* **2020**, *8*, 220. [CrossRef]
25. Sajjadi, S.S.; Baleanu, D.; Jajarmi, A.; Pirouz, H.M. A new adaptive synchronization and hyperchaos control of a biological snap oscillator. *Chaos Solitons Fractals* **2020**, *138*, 109919. [CrossRef]
26. Baleanu, D.; Jajarmi, A.; Sajjadi, S.S.; Asad, J.H. The fractional features of a harmonic oscillator with position-dependent mass. *Commun. Theor. Phys.* **2020**, *72*, 055002. [CrossRef]
27. Jajarmi, A.; Baleanu, D. On the fractional optimal control problems with a general derivative operator. *Asian J. Control* **2021**, *23*, 1062–1071. [CrossRef]
28. Azhar, I.; Siddiqui, M.J.; Muhi, I.; Abbas, M.; Akram, T. Nonlinear waves propagation and stability analysis for planar waves at far field using quintic B-spline collocation method. *Alex. Eng. J.* **2020**, *59*, 2695–2703.
29. Khalid, N.; Abbas, M.; Iqbal, M.K.; Singh, J.; Ismail, A.I. A computational approach for solving time fractional differential equation via spline functions. *Alex. Eng. J.* **2020**, *59*, 3061–3078. [CrossRef]
30. Akram, T.; Abbas, M.; Iqbal, A.; Baleanu, D.; Asad, J.H. Novel Numerical Approach Based on Modified Extended Cubic B-Spline Functions for Solving Non-Linear Time-Fractional Telegraph Equation. *Symmetry* **2020**, *12*, 1154. [CrossRef]
31. Din, A.; Li, Y. Controlling heroin addiction via age-structured modeling. *Adv. Differ. Equ.* **2020**, *2020*, 1–17. [CrossRef]
32. Akram, T.; Abbas, M.; Ali, A.; Iqbal, A.; Baleanu, D. A Numerical Approach of a Time Fractional Reaction–Diffusion Model with a Non-Singular Kernel. *Symmetry* **2020**, *12*, 1653. [CrossRef]
33. Amin, M.; Abbas, M.; Iqbal, M.K.; Baleanu, D. Numerical Treatment of Time-Fractional Klein–Gordon Equation Using Redefined Extended Cubic B-Spline Functions. *Front. Phys.* **2020**, *8*, 288. [CrossRef]
34. Amin, M.; Abbas, M.; Iqbal, M.K.; Ismail, A.I.M.; Baleanu, D. A fourth order non-polynomial quintic spline collocation technique for solving time fractional superdiffusion equations. *Adv. Differ. Equ.* **2019**, *2019*, 1–21. [CrossRef]
35. Khalid, N.; Abbas, M.; Iqbal, M.K. Non-polynomial quintic spline for solving fourth-order fractional boundary value problems involving product terms. *Appl. Math. Comput.* **2019**, *349*, 393–407. [CrossRef]
36. Iqbal, M.K.; Abbas, M.; Wasim, I. New cubic B-spline approximation for solving third order Emden-Flower type equations. *Appl. Math. Comput.* **2018**, *331*, 319–333. [CrossRef]
37. Akram, T.; Abbas, M.; Riaz, M.B.; Ismail, A.I.; Norhashidah, M.A. An efficient numerical technique for solving time fractional Burgers equation. *Alex. Eng. J.* **2020**, *59*, 2201–2220. [CrossRef]
38. Khalid, N.; Abbas, M.; Iqbal, M.K.; Baleanu, D. A numerical investigation of Caputo time fractional Allen-Cahn equation using redefined cubic B-spline functions. *Adv. Differ. Equ.* **2020**, *2020*, 1–22. [CrossRef]
39. Ndaïrou, F.; Torres, D.F.M. Pontryagin Maximum Principle for Distributed-Order Fractional Systems. *Mathematics* **2021**, *9*, 1883. [CrossRef]
40. Sabatier, J.; Agrawal, O.P.; Machado, J.A.T. *Advances in Fractional Calculus*; Springer: Dordrecht, The Netherlands, 2007. [CrossRef]
41. Baleanu, D.; Machado, J.A.T.; Albert, C.J. *Fractional Dynamics and Control*; Springer Science & Business Media: Berlin/Heidelberg, Germany, 2011.
42. Torres, D.F.M. Cauchy's formula on nonempty closed sets and a new notion of Riemann-Liouville fractional integral on time scales. *Appl. Math. Lett.* **2021**, *121*, 107407. [CrossRef]

43. Al-Refai, M.; Abdeljawad, T. Analysis of the fractional diffusion equations with fractional derivative of non-singular kernel. *Adv. Differ. Equ.* **2017**, *2017*, 1–12. [CrossRef]
44. Mozyrska, D.; Torres, D.F.M.; Wyrwas, M. Solutions of systems with the Caputo-Fabrizio fractional delta derivative on time scales. *Nonlinear Anal. Hybrid Syst.* **2019**, *32*, 168–176. [CrossRef]
45. Abdeljawad, T. Fractional operators with exponential kernels and a Lyapunov type inequality. *Adv. Differ. Equ.* **2017**, *2017*, 1–11. [CrossRef]
46. Hasan, S.; El-Ajou, A.; Hadid, S.; Al-Smadi, M.; Momani, S. Atangana-Baleanu fractional framework of reproducing kernel technique in solving fractional population dynamics system. *Chaos Solitons Fractals* **2020**, *133*, 109624. [CrossRef]
47. Khan, S.A.; Shah, K.; Zaman, G.; Jarad, F. Existence theory and numerical solutions to smoking model under Caputo-Fabrizio fractional derivative. *Chaos* **2019**, *29*, 013128. [CrossRef] [PubMed]
48. Sidi Ammi, M.R.; Tahiri, M.; Torres, D.F.M. Necessary optimality conditions of a reaction-diffusion SIR model with ABC fractional derivatives. *arXiv* **2021**, arXiv:2106.15055.
49. Din, A.; Li, Y.; Liu, Q. Viral dynamics and control of hepatitis B virus (HBV) using an epidemic model. *Alex. Eng. J.* **2020**, *59*, 667–679. [CrossRef]
50. Din, A.; Liang, J.; Zhou, T. Detecting critical transitions in the case of moderate or strong noise by binomial moments. *Phys. Rev. E* **2018**, *98*, 012114. [CrossRef]
51. Lai, C.C.; Shih, T.P.; Ko, W.C.; Tang, H.J.; Hsueh, P.R. Severe acute respiratory syndrome coronavirus 2 (SARS-CoV-2) and coronavirus disease-2019 (COVID-19): The epidemic and the challenges. *Int. J. Antimicrob. Agents* **2020**, *55*, 105924. [CrossRef] [PubMed]
52. Din, A.; Li, Y. Lévy noise impact on a stochastic hepatitis B epidemic model under real statistical data and its fractal–fractional Atangana–Baleanu order model. *Phys. Scr.* **2021**, *96*, 124008. [CrossRef]
53. Wang, J.; Shah, K.; Ali, A. Existence and Hyers-Ulam stability of fractional nonlinear impulsive switched coupled evolution equations. *Math. Methods Appl. Sci.* **2018**, *41*, 2392–2402. [CrossRef]
54. Ahmed, E.; El-Sayed, A.M.A.; El-Saka, H.A.A.; Ashry, G.A. On applications of Ulam-Hyers stability in biology and economics. *arXiv* **2010**, arXiv:1004.1354.
55. Pakistan, G. COVID-19 Situation. Know about COVID-19, See the Realtime Pakistan and Worldwide. Available online: https://covid.gov.pk (accessed on 28 October 2021).
56. Rodrigues, H.S.; Monteiro, M.T.T.; Torres, D.F.M. Seasonality effects on dengue: Basic reproduction number, sensitivity analysis and optimal control. *Math. Methods Appl. Sci.* **2016**, *39*, 4671–4679. [CrossRef]
57. Rosa, S.; Torres, D.F.M. Parameter estimation, sensitivity analysis and optimal control of a periodic epidemic model with application to HRSV in Florida. *Stat. Optim. Inf. Comput.* **2018**, *6*, 139–149. [CrossRef]

MDPI

Article

On Periodic Fractional (p, q)-Integral Boundary Value Problems for Sequential Fractional (p, q)-Integrodifference Equations

Jarunee Soontharanon [1] and Thanin Sitthiwirattham [2,*]

[1] Department of Mathematics, Faculty of Applied Science, King Mongkut's University of Technology North Bangkok, Bangkok 10800, Thailand; jarunee.s@sci.kmutnb.ac.th

[2] Mathematics Department, Faculty of Science and Technology, Suan Dusit University, Bangkok 10300, Thailand

* Correspondence: thanin_sit@dusit.ac.th

Abstract: We study the existence results of a fractional (p, q)-integrodifference equation with periodic fractional (p, q)-integral boundary condition by using Banach and Schauder's fixed point theorems. Some properties of (p, q)-integral are also presented in this paper as a tool for our calculations.

Keywords: fractional (p, q)-integral; fractional (p, q)-difference; periodic boundary value problems; existence

Citation: Soontharanon, J.; Sitthiwirattham, T. On Periodic Fractional (p, q)-Integral Boundary Value Problems for Sequential Fractional (p, q)-Integrodifference Equations. *Axioms* **2021**, *10*, 264. https://doi.org/10.3390/axioms10040264

Academic Editor: Natália Martins

Received: 8 September 2021
Accepted: 18 October 2021
Published: 19 October 2021

1. Introduction

The studies of quantum calculus with integer order were presented in the last three decades, and many researchers extensively studied calculus without a limit that deals with a set of nondifferentiable functions, the so-called quantum calculus. Many types of quantum difference operators are employed in several applications of mathematical areas, such as the calculus of variations, particle physics, quantum mechanics, and theory of relativity. The q-calculus, one type of quantum initiated by Jackson [1–5], was employed in several fields of applied sciences and engineering such as physical problems, dynamical system, control theory, electrical networks, economics, and so on [6–14].

For fractional quantum calculus, Agarwal [15] and Al-Salam [16] proposed fractional q-calculus, and Díaz and Osler [17] proposed fractional difference calculus. In 2017, Brikshavana and Sitthiwirattham [18] introduced fractional Hahn difference calculus. In 2019, Patanarapeelert and Sitthiwirattham [19] studied fractional symmetric Hahn difference calculus.

Later, the motivation of quantum calculus based on two parameters (p, q)-integer was presented. The (p, q)-calculus (postquantum calculus) was introduced by Chakrabarti and Jagannathan [20]. This calculus was used in many fields such as special functions, approximation theory, physical sciences, Lie group, hypergeometric series, Bézier curves, and surfaces. For some recent papers about (p, q)-differenceequations, we refer to [21–33] and the references therein. For example, the fundamental theorems of (p, q)-calculus and some (p, q)-Taylor formulas were studied in [21]. In [32], the (p, q)-Melin transform and its applications were studied. The Picard and Gauss–Weierstrass singular integral in (p, q)-calculus were introduced in [33]. For the boundary value problem for (p,q)-difference equations were studied in [34–36]. For example, the nonlocal boundary value problems for first-order (p, q)-difference equations were studied in [34]. The second-order (p, q)-difference equations with separated boundary conditions were studied in [35]. In [36], the authors studied the first-order and second-order (p, q)-difference equations with impulse.

Recently, Soontharanon and Sitthiwirattham [37] introduced the fractional (p, q)-difference operators and its properties. Now, this calculus was used in the inequalities [38,39] and the boundary value problems [40–42]. However, the study of the boundary value problems for fractional (p, q)-difference equation in the beginning, there are a few literature on this knowledge. In [40], the existence results of a fractional (p, q)-integrodifference equation with Robin boundary condition were studied in 2020. In 2021 [41], the authors investigated the boundary value problem of a class of fractional (p, q)-difference Schrödinger equations. In the same year, the existence results of solution and positive solution for the boundary value problem of a class of fractional (p, q)-difference equations involving the Riemann–Liouville fractional derivative [42] were studied.

Motivated by the above papers, we seek to enrich the contributions in this new research area. In this paper, we introduce and study the boundary value problem involving function F, which depends on fractional (p, q)-integral and fractional (p, q)-difference, and the boundary condition is nonlocal. Our problem is sequential fractional (p, q)-integrodifference equation with periodic fractional (p, q)-integral boundary conditions of the form

$$\begin{aligned} &D^{\alpha}_{p,q}D^{\beta}_{p,q}u(t) = F\Big[t, u(t), \Psi^{\gamma}_{p,q}u(t), D^{\nu}_{p,q}u(t)\Big], \quad t \in I^{T}_{p,q}, \\ &u(0) = u\left(\frac{T}{p}\right) \\ &\mathcal{I}^{\theta}_{p,q}g(\eta)u(\eta) = \varphi(u), \quad \eta \in I^{T}_{p,q} - \{0, T\}, \end{aligned} \tag{1}$$

where $I^{T}_{p,q} := \left\{\left(\frac{q}{p}\right)^{k}\frac{T}{p} : k \in \mathbb{N}_0\right\} \cup \{0\}$; $0 < q < p \leq 1$; $\alpha, \beta, \gamma, \nu, \theta \in (0, 1]$; $F \in C\big(I^{T}_{p,q} \times \mathbb{R} \times \mathbb{R} \times \mathbb{R}, \mathbb{R}\big)$, $g \in \big(I^{T}_{p,q}, \mathbb{R}^{+}\big)$ are given functions; $\varphi : C\big(I^{T}_{p,q}, \mathbb{R}\big) \to \mathbb{R}$ is given functional; and for $\phi \in C\big(I^{T}_{p,q} \times I^{T}_{p,q}, [0, \infty)\big)$, we define an operator of the (p, q)-integral of the product of functions ϕ and u as

$$\Psi^{\gamma}_{p,q}u(t) := \left(\mathcal{I}^{\gamma}_{p,q}\phi\, u\right)(t) = \frac{1}{p^{\binom{\gamma}{2}}\Gamma_{p,q}(\gamma)}\int_0^t (t - qs)^{\underline{\gamma-1}}_{p,q}\phi\left(t, \frac{s}{p^{\gamma-1}}\right)u\left(\frac{s}{p^{\gamma-1}}\right)d_{p,q}s.$$

We aim to show the existence results to the problem (1). Firstly, we convert the given nonlinear problem (1) into a fixed point problem related to (1), by considering a linear variant of the problem at hand. Once the fixed point operator is available, we make use the classical Banach's and Schauder's fixed point theorems to establish existence results.

The paper is organized as follows: Section 2 contains some preliminary concepts related to our problem. We present the existence and uniqueness result in Section 2, and the existence of at least one solution in Section 4. To illustrate our results, we provide some examples in Section 5. Finally, Section 6 discusses our conclusions.

2. Preliminaries

In this section, we provide some basic definitions, notations, and lemmas as follows. For $0 < q < p \leq 1$, we define

$$[k]_q := \begin{cases} \dfrac{1 - q^k}{1 - q}, & k \in \mathbb{N} \\ 1, & k = 0, \end{cases}$$

$$[k]_{p,q} := \begin{cases} \dfrac{p^k - q^k}{p - q} = p^{k-1}[k]_{\frac{q}{p}}, & k \in \mathbb{N} \\ 1, & k = 0, \end{cases}$$

$$[k]_{p,q}! := \begin{cases} [k]_{p,q}[k-1]_{p,q}\cdots[1]_{p,q} = \displaystyle\prod_{i=1}^{k}\frac{p^i - q^i}{p - q}, & k \in \mathbb{N} \\ 1, & k = 0. \end{cases}$$

The (p, q)-forward jump and the (p, q)-backward jump operators are defined as

$$\sigma^{k}_{p,q}(t) := \left(\frac{q}{p}\right)^{k}t \quad \text{and} \quad \rho^{k}_{p,q}(t) := \left(\frac{p}{q}\right)^{k}t, \quad \text{for } k \in \mathbb{N}, \text{ respectively.}$$

The q-analogue of the power function $(a - b)^{\underline{n}}_{q}$ with $n \in \mathbb{N}_0 := \{0, 1, 2, \dots\}$ is given by

$$(a-b)_q^{\underline{0}} := 1, \qquad (a-b)_q^{\underline{n}} := \prod_{i=0}^{n-1}(a-bq^i), \qquad a,b \in \mathbb{R}.$$

The (p,q)-analogue of the power function $(a-b)_{p,q}^{\underline{n}}$ with $n \in \mathbb{N}_0$ is given by

$$(a-b)_{p,q}^{\underline{0}} := 1, \qquad (a-b)_{p,q}^{\underline{n}} := \prod_{k=0}^{n-1}(ap^k - bq^k), \qquad a,b \in \mathbb{R}.$$

Generally, for $\alpha \in \mathbb{R}$, we define

$$(a-b)_q^{\underline{\alpha}} = a^{\alpha}\prod_{i=0}^{\infty}\frac{1-\left(\frac{b}{a}\right)q^i}{1-\left(\frac{b}{a}\right)q^{\alpha+i}}, \ a \neq 0.$$

$$(a-b)_{p,q}^{\underline{\alpha}} = p^{\binom{\alpha}{2}}(a-b)_{\frac{q}{p}}^{\underline{\alpha}} = a^{\alpha}\prod_{i=0}^{\infty}\frac{1}{p^{\alpha}}\left[\frac{1-\frac{b}{a}\left(\frac{q}{p}\right)^i}{1-\frac{b}{a}\left(\frac{q}{p}\right)^{i+\alpha}}\right], \ a \neq 0.$$

In particular, $a_q^{\underline{\alpha}} = a^{\alpha}$, $a_{p,q}^{\underline{\alpha}} = \left(\frac{a}{p}\right)^{\alpha}$ and $(0)_q^{\underline{\alpha}} = (0)_{p,q}^{\underline{\alpha}} = 0$ for $\alpha > 0$.

The (p,q)-gamma and (p,q)-beta functions are defined by

$$\Gamma_{p,q}(x) := \begin{cases} \frac{(p-q)_{p,q}^{\underline{x-1}}}{(p-q)^{x-1}} = \frac{\left(1-\frac{q}{p}\right)_{p,q}^{\underline{x-1}}}{\left(1-\frac{q}{p}\right)^{x-1}}, & x \in \mathbb{R}\setminus\{0,-1,-2,\ldots\} \\ [x-1]_{p,q}!, & x \in \mathbb{N}, \end{cases}$$

$$B_{p,q}(x,y) := \int_0^1 t^{x-1}(1-qt)_{p,q}^{\underline{y-1}}d_{p,q}t = p^{\frac{1}{2}(y-1)(2x+y-2)}\frac{\Gamma_{p,q}(x)\Gamma_q p,q(y)}{\Gamma_{p,q}(x-y)},$$

respectively.

Definition 1. *For $0<q<p\leq 1$ and $f:[0,T]\to\mathbb{R}$, we define the (p,q)-difference of f as*

$$D_{p,q}f(t) := \begin{cases} \dfrac{f(pt)-f(qt)}{(p-q)(t)}, & \text{for } t \neq 0 \\ f'(0), & \text{for } t = 0 \end{cases}$$

provided that f is differentiable at 0 and f is called (p,q)-differentiable on $I_{p,q}^T$ if $D_{p,q}f(t)$ exists for all $t \in I_{p,q}^T$.

Observe that the function $g(t) = D_{p,q}f(t)$ is defined on $[0, T/p]$.

Definition 2. *Let I be any closed interval of $\mathbb{R}$ containing a, b and 0. Assuming that $f: I \to \mathbb{R}$ is a given function, we define (p,q)-integral of f from a to b by*

$$\int_a^b f(t)d_{p,q}t := \int_0^b f(t)d_{p,q}t - \int_0^a f(t)d_{p,q}t,$$

where

$$\mathcal{I}_{p,q}f(x) = \int_0^x f(t)d_{p,q}t = (p-q)x\sum_{k=0}^{\infty}\frac{q^k}{p^{k+1}}f\left(\frac{q^k}{p^{k+1}}x\right), \quad x \in I,$$

provided that the series converges at $x = a$ and $x = b$ and f is called (p,q)-integrable on $[a,b]$ if it is (p,q)-integrable on $[a,b]$ for all $a,b \in I$.

An operator $\mathcal{I}^N_{p,q}$ is defined as

$$\mathcal{I}^0_{p,q}f(x) = f(x) \text{ and } \mathcal{I}^N_{p,q}f(x) = \mathcal{I}_{p,q}\mathcal{I}^{N-1}_{p,q}f(x), N \in \mathbb{N}.$$

The relations between (p,q)-difference and (p,q)-integral operators are given by

$$D_{p,q}\mathcal{I}_{p,q}f(x) = f(x) \text{ and } \mathcal{I}_{p,q}D_{p,q}f(x) = f(x) - f(0).$$

Fractional (p,q)-integral and fractional (p,q)-difference of Riemann–Liouville type are defined as follows.

Definition 3. *For $\alpha > 0$, $0 < q < p \leq 1$ and f defined on $I^T_{p,q}$, the fractional (p,q)-integral is defined by*

$$\begin{aligned}\mathcal{I}^\alpha_{p,q}f(t) &:= \frac{1}{p^{\binom{\alpha}{2}}\Gamma_{p,q}(\alpha)}\int_0^t (t-qs)^{\underline{\alpha-1}}_{p,q} f\left(\frac{s}{p^{\alpha-1}}\right) d_{p,q}s \\ &= \frac{(p-q)t}{p^{\binom{\alpha}{2}}\Gamma_{p,q}(\alpha)}\sum_{k=0}^{\infty}\frac{q^k}{p^{k+1}}\left(t-\left(\frac{q}{p}\right)^{k+1}t\right)^{\underline{\alpha-1}}_{p,q} f\left(\frac{q^k}{p^{k+\alpha}}t\right),\end{aligned}$$

and $(\mathcal{I}^0_{p,q}f)(t) = f(t)$.

Definition 4. *For $\alpha > 0$, $0 < q < p \leq 1$ and f defined on $I^T_{p,q}$, the fractional (p,q)-difference operator of Riemann–Liouville type of order α is defined by*

$$\begin{aligned}D^\alpha_{p,q}f(t) &:= D^N_{p,q}\mathcal{I}^{N-\alpha}_{p,q}f(t) \\ &= \frac{1}{p^{\binom{-\alpha}{2}}\Gamma_{p,q}(-\alpha)}\int_0^t (t-qs)^{\underline{-\alpha-1}}_{p,q} f\left(\frac{s}{p^{-\alpha-1}}\right) d_{p,q}s,\end{aligned}$$

and $D^0_{p,q}f(t) = f(t)$, where $N-1 < \alpha < N$, $N \in \mathbb{N}$.

Lemma 1 ([37]). *Let $\alpha \in (N-1,N)$, $N \in \mathbb{N}$, $0 < q < p \leq 1$ and $f : I^T_{p,q} \to \mathbb{R}$. Then,*

$$\mathcal{I}^\alpha_{p,q}D^\alpha_{p,q}f(t) = f(t) + C_1t^{\alpha-1} + C_2t^{\alpha-2} + \cdots + C_Nt^{\alpha-N}$$

for some $C_i \in \mathbb{R}$, $i = 1,2,\ldots,N$.

Lemma 2 ([37]). *Let $0 < q < p \leq 1$ and $f : I^T_{p,q} \to \mathbb{R}$ be continuous at 0. Then,*

$$\int_0^x\int_0^s f(\tau)\, d_{p,q}\tau\, d_{p,q}s = \int_0^{\frac{x}{p}}\int_{pq\tau}^x f(\tau)\, d_{p,q}s\, d_{p,q}\tau.$$

Lemma 3 ([37]). *Let $\alpha,\beta > 0$, $0 < q < p \leq 1$. Then,*

$$\begin{aligned}(a) \quad &\int_0^t (t-qs)^{\underline{\alpha-1}}_{p,q} s^\beta\, d_{p,q}s = t^{\alpha+\beta}B_{p,q}(\beta+1,\alpha), \\ (b) \quad &\int_0^t\int_0^x (t-qx)^{\underline{\alpha-1}}_{p,q}(x-qs)^{\underline{\beta-1}}_{p,q}\, d_{p,q}s\, d_{p,q}x = \frac{B_{p,q}(\beta+1,\alpha)}{[\beta]_{p,q}}t^{\alpha+\beta}.\end{aligned}$$

Lemma 4 ([40]). *Let $\alpha,\beta > 0$, $0 < q < p \leq 1$ and $n \in \mathbb{Z}$. Then,*

$$(a)\quad \int_0^t (t-qs)_{p,q}^{\underline{\alpha-1}}\, d_{p,q}s \;=\; p^{\binom{\alpha}{2}}\frac{\Gamma_{p,q}(\alpha)}{\Gamma_{p,q}(\alpha+1)}\, t^{\alpha},$$

$$(b)\quad \int_0^t \int_0^{\frac{x}{p^{-\beta-1}}} (t-qx)_{p,q}^{\underline{-\beta-1}} \left(\frac{x}{p^{-\beta-1}} - qs\right)_{p,q}^{\underline{\alpha-1}} d_{p,q}s\, d_{p,q}x \;=\; p^{\binom{\alpha}{2}+\binom{-\beta}{2}}\frac{\Gamma_{p,q}(\alpha)}{\Gamma_{p,q}(\alpha+1)}\, t^{\alpha+\beta},$$

$$(c)\quad \int_0^t (t-qs)_{p,q}^{\underline{-\beta-1}} \left(\frac{s}{p^{-\beta-1}}\right)^{\alpha-n} d_{p,q}s \;=\; p^{\binom{-\beta}{2}}\frac{\Gamma_{p,q}(\alpha-n+1)\Gamma_{p,q}(-\beta)}{\Gamma_{p,q}(\alpha-\beta-n+1)}\, t^{\alpha-\beta-n}.$$

Lemma 5. *Let $\alpha,\beta,\theta > 0$, $0 < q < p \le 1$ and $n \in \mathbb{Z}$. Then,*

$$(a)\quad \int_0^t \int_0^{\frac{x}{p^{\theta-1}}} (t-qx)_{p,q}^{\underline{\theta-1}} \left(\frac{x}{p^{\theta-1}} - qs\right)_{p,q}^{\underline{\beta-1}} \left(\frac{s}{p^{\alpha-1}}\right)_{p,q}^{\underline{\alpha-n}} d_{p,q}s\, d_{p,q}x$$
$$= p^{\binom{\beta}{2}+\binom{\theta}{2}}\frac{\Gamma_{p,q}(\alpha-n+1)\Gamma_{p,q}(\beta)\Gamma_{p,q}(\theta)}{\Gamma_{p,q}(\alpha+\beta+\theta-n+1)}\, t^{\alpha+\beta+\theta-n},$$

$$(b)\quad \int_0^t \int_0^{\frac{y}{p^{\theta-1}}} \int_0^{\frac{x}{p^{\beta-1}}} (t-qy)_{p,q}^{\underline{\theta-1}} \left(\frac{y}{p^{\theta-1}} - qx\right)_{p,q}^{\underline{\beta-1}} \left(\frac{x}{p^{\beta-1}} - qs\right)_{p,q}^{\underline{\alpha-1}} d_{p,q}s\, d_{p,q}x\, d_{p,q}y$$
$$= p^{\binom{\alpha}{2}+\binom{\beta}{2}+\binom{\theta}{2}}\frac{\Gamma_{p,q}(\alpha)\Gamma_{p,q}(\beta)\Gamma_{p,q}(\theta)}{\Gamma_{p,q}(\alpha+\beta+\theta+1)}\, t^{\alpha+\beta+\theta}.$$

Proof. By Lemmas 2, 3 and 4 and definition of the (p,q)-beta function, we have

$$(a)\quad \int_0^t \int_0^{\frac{x}{p^{\theta-1}}} (t-qx)_{p,q}^{\underline{\theta-1}} \left(\frac{x}{p^{\theta-1}} - qs\right)_{p,q}^{\underline{\beta-1}} \left(\frac{s}{p^{\alpha-1}}\right)_{p,q}^{\underline{\alpha-n}} d_{p,q}s\, d_{p,q}x$$
$$= \int_0^t (t-qx)_{p,q}^{\underline{\theta-1}} \left[\int_0^{\frac{x}{p^{\theta-1}}} \left(\frac{x}{p^{\theta-1}} - qs\right)_{p,q}^{\underline{\beta-1}} \left(\frac{s}{p^{\alpha-1}}\right)_{p,q}^{\underline{\alpha-n}} d_{p,q}s\right] d_{p,q}x$$
$$= \frac{p^{\binom{\beta}{2}}}{p^{(\theta-1)(\alpha+\beta-n)}}\cdot\frac{\Gamma_{p,q}(\alpha-n+1)\Gamma_{p,q}(\beta)}{\Gamma_{p,q}(\alpha+\beta-n+1)}\int_0^t (t-qx)_{p,q}^{\underline{\theta-1}} x^{\alpha+\beta-n}\, d_{p,q}x$$
$$= p^{\binom{\beta}{2}+\binom{\theta}{2}}\frac{\Gamma_{p,q}(\alpha-n+1)\Gamma_{p,q}(\beta)\Gamma_{p,q}(\theta)}{\Gamma_{p,q}(\alpha+\beta+\theta-n+1)}\, t^{\alpha+\beta+\theta-n},$$

$$(b)\quad \int_0^t \int_0^{\frac{y}{p^{\theta-1}}} \int_0^{\frac{x}{p^{\beta-1}}} (t-qy)_{p,q}^{\underline{\theta-1}} \left(\frac{y}{p^{\theta-1}} - qx\right)_{p,q}^{\underline{\beta-1}} \left(\frac{x}{p^{\beta-1}} - qs\right)_{p,q}^{\underline{\alpha-1}} d_{p,q}s\, d_{p,q}x\, d_{p,q}y$$
$$= \int_0^t \int_0^{\frac{y}{p^{\theta-1}}} (t-qy)_{p,q}^{\underline{\theta-1}} \left(\frac{y}{p^{\theta-1}} - qx\right)_{p,q}^{\underline{\beta-1}} \left[\int_0^{\frac{x}{p^{\beta-1}}} \left(\frac{x}{p^{\beta-1}} - qs\right)_{p,q}^{\underline{\alpha-1}} d_{p,q}s\right] d_{p,q}x\, d_{p,q}y$$
$$= p^{\binom{\alpha}{2}}\frac{\Gamma_{p,q}(\alpha)}{\Gamma_{p,q}(\alpha+1)}\int_0^t \int_0^{\frac{y}{p^{\theta-1}}} (t-qy)_{p,q}^{\underline{\theta-1}} \left(\frac{y}{p^{\theta-1}} - qx\right)_{p,q}^{\underline{\beta-1}} \left(\frac{x^{\alpha}}{p^{\beta-1}}\right) d_{p,q}x\, d_{p,q}y$$
$$= p^{\binom{\alpha}{2}+\binom{\beta}{2}+\binom{\theta}{2}}\frac{\Gamma_{p,q}(\alpha)\Gamma_{p,q}(\beta)\Gamma_{p,q}(\theta)}{\Gamma_{p,q}(\alpha+\beta+\theta+1)}\, t^{\alpha+\beta+\theta}.$$

The proof is complete. □

The following lemma, dealing with a linear variant of problem (1), plays an important role in the forthcoming analysis.

Lemma 6. *Let $\Omega \neq 0$, $\alpha,\beta,\theta \in (0,1]$, $0 < q < p \le 1$, $h \in C(I_{p,q}^T, \mathbb{R})$ and $g \in C(I_{p,q}^T, \mathbb{R}^+)$ be given functions, $\varphi : C(I_{p,q}^T, \mathbb{R}) \to \mathbb{R}$ be given functional. Then, the problem*

$$D^{\alpha}_{p,q}D^{\beta}_{p,q}u(t) = h(t), \quad t \in I^{T}_{p,q}, \tag{2}$$

$$u(0) = u\left(\frac{T}{p}\right) \tag{3}$$

$$\mathcal{I}^{\theta}_{p,q}g(\eta)u(\eta) = \varphi(u), \quad \eta \in I^{T}_{p,q} - \left\{0, \frac{T}{p}\right\} \tag{4}$$

has the unique solution:

$$\begin{aligned} u(t) = & \frac{1}{p^{\binom{\alpha}{2}+\binom{\beta}{2}}\Gamma_{p,q}(\alpha)\Gamma_{p,q}(\beta)} \int_0^t \int_0^{\frac{x}{p^{\beta-1}}} (t-qx)^{\underline{\beta-1}}_{p,q}\left(\frac{x}{p^{\beta-1}} - qs\right)^{\underline{\alpha-1}}_{p,q} h\left(\frac{s}{p^{\alpha-1}}\right) d_{p,q}s\, d_{p,q}x \\ & - \frac{t^{\beta-1}}{\Omega}\left\{\boldsymbol{B}_\eta \mathbb{P}[h] + \boldsymbol{A}_T(\varphi(u) - \mathbb{Q}[h])\right\} \\ & + \frac{t^{\alpha+\beta-1}}{\Omega\Gamma_{p,q}(\alpha+\beta)}\left\{\boldsymbol{A}_\eta \mathbb{P}[h] + \left(\frac{T}{p}\right)^{\beta-1}(\varphi(u) - \mathbb{Q}[h])\right\} \end{aligned} \tag{5}$$

where the functionals $\mathbb{P}[h]$ *and* $\mathbb{Q}[h]$ *are defined by*

$$\begin{aligned} \mathbb{P}[h] := & \frac{1}{p^{\binom{\alpha}{2}+\binom{\beta}{2}}\Gamma_{p,q}(\alpha)\Gamma_{p,q}(\beta)} \int_0^{\frac{T}{p}} \int_0^{\frac{x}{p^{\beta-1}}} \left(\frac{T}{p} - qx\right)^{\underline{\beta-1}}_{p,q}\left(\frac{x}{p^{\beta-1}} - qs\right)^{\underline{\alpha-1}}_{p,q} \times \\ & h\left(\frac{s}{p^{\alpha-1}}\right) d_{p,q}s\, d_{p,q}x \end{aligned} \tag{6}$$

$$\begin{aligned} \mathbb{Q}[h] := & \frac{1}{p^{\binom{\alpha}{2}+\binom{\beta}{2}+\binom{\theta}{2}}\Gamma_{p,q}(\alpha)\Gamma_{p,q}(\beta)\Gamma_{p,q}(\theta)} \int_0^{\eta} \int_0^{\frac{y}{p^{\theta-1}}} \int_0^{\frac{x}{p^{\beta-1}}} (\eta - qy)^{\underline{\theta-1}}_{p,q} \times \\ & \left(\frac{y}{p^{\theta-1}} - qx\right)^{\underline{\beta-1}}_{p,q}\left(\frac{x}{p^{\beta-1}} - qs\right)^{\underline{\alpha-1}}_{p,q} g\left(\frac{y}{p^{\theta-1}}\right) h\left(\frac{s}{p^{\alpha-1}}\right) d_{p,q}s\, d_{p,q}x\, d_{p,q}y \end{aligned} \tag{7}$$

and the constants $\boldsymbol{A}_T, \boldsymbol{A}_\eta, \boldsymbol{B}_\eta$ *and* Ω *are defined by*

$$\boldsymbol{A}_T := \frac{1}{p^{\binom{\beta}{2}}\Gamma_{p,q}(\beta)} \int_0^{\frac{T}{p}} \left(\frac{T}{p} - qs\right)^{\underline{\beta-1}}_{p,q}\left(\frac{s}{p^{\theta-1}}\right)^{\underline{\alpha-1}} d_{p,q}s = \frac{\left(\frac{T}{p}\right)^{\alpha+\beta-1}}{\Gamma_{p,q}(\alpha+\beta)} \tag{8}$$

$$\boldsymbol{A}_\eta := \frac{1}{p^{\binom{\theta}{2}}\Gamma_{p,q}(\theta)} \int_0^{\eta} (\eta - qs)^{\underline{\theta-1}}_{p,q} g\left(\frac{s}{p^{\theta-1}}\right)\left(\frac{s}{p^{\theta-1}}\right)^{\underline{\beta-1}} d_{p,q}s \tag{9}$$

$$\begin{aligned} \boldsymbol{B}_\eta := & \frac{1}{p^{\binom{\beta}{2}+\binom{\theta}{2}}\Gamma_{p,q}(\beta)\Gamma_{p,q}(\theta)} \int_0^{\eta} \int_0^{\frac{x}{p^{\theta-1}}} (\eta - qx)^{\underline{\theta-1}}_{p,q}\left(\frac{x}{p^{\theta-1}} - qs\right)^{\underline{\beta-1}}_{p,q} g\left(\frac{x}{p^{\theta-1}}\right) \times \\ & \left(\frac{s}{p^{\alpha-1}}\right)^{\underline{\alpha-1}} d_{p,q}s\, d_{p,q}x \end{aligned} \tag{10}$$

$$\Omega := \left(\frac{T}{p}\right)^{\beta-1}\boldsymbol{B}_\eta - \boldsymbol{A}_T\boldsymbol{A}_\eta. \tag{11}$$

Proof. Taking fractional (p,q)-integral of order α for (2) and using Lemma 1, we then have

$$\begin{aligned} D^{\beta}_{p,q}u(t) = & C_1 t^{\alpha-1} + \mathcal{I}^{\alpha}_{p,q}h(t) \\ = & C_1 t^{\alpha-1} + \frac{1}{p^{\binom{\alpha}{2}}\Gamma_{p,q}(\alpha)} \int_0^t (t-qs)^{\underline{\alpha-1}}_{p,q} h\left(\frac{s}{p^{\alpha-1}}\right) d_{p,q}s. \end{aligned} \tag{12}$$

Next, taking fractional (p,q)-difference of order β for (12), we have

$$u(t) = C_0 t^{\beta-1} + C_1 \frac{t^{\alpha+\beta-1}}{\Gamma_{p,q}(\alpha+\beta)} + \frac{1}{p^{\binom{\alpha}{2}+\binom{\beta}{2}}\Gamma_{p,q}(\alpha)\Gamma_{p,q}(\beta)} \times$$
$$\int_0^t \int_0^{\frac{x}{p^{\beta-1}}} (t-qx)_{p,q}^{\underline{\beta-1}} \left(\frac{x}{p^{\beta-1}} - qs\right)_{p,q}^{\underline{\alpha-1}} h\left(\frac{s}{p^{\alpha-1}}\right) d_{p,q}s\, d_{p,q}x. \tag{13}$$

Substituting $t = 0, \frac{T}{p}$ into (13) and employing the condition (3), we have

$$C_0\left(\frac{T}{p}\right)^{\beta-1} + C_1 \mathbf{A}_T = -\mathbb{P}[h]. \tag{14}$$

By taking fractional (p, q)-integral of order θ for (13), we have

$$\mathcal{I}^{\theta}_{p,q}u(t) = C_0 \frac{t^{\beta+\theta-1}}{\Gamma_{p,q}(\beta+\theta)} + \frac{C_1}{p^{\binom{\beta}{2}+\binom{\theta}{2}}\Gamma_{p,q}(\beta)\Gamma_{p,q}(\theta)} \times$$
$$\int_0^t \int_0^{\frac{x}{p^{\theta-1}}} (t-qx)_{p,q}^{\underline{\theta-1}} \left(\frac{x}{p^{\theta-1}} - qs\right)_{p,q}^{\underline{\beta-1}} \left(\frac{s}{p^{\alpha-1}}\right)^{\underline{\alpha-1}} d_{p,q}s\, d_{p,q}x$$
$$+ \frac{1}{p^{\binom{\alpha}{2}+\binom{\beta}{2}+\binom{\theta}{2}}\Gamma_{p,q}(\alpha)\Gamma_{p,q}(\beta)\Gamma_{p,q}(\theta)} \int_0^t \int_0^{\frac{y}{p^{\theta-1}}} \int_0^{\frac{x}{p^{\beta-1}}} (t-qy)_{p,q}^{\underline{\theta-1}} \times$$
$$\left(\frac{y}{p^{\theta-1}} - qx\right)_{p,q}^{\underline{\beta-1}} \left(\frac{x}{p^{\beta-1}} - qs\right)_{p,q}^{\underline{\alpha-1}} h\left(\frac{s}{p^{\alpha-1}}\right) d_{p,q}s\, d_{p,q}x\, d_{p,q}y. \tag{15}$$

From the condition (4) we have

$$C_0 \mathbf{A}_\eta + C_1 \mathbf{B}_\eta = \varphi(u) - \mathbb{Q}[h] \tag{16}$$

Solving the system of linear Equations (14) and (16),we obtain

$$C_0 = \frac{-\mathbf{B}_\eta \mathbb{P}[h] - \mathbf{A}_T(\varphi(u) - \mathbb{Q}[h])}{\Omega} \quad \text{and} \quad C_1 = \frac{\left(\frac{T}{p}\right)^{\beta-1}(\varphi(u) - \mathbb{Q}[h]) + \mathbf{A}_\eta \mathbb{P}[h]}{\Omega},$$

where $\mathbb{P}[h], \mathbb{Q}[h], \mathbf{A}_T, \mathbf{A}_\eta, \mathbf{B}_\eta$ and Ω are defined by (6)–(11), respectively.

After substituting C_0, C_1 into (13), we obtain (5). We can prove the converse by direct computation. The proof is complete. □

3. Existence and Uniqueness Result

In this section, we prove the existence and uniqueness result for problem (1) by using Banach fixed point theorem as follows.

Lemma 7 ([43] **Banach fixed point theorem).** *Let a nonempty closed subset C of a Banach space X, then there is a unique fixed point for any contraction mapping P of C into itself.*

Let $\mathcal{C} = C\left(I^T_{p,q}, \mathbb{R}\right)$ be a Banach space of all function u with the norm defined by

$$\|u\|_{\mathcal{C}} = \max\left\{\|u\|, \left\|D^{\nu}_{p,q}u\right\|\right\},$$

where $\|u\| = \max_{t\in I^T_{p,q}}\left\{|u(t)|\right\}$ and $\|D^{\nu}_{p,q}u\|_{\mathcal{C}} = \max_{t\in I^T_{p,q}}\left\{\left|D^{\nu}_{p,q}u(t)\right|\right\}$.

By Lemma 6, replacing $h(t)$ by $F\left[t, u(t), \Psi^{\gamma}_{p,q}u(t), D^{\nu}_{p,q}u(t)\right]$, we define an operator $\mathcal{A} : \mathcal{C} \to \mathcal{C}$ by

$$
\begin{aligned}
(\mathcal{A}u)(t) :=\; & \frac{\varphi(u)}{\Omega}\left[\frac{(\frac{T}{p})^{\beta-1}}{\Gamma_{p,q}(\alpha+\beta)}t^{\alpha+\beta-1} - \mathbf{A}_T t^{\beta-1}\right] \\
& + \frac{\mathbb{Q}^*[F_u]}{\Omega}\left[\mathbf{A}_T t^{\beta-1} - \frac{(\frac{T}{p})^{\beta-1}}{\Gamma_{p,q}(\alpha+\beta)}t^{\alpha+\beta-1}\right] \\
& + \frac{\mathbb{P}^*[F_u]}{\Omega}\left[\mathbf{A}_\eta \frac{t^{\alpha+\beta-1}}{\Gamma_{p,q}(\alpha+\beta)} - \mathbf{B}_\eta t^{\beta-1}\right] \\
& + \frac{1}{p^{\binom{\alpha}{2}+\binom{\beta}{2}}\Gamma_{p,q}(\alpha)\Gamma_{p,q}(\beta)}\int_0^t\int_0^{\frac{x}{p^{\beta-1}}}(t-qx)_{p,q}^{\underline{\beta-1}}\left(\frac{x}{p^{\beta-1}}-qs\right)_{p,q}^{\underline{\alpha-1}}\times \\
& F\left[\frac{s}{p^{\alpha-1}}, u\left(\frac{s}{p^{\alpha-1}}\right), \Psi_{p,q}^{\gamma}u\left(\frac{s}{p^{\alpha-1}}\right), D_{p,q}^{\nu}u\left(\frac{s}{p^{\alpha-1}}\right)\right]d_{p,q}s\, d_{p,q}x
\end{aligned} \tag{17}
$$

where the functionals $\mathbb{P}^*[F_u]$ and $\mathbb{Q}^*[F_u]$ are defined by

$$
\begin{aligned}
\mathbb{P}^*[F_u] :=\; & \frac{1}{p^{\binom{\alpha}{2}+\binom{\beta}{2}}\Gamma_{p,q}(\alpha)\Gamma_{p,q}(\beta)}\int_0^{\frac{T}{p}}\int_0^{\frac{x}{p^{\beta-1}}}\left(\frac{T}{p}-qx\right)_{p,q}^{\underline{\beta-1}}\left(\frac{x}{p^{\beta-1}}-qs\right)_{p,q}^{\underline{\alpha-1}}\times \\
& F\left[\frac{s}{p^{\alpha-1}}, u\left(\frac{s}{p^{\alpha-1}}\right), \Psi_{p,q}^{\gamma}u\left(\frac{s}{p^{\alpha-1}}\right), D_{p,q}^{\nu}u\left(\frac{s}{p^{\alpha-1}}\right)\right]d_{p,q}s\, d_{p,q}x
\end{aligned} \tag{18}
$$

$$
\begin{aligned}
\mathbb{Q}^*[F_u] :=\; & \frac{1}{p^{\binom{\alpha}{2}+\binom{\beta}{2}+\binom{\theta}{2}}\Gamma_{p,q}(\alpha)\Gamma_{p,q}(\beta)\Gamma_{p,q}(\theta)}\times \\
& \int_0^{\eta}\int_0^{\frac{y}{p^{\theta-1}}}\int_0^{\frac{x}{p^{\beta-1}}}(\eta-qy)_{p,q}^{\underline{\theta-1}}\left(\frac{y}{p^{\theta-1}}-qx\right)_{p,q}^{\underline{\beta-1}}\left(\frac{x}{p^{\beta-1}}-qs\right)_{p,q}^{\underline{\alpha-1}} g\left(\frac{y}{p^{\theta-1}}\right)\times \\
& F\left[\frac{s}{p^{\alpha-1}}, u\left(\frac{s}{p^{\alpha-1}}\right), \Psi_{p,q}^{\gamma}u\left(\frac{s}{p^{\alpha-1}}\right), D_{p,q}^{\nu}u\left(\frac{s}{p^{\alpha-1}}\right)\right]d_{p,q}s\, d_{p,q}x\, d_{p,q}y
\end{aligned} \tag{19}
$$

and the constants $\mathbf{A}_T, \mathbf{A}_\eta, \mathbf{B}_\eta$ and Ω are defined by (8)–(11), respectively.

We see that the problem (1) has solution if and only if the operator $\mathcal{A}$ has fixed point.

Theorem 1. *Assume that $F : I_{p,q}^T \times \mathbb{R} \times \mathbb{R} \times \mathbb{R} \to \mathbb{R}$ is continuous, $\phi : I_{p,q}^T \times I_{p,q}^T \to [0,\infty)$ is continuous with $\phi_0 = \max\left\{\phi(t,s) : (t,s) \in I_{p,q}^T \times I_{p,q}^T\right\}$, and $\varphi : C(I_{p,q}^T, \mathbb{R}) \to \mathbb{R}$ is given functional. Suppose that the following conditions hold:*

(H_1) *There exist positive constants L_1, L_2, L_3 such that for each $t \in I_{p,q}^T$ and $u_i, v_i \in \mathbb{R}$, $i = 1,2,3$,*

$$\Big|F[t,u_1,u_2,u_3] - F[t,v_1,v_2,v_3]\Big| \le L_1|u_1-v_1| + L_2|u_2-v_2| + L_3|u_3-v_3|.$$

(H_2) *There exists a positive constant ω such that for each $u, v \in \mathcal{C}$,*

$$|\varphi(u) - \varphi(v)| \le \omega\|u-v\|_{\mathcal{C}}\,.$$

(H_3) *For each $t \in I_{p,q}^T$, $0 < g < g(t) < G$.*

(H_4) $\mathcal{X} := \omega\mathbf{O}_T + (\mathcal{L}+L_3)\Theta \le 1$,

where

$$\mathcal{L} := L_1 + L_2 \frac{\phi_0 (\frac{T}{p})^{\gamma}}{\Gamma_{p,q}(\gamma+1)}, \tag{20}$$

$$\Theta := \frac{O_T G\eta^{\alpha+\beta+\theta}}{\Gamma_{p,q}(\alpha+\beta+\theta-1)} + \frac{(\frac{T}{p})^{\alpha+\beta}}{\Gamma_{p,q}(\alpha+\beta+1)}(O_\eta+1), \tag{21}$$

$$O_T := \left[\frac{\left(\frac{T}{p}\right)^{\alpha+\beta-1}}{\Gamma_{p,q}(\alpha+\beta)} + A_T \right] \frac{\left(\frac{T}{p}\right)^{\beta-1}}{\min|\Omega|}, \tag{22}$$

$$O_\eta := \left[\frac{\left(\frac{T}{p}\right)^{\alpha}}{\Gamma_{p,q}(\alpha+\beta)} \max A_\eta + \max B_\eta \right] \frac{\left(\frac{T}{p}\right)^{\beta-1}}{\min|\Omega|}. \tag{23}$$

Then, problem (1) has a unique solution in $I^T_{p,q}$.

Proof. For each $t \in I^T_{p,q}$ and $u, v \in \mathcal{C}$,

$$\begin{aligned}
\left|\Psi^{\gamma}_{p,q}u(t) - \Psi^{\gamma}_{p,q}v(t)\right| &\le \frac{\phi_0}{p^{\binom{\gamma}{2}}\Gamma_{p,q}(\gamma)} \int_0^t (t-qs)^{\underline{\gamma-1}}_{p,q} \left| u\left(\frac{s}{p^{\gamma-1}}\right) - v\left(\frac{s}{p^{\gamma-1}}\right)\right| d_{p,q}s \\
&\le \frac{\phi_0}{p^{\binom{\gamma}{2}}\Gamma_{p,q}(\gamma)} \|u-v\| \int_0^{\frac{T}{p}} \left(\frac{T}{p} - qs\right)^{\underline{\gamma-1}}_{p,q} d_{p,q}s \\
&= \frac{\phi_0 \left(\frac{T}{p}\right)^{\gamma}}{\Gamma_{p,q}(\gamma+1)} \|u-v\|.
\end{aligned}$$

Denote that

$$\mathcal{F}|u-v|(t) := \left| F\left[t, u(t), \Psi^{\gamma}_{p,q}u(t), D^{\nu}_{p,q}u(t)\right] - F\left[t, v(t), \Psi^{\gamma}_{p,q}v(t), D^{\nu}_{p,q}v(t)\right]\right|.$$

By using Lemma 5(a), we obtain

$$\begin{aligned}
&\left|\mathbb{P}^*[F_u] - \mathbb{P}^*[F_v]\right| \\
&\le \frac{1}{p^{\binom{\alpha}{2}+\binom{\beta}{2}}\Gamma_{p,q}(\alpha)\Gamma_{p,q}(\beta)} \int_0^{\frac{T}{p}} \int_0^{\frac{x}{p^{\beta-1}}} \left(\frac{T}{p} - qx\right)^{\underline{\beta-1}}_{p,q} \left(\frac{x}{p^{\beta-1}} - qs\right)^{\underline{\alpha-1}}_{p,q} \times \\
&\quad \mathcal{F}|u-v|\left(\frac{s}{p^{\alpha-1}}\right) d_{p,q}s\, d_{p,q}x \\
&\le \frac{\left[L_1|u-v| + L_2\left|\Psi^{\gamma}_{p,q}u - \Psi^{\gamma}_{p,q}v\right| + L_3\left|D^{\nu}_{p,q}u - D^{\nu}_{p,q}v\right|\right]}{p^{\binom{\alpha}{2}+\binom{\beta}{2}}\Gamma_{p,q}(\alpha)\Gamma_{p,q}(\beta)} \int_0^{\frac{T}{p}} \int_0^{\frac{x}{p^{\beta-1}}} \left(\frac{T}{p} - qx\right)^{\underline{\beta-1}}_{p,q} \times \\
&\quad \left(\frac{x}{p^{\beta-1}} - qs\right)^{\underline{\alpha-1}}_{p,q} \left(\frac{s}{p^{\alpha-1}}\right) d_{p,q}s\, d_{p,q}x \\
&\le \left(\left[L_1 + L_2\phi_0 \frac{\left(\frac{T}{p}\right)^{\gamma}}{\Gamma_{p,q}(\gamma+1)} \right] |u-v| + L_3\left|D^{\nu}_{p,q}u - D^{\nu}_{p,q}v\right| \right) \frac{\left(\frac{T}{p}\right)^{\alpha+\beta}}{\Gamma_{p,q}(\alpha+\beta+1)} \\
&\le \frac{(\mathcal{L}+L_3)\left(\frac{T}{p}\right)^{\alpha+\beta}}{\Gamma_{p,q}(\alpha+\beta+1)} \|u-v\|_{\mathcal{C}},
\end{aligned} \tag{24}$$

and by using Lemma 5(b), we have

$$
\begin{aligned}
&\left|\mathbb{Q}^*[F_u] - \mathbb{Q}^*[F_v]\right| \\
&\leq \frac{G}{p^{\binom{\alpha}{2}+\binom{\beta}{2}+\binom{\theta}{2}}\Gamma_{p,q}(\alpha)\Gamma_{p,q}(\beta)\Gamma_{p,q}(\theta)} \int_0^\eta \int_0^{\frac{y}{p^{\theta-1}}} \int_0^{\frac{x}{p^{\beta-1}}} (\eta - qy)_{p,q}^{\underline{\theta-1}} \left(\frac{y}{p^{\theta-1}} - qx\right)_{p,q}^{\underline{\beta-1}} \times \\
&\quad \left(\frac{x}{p^{\beta-1}} - qs\right)_{p,q}^{\underline{\alpha-1}} \mathcal{F}|u-v|\left(\frac{s}{p^{\alpha-1}}\right) d_{p,q}s\, d_{p,q}x\, d_{p,q}y \\
&\leq \frac{G\left[L_1|u-v| + L_2\left|\Psi_{p,q}^\gamma u - \Psi_{p,q}^\gamma v\right| + L_3\left|D_{p,q}^\nu u - D_{p,q}^\nu v\right|\right]}{p^{\binom{\alpha}{2}+\binom{\beta}{2}+\binom{\theta}{2}}\Gamma_{p,q}(\alpha)\Gamma_{p,q}(\beta)\Gamma_{p,q}(\theta)} \times \\
&\quad \int_0^\eta \int_0^{\frac{y}{p^{\theta-1}}} \int_0^{\frac{x}{p^{\beta-1}}} (\eta - qy)_{p,q}^{\underline{\theta-1}} \left(\frac{y}{p^{\theta-1}} - qx\right)_{p,q}^{\underline{\beta-1}} \left(\frac{x}{p^{\beta-1}} - qs\right)_{p,q}^{\underline{\alpha-1}} d_{p,q}s\, d_{p,q}x\, d_{p,q}y \\
&\leq \left[\left(L_1 + L_2 \frac{\phi_0\left(\frac{T}{p}\right)^\gamma}{\Gamma_{p,q}(\gamma+1)}\right)|u-v| + L_3\left|D_{p,q}^\nu u - D_{p,q}^\nu v\right|\right] \frac{G\eta^{\alpha+\beta+\theta}}{\Gamma_{p,q}(\alpha+\beta+\theta+1)} \\
&\leq \frac{G(\mathcal{L}+L_3)\eta^{\alpha+\beta+\theta}}{\Gamma_{p,q}(\alpha+\beta+\theta+1)} \|u-v\|_{\mathcal{C}}.
\end{aligned} \tag{25}
$$

Then,

$$
\begin{aligned}
&|(\mathcal{A}u)(t) - (\mathcal{A}v)(t)| \\
&\quad\leq \frac{\omega\|u-v\|_{\mathcal{C}}\left(\frac{T}{p}\right)^{\beta-1}}{|\Omega|}\left[\frac{\left(\frac{T}{p}\right)^{\alpha+\beta-1}}{\Gamma_{p,q}(\alpha+\beta)} + \mathbf{A}_T\right] \\
&\qquad + (\mathcal{L}+L_3)\|u-v\|_{\mathcal{C}} \frac{G\,\eta^{\alpha+\beta+\theta}}{|\Omega|\Gamma_{p,q}(\alpha+\beta+\theta-1)} \cdot \left(\frac{T}{p}\right)^{\beta-1}\left[\frac{\left(\frac{T}{p}\right)^{\alpha+\beta-1}}{\Gamma_{p,q}(\alpha+\beta)} + \mathbf{A}_T\right] \\
&\qquad + (\mathcal{L}+L_3)\|u-v\|_{\mathcal{C}} \frac{\left(\frac{T}{p}\right)^{\alpha+\beta}}{|\Omega|\Gamma_{p,q}(\alpha+\beta+1)} \cdot \left(\frac{T}{p}\right)^{\beta-1}\left[\frac{\left(\frac{T}{p}\right)^{\alpha}}{\Gamma_{p,q}(\alpha+\beta)}\mathbf{A}_\eta + \mathbf{B}_\eta\right] \\
&\qquad + \frac{(\mathcal{L}+L_3)\|u-v\|_{\mathcal{C}}}{p^{\binom{\alpha}{2}+\binom{\beta}{2}}\Gamma_{p,q}(\alpha)\Gamma_{p,q}(\beta)} \int_0^{\frac{T}{p}} \int_0^{\frac{x}{p^{\beta-1}}} \left(\frac{T}{p} - qx\right)_{p,q}^{\underline{\beta-1}} \left(\frac{x}{p^{\beta-1}} - qs\right)_{p,q}^{\underline{\alpha-1}} d_{p,q}s\, d_{p,q}x \\
&\quad\leq \left\{\mathbf{O}_T\left[\omega + (\mathcal{L}+L_3)\frac{G\,\eta^{\alpha+\beta+\theta}}{\Gamma_{p,q}(\alpha+\beta+\theta-1)}\right]\right. \\
&\qquad \left. + \mathbf{O}_\eta\left[(\mathcal{L}+L_3)\frac{\left(\frac{T}{p}\right)^{\alpha+\beta}}{\Gamma_{p,q}(\alpha+\beta+1)}\right] + (\mathcal{L}+L_3)\frac{\left(\frac{T}{p}\right)^{\alpha+\beta}}{\Gamma_{p,q}(\alpha+\beta+1)}\right\}\|u-v\|_{\mathcal{C}} \\
&\quad= \mathcal{X}\|u-v\|_{\mathcal{C}}.
\end{aligned} \tag{26}
$$

Taking fractional (p, q)-difference of order ν for (17), we get

$$
\begin{aligned}
&(D^{\nu}_{p,q}\mathcal{A}u)(t)\\
&= \frac{\varphi(u)}{\Omega}\left[\frac{(\frac{T}{p})^{\beta-1}\Gamma_{p,q}(\alpha+\beta)}{\Gamma_{p,q}(\alpha+\beta)\Gamma_{p,q}(\alpha+\beta-\nu)}t^{\alpha+\beta-\nu-1} - \mathbf{A}_T\frac{\Gamma_{p,q}(\beta)}{\Gamma_{p,q}(\beta-\nu)}t^{\beta-\nu-1}\right]\\
&\quad + \frac{\mathbb{Q}^*[F_u]}{\Omega}\left[\mathbf{A}_T\frac{\Gamma_{p,q}(\beta)}{\Gamma_{p,q}(\beta-\nu)}t^{\beta-\nu-1} - \frac{(\frac{T}{p})^{\beta-1}\Gamma_{p,q}(\alpha+\beta)}{\Gamma_{p,q}(\alpha+\beta)\Gamma_{p,q}(\alpha+\beta-\nu)}t^{\alpha+\beta-\nu-1}\right]\\
&\quad + \frac{\mathbb{P}^*[F_u]}{\Omega}\left[\mathbf{A}_\eta\frac{\Gamma_{p,q}(\alpha+\beta)}{\Gamma_{p,q}(\alpha+\beta-\nu)\Gamma_{p,q}(\alpha+\beta)}t^{\alpha+\beta-\nu-1} - \mathbf{B}_\eta\frac{\Gamma_{p,q}(\beta)}{\Gamma_{p,q}(\beta-\nu)}t^{\beta-\nu-1}\right]\\
&\quad + \frac{1}{p^{\binom{\alpha}{2}+\binom{\beta}{2}+\binom{-\nu}{2}}\Gamma_{p,q}(\alpha)\Gamma_{p,q}(\beta)\Gamma_{p,q}(-\nu)}\times\\
&\quad \int_0^t\int_0^{\frac{y}{p^{-\nu-1}}}\int_0^{\frac{x}{p^{\beta-1}}}(t-qy)_{p,q}^{\underline{-\nu-1}}\left(\frac{y}{p^{-\nu-1}}-qx\right)_{p,q}^{\underline{\beta-1}}\left(\frac{x}{p^{\beta-1}}-qs\right)_{p,q}^{\underline{\alpha-1}}\times\\
&\quad F\left[\frac{s}{p^{\alpha-1}},u\left(\frac{s}{p^{\alpha-1}}\right),\Psi^{\gamma}_{p,q}u\left(\frac{s}{p^{\alpha-1}}\right),D^{\nu}_{p,q}u\left(\frac{s}{p^{\alpha-1}}\right)\right]d_{p,q}s\,d_{p,q}x\,d_{p,q}y. \qquad (27)
\end{aligned}
$$

Thus,

$$
\begin{aligned}
&\left|(D^{\nu}_{p,q}\mathcal{A}u)(t)-(D^{\nu}_{p,q}\mathcal{A}v)(t)\right|\\
&\quad\le \omega\|u-v\|_{\mathcal{C}}\left(\frac{T}{p}\right)^{-\nu}\frac{\Gamma_{p,q}(\alpha+\beta)}{\Gamma_{p,q}(\alpha+\beta-\nu)}\mathbf{O}_T\\
&\quad\le (\mathcal{L}+L_3)\|u-v\|_{\mathcal{C}}\frac{G\,\eta^{\alpha+\beta+\theta}}{\Gamma_{p,q}(\alpha+\beta+\theta-1)}\cdot\left(\frac{T}{p}\right)^{-\nu}\frac{\Gamma_{p,q}(\alpha+\beta)}{\Gamma_{p,q}(\alpha+\beta-\nu)}\mathbf{O}_T\\
&\qquad + (\mathcal{L}+L_3)\|u-v\|_{\mathcal{C}}\frac{\left(\frac{T}{p}\right)^{\alpha+\beta}}{\Gamma_{p,q}(\alpha+\beta+1)}\cdot\left(\frac{T}{p}\right)^{-\nu}\frac{\Gamma_{p,q}(\alpha+\beta)}{\Gamma_{p,q}(\alpha+\beta-\nu)}\mathbf{O}_\eta\\
&\qquad + (\mathcal{L}+L_3)\|u-v\|_{\mathcal{C}}\frac{(\frac{T}{p})^{\alpha+\beta-\nu}}{\Gamma_{p,q}(\alpha+\beta-\nu+1)}\\
&\quad\le \left\{\mathbf{O}_T\left[\omega+(\mathcal{L}+L_3)\frac{G\,\eta^{\alpha+\beta+\theta}}{\Gamma_{p,q}(\alpha+\beta+\theta-1)}\right]\frac{\left(\frac{T}{p}\right)^{-\nu}\Gamma_{p,q}(\alpha+\beta)}{\Gamma_{p,q}(\alpha+\beta-\nu)}\right.\\
&\qquad + \mathbf{O}_\eta\left[(\mathcal{L}+L_3)\frac{\left(\frac{T}{p}\right)^{\alpha+\beta}}{\Gamma_{p,q}(\alpha+\beta+1)}\right]\frac{\left(\frac{T}{p}\right)^{-\nu}\Gamma_{p,q}(\alpha+\beta)}{\Gamma_{p,q}(\alpha+\beta-\nu)}\\
&\qquad \left.+ (\mathcal{L}+L_3)\frac{\left(\frac{T}{p}\right)^{\alpha+\beta-\nu}}{\Gamma_{p,q}(\alpha+\beta-\nu+1)}\right\}\|u-v\|_{\mathcal{C}}\\
&\quad< \mathcal{X}\|u-v\|_{\mathcal{C}}. \qquad (28)
\end{aligned}
$$

From (26) and (28), we have

$$\|\mathcal{A}u-\mathcal{A}v\|_{\mathcal{C}}\le\mathcal{X}\|u-v\|_{\mathcal{C}}.$$

By (H_4), we can conclude that $\mathcal{A}$ is a contraction. Thus, by using Banach fixed point theorem in lemma 7, $\mathcal{A}$ has a fixed point, which is a unique solution of problem (1) on $I^T_{p,q}$. □

4. Existence of at Least One Solution

In this section, we prove the existence of at least one solution to (1). The following lemmas reviewing the Schauder's fixed point theorem are also provided.

Lemma 8 ([43] Arzelá-Ascoli theorem). *A collection of functions in $C[a,b]$ with the sup norm, is relatively compact if and only if it is uniformly bounded and equicontinuous on $[a,b]$.*

Lemma 9 ([43]). *If a set is closed and relatively compact, then it is compact.*

Lemma 10 ([44] Schauder's fixed point theorem). *Let (D,d) be a complete metric space, U be a closed convex subset of D, and $T : D \to D$ be the map such that the set $Tu : u \in U$ is relatively compact in D. Then, the operator T has at least one fixed point $u^* \in U$: $Tu^* = u^*$.*

Theorem 2. *Assume that $F : I^T_{p,q} \times \mathbb{R} \times \mathbb{R} \times \mathbb{R} \to \mathbb{R}$ is continuous, and $\varphi : C(I^T_{p,q}, \mathbb{R}) \to \mathbb{R}$ is given functional. Suppose that the following conditions hold:*

(H_5) There exists a positive constant M such that for each $t \in I^T_{p,q}$ and $u_i \in \mathbb{R}$, $i = 1,2,3$,

$$\Big|F[t, u_1, u_2, u_3]\Big| \leq M.$$

(H_6) There exists a positive constant N such that for each $u \in \mathcal{C}$,

$$|\varphi(u)| \leq N.$$

Then, problem (1) has at least one solution on $I^T_{p,q}$.

Proof. To prove this theorem, we proceed as follows.

Step I. Verify $\mathcal{A}$ maps bounded sets into bounded sets in $B_R = \{u \in \mathcal{C} : \|u\|_{\mathcal{C}} \leq R\}$. Let us prove that for any $R > 0$, there exists a positive constant L such that for each $x \in B_R$, we have $\|\mathcal{A}u\|_{\mathcal{C}} \leq L$. By using Lemma 5, for each $t \in I^T_{p,q}$ and $u \in B_R$, we have

$$\begin{aligned}
&\Big|\mathbb{P}^*[F_u]\Big| \\
&\leq \frac{M}{p^{\binom{\alpha}{2}+\binom{\beta}{2}}\Gamma_{p,q}(\alpha)\Gamma_{p,q}(\beta)} \int_0^{\frac{T}{p}} \int_0^{\frac{x}{p^{\beta-1}}} \left(\frac{T}{p} - qx\right)_{p,q}^{\underline{\beta-1}} \left(\frac{x}{p^{\beta-1}} - qs\right)_{p,q}^{\underline{\alpha-1}} d_{p,q}s\, d_{p,q}x \\
&\leq \frac{M\left(\frac{T}{p}\right)^{\alpha+\beta}}{\Gamma_{p,q}(\alpha+\beta+1)},
\end{aligned} \tag{29}$$

$$\begin{aligned}
&\Big|\mathbb{Q}^*[F_u]\Big| \\
&\leq \frac{GM}{p^{\binom{\alpha}{2}+\binom{\beta}{2}+\binom{\theta}{2}}\Gamma_{p,q}(\alpha)\Gamma_{p,q}(\beta)\Gamma_{p,q}(\theta)} \int_0^{\eta} \int_0^{\frac{y}{p^{\theta-1}}} \int_0^{\frac{x}{p^{\beta-1}}} (\eta - qy)_{p,q}^{\underline{\theta-1}} \left(\frac{y}{p^{\theta-1}} - qx\right)_{p,q}^{\underline{\beta-1}} \times \\
&\left(\frac{x}{p^{\beta-1}} - qs\right)_{p,q}^{\underline{\alpha-1}} d_{p,q}s\, d_{p,q}x\, d_{p,q}y \\
&\leq \frac{GM\eta^{\alpha+\beta+\theta}}{\Gamma_{p,q}(\alpha+\beta+\theta+1)}.
\end{aligned} \tag{30}$$

From (29) and (30), we have

$$
\begin{aligned}
\left|(\mathcal{A}u)(t)\right| \leq\ & N\mathbf{O}_T + \frac{GM\eta^{\alpha+\beta+\theta}}{\Gamma_{p,q}(\alpha+\beta+\theta-1)}\mathbf{O}_T \\
& + \frac{M\left(\frac{T}{p}\right)^{\alpha+\beta}}{\Gamma_{p,q}(\alpha+\beta+1)}\mathbf{O}_\eta \\
& + \frac{M}{p^{\binom{\alpha}{2}+\binom{\beta}{2}}\Gamma_{p,q}(\alpha)\Gamma_{p,q}(\beta)}\int_0^{\frac{T}{p}}\int_0^{\frac{x}{p^{\beta-1}}}\left(\frac{T}{p}-qx\right)_{p,q}^{\underline{\beta-1}}\left(\frac{x}{p^{\beta-1}}-qs\right)_{p,q}^{\underline{\alpha-1}} d_{p,q}s\, d_{p,q}x \\
\leq\ & N\mathbf{O}_T + M\left[\frac{\mathbf{O}_T\, G\eta^{\alpha+\beta+\theta}}{\Gamma_{p,q}(\alpha+\beta+\theta-1)} + \frac{\left(\frac{T}{p}\right)^{\alpha+\beta}}{\Gamma_{p,q}(\alpha+\beta+1)}(\mathbf{O}_\eta+1)\right] \\
\leq\ & N\mathbf{O}_T + M\Theta := L.
\end{aligned} \tag{31}
$$

We find that

$$
\begin{aligned}
\left|\left(D_{p,q}^{\nu}\mathcal{A}u\right)(t)\right| \leq\ & N\mathbf{O}_T\left(\frac{T}{p}\right)^{-\nu}\frac{\Gamma_{p,q}(\alpha+\beta)}{\Gamma_{p,q}(\alpha+\beta-\nu)} \\
& + M\left[\frac{\mathbf{O}_T\, G\eta^{\alpha+\beta+\theta}}{\Gamma_{p,q}(\alpha+\beta+\theta-1)}\left(\frac{T}{p}\right)^{-\nu}\frac{\Gamma_{p,q}(\alpha+\beta)}{\Gamma_{p,q}(\alpha+\beta-\nu)}\right. \\
& \left. + \frac{\left(\frac{T}{p}\right)^{\alpha+\beta}}{\Gamma_{p,q}(\alpha+\beta+1)}(\mathbf{O}_\eta+1)\left(\frac{T}{p}\right)^{-\nu}\frac{\Gamma_{p,q}(\alpha+\beta)}{\Gamma_{p,q}(\alpha+\beta-\nu)}\right] \\
<\ & L.
\end{aligned} \tag{32}
$$

Thus, $\|(\mathcal{A}u)\|_{\mathcal{C}} \leq L$, which implies that $\mathcal{A}$ is uniformly bounded.

Step II. Since F is continuous, we can conclude that the operator $\mathcal{A}$ is continuous on B_R.

Step III. For any $t_1, t_2 \in I_{p,q}^T$ with $t_1 < t_2$, we find that

$$
\begin{aligned}
\left|(\mathcal{A}u)(t_1) - (\mathcal{A}u)(t_2)\right| \leq\ & \frac{\left|t_2^{\beta-1} - t_1^{\beta-1}\right|}{|\Omega|}\Big[\mathbf{A}_T(N+\mathbb{Q}^*[F_u]) + \mathbf{B}_\eta\mathbb{P}^*[F_u]\Big] \\
& + \frac{\left|t_2^{\alpha+\beta-1} - t_1^{\alpha+\beta-1}\right|}{|\Omega|\Gamma_{p,q}(\alpha+\beta)}\left[\left(\frac{T}{p}\right)^{\beta-1}(N+\mathbb{Q}^*[F_u]) + \mathbf{A}_\eta\mathbb{P}^*[F_u]\right] \\
& + \frac{M}{\Gamma_{p,q}(\alpha+\beta)}\left|t_2^{\alpha+\beta} - t_1^{\alpha+\beta}\right|,
\end{aligned} \tag{33}
$$

and

$$
\begin{aligned}
& \left|(D_{p,q}{}^{\nu}\mathcal{A}u)(t_2) - (D_{p,q}^{\nu}\mathcal{A}u)(t_1)\right| \\
& \leq \frac{\left|t_2^{\alpha+\beta} - t_1^{\alpha+\beta}\right|}{|\Omega|\Gamma_{p,q}(\beta-\nu)}\Big[\mathbf{A}_T(N+\mathbb{Q}^*[F_u]) + \mathbf{B}_\eta\mathbb{P}^*[F_u]\Big] + \frac{\left|t_2^{\alpha+\beta-\nu-1} - t_1^{\alpha+\beta-\nu-1}\right|}{|\Omega|\Gamma_{p,q}(\alpha+\beta-\nu)}\times \\
& \left[\left(\frac{T}{p}\right)^{\beta-1}(N+\mathbb{Q}^*[F_u]) + \mathbf{A}_\eta\mathbb{P}^*[F_u]\right] + \frac{M}{\Gamma_{p,q}(\alpha+\beta-\nu+1)}\left|t_2^{\alpha+\beta-\nu} - t_1^{\alpha+\beta-\nu}\right|.
\end{aligned} \tag{34}
$$

We see that the right-hand side of (33) and (34) tends to be zero when $|t_2 - t_1| \to 0$. Thus, $\mathcal{A}$ is relatively compact on B_R. This implies that $\mathcal{A}(B_R)$ is an equicontinuous set. By Arzelá-Ascoli theorem in Lemma 8, Lemma 9, and the above steps, we see that $\mathcal{A} : \mathcal{C} \to \mathcal{C}$

is completely continuous. Hence, we can conclude from Schauder fixed point theorem in Lemma 10 that problem (1) has at least one solution. □

5. Examples

In this section, to illustrate our results, we consider some examples.

Example 1. *Consider the following fractional (p,q)-integrodifference equation as*

$$D^{\frac{3}{4}}_{\frac{2}{3},\frac{1}{2}} D^{\frac{1}{2}}_{\frac{2}{3},\frac{1}{2}} u(t) = \frac{1}{(100e^2+t^3)(1+|u(t)|)} \left[e^{-3t}\left(u^2+2|u|\right) + e^{-(\pi+\sin^2\pi t)} \left| \Psi^{\frac{1}{3}}_{\frac{2}{3},\frac{1}{2}} u(t) \right| \right.$$
$$\left. + e^{-(2\pi+\cos^2\pi t)} \left| D^{\frac{1}{4}}_{\frac{2}{3},\frac{1}{2}} u(t) \right| \right], \; t \in I^{10}_{\frac{2}{3},\frac{1}{2}} = \left\{ \frac{10\left(\frac{1}{2}\right)^k}{\left(\frac{2}{3}\right)^{k+1}} : k \in \mathbb{N}_0 \right\} \cup \{0\} \tag{35}$$

with periodic fractional (p,q)-integral boundary condition

$$\begin{aligned} u(0) &= u(15) \\ \mathcal{I}^{\frac{2}{3}}_{\frac{2}{3},\frac{1}{2}} \left(2e + \sin\left(\frac{1215}{256}\right)\right)^2 u\left(\frac{1215}{256}\right) &= \sum_{i=0}^{\infty} \frac{C_i|u(t_i)|}{1+|u(t_i)|}, \; t_i = \sigma^i_{\frac{2}{3},\frac{1}{2}}(10), \end{aligned} \tag{36}$$

where C_i is given constants with $\frac{1}{1000} \le \sum_{i=0}^{\infty} C_i \le \frac{e}{1000}$ and $\phi(t,s) = \frac{e^{-|t-s|}}{(t+e)^3}$.

Letting $\alpha = \frac{3}{4}$, $\beta = \frac{1}{2}$, $\gamma = \frac{1}{3}$, $\nu = \frac{1}{4}$, $\theta = \frac{2}{3}$, $p = \frac{2}{3}$, $q = \frac{1}{2}$, $T = 10$, $\eta = \sigma^4_{\frac{2}{3},\frac{1}{2}}(10) = \frac{1215}{256}$, $g(t) = (20e + \sin t)^2$ and $F\left[t, u(t), \Psi^{\gamma}_{p,q}u(t), D^{\nu}_{p,q}u(t)\right] = \frac{1}{(100e^2+t^3)(1+|u(t)|)} \times \left[e^{-3t}(u^2+2|u|) + e^{-(\pi+\sin^2\pi t)} \left| \Psi^{\frac{1}{3}}_{\frac{2}{3},\frac{1}{2}} u(t) \right| + e^{-(2\pi+\cos^2\pi t)} \left| D^{\frac{1}{4}}_{\frac{2}{3},\frac{1}{2}} u(t) \right| \right]$.

Using above values, we find that

$$\phi_0 = 0.0498, \quad |\mathbf{A}_T| = 2.06344, \quad |\mathbf{A}_\eta| \le 264.588, \quad |\mathbf{B}_\eta| \le 196.777 \text{ and } |\Omega| \ge 283.525.$$

For all $t \in I^{10}_{\frac{2}{3},\frac{1}{2}}$ and $u, v \in \mathbb{R}$, we find that

$$\begin{aligned} &\left| F\left[t, u, \Psi^{\gamma}_{p,q}u, D^{\nu}_{p,q}u\right] - F\left[t, v, \Psi^{\gamma}_{p,q}v, D^{\nu}_{p,q}v\right] \right| \\ &\le \frac{1}{100e^2}|u-v| + \frac{1}{1000e^{2+\pi}}\left|\Psi^{\gamma}_{p,q}u - \Psi^{\gamma}_{p,q}v\right| + \frac{1}{100e^{2+2\pi}}\left|D^{\nu}_{p,q}u - D^{\nu}_{p,q}v\right|. \end{aligned}$$

Thus, (H_1) holds with $L_1 = 0.001353$, $L_2 = 5.848 \times 10^{-5}$ and $L_3 = 2.5273 \times 10^{-6}$. So $\mathcal{L} = 0.00136$.

For all $u, v \in \mathcal{C}$,

$$|\varphi(u) - \varphi(v)| \le \frac{e}{1000}\|u-v\|_{\mathcal{C}}.$$

Thus, (H_2) holds with $\omega = 0.002718$.

In addition, (H_3) holds with $g = 19.6831$, $G = 41.42935$.

Since

$$\mathbf{O}_T = 0.001885 \quad \mathbf{O}_\eta = 2.10484 \text{ and } \Theta = 89.5277,$$

therefore, (H_4) holds with

$$\mathcal{X} = 0.121989 < 1.$$

Hence, by Theorem 1 this problem has a unique solution.

Example 2. *Consider the following fractional (p,q)-integrodifference equation as*

$$D^{\frac{3}{4}}_{\frac{2}{3},\frac{1}{2}} D^{\frac{1}{2}}_{\frac{2}{3},\frac{1}{2}} u(t) = \frac{1}{10}\left(t+\frac{1}{3}\right) e^{-(t+\frac{1}{10})\left[u(t)+\left|\Psi^{\frac{1}{3}}_{\frac{2}{3},\frac{1}{2}} u(t)\right| + \left|D^{\frac{1}{4}}_{\frac{2}{3},\frac{1}{2}} u(t)\right|\right]}, \; t \in I^{10}_{\frac{2}{3},\frac{1}{2}} \tag{37}$$

with periodic fractional (p,q)-integral boundary condition

$$\begin{aligned} u(0) &= u(15) \\ \mathcal{I}^{\frac{2}{3}}_{\frac{2}{3},\frac{1}{2}} \left(2e + \sin\left(\frac{1215}{256}\right)\right)^2 u\left(\frac{1215}{256}\right) &= \sum_{i=0}^{\infty} C_i e^{-|u(t_i)|}, \;\; t_i = \sigma^i_{\frac{2}{3},\frac{1}{2}}(10), \end{aligned} \tag{38}$$

where D_i is given constants with $\frac{1}{500} \leq \sum_{i=0}^{\infty} D_i \leq \frac{e}{500}$.

Letting $\alpha = \frac{3}{4}$, $\beta = \frac{1}{2}$, $\gamma = \frac{1}{3}$, $\nu = \frac{1}{4}$, $\theta = \frac{2}{3}$, $p = \frac{2}{3}$, $q = \frac{1}{2}$, $T = 10$, $\eta = \frac{1215}{256}$. It is clear that $\left|F\left[t, u, \Psi^{\gamma}_{p,q} u, D^{\nu}_{p,q} u\right]\right| \leq \frac{23}{15} = M$ for $t \in I^{10}_{\frac{2}{3},\frac{1}{2}}$, and $|\varphi(u)| \leq \frac{e}{500} = N$ for $u \in \mathcal{C}$. Thus, we can conclude from Theorem 2 that our problem has at least one solution.

6. Conclusions

A fractional (p, q)-integrodifference equation with periodic fractional (p, q)-integral boundary condition (1) is studied. Our problem contains three fractional (p, q)-difference operators, and two fractional (p, q)-integral operators. We establish the conditions for the existence and uniqueness of solution for problem (1) by using the Banach fixed point theorem, and this result is shown in Theorem 1. We also established the conditions of at least one solution by using the Schauder's fixed point theorem, and this result is shown in Theorem 2. The choice to use of Theorems 1 or 2 depends on the conditions of the assumptions. The main results are illustrated by a numerical example. Some properties of fractional (p, q)-integral needed in our study are also discussed. The results of the paper are new and enrich the subject of boundary value problems for fractional (p, q)-difference equations. In the future work, we may extend this work by considering new boundary value problems.

Author Contributions: Conceptualization, J.S. and T.S.; methodology, J.S. and T.S.; formal analysis, J.S. and T.S.; funding acquisition, J.S and T.S. All authors have read and agreed to the published version of the manuscript.

Funding: This research was funded by King Mongkut's University of Technology North Bangkok. Contract no.KMUTNB-62-KNOW-22.

Institutional Review Board Statement: Not applicable.

Informed Consent Statement: Not applicable.

Data Availability Statement: Not applicable.

Acknowledgments: The authors would like to express their gratitude to anonymous referees for very helpful suggestions and comments which led to improvements of our original manuscript.

Conflicts of Interest: The authors declare no conflict of interest.

References

1. Jackson, F.H. On q-difference equations. *Am. J. Math.* **1910**, *32*, 305–314. [CrossRef]
2. Jackson, F.H. On q-difference integrals. *Q. J. Pure Appl. Math.* **1910**, *41*, 193–203.
3. Ernst, T. *A Comprehensive Treatment of q-Calculus*; Springer: Basel, Switzerland, 2012.
4. Kac, V.; Cheung, P. *Quantum Calculus*; Springer: New York, NY, USA, 2002.
5. Ernst, T. *A New Notation for q-Calculus and a New q-Taylor Formula*; U.U.D.M. Report 1999:25; Department of Mathematics, Uppsala University: Uppsala, Sweden, 1999; ISSN 1101-3591.
6. Finkelstein, R.J. Symmetry group of the hydrogen atom. *J. Math. Phys.* **1967**, *8*, 443–449. [CrossRef]
7. Finkelstein, R.J. q-gauge theory. *Int. J. Mod. Phys. A* **1996**, *11*, 733–746. [CrossRef]

8. Ilinski, K.N.; Kalinin, G.V.; Stepanenko, A.S. q-functional field theory for particles with exotic statistics. *Phys. Lett. A* **1997**, *232*, 399–408. [CrossRef]
9. Finkelstein, R.J. q-field theory. *Lett. Math. Phys.* **1995**, *34*, 169–176. [CrossRef]
10. Finkelstein, R.J. The q-Coulomb problem. *J. Math. Phys.* **1996**, *37*, 2628–2636. [CrossRef]
11. Cheng, H. Canonical Quantization of Yang-Mills theories. *Perspectives in Mathematical Physics*; International Press: Boston, MA, USA, 1996.
12. Finkelstein, R.J. q-gravity. *Lett. Math. Phys.* **1996**, *38*, 53–62. [CrossRef]
13. Negadi, T.; Kibler, M. A q-deformed Aufbau Prinzip. *J. Phys. A Math. Gen.* **1992**, *25*, 157–160. [CrossRef]
14. Siegel, W. *Introduction to String Field Theory, Volume 8 of Advanced Series in Mathematical Physics*; World Scientific Publishing: Teaneck, NJ, USA, 1988.
15. Agarwal, R.P. Certain fractional q-integrals and q-derivatives. *Proc. Camb. Phil. Soc.* **1969**, *66*, 365–370. [CrossRef]
16. Al-Salam, W.A. Some fractional q-integrals and q-derivatives. *Proc. Edin. Math. Soc.* **1966**, *15*, 135–140. [CrossRef]
17. Díaz, J.B.; Osler, T.J. Differences of fractional order. *Math. Comp.* **1974**, *28*, 185–202. [CrossRef]
18. Brikshavana, T.; Sitthiwirattham, T. On Fractional Hahn Calculus. *Adv. Differ. Equ.* **2017**, *354*, 1–15. [CrossRef]
19. Patanarapeelert, N.; Sitthiwirattham, T. On Fractional Symmetric Hahn Calculus. *Mathematics* **2019**, *7*, 873. [CrossRef]
20. Chakrabarti, R.; Jagannathan, R. A (p,q)-oscillator realization of two-parameter quantum algebras. *J. Math. Phys. Math. Gen.* **1991**, *24*, 5683–5701. [CrossRef]
21. Sadjang, P.N. On the fundamental theorem of (p,q)-calculus and some (p,q)-Taylor formulas. *Results Math.* **2018**, *73*, 1–21.
22. Mursaleen, M.; Ansari, K.J.; Khan, A. On (p,q)-analogue of Bernstein operators. *Appl. Math. Comput.* **2015**, *266*, 874–882. [CrossRef]
23. Mursaleen, M.; Ansari, K.J.; Khan, A. Some approximation results by (p,q)-analogue of Bernstein-Stancu operators. *Appl. Math. Comput.* **2015**, *264*, 392–402; Corrigendum in *Appl. Math. Comput.* **2015**, *269*, 744–746. [CrossRef]
24. Mursaleen, M.; Khan, F.; Khan, A. Approximation by (p,q)-Lorentz polynomials on a compact disk. *Complex Anal. Oper. Theory* **2016**, *10*, 1725–1740. [CrossRef]
25. Rahman, S.; Mursaleen, M.; Alkhaldi, A.H. Convergence of iterates of q-Bernstein and (p,q)-Bernstein operators and the Kelisky-Rivlin type theorem. *Filomat* **2018**, *32*, 4351–4364. [CrossRef]
26. Jebreen, H.B.; Mursaleen, M.; Ahasan, M. On the convergence of Lupas (p,q)-Bernstein operators via contraction principle. *J. Ineq. Appl.* **2019**, *2019*. [CrossRef]
27. Khan, A.; Sharma, V. Statistical approximation by (p,q)-analogue of Bernstein-Stancu operators. *Azerbaijan J. Math.* **2018**, *8*, 100–121.
28. Khan, K.; Lobiyal, D.K. Bezier curves based on Lupas (p,q)-analogue of Bernstein functions in CAGD. *J. Comput. Appl. Math.* **2017**, *317*, 458–477. [CrossRef]
29. Milovanovic, G.V.; Gupta, V.; Malik, N. (p,q)-Beta functions and applications in approximation. *Bol. Soc. Mat. Mex.* **2018**, *24*, 219–237. [CrossRef]
30. Cheng, W.T.; Zhang, W.H.; Cai, Q.B. (p,q)-gamma operators which preserve x^2. *J. Inequal. Appl.* **2019**, *2019*, 1–14. [CrossRef]
31. Nasiruzzaman, M.; Mukheimer, A.; Mursaleen, M. Some Opial-type integral inequalities via (p,q)-calculus. *J. Ineq. Appl.* **2019**, *2019*, 1–15. [CrossRef]
32. Jain, P.; Basu, C.; Panwar, V. On the (p,q)-Melin Transform and Its Applications. *Acta Math. Sci. Ser. B* **2021**, *41*, 1719–1732. [CrossRef]
33. Aral, A.; Deniz, E.; Erbay, H. The Picard and Gauss-Weierstrass singular integral in (p,q)-calculus. *Bull. Malays. Math. Sci. Soc.* **2020**, *43*, 1569–1583. [CrossRef]
34. Kamsrisuk, N.; Promsakon, C.; Ntouyas, S.K.; Tariboon, J. Nonlocal boundary value problems for (p,q)-difference equations. *Differ. Equ. Appl.* **2018**, *10*, 183–195. [CrossRef]
35. Promsakon, C.; Kamsrisuk, N.; Ntouyas, S.K.; Tariboon, J. On the second-order quantum (p,q)-difference equations with separated boundary conditions. *Adv. Math. Phys.* **2018**, *2018*, 9089865. [CrossRef]
36. Nuntigrangjana, T.; Putjuso, S.N.; Ntouyas, S.K.; Tariboon, J. Impulsive quantum (p,q)-difference equations. *Adv. Differ. Equ.* **2020**, *2020*, 1–20. [CrossRef]
37. Soontharanon, J.; Sitthiwirattham, T. On Fractional (p,q)-Calculus. *Adv. Differ. Equ.* **2020**, *2020*, 1–18. [CrossRef]
38. Pheak, N.; Nonlaopon, K.; Tariboon, J.; Ntouyas, S.K. Fractional (p,q)-Calculus on Finite Intervals and Some Integral Inequalities. *Symmetry* **2021**, *13*, 504.
39. Pheak, N.; Nonlaopon, K.; Tariboon, J.; Ntouyas, S.K. Praveen Agarwal, Some trapezoid and midpoint type inequalities via fractional (p,q)-calculus. *Adv. Differ. Equ.*, **2021**, *2021*, 1–22.
40. Soontharanon, J.; Sitthiwirattham, T. Existence Results of Nonlocal Robin Boundary Value Problems for Fractional (p,q)-Integrodifference Equations. *Adv. Differ. Equ.* **2020**, *2020*, 1–17. [CrossRef]
41. Qin, Z.; Sun, S. On a nonlinear fractional (p,q)-difference Schrödinger equation. *J. Appl. Math. Comput.* **2021**. [CrossRef]
42. Qin, Z.; Sun, S. Positive solutions for fractional (p,q)-difference boundary value problems. *J. Appl. Math. Comput.* **2021**. [CrossRef]
43. Griffel, D.H. *Applied Functional Analysis*; Ellis Horwood Publishers: Chichester, UK, 1981.
44. Guo, D.; Lakshmikantham, V. *Nonlinear Problems in Abstract Cone*; Academic Press: Orlando, FL, USA, 1988.

MDPI

Article

Global Stability Condition for the Disease-Free Equilibrium Point of Fractional Epidemiological Models †

Ricardo Almeida ‡, Natália Martins *,‡ and Cristiana J. Silva ‡

Center for Research and Development in Mathematics and Applications (CIDMA), Department of Mathematics, University of Aveiro, 3810-193 Aveiro, Portugal; ricardo.almeida@ua.pt (R.A.); cjoaosilva@ua.pt (C.J.S.)

* Correspondence: natalia@ua.pt

† Dedicated to Professor Delfim F. M. Torres on the occasion of his 50th birthday.

‡ These authors contributed equally to this work.

Abstract: In this paper, we present a new result that allows for studying the global stability of the disease-free equilibrium point when the basic reproduction number is less than 1, in the fractional calculus context. The method only involves basic linear algebra and can be easily applied to study global asymptotic stability. After proving some auxiliary lemmas involving the Mittag–Leffler function, we present the main result of the paper. Under some assumptions, we prove that the disease-free equilibrium point of a fractional differential system is globally asymptotically stable. We then exemplify the procedure with some epidemiological models: a fractional-order SEIR model with classical incidence function, a fractional-order SIRS model with a general incidence function, and a fractional-order model for HIV/AIDS.

Keywords: epidemiology; mathematical modeling; fractional calculus; equilibrium; stability

Citation: Almeida, R.; Martins, N.; Silva, C.J. Global Stability Condition for the Disease-Free Equilibrium Point of Fractional Epidemiological Models. *Axioms* **2021**, *10*, 238. https://doi.org/10.3390/axioms10040238

Academic Editor: Ioannis Dassios

Received: 23 July 2021
Accepted: 17 September 2021
Published: 25 September 2021

1. Introduction

Fractional differential equations play an important role in modeling real-life phenomena. By replacing an integer-order derivative with a real-order fractional derivative, often we can fit the system of equations to the real data more efficiently because many dynamical systems cannot be completely described by ODEs. To mention a few of them, we refer applications to bioengineering [1,2], biology [3], Lévy motion [4], harmonic oscillators with damping [5], economy [6,7], and engineering [8,9]. In this work, we are particularly interested in applications of fractional calculus in epidemiological models. This topic has been intensively studied in the recent past, from the well formulation of the problem, the existence of equilibrium points, modeling, and forecasting of epidemiological systems. For example, Refs. [10–12] proposed fractional epidemiological models to study the spread of COVID-19 in different countries, Refs. [13,14] investigated the HIV infection, in Ref. [15], a varicella outbreak in China was considered, the spread of dengue fever outbreak in the Cape Verde islands was studied in [16], and in [17] a fractional measles model was proposed. Stability studies were given in e.g., [13,18–21] and numerical methods in e.g., [22–24]. We also refer to [25] where a review of several fractional epidemiological models was carried out.

An important problem is the study of the global stability of the equilibrium points, in order to better understand the evolution of the disease over time. That is, the system will evolve to the equilibrium point, independently of the starting points. The study of local stability is a relatively simple matter, as it usually involves finding the eigenvalues of the Jacobian matrix and studying their sign. However, the question of global stability is not, in many cases, simple to answer as it usually involves constructing suitable Lyapunov-like functions and there is no routine on how to find them. We emphasize here the fact that the use of the Lyapunov stability theory to establish the global asymptotic stability for fractional differential equations is more complicated than for the ODEs (see, e.g., [26–29]).

This was the motivation to investigate a new method to study global asymptotic stability for dynamical systems described by fractional differential equations. In [30], a novel method was presented. By writing the system in a matricial form and analyzing the matrices involved is such procedure, under some simple assumptions, we can ensure that the equilibrium point is globally stable if the basic reproduction number $\mathcal{R}_0$ is less than 1. The aim of this work is to generalize the main result of [30] to the fractional setting. To our knowledge, this is the first work on global stability following this last approach to the problem.

The paper is organized as follows. In Section 2, we present some concepts and known results needed for this work. Section 3 presents the new contributions of this paper. After deducing some auxiliary lemmas, we prove the main result of this work, Theorem 2. Under some assumptions that can be easily verified for a wide range of epidemiological models given by fractional differential systems, we prove that, if the basic reproduction number is less than 1, then the equilibrium point is globally asymptotically stable. Lastly, in Section 4, we present three examples to show the utility of our research.

2. Preliminaries

We begin this section with some basic definitions and results of the fractional calculus needed in this work. For more details, we refer the reader to [31,32].

Throughout the text, $\alpha \in]0,1[$ and $\Gamma(z) = \int_0^\infty t^{z-1}e^t\,dt$, $z > 0$, is the Gamma function.

Definition 1. *Let $f : \mathbb{R}_0^+ \to \mathbb{R}$ be an integrable function. The (left-sided) Riemann–Liouville fractional integral of function f of order α is given by*

$$\mathbb{I}_{0+}^{\alpha} f(t) := \frac{1}{\Gamma(\alpha)} \int_0^t (t-\tau)^{\alpha-1} f(\tau) d\tau, \quad t > 0.$$

Definition 2. *The (left-sided) Caputo fractional derivative of order α of function $f \in C^1(\mathbb{R}_0^+, \mathbb{R})$ is defined by*

$${}^{C}\mathbb{D}_{0+}^{\alpha} f(t) := \frac{1}{\Gamma(1-\alpha)} \int_0^t (t-\tau)^{-\alpha} f'(\tau)\,d\tau, \quad t > 0.$$

Next, we recall the definition of the generalized Mittag–Leffler function, which is a special function that generalizes the standard exponential function. The Mittag–Leffler function is of great importance in fractional calculus because it arises naturally in the solution of fractional-order differential and integral equations.

Definition 3. *The Mittag–Leffler function with two parameters is defined by*

$$E_{\alpha,\beta}(t) := \sum_{k=0}^{\infty} \frac{t^k}{\Gamma(\alpha k + \beta)}, \quad t \in \mathbb{C},$$

with $\alpha, \beta \geq 0$. When $\beta = 1$, we define the one parameter Mittag–Leffler function $E_\alpha(t) := E_{\alpha,1}(t)$.

To understand the theory of fractional differential equations, one needs to know properties of these special functions. Its main properties and applications can be found, for example, in [33]. We emphasize here the fact that $E_{\alpha,\beta}(t)$ can take negative values (cf. [34]).

Recently, we have observed an increasing interest in the Mittag–Leffler function for matrix arguments, since the solution of many systems of differential equations of noninteger order can be expressed using this matrix function. For theoretical properties and a survey on numerical approximation of the matrix Mittag–Leffler function, we recommend the recent paper [35] and the references cited therein.

Definition 4. *Given $A \in \mathbb{C}^{n\times n}$, the matrix Mittag–Leffler function with two parameters is defined through the convergent series*

$$E_{\alpha,\beta}(A) := \sum_{k=0}^{\infty} \frac{A^k}{\Gamma(\alpha k+\beta)}$$

where $\alpha, \beta \geq 0$. If $\beta = 1$, we define the one parameter matrix Mittag–Leffler function $E_\alpha(A) := E_{\alpha,1}(A)$.

Remark 1. *For $\alpha = \beta = 1$, the matrix Mittag–Leffler function is the matrix exponential, that is, $E_{1,1}(A) = \exp(A) = \sum_{k=0}^{\infty} \frac{A^k}{k!}$. Unfortunately, as noticed in [36], there are several works where some properties of the matrix exponential were incorrectly extended to the matrix Mittag–Leffler function and then used to solve certain linear matrix fractional differential equations. One of the properties that cannot be extended to the matrix Mittag–Leffler function is the semigroup property: for given commutating matrices A and B, in general, we have $E_\alpha(A+B) \neq E_\alpha(A)\cdot E_\alpha(B)$. We note, however, that, if matrices A and B commute and $\alpha \approx 1$, then $E_\alpha(A+B) \approx E_\alpha(A)\cdot E_\alpha(B)$.*

We recall now two properties of the matrix Mittag–Leffler function that are useful in the present work (see [35]):

1. if $A = \mathrm{diag}(a_{11},\ldots,a_{nn})$, then $E_{\alpha,\beta}(A) = \mathrm{diag}(E_{\alpha,\beta}(a_{11}),\ldots,E_{\alpha,\beta}(a_{nn}))$;
2. if there exists a non-singular matrix P such that $A = PDP^{-1}$, then $E_{\alpha,\beta}(A) = PE_{\alpha,\beta}(D)P^{-1}$.

To finalize this section, we review some concepts on matrix theory.

Definition 5. *We say that a square matrix A is an M-matrix if the off-diagonal entries are nonpositive and the real parts of all eigenvalues are nonnegative.*

Given a square matrix A, the set of eigenvalues of A is denoted by $\sigma(A)$. The spectral bound of matrix A is defined as $m(A) = \max\{\mathrm{Re}(\lambda) : \lambda \in \sigma(A)\}$, where $\mathrm{Re}(\lambda)$ denotes the real part of λ, and the spectral radius of A is defined as $\rho(A) = \max\{|\lambda| : \lambda \in \sigma(A)\}$.

The following result is a fundamental tool in the proof of Lemma 4.

Lemma 1 ([36]). *Let $A = \begin{bmatrix} a & b \\ c & d \end{bmatrix}$ be a diagonalizable matrix of order 2 with eigenvalues $\lambda_1 = \frac{(a+d)-\Omega}{2}$ and $\lambda_2 = \frac{(a+d)+\Omega}{2}$ where $\Omega := \sqrt{(a-d)^2+4bc}$. Let $e_1 := E_{\alpha,\beta}(\lambda_1)$ and $e_2 := E_{\alpha,\beta}(\lambda_2)$. If Ω and c are not zero, the matrix Mittag–Leffler function of matrix A is*

$$E_{\alpha,\beta}(A) = \frac{1}{2\Omega}\begin{bmatrix} (d-a)(e_1-e_2)+\Omega(e_1+e_2) & -2b(e_1-e_2) \\ -2c(e_1-e_2) & -(d-a)(e_1-e_2)+\Omega(e_1+e_2) \end{bmatrix}.$$

We remark that, in Lemma 1, if $\Omega = 0$ or $c = 0$, a simple formula for the Mittag–Leffler function of a diagonalizable matrix of order 2 can be easily obtained.

Theorem 1 (cf. [37]). *Let $A \in \mathbb{R}^{n\times n}$. If the spectrum of A satisfies the relation*

$$\sigma(A) \subseteq \left\{\lambda \in \mathbb{C}\setminus\{0\} : |\arg(\lambda)| > \frac{\alpha\pi}{2}\right\},$$

then $\lim_{t\to\infty} \|E_\alpha(At^\alpha)\| = 0$.

3. Main Results

Suppose that the epidemiological model under study is described by the fractional differential system

$$\begin{cases} {}^{C}\mathbb{D}^{\alpha}_{0+}X(t) = F(X, I) \\ {}^{C}\mathbb{D}^{\alpha}_{0+}I(t) = G(X, I) \\ G(X, 0) = 0 \end{cases} \quad (1)$$

with nonnegative initial conditions $X(0) = X_0 \in \mathbb{R}^m$ and $I(0) = I_0 \in \mathbb{R}^n$, where the components of the vector X denote the number of uninfected individuals (e.g., susceptible, recovered, vaccinated, etc.) and the components of I denote the number of the infected and infectious (the ones that can transmit the disease, such as the asymptomatic but infectious and active infected). In addition, we assume that function F is continuous, G is of class C^1, and the fractional differential system (1) with initial conditions $X(0) = X_0$ and $I(0) = I_0$ admits a unique solution.

Throughout this paper, we denote by $U_0 = (X^\star, 0) \in \mathbb{R}^{m+n}$ the disease-free equilibrium (DFE) point of the system (1), that is, $F(X^\star, 0) = G(X^\star, 0) = 0$.

Let $A := \frac{\partial G}{\partial I}(X^\star, 0)$ and assume that matrix A can be written in the form $A = M - D$, where M, D are two square matrices with $M \geq 0$ (all entries are nonnegative), and $D > 0$ is a diagonal matrix. The following result was proven in [38]:

$$m(A) < 0 \quad \text{if and only if} \quad \rho(MD^{-1}) < 1,$$

or

$$m(A) > 0 \quad \text{if and only if} \quad \rho(MD^{-1}) > 1.$$

The value

$$\mathcal{R}_0 := \rho(MD^{-1})$$

plays an important role in epidemiological models, and it is known as the basic reproduction number. This number gives the average number of secondary cases produced by one infected individual in a population where all individuals are susceptible to the infection.

The following result is well known in the literature. For the convenience of the reader, we present here one possible proof that follows from the fact that the scalar Mittag–Leffler function is completely monotonic ([39,40]).

Lemma 2. *For $0 < \alpha < 1$, $E_{\alpha,\alpha}(t) \geq 0$, for all $t \in \mathbb{R}$.*

Proof. Clearly, $E_{\alpha,\alpha}(t) \geq 0$, for all $t \geq 0$. To prove that $E_{\alpha,\alpha}(-t) \geq 0$, for all $t > 0$, we use the fact that the scalar Mittag–Leffler function $E_\alpha(-t)$, $t \geq 0$, is completely monotonic, that is,

$$(-1)^m \frac{d^m}{dt^m} E_\alpha(-t) \geq 0, \quad \forall m \in \mathbb{N}. \quad (2)$$

Since

$$\alpha \frac{d}{dt} E_\alpha(-t) = -E_{\alpha,\alpha}(-t), \quad t \geq 0,$$

it follows from (2) that

$$E_{\alpha,\alpha}(-t) \geq 0, \quad t \geq 0.$$

This completes the proof. □

The following result is also useful in this work.

Lemma 3. *For $0 < \alpha < 1$, $E_{\alpha,\alpha} : \mathbb{R} \to \mathbb{R}$ is an increasing function.*

Proof. It is clear that $E_{\alpha,\alpha}$ is an increasing function on $\mathbb{R}_0^+$. Now, we prove that $\frac{d}{dt}E_{\alpha,\alpha}(t) \geq 0$, for all $t \in \mathbb{R}_0^-$. Since

$$\frac{d}{dt}E_{\alpha,\alpha}(t) = \alpha\frac{d^2}{dt^2}E_\alpha(t) = \sum_{k=1}^{\infty}\frac{kt^{k-1}}{\Gamma(k\alpha+\alpha)},$$

from (2), we conclude that $\frac{d}{dt}E_{\alpha,\alpha}(t) \geq 0$, proving the desired result. □

Now, we prove the following lemma that shows the applicability of our main result (Theorem 2).

Lemma 4. *Let $A \in \mathbb{R}^{2\times 2}$ be a matrix and $0 < \alpha < 1$. If matrix A is diagonalizable and $-A$ is an M-matrix, then $E_{\alpha,\alpha}(A) \geq 0$.*

Proof. With our assumptions and using the notations from Lemma 1, we have that $b, c, \Omega \in \mathbb{R}_0^+$, $\lambda_1, \lambda_2 \in \mathbb{R}_0^-$, and $\lambda_2 \geq \lambda_1$. First, suppose that $\Omega \neq 0$ and $c \neq 0$. Hence, from Lemmas 2 and 3, we conclude that $e_2 := E_{\alpha,\alpha}(\lambda_2) \geq e_1 := E_{\alpha,\alpha}(\lambda_1) \geq 0$. It remains to be proved that

$$\Omega(e_1+e_2) \geq -(d-a)(e_1-e_2) \quad \text{and} \quad \Omega(e_1+e_2) \geq (d-a)(e_1-e_2).$$

Suppose that $d \geq a$ (the other case is similar). Then, we just need to prove the first inequality. Since both sides of the inequality are nonnegative, we have that

$$[(a-d)^2+4bc](e_1+e_2)^2 \geq (d-a)^2(e_1-e_2)^2,$$

which is equivalent to

$$\begin{aligned}(a-d)^2e_1^2+(a-d)^2e_2^2+2(a-d)^2e_1e_2+4bc(e_1+e_2)^2 \\ \geq (a-d)^2e_1^2+(a-d)^2e_2^2-2(a-d)^2e_1e_2\end{aligned}$$

proving the desired. Now, we suppose that $c = 0$ and $a \neq d$. If $a < d$, then we get

$$E_{\alpha,\beta}(A) = \begin{bmatrix} e_1 & \frac{b}{a-d}(e_1-e_2) \\ 0 & e_2 \end{bmatrix} \geq 0,$$

and, if $a > d$, then

$$E_{\alpha,\beta}(A) = \begin{bmatrix} e_2 & \frac{b}{a-d}(e_2-e_1) \\ 0 & e_1 \end{bmatrix} \geq 0.$$

If $c = 0$ and $a = d$, then

$$E_{\alpha,\beta}(A) = \begin{bmatrix} e_1 & 0 \\ 0 & e_1 \end{bmatrix} \geq 0$$

since $b = 0$ (otherwise, A is not diagonalizable). If $\Omega = 0$, the proof is trivial since in this case A is diagonalizable iff $a = d$ and $c = d = 0$. □

The following result is a fundamental tool in the proof of Theorem 2.

Lemma 5. *Let $B \in \mathbb{R}^{n\times n}$ be an invertible matrix and $H : \mathbb{R}^{m+n} \to \mathbb{R}^n$ be a continuous function. Suppose that the fractional differential equation*

$${}^C\mathbb{D}_{0+}^{\alpha}I(t) = B \cdot I(t) - H(X(t), I(t))$$

with initial condition $I(0) = I_0 \in \mathbb{R}^n$, has a unique solution. Then, the solution of this initial value problem satisfies

$$I(t) = E_\alpha(Bt^\alpha) \cdot I_0 - \int_0^t B^{-1} \cdot \frac{d}{ds} E_\alpha(Bs^\alpha) \cdot H(X(t-s), I(t-s))\, ds.$$

Proof. The proof follows the ideas from [41] (Theorem 7.2). First, observe that

$${}^C\mathbb{D}^\alpha_{0+}(E_\alpha(Bt^\alpha) \cdot I_0) = B \cdot E_\alpha(Bt^\alpha) \cdot I_0.$$

To compute

$${}^C\mathbb{D}^\alpha_{0+}\left(\int_0^t B^{-1} \cdot \frac{d}{ds} E_\alpha(Bs^\alpha) \cdot H(X(t-s), I(t-s))\, ds\right),$$

let

$$\overline{y}(t) = \int_0^t B^{-1} \cdot \frac{d}{ds} E_\alpha(Bs^\alpha) \cdot H(X(t-s), I(t-s))\, ds.$$

Since

$$B^{-1} \cdot \frac{d}{ds} E_\alpha(Bs^\alpha) = \alpha s^{\alpha-1} E'_\alpha(Bs^\alpha) = \alpha s^{\alpha-1} \sum_{k=1}^\infty \frac{k(Bs^\alpha)^{k-1}}{\Gamma(k\alpha+1)} = \sum_{k=1}^\infty \frac{B^{k-1}s^{\alpha k-1}}{\Gamma(k\alpha)},$$

we get the following:

$$\begin{aligned}
\overline{y}(t) &= \sum_{k=1}^\infty \frac{B^{k-1}}{\Gamma(k\alpha)} \int_0^t s^{\alpha k-1} H(X(t-s), I(t-s))\, ds \\
&= \sum_{k=1}^\infty \frac{B^{k-1}}{\Gamma(k\alpha)} \int_0^t (t-\tau)^{\alpha k-1} H(X(\tau), I(\tau))\, d\tau \\
&= \sum_{k=1}^\infty B^{k-1} \cdot \mathbb{I}^{\alpha k}_{0+} H(X(t), I(t)).
\end{aligned}$$

Thus,

$$\begin{aligned}
{}^C\mathbb{D}^\alpha_{0+}\overline{y}(t) &= \sum_{k=1}^\infty B^{k-1} \cdot {}^C\mathbb{D}^\alpha_{0+}\mathbb{I}^{\alpha k}_{0+} H(X(t), I(t)) = \sum_{k=1}^\infty B^{k-1} \cdot \mathbb{I}^{\alpha(k-1)}_{0+} H(X(t), I(t)) \\
&= \sum_{k=0}^\infty B^k \cdot \mathbb{I}^{\alpha k}_{0+} H(X(t), I(t)) = H(X(t), I(t)) + \sum_{k=1}^\infty B^k \cdot \mathbb{I}^{\alpha k}_{0+} H(X(t), I(t)) \\
&= H(X(t), I(t)) + B \cdot \overline{y}(t).
\end{aligned}$$

Hence, we may conclude that

$$\begin{aligned}
{}^C\mathbb{D}^\alpha_{0+} I(t) &= B \cdot E_\alpha(Bt^\alpha) \cdot I_0 - H(X(t), I(t)) - B \cdot \overline{y}(t) \\
&= B \cdot \left(E_\alpha(Bt^\alpha) \cdot I_0 - \overline{y}(t)\right) - H(X(t), I(t)) \\
&= B \cdot I(t) - H(X(t), I(t)),
\end{aligned}$$

proving the desired result. □

We are now in conditions to present a new global stability condition for the DFE of system (1) when $\mathcal{R}_0 < 1$. Knowing that an equilibrium point is globally asymptotically stable with respect to the system that describes the evolution of the uninfected individuals, and with some extra assumptions related to the matrices involved in the system associated with infected and infectious individuals, we can conclude that the equilibrium point is in fact globally asymptotically stable with respect to the complete system. Although the result imposes some restrictions in order to be applied, for many epidemiological models, it can be easily used, as we will illustrate in Section 4.

Theorem 2. *Suppose that*

1. *For* ${}^{C}\mathbb{D}^{\alpha}_{0+}X(t) = F(X,0)$, *the vector* $X^\star$ *is globally asymptotically stable;*
2. *Function* G *can be written as* $G(X,I) = A \cdot I - \hat{G}(X,I)$, *where* $\hat{G}(X,I) \geq 0$ *for all* (X,I), *and* $A = \frac{\partial G}{\partial I}(X^\star,0)$ *can be written as* $A = M - D$, *where* $M \geq 0$ *and* $D > 0$ *is a diagonal matrix;*
3. *Matrix* A *is diagonalizable, the real parts of the eigenvalues of* A *are nonpositive, and* $E_{\alpha,\alpha}(A) \geq 0$;
4. $I(t) \geq 0$ *for all* $t > 0$ *(nonnegativity of solutions).*

If $\mathcal{R}_0 < 1$, *then the DFE,* $U_0 = (X^\star,0)$, *is a globally asymptotically stable equilibrium of system* (1), *for all* $0 < \alpha < 1$.

Proof. First, observe that, since $\mathcal{R}_0 < 1$, then $m(A) < 0$ (see [38]), and so matrix A is invertible. Since

$${}^{C}\mathbb{D}^{\alpha}_{0+}I(t) = G(X(t),I(t)) = A \cdot I(t) - \hat{G}(X(t),I(t)),$$

then, by Lemma 5, we get

$$0 \leq I(t) = E_\alpha(At^\alpha) \cdot I_0 - \int_0^t A^{-1}\frac{d}{ds}E_\alpha(As^\alpha) \cdot \hat{G}(X(t-s),I(t-s))\,ds \leq E_\alpha(At^\alpha) \cdot I_0,$$

since

$$A^{-1}\frac{d}{ds}E_\alpha(As^\alpha) = \alpha s^{\alpha-1}E'_\alpha(As^\alpha) = s^{\alpha-1}E_{\alpha,\alpha}(As^\alpha) \geq 0.$$

Since the real parts of the eigenvalues of matrix A are negative, from Theorem 1, we get

$$\lim_{t\to\infty} \|E_\alpha(At^\alpha)\| = 0,$$

and hence

$$\lim_{t\to\infty} I(t) = 0.$$

Since $X^\star$ is globally asymptotically stable with respect to ${}^{C}\mathbb{D}^{\alpha}_{0+}X(t) = F(X,0)$, which in turn is the limiting system of ${}^{C}\mathbb{D}^{\alpha}_{0+}X(t) = F(X,I)$, then we get

$$\lim_{t\to\infty} X(t) = X^\star(t),$$

which completes the proof. □

Remark 2. *Note that the assumption* $E_{\alpha,\alpha}(A) \geq 0$ *in Theorem 2 is trivially satisfied if matrix* A *has dimensions 1 or 2 (by Lemmas 2 and 4, respectively). In addition, if* $\alpha \approx 1$, *since* $E_{\alpha,\alpha}(A) \approx \exp(A)$, *the above condition also holds for any matrix* $A \in \mathbb{R}^{n\times n}$.

4. Examples

In this section, we illustrate our main result, Theorem 2, by considering three Caputo fractional-order compartmental models and show that the disease free equilibrium is globally stable in all the cases, whenever $\mathcal{R}_0 < 1$. We stress that usually (see e.g., [21]) the global stability of the disease free equilibrium is proved by considering an appropriate Lyapunov function and LaSalle's invariance principle [42], which are often difficult to apply especially when the model has a considerable number of variables and parameters. Our main result allows us to prove the global stability of the disease free equilibrium in an easier and simpler way. For the numerical implementation of the fractional derivatives, we have used the Adams–Bashforth–Moulton scheme, which has been implemented in the Matlab code fde12 by Garrappa [43].

The code implements the predictor–corrector PECE method of Adams–Bashforth–Moulton type described in [44]. We fixed a time step size of $h = 2^{-6}$ and consider, without loss of generality, the fractional-order derivatives $\alpha \in \{0.8, 0.85, 0.9, 0.95, 1.0\}$.

4.1. A Fractional SEIR Model with Traditional Incidence Rate

We start by considering a Caputo fractional-order version of the classical SEIR model that has been applied to describe the transmission dynamics of infectious diseases where there exists a significant latency period during which the individuals are infected but not yet infectious. During this period, the individual is in the so-called *exposed* compartment E, see e.g., [45]. The other compartments of the model are *susceptible* S, *infected* I, and *recovered* R, and each of them denotes a fraction of the total population. The following assumptions are considered: the birth and death rates are assumed to be equal, and denoted by μ; the incidence rate is the traditional one, given by βSI, where β represents the transmission rate; the latent period is denoted by ε; infected individuals recover at a rate γ and remain recovered with permanent immunity. All parameters are assumed to be positive. The model is given by the following system:

$$\begin{cases} {}^{C}\mathbb{D}^{\alpha}_{0+} S(t) = \mu - \mu S(t) - \beta S(t) I(t), \\ {}^{C}\mathbb{D}^{\alpha}_{0+} E(t) = -(\varepsilon + \mu) E(t) + \beta S(t) I(t), \\ {}^{C}\mathbb{D}^{\alpha}_{0+} I(t) = \varepsilon E(t) - (\gamma + \mu) I(t), \\ {}^{C}\mathbb{D}^{\alpha}_{0+} R(t) = \gamma I(t) - \mu R(t). \end{cases} \tag{3}$$

The disease-free equilibrium of the model (3) is given by

$$\Sigma_0 = \left(S^0, E^0, I^0, R^0\right) = (1, 0, 0, 0).$$

Following the notation from Section 3, we have

$$A = \begin{bmatrix} -\varepsilon - \mu & \beta \\ \varepsilon & -\gamma - \mu \end{bmatrix}.$$

The matrix A can be written as $A = M - D$ with

$$M = \begin{bmatrix} 0 & \beta \\ \varepsilon & 0 \end{bmatrix}$$

and

$$D = \begin{bmatrix} \varepsilon + \mu & 0 \\ 0 & \gamma + \mu \end{bmatrix}.$$

The point $X^\star = (1, 0)$ is globally asymptotically stable for the system of uninfected individuals:

$$\begin{cases} {}^{C}\mathbb{D}^{\alpha}_{0+} S(t) = \mu - \mu S(t), \\ {}^{C}\mathbb{D}^{\alpha}_{0+} R(t) = -\mu R(t). \end{cases} \tag{4}$$

It is easy to verify that the function

$$R(t) = R_0 \, E_\alpha(-\mu t^\alpha)$$

satisfies the second equation of (4). From [41] (Theorem 7.2), the solution of the first equation of (4) is the function

$$S(t) = S_0\, E_\alpha(-\mu t^\alpha) + \int_0^t \mu s^{\alpha-1} E_{\alpha,\alpha}(-\mu s^\alpha)\, ds.$$

Simple computations lead to

$$S(t) = S_0\, E_\alpha(-\mu t^\alpha) - E_\alpha(-\mu t^\alpha) + 1.$$

Thus,

$$(S(t), R(t)) \to (1, 0) \quad \text{as} \quad t \to \infty.$$

In addition, by Lemma 4, $E_{\alpha,\alpha}(A)$ is nonnegative and so, by Theorem 2, the disease-free equilibrium of the model (9) is globally asymptotically stable.

Consider the following parameter values: $\mu = 1/80$, $\beta = 0.05$ and $\gamma = 1$, $\varepsilon = 1$. Then, the eigenvalues of the matrix A are -0.7888 and -1.2361; therefore, $m(A) < 0$. Moreover, we confirm that $\mathcal{R}_0 := \rho(MD^{-1}) = 0.2893 < 1$.

Through adequate numerical simulations, we illustrate the global stability of the disease free equilibrium, whenever $\mathcal{R}_0 < 1$, considering different values of α and initial conditions. In Figure 1, we consider different values for α and the initial condition x_0 from (5).

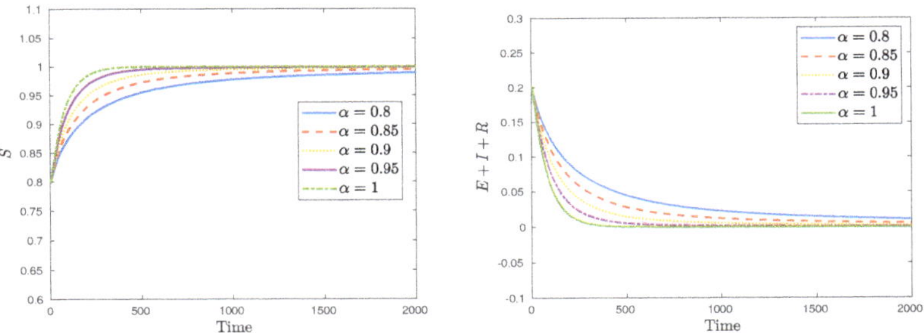

Figure 1. Stability of the disease free equilibrium $\Sigma_0 = (1, 0, 0, 0)$, for the SEIR fractional model (3), considering different values of $\alpha \in \{0.8, 0.85, 0.9, 0.95, 1.0\}$. On the left: S. On the right: $E + I + R$.

The global stability of the disease free equilibrium $\Sigma_0 = (1, 0, 0, 0)$ is illustrated in Figure 2, considering different initial conditions x_i, $i = 0, \ldots, 7$, given by (5):

$$\begin{aligned}
x_0 &= (S_{0,0}, E_{0,0}, I_{0,0}, R_{0,0}) = (0.3, 0.5, 0.1, 0.1)\,,\\
x_1 &= (S_{0,1}, E_{0,1}, I_{0,1}, R_{0,1}) = (0.4, 0.1, 0.3, 0.2)\,,\\
x_2 &= (S_{0,2}, E_{0,2}, I_{0,2}, R_{0,2}) = (0.5, 0.05, 0.4, 0.05)\,,\\
x_3 &= (S_{0,3}, E_{0,3}, I_{0,3}, R_{0,3}) = (0.6, 0.1, 0.2, 0.1)\,,\\
x_4 &= (S_{0,4}, E_{0,4}, I_{0,4}, R_{0,4}) = (0.7, 0.05, 0.1, 0.15)\,,\\
x_5 &= (S_{0,5}, E_{0,5}, I_{0,5}, R_{0,5}) = (0.8, 0.1, 0.1, 0.0)\,,\\
x_6 &= (S_{0,6}, E_{0,6}, I_{0,6}, R_{0,6}) = (0.85, 0.05, 0.1, 0.0)\,,\\
x_7 &= (S_{0,7}, E_{0,7}, I_{0,7}, R_{0,7}) = (0.95, 0.025, 0.025, 0.0)\,.
\end{aligned} \tag{5}$$

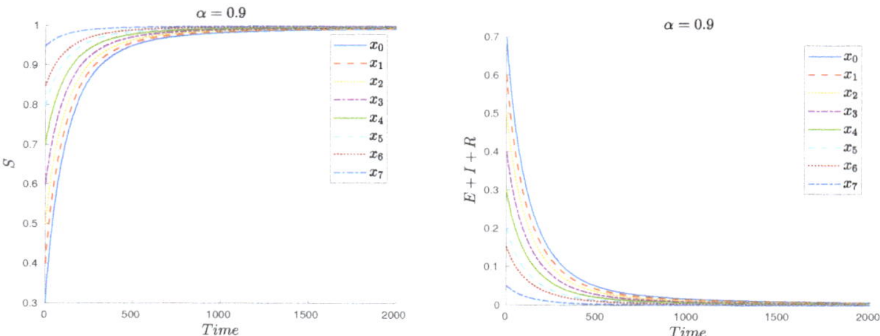

Figure 2. Global stability of the disease free equilibrium $\Sigma_0 = (1, 0, 0, 0)$, for the fractional model (3), considering $\alpha = 0.9$ and different initial conditions x_i, $i = 0, \ldots, 7$, from (5). On the left S and on the right $E + I + R$.

4.2. A Fractional SIRS Model with General Incidence Rate

In the second example, we consider the Caputo fractional-order version of the classical $SIRS$ model from [21], given by the following system:

$$\begin{cases} {}^C\mathbb{D}^{\alpha}_{0+} S(t) = \Lambda - \mu S(t) - \frac{\beta I(t) S(t)}{1 + k_1 S(t) + k_2 I(t) + k_3 S(t) I(t)} + \lambda R(t), \\ {}^C\mathbb{D}^{\alpha}_{0+} I(t) = \frac{\beta I(t) S(t)}{1 + k_1 S(t) + k_2 I(t) + k_3 S(t) I(t)} - (\mu + r) I(t), \\ {}^C\mathbb{D}^{\alpha}_{0+} R(t) = r I(t) - (\mu + \lambda) R(t). \end{cases} \tag{6}$$

The model considers a homogeneous population divided into three subgroups: susceptible individuals $S(t)$, infected and infectious individuals $I(t)$, and recovered $R(t)$, individuals at time t. The parameters Λ, β, μ, and r, represent the recruitment rate of the population, the infection rate, the natural death rate, and the recovery rate of the infected individuals, respectively. The rate that recovered individuals lose immunity and return to the susceptible class is represented by λ. While contacting with infected individuals, the susceptible become infected at the incidence rate

$$\frac{\beta S I}{1 + k_1 S + k_2 I + k_3 S I},$$

where k_1, k_2, and k_3 are nonnegative constants [21]. We remark that system (6) admits a unique positive solution (see [21] (Theorem 7)).

The disease-free equilibrium of the model (6) is given by

$$\Sigma_0 = \left(S^0, I^0, R^0\right) = \left(\frac{\Lambda}{\mu}, 0, 0\right).$$

In this case, following the notation from Section 3,

$$A = M - D = \left[\frac{\Lambda\beta}{\Lambda k_1 + \mu} - \mu - r\right]$$

with

$$M = \left[\frac{\Lambda\beta}{\Lambda k_1 + \mu}\right] \quad \text{and} \quad D = [\mu + r].$$

Hence,

$$\mathcal{R}_0 = \rho(MD^{-1}) = \frac{\Lambda\beta}{(\mu + r)(\Lambda k_1 + \mu)}.$$

We easily conclude that $A < 0$ whenever $\mathcal{R}_0 < 1$.

In what follows, we prove that the first condition of Theorem 2 holds, that is, $X^\star = (\Lambda/\mu, 0)$ is globally asymptotically stable for the system of uninfected individuals:

$$\begin{cases} {}^C\mathbb{D}^\alpha_{0+} S(t) = \Lambda - \mu S(t) + \lambda R(t), \\ {}^C\mathbb{D}^\alpha_{0+} R(t) = -(\mu + \lambda) R(t). \end{cases} \tag{7}$$

The solution of the fractional differential equations (7) is the functions

$$R(t) = R_0\, E_\alpha(-(\mu+\lambda)t^\alpha)$$

and

$$S(t) = S_0\, E_\alpha(-\mu t^\alpha) + \int_0^t \Big[\Lambda + \lambda R_0\, E_\alpha(-(\mu+\lambda)(t-s)^\alpha)\Big] s^{\alpha-1} E_{\alpha,\alpha}(-\mu s^\alpha)\, ds.$$

Obviously $R(t) \to 0$, as t goes to infinity. Now, we prove that $S(t) \to \Lambda/\mu$, as $t \to \infty$. First, observe that

$$\begin{aligned} \int_0^t s^{\alpha-1} E_{\alpha,\alpha}(-\mu s^\alpha)\, ds &= \sum_{k=0}^\infty \frac{(-\mu)^k}{\Gamma(k\alpha+\alpha)} \int_0^t s^{k\alpha+\alpha-1}\, ds = \sum_{k=0}^\infty \frac{(-\mu)^k}{\Gamma(k\alpha+\alpha+1)} t^{k\alpha+\alpha} \\ &= -\frac{1}{\mu} \sum_{k=0}^\infty \frac{(-\mu)^{k+1}}{\Gamma((k+1)\alpha+1)} t^{(k+1)\alpha} = -\frac{1}{\mu}(E_\alpha(-\mu t^\alpha) - 1). \end{aligned}$$

For the other term inside the integral, we get

$$\begin{aligned} &\int_0^t E_\alpha(-(\mu+\lambda)(t-s)^\alpha) s^{\alpha-1} E_{\alpha,\alpha}(-\mu s^\alpha)\, ds \\ &\qquad = \sum_{m=0}^\infty \sum_{k=0}^\infty \frac{(-(\mu+\lambda))^m (-\mu)^k}{\Gamma(m\alpha+1)\Gamma(k\alpha+\alpha)} \int_0^t (t-s)^{m\alpha} s^{k\alpha+\alpha-1}\, ds. \end{aligned}$$

To evaluate this integral, we use the known formula involving the Beta function:

$$B(x,y) = \frac{\Gamma(x)\Gamma(y)}{\Gamma(x+y)}, \quad x, y > 0.$$

With the change of variable $u = s/t$, we get

$$\begin{aligned} \int_0^t (t-s)^{m\alpha} s^{k\alpha+\alpha-1}\, ds &= t^{m\alpha} \int_0^t (1-s/t)^{m\alpha} s^{k\alpha+\alpha-1}\, ds = t^{m\alpha+k\alpha+\alpha} \int_0^1 (1-u)^{m\alpha} u^{k\alpha+\alpha-1}\, du \\ &= t^{m\alpha+k\alpha+\alpha} B(k\alpha+\alpha, m\alpha+1) = t^{m\alpha+k\alpha+\alpha} \frac{\Gamma(k\alpha+\alpha)\Gamma(m\alpha+1)}{\Gamma(m\alpha+k\alpha+\alpha+1)}. \end{aligned}$$

Thus, we prove that the solution $S(\cdot)$ is given by

$$S(t) = S_0\, E_\alpha(-\mu t^\alpha) + \frac{\Lambda}{\mu}(1 - E_\alpha(-\mu t^\alpha)) + \lambda R_0 \sum_{m=0}^\infty \sum_{k=0}^\infty \frac{(-(\mu+\lambda))^m (-\mu)^k}{\Gamma(m\alpha+k\alpha+\alpha+1)} t^{m\alpha+k\alpha+\alpha}.$$

Observe that, as t goes to infinity,

$$S_0\, E_\alpha(-\mu t^\alpha) \to 0 \quad \text{and} \quad \frac{\Lambda}{\mu}(1 - E_\alpha(-\mu t^\alpha)) \to \frac{\Lambda}{\mu}.$$

To sum up, it remains to prove that the double sum converges to zero. For that purpose, we recall the concept of the Mittag–Leffler function of two variables (cf. [46]) $E_{\alpha,\beta}(x,y,\cdot)$. With such notation, we can write

$$\sum_{m=0}^{\infty}\sum_{k=0}^{\infty}\frac{(-(\mu+\lambda))^m(-\mu)^k}{\Gamma(m\alpha+k\alpha+\alpha+1)}t^{m\alpha+k\alpha+\alpha}=t^\alpha E_{\alpha,\alpha}(-(\mu+\lambda)t^\alpha,-\mu t^\alpha,\alpha+1)$$

which converges to zero, as t goes to infinity, by [46] (Theorem 3.1). This proves the desired conclusion. We also remark that, by Lemma 2, $E_{\alpha,\alpha}(A)$ is nonnegative. Therefore, by Theorem 2, the disease-free equilibrium of the model (6) is globally asymptotically stable.

Considering the parameter values from [21], $\Lambda = 0.8$, $\mu = 0.1$, $\lambda = 0.5$, $\beta = 0.1$, $r = 0.5$, $k_1 = 0.1$, $k_2 = 0.02$, and $k_3 = 0.003$, we have $\mathcal{R}_0 = 0.7407 < 1$. For initial conditions, we consider the following ones and without any specific criteria:

$$\begin{aligned} &y_0=(S_{0,0},I_{0,0},R_{0,0})=(10,1,1)\,, \quad y_1=(S_{0,1},I_{0,1},R_{0,1})=(100,10,5)\,,\\ &y_2=(S_{0,2},I_{0,2},R_{0,2})=(200,20,10)\,, \quad y_3=(S_{0,3},I_{0,3},R_{0,3})=(300,30,20)\,,\\ &y_4=(S_{0,4},I_{0,4},C_{0,4},A_{0,5})=(400,40,50)\,. \end{aligned} \tag{8}$$

The stability of the disease free equilibrium for model (6) is illustrated in Figure 3.

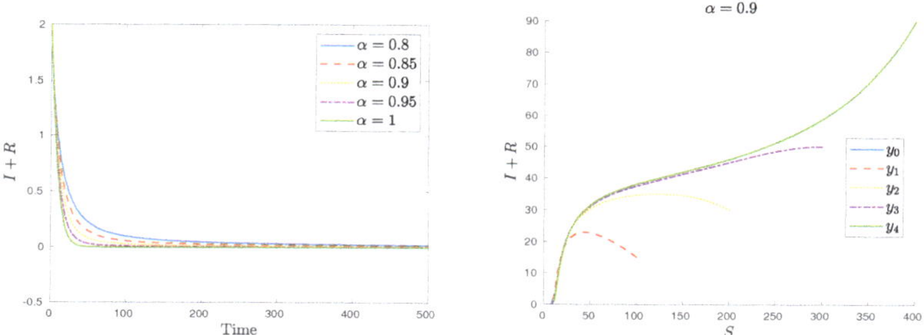

Figure 3. Stability of the disease free equilibrium $\Sigma_0 = (\frac{\Lambda}{\mu} = 8, 0, 0)$, for the SIRS fractional model (6). In the left: $I + R$, considering different values of $\alpha \in \{0.8, 0.85, 0.9, 0.95, 1.0\}$ and initial condition y_1 from (8). On the right: S and $I + R$, considering the initial conditions y_i, $i = 0, \dots, 4$ from (8) and fixed $\alpha = 0.9$.

4.3. A Modified Fractional SICA Model for HIV/AIDS

In this example, we consider a modified Caputo fractional-order model for HIV/AIDS, based on the model proposed in [14,47]. We show that this fractional model satisfies the conditions of Theorem 2 and, through some numerical simulations, we illustrate the global stability of the disease free equilibrium, when $\mathcal{R}_0 < 1$.

In this model, the total population is assumed to be homogeneous and divided into four mutually-exclusive compartments: susceptible individuals (S); HIV-infected individuals with no clinical symptoms of AIDS but able to transmit HIV to other individuals (I); HIV-infected individuals under antiretroviral (ART) treatment (the so called chronic stage) with a viral load remaining low (C); and HIV-infected individuals with AIDS clinical symptoms (A). Analogously to the assumption made in [47], we consider that individuals in the chronic stage C have a very low viral load and do not transmit HIV infection [48], but, differently from [47], we assume that individuals with AIDS A, due to their higher viral load, may transmit HIV virus, at a rate $\eta_A\,\beta$ with $\eta_A > 1$. Therefore, effective contact with people infected with HIV is at a rate λ, given by

$$\lambda = \beta(I + \eta_A A),$$

where β is the effective contact rate for HIV transmission. We assume that the recruitment rate is equal to the natural death rate and is denoted by μ. The following assumptions are the same as in [14]. HIV-infected individuals with no AIDS symptoms I progress to the class of individuals with HIV infection under ART treatment C, at a rate ϕ, and HIV-infected individuals with AIDS symptoms are treated for HIV at rate γ. Individuals in the class C leave for the class I, at a rate ω. HIV-infected individuals with AIDS symptoms A that start treatment move to the class of HIV-infected individuals I, moving to the chronic class C only if the treatment is maintained. HIV-infected individuals with no AIDS symptoms I that do not take ART treatment progress to the AIDS class A, at rate ρ. Only HIV-infected individuals with AIDS symptoms A suffer from an AIDS induced death, at a rate d. The total population at time t, denoted by $N(t)$, is given by $N(t) = S(t) + I(t) + C(t) + A(t)$. The Caputo fractional-order system that describes the previous assumptions is given by

$$\begin{cases} {}^{C}\mathbb{D}^{\alpha}_{0+}S(t) = \mu - \beta(I(t) + \eta_A A(t))S(t) - \mu S(t), \\ {}^{C}\mathbb{D}^{\alpha}_{0+}I(t) = \beta(I(t) + \eta_A A(t))S(t) - (\rho + \phi + \mu)I(t) + \omega C(t) + \gamma A(t), \\ {}^{C}\mathbb{D}^{\alpha}_{0+}C(t) = \phi I(t) - (\omega + \mu)C(t), \\ {}^{C}\mathbb{D}^{\alpha}_{0+}A(t) = \rho\, I(t) - (\gamma + \mu + d)A(t)\,. \end{cases} \tag{9}$$

The disease free equilibrium of system (9) is given by

$$\Sigma_0 = \left(S^0, I^0, C^0, A^0\right) = (1, 0, 0, 0). \tag{10}$$

Using the notation from Section 3, we have

$$A = \begin{bmatrix} \beta - \rho - \phi - \mu & \beta\eta_A + \gamma \\ \rho & -\gamma - \mu - d \end{bmatrix}.$$

The matrix A can be written as $A = M - D$ with

$$M = \begin{bmatrix} \beta & \beta\eta_A + \gamma \\ \rho & 0 \end{bmatrix}$$

and

$$D = \begin{bmatrix} \rho + \phi + \mu & 0 \\ 0 & \gamma + \mu + d \end{bmatrix}.$$

In this case, we will prove that $X^\star = (1, 0)$ is globally asymptotically stable for the system of uninfected individuals:

$$\begin{cases} {}^{C}\mathbb{D}^{\alpha}_{0+}S(t) = \mu - \mu S(t), \\ {}^{C}\mathbb{D}^{\alpha}_{0+}C(t) = -(\omega + \mu)C(t). \end{cases} \tag{11}$$

The solution of (11) is

$$C(t) = C_0\, E_\alpha(-(\omega + \mu)t^\alpha)$$

and

$$S(t) = S_0\, E_\alpha(-\mu t^\alpha) + \int_0^t \mu s^{\alpha-1} E_{\alpha,\alpha}(-\mu s^\alpha)\, ds.$$

Similarly to Section 4.1, we can prove that the disease-free equilibrium of the model (9) is globally asymptotically stable.

Let us consider the parameter values from Table 1.

Table 1. Parameter values of model (9) corresponding to $\mathcal{R}_0 = 0.1863 < 1$.

Symbol	Description	Value
μ	Recruitment rate/natural death rate	1/69.54
β	HIV transmission rate	0.05
η_A	Modification parameter	1.3
ϕ	HIV treatment rate for I individuals	1
ρ	Default treatment rate for I individuals	0.1
γ	AIDS treatment rate	0.33
ω	Default treatment rate for C individuals	0.09
d	AIDS induced death rate	1

Then, the eigenvalues of the matrix A are -0.9612 and -1.4474; therefore, $m(A) < 0$. Moreover, we confirm that $\mathcal{R}_0 := \rho(MD^{-1}) = 0.1863 < 1$.

We now show, using numerical simulations, that for different values of α and initial conditions, the global stability of the disease free equilibrium holds, whenever $\mathcal{R}_0 < 1$.

In Figure 4, we consider different values for α and the initial conditions x_0 from (12).

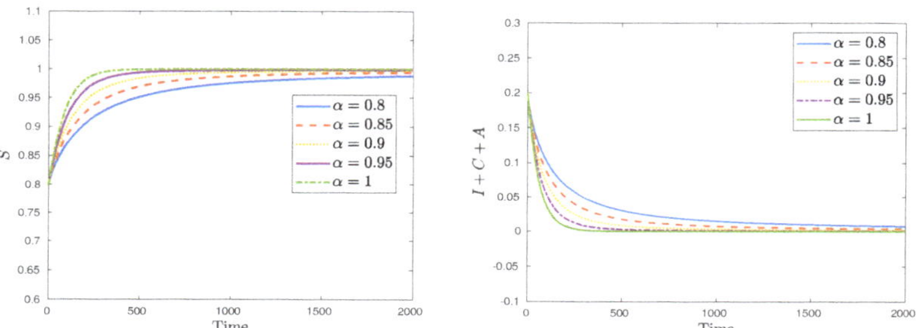

Figure 4. Stability of the disease free equilibrium $\Sigma_0 = (1, 0, 0, 0)$, for the SICA fractional model (9), considering different values of $\alpha \in \{0.8, 0.85, 0.9, 0.95, 1.0\}$. On the left: S. On the right: $I + C + A$.

The global stability of the disease free equilibrium (10) is illustrated in Figure 5, considering different initial conditions x_i, $i = 0, \ldots, 7$, given by (12).

$$\begin{aligned}
x_0 &= (S_{0,0}, I_{0,0}, C_{0,0}, A_{0,0}) = (0.8, 0.1, 0.1, 0)\,, \\
x_1 &= (S_{0,1}, I_{0,1}, C_{0,1}, A_{0,1}) = (0.4, 0.2, 0.2, 0.2)\,, \\
x_2 &= (S_{0,2}, I_{0,2}, C_{0,2}, A_{0,2}) = (0.7, 0.1, 0.1, 0.1)\,, \\
x_3 &= (S_{0,3}, I_{0,3}, C_{0,3}, A_{0,3}) = (0.5, 0.1, 0.2, 0.2)\,, \\
x_4 &= (S_{0,4}, I_{0,4}, C_{0,4}, A_{0,4}) = (0.9, 0.05, 0.05, 0)\,, \\
x_5 &= (S_{0,5}, I_{0,5}, C_{0,5}, A_{0,5}) = (0.6, 0.2, 0.1, 0.1)\,, \\
x_6 &= (S_{0,6}, I_{0,6}, C_{0,6}, A_{0,6}) = (0.55, 0.25, 0.1, 0.1)\,, \\
x_7 &= (S_{0,7}, I_{0,7}, C_{0,7}, A_{0,7}) = (0.75, 0.1, 0.1, 0.05)\,.
\end{aligned} \tag{12}$$

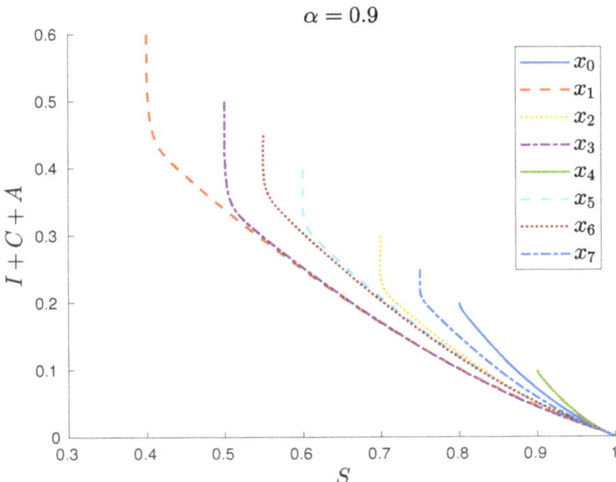

Figure 5. Global stability of the disease free equilibrium $\Sigma_0 = (1, 0, 0, 0)$, for the fractional model (9), considering $\alpha = 0.9$ and different initial conditions x_i, $i = 0, \ldots, 7$, from (12), on the x-axis S and on the y-axis $I + C + A$.

5. Conclusions

A new and simple result for the global stability for the disease-free equilibrium of fractional epidemiological models is presented. We highlight here the fact that the approach available in the literature so far involves the determination of an appropriate Lyapunov function, very laborious computations and, in the end, the application of LaSalle's invariance principle. Our new method uses only basic results from matrix theory and some well-known results from fractional-order differential equations.

We also remark that the applicability of our main result, Theorem 2, is only possible if the matrix A satisfies the condition $E_{\alpha,\alpha}(A) \geq 0$. We proved that, under the assumptions of Theorem 2, this condition holds if matrix A has dimensions 1 or 2. It would be interesting to check under what conditions we can guarantee that $E_{\alpha,\alpha}(A) \geq 0$ if the matrix A has dimensions greater than 2. Since many epidemiological models divide the population into subpopulations of epidemiological significance where the number of the infectious compartments are at most 2, our main result can be applied to a wide variety of epidemiological models.

Author Contributions: Conceptualization, R.A., N.M., and C.J.S.; methodology, R.A., N.M., and C.J.S.; formal analysis, R.A., N.M., and C.J.S.; writing—original draft preparation, R.A., N.M., and C.J.S.; writing—review and editing, R.A., N.M., and C.J.S. All authors have read and agreed to the published version of the manuscript.

Funding: Work is supported by Portuguese funds through the CIDMA-Center for Research and Development in Mathematics and Applications, and the Portuguese Foundation for Science and Technology (FCT-Fundação para a Ciência e a Tecnologia), within project UIDB/04106/2020. Cristiana J. Silva is also supported by FCT via the FCT Researcher Program CEEC Individual 2018 with reference CEECIND/00564/2018.

Institutional Review Board Statement: Not applicable.

Informed Consent Statement: Not applicable.

Data Availability Statement: Not applicable.

Acknowledgments: The authors are grateful to the two reviewer's valuable comments that improved the manuscript.

Conflicts of Interest: The authors declare no conflict of interest.

References

1. Ionescu, C.; Lopes, A.; Copot, D.; Machado, J.A.T.; Bates, J.H.T. The role of fractional calculus in modeling biological phenomena: A review. *Commun. Nonlinear Sci. Numer. Simul.* **2017**, *51*, 141–159. [CrossRef]
2. Jesus, I.S.; Machado, J.A.T.; Cunha, J.B. Fractional electrical impedances in botanical elements. *J. Vib. Control* **2008**, *14*, 1389–1402. [CrossRef]
3. Boukhouima, A.; Hattaf, K.; Lotfi, E.M.; Mahrouf, M.; Torres, D.F.M.; Yousfi, N. Lyapunov functions for fractional-order systems in biology: Methods and applications. *Chaos Solitons Fractals* **2020**, *140*, 110224. [CrossRef]
4. Benson, D.A.; Wheatcraft, S.W.; Meerschaert, M.M. The fractional-order governing equation of Lévy motion. *Water Resour. Res.* **2000**, *36*, 1413–1423. [CrossRef]
5. Stanislavsky, A.A. Fractional oscillator. *Phys. Rev. E* **2004**, *70*, 051103. [CrossRef]
6. Skovranek, T. The Mittag–Leffler fitting of the Phillips curve. *Mathematics* **2019**, *7*, 589. [CrossRef]
7. Xie, Y.; Wang, Z.; Meng, B. Stability and bifurcation of a delayed time-fractional order business cycle model with a general liquidity preference function and investment function. *Mathematics* **2019**, *7*, 846. [CrossRef]
8. Maalej, B.; Khlif, R.J.; Mhiri, C.; Elleuch, M.H.; Derbel, N. Adaptive fractional control optimized by genetic algorithms with application to polyarticulated robotic systems. *Math. Probl. Eng.* **2021**, *2021*, 5579541. [CrossRef]
9. Paola, M.; Pinnola, F.P.; Zingales, M. Fractional differential equations and related exact mechanical models. *Comput. Math. Appl.* **2013**, *66*, 608–620. [CrossRef]
10. Barros, L.C.; Lopes, M.M.; Pedro, F.S.; Esmi, E.; Santos, J.P.C.; Sánchez, D.E. The memory effect on fractional calculus: An application in the spread of COVID-19. *Comput. Appl. Math.* **2021**, *40*, 72. [CrossRef]
11. Bushnaq, S.; Saeed, T.; Torres, D.F.M.; Zeb, A. Control of COVID-19 dynamics through a fractional-order model. *Alex. Eng. J.* **2021**, *60*, 3587–3592. [CrossRef]
12. Ndaïrou, F.; Area, I.; Nieto, J.J.; Silva, C.J.; Torres, D.F.M. Fractional model of COVID-19 applied to Galicia, Spain and Portugal. *Chaos Solitons Fractals* **2021**, *144*, 110652. [CrossRef] [PubMed]
13. Liu, Z.; Lu, P. Stability analysis for HIV infection of CD4+ T-cells by a fractional differential time-delay model with cure rate. *Adv. Differ. Equ.* **2014**, *2014*, 298. [CrossRef]
14. Silva, C.J.; Torres, D.F.M. Stability of a fractional HIV/AIDS model. *Math. Comput. Simul.* **2019**, *164*, 180–190. [CrossRef]
15. Almeida, R.; Cruz, A.M.C.B.; Martins, N.; Monteiro, M.T.T. An epidemiological MSEIR model described by the Caputo fractional derivative. *Int. J. Dyn. Control* **2019**, *7*, 776–784. [CrossRef]
16. Diethelm, K. A fractional calculus based model for the simulation of an outbreak of dengue fever. *Nonlinear Dyn.* **2013**, *71*, 613–619. [CrossRef]
17. Qureshi, S.; Jan, R. Modeling of measles epidemic with optimized fractional order under Caputo differential operator. *Chaos Solitons Fractals* **2021**, *145*, 110766. [CrossRef]
18. Goufo, E.F.D.; Maritz, R.; Munganga, J. Some properties of the Kermack-McKendrick epidemic model with fractional derivative and nonlinear incidence. *Adv. Differ. Equ.* **2014**, *2014*, 278. [CrossRef]
19. Khan, M.A.; Ullah, S.; Ullah, S.; Farhan, M. Fractional order SEIR model with generalized incidence rate. *AIMS Math.* **2020**, *5*, 2843–2857. [CrossRef]
20. Özalp, N.; Demirci, E. A fractional order SEIR model with vertical transmission. *Appl. Math. Model.* **2011**, *54*, 1–6. [CrossRef]
21. Ammi, M.R.S.; Tahiri, M.; Torres, D.F.M. Global Stability of a Caputo Fractional SIRS Model with General Incidence Rate. *Math. Comput. Sci.* **2021**, *15*, 91–105. [CrossRef]
22. Günay, B.; Agarwal, P.; Guirao, J.L.G.; Momani, S. A fractional approach to a computational eco-epidemiological model with holling type-II functional response. *Symmetry* **2021**, *13*, 1159. [CrossRef]
23. Hassouna, M.; Ouhadan, A.; Kinani, E.H.E. On the solution of fractional order SIS epidemic model. *Chaos Solitons Fractals* **2018**, *117*, 168–174. [CrossRef] [PubMed]
24. Seo, Y.; Zeb, A.; Zaman, G.; Jung, I. Square-root dynamics of a SIR-model in fractional order. *Appl. Math.* **2012**, *3*, 1882–1887. [CrossRef]
25. Chen, Y.; Liu, F.; Yu, Q.; Li, T. Review of fractional epidemic models. *Appl. Math. Model.* **2021**, *97*, 281–307. [CrossRef]
26. Hattaf, K.; Yousfi, N. Global stability for fractional diffusion equations in biological systems. *Complexity* **2020**, *2020*, 5476842. [CrossRef]
27. Nabti, A.; Ghanbari, B. Global stability analysis of a fractional SVEIR epidemic model. *Math. Methods Appl. Sci.* **2021**, *44*, 8577–8597. [CrossRef]
28. Salahshour, S.; Ahmadian, A.; Salimi, M.; Pansera, B.A.; Ferrara, M. A new Lyapunov stability analysis of fractional-order systems with nonsingular kernel derivative. *Alex. Eng. J.* **2020**, *59*, 2985–2990. [CrossRef]
29. Xu, Q.; Zhuang, S.; Xu, X.; Che, C.; Xia, Y. Stabilization of a class of fractional-order nonautonomous systems using quadratic Lyapunov functions. *Adv. Differ. Equ.* **2018**, *2018*, 14. [CrossRef]
30. Castillo–Chavez, C.; Feng, Z.; Huang, W. On the computation of R_0 and its role in global stability. *IMA Vol. Math. Its Appl.* **2002**, *125*, 229–250.
31. Podlubny, I. *Fractional Differential Equations, Mathematics in Science and Engineering, 198*; Academic Press Inc.: San Diego, CA, USA, 1999.
32. Samko, S.G.; Kilbas, A.A.; Marichev, O.I. *Fractional Integrals and Derivatives, Translated from the 1987 Russian Original*; Gordon and Breach: Yverdon, Switzerland, 1993.

33. Haubold, H.J.; Mathai, A.M.; Saxena, R.K. Mittag–Leffler functions and their applications. *J. Appl. Math.* **2011**, *2011*, 298628. [CrossRef]
34. Hilfer, R.; Seybold, H.J. Computation of the generalized Mittag–Leffler function and its inverse in the complex plane. *Integral Transform. Spec. Funct.* **2006**, *17*, 637–652. [CrossRef]
35. Popolizio, M. On the Matrix Mittag–Leffler Function: Theoretical Properties and Numerical Computation. *Mathematics* **2019**, *7*, 1140. [CrossRef]
36. Sadeghi, A.; Cardoso, J.R. Some notes on properties of the matrix Mittag–Leffler function. *Appl. Math. Comput.* **2018**, *338*, 733–738. [CrossRef]
37. Cong, N.D.; Doan, T.S.; Tuan, H.T. Asymptotic stability of linear fractional systems with constant coefficients and small time-dependent perturbations. *Vietnam J. Math.* **2018**, *46*, 665–680. [CrossRef]
38. Diekmann, O.; Heesterbeek, J.A.P.; Metz, J.A.J. On the definition and the computation of the basic reproduction ratio R_0 in models for infectious diseases in heterogeneous population. *J. Math. Biol.* **1990**, *28*, 365–382. [CrossRef]
39. Miller, K.S.; Samko, S.G. A note on the complete monotonicity of the generalized Mittag–Leffler function. *Real Anal. Exch.* **1997**, *23*, 753–755. [CrossRef]
40. Pollard, H. The completely monotonic character of the Mittag-Leffer function $E_\alpha(-x)$. *Bull. Am. Math. Soc.* **1948**, *54*, 1115–1116. [CrossRef]
41. Diethelm, K. *The Analysis of Fractional Differential Equations, An Application-Oriented Exposition Using Differential Operators of Caputo Type*; LNM Springer: Berlin, Germany, 2010; Volume 2004.
42. LaSalle, J.P. *The Stability of Dynamical Systems, Regional Conferences Series in Applied Mathematics*; SIAM: Philadelphia, PA, USA, 1976.
43. Garrappa, R. Predictor-Corrector PECE Method for Fractional Differential Equations, MATLAB Central File Exchange, File ID: 32918. 2012. Available online: https://www.mathworks.com/matlabcentral/fileexchange/32918-predictor-corrector-pece-method-for-fractional-differential-equations (accessed on 13 July 2021).
44. Diethelm, K.; Freed, A.D. The Frac PECE subroutine for the numerical solution of differential equations of fractional order. In *Forschung und Wissenschaftliches Rechnen 1998*; Heinzel, S., Plesser, T., Eds.; Gessellschaft fur Wissenschaftliche Datenverarbeitung: Gottingen, Germany, 1999; pp. 57–71.
45. Li, M.Y.; Muldowney, J.S. Global stability for the SEIR model in epidemiology. *Math. Biosci.* **1995**, *125*, 155–164. [CrossRef]
46. Lavault, C. Integral representations and asymptotic behaviour of a Mittag–Leffler type function of two variables. *Adv. Oper. Theory* **2018**, *3*, 365–373. [CrossRef]
47. Silva, C.J.; Torres, D.F.M. A SICA compartmental model in epidemiology with application to HIV/AIDS in Cape Verde. *Ecol. Complex.* **2017**, *30*, 70–75. [CrossRef]
48. Pebody, R. Undetectable Viral Load and Transmission–Information for People with HIV. November 2020. Available online: https://www.aidsmap.com/about-hiv/undetectable-viral-load-and-transmission-information-people-hiv (accessed on 13 July 2021).

Article

On a Non-Newtonian Calculus of Variations

Delfim F. M. Torres

Center for Research and Development in Mathematics and Applications (CIDMA), Department of Mathematics, University of Aveiro, 3810-193 Aveiro, Portugal; delfim@ua.pt

Abstract: The calculus of variations is a field of mathematical analysis born in 1687 with Newton's problem of minimal resistance, which is concerned with the maxima or minima of integral functionals. Finding the solution of such problems leads to solving the associated Euler–Lagrange equations. The subject has found many applications over the centuries, e.g., in physics, economics, engineering and biology. Up to this moment, however, the theory of the calculus of variations has been confined to Newton's approach to calculus. As in many applications negative values of admissible functions are not physically plausible, we propose here to develop an alternative calculus of variations based on the non-Newtonian approach first introduced by Grossman and Katz in the period between 1967 and 1970, which provides a calculus defined, from the very beginning, for positive real numbers only, and it is based on a (non-Newtonian) derivative that permits one to compare relative changes between a dependent positive variable and an independent variable that is also positive. In this way, the non-Newtonian calculus of variations we introduce here provides a natural framework for problems involving functions with positive images. Our main result is a first-order optimality condition of Euler–Lagrange type. The new calculus of variations complements the standard one in a nontrivial/multiplicative way, guaranteeing that the solution remains in the physically admissible positive range. An illustrative example is given.

Keywords: calculus of variations; non-Newtonian calculus; multiplicative integral functionals; multiplicative Euler–Lagrange equations; admissible positive functions

MSC: 26A24; 49K05

Citation: Torres, D.F.M. On a Non-Newtonian Calculus of Variations. *Axioms* **2021**, *10*, 171. https://doi.org/10.3390/axioms10030171

Academic Editor: Giampiero Palatucci

Received: 12 June 2021
Accepted: 26 July 2021
Published: 29 July 2021

1. Introduction

A popular method of creating a new mathematical system is to vary the axioms of a known one. Non-Newtonian calculi provide alternative approaches to the usual calculus of Newton (1643–1727) and Leibniz (1646–1716), which were first introduced by Grossman and Katz (1933–2010) in the period between 1967 and 1970 [1]. The two most popular non-Newtonian calculi are the multiplicative and bigeometric calculi, which in fact are modifications of each other: in these calculi, the addition and subtraction are changed to multiplication and division [2]. Since such multiplicative calculi are variations on the usual calculus, the traditional one is sometimes called the additive calculus [3].

Recently, it has been shown that non-Newtonian/multiplicative calculi are more suitable than the ordinary Newtonian/additive calculus for some problems, e.g., in actuarial science, finance, economics, biology, demography, pattern recognition in images, signal processing, thermostatistics and quantum information theory [3–7]. This is explained by the fact that while the basis for the standard/additive calculus is the representation of a function as locally linear, the basis of a multiplicative calculus is the representation of a function as locally exponential [1,3,7]. In fact, the usefulness of product integration goes back to Volterra (1860–1940), who introduced in 1887 the notion of a product integral and used it to study solutions of differential equations [8,9]. For readers not familiar with product integrals, we refer to the book [10], which contains short biographical sketches of Volterra, Schlesinger and other mathematicians involved in the development of product integrals, and an extensive list of references, offering a gentle opportunity to become acquainted with the subject of non-Newtonian integration. For our purposes, it is enough to understand that a non-Newtonian calculus is a methodology that allows one to have a different look at problems that can be investigated via calculus: it provides differentiation and integration tools, based on multiplication instead of addition, and in some cases—mainly problems of price elasticity, multiplicative growth, etc.—the use of such multiplicative calculi is preferable to the traditional Newtonian calculus [11–14]. Moreover, a non-Newtonian calculus is a self-contained system, independent of any other system of calculus [15].

The main aim of our work was to obtain, for the first time in the literature, a non-Newtonian calculus of variations that involves the minimization of a functional defined by a non-Newtonian integral with an integrand/Lagrangian depending on the non-Newtonian derivative. The calculus of variations is a field of mathematical analysis that uses, as the name indicates, variations, which are small changes in functions, to find maxima and minima of the considered functionals: mappings from a set of functions to the real numbers. In the non-Newtonian framework, instead of the classical variations of the form $y(\cdot) + \epsilon h(\cdot)$, proposed by Lagrange (1736–1813) and still used nowadays in all recent formulations of the calculus of variations [16–18], for example, in the fractional calculus of variations [19,20], quantum variational calculus [21,22] and the calculus of variations on time scales [23,24], we propose here to use "multiplicative variations". More precisely, in contrast with the calculi of variations found in the literature, we show here, for the first time, how to consider variations of the form $y(\cdot) \cdot \epsilon^{\ln h(\cdot)}$. The functionals of the calculus of variations are expressed as definite integrals, here in a non-Newtonian sense, involving functions and their derivatives, in a non-Newtonian sense here. The functions that maximize or minimize the functionals of the calculus of variations are found using the Euler–Lagrange equation, which we prove here in the non-Newtonian setting. Given the importance of the calculus of variations in applications, for example, in physics [25,26], economics [27,28] and biology [29,30], and the importance that non-Newtonian calculus already has in these areas, we trust that the calculus of variations initiated here will call attention to the research community. We give credit to the citation found in the 1972 book of Grossman and Katz [1]: "for each successive class of phenomena, a new calculus or a new geometry".

2. Materials and Methods

From 1967 till 1970, Grossman and Katz gave definitions of new kinds of derivatives and integrals, converting the roles of subtraction and addition into division and multiplication, respectively, and established a new family of calculi, called non-Newtonian calculi [1,31,32], which are akin to the classical calculus developed by Newton and Leibniz three centuries ago. Non-Newtonian calculi use different types of arithmetic and their generators. Let α be a bijection between subsets X and Y of the set of real numbers $\mathbb{R}$, and endow Y, with the induced operations sum and multiplication and the ordering given by the inverse map α^{-1}. Then the α-arithmetic is a field with the order topology [33]. In concrete, given a bijection $\alpha : X \to Y \subseteq \mathbb{R}$, called a generator [15], we say that α defines an arithmetic if the following four operations are defined:

$$\begin{aligned} x \oplus y &= \alpha\big(\alpha^{-1}(x) + \alpha^{-1}(y)\big), \\ x \ominus y &= \alpha\big(\alpha^{-1}(x) - \alpha^{-1}(y)\big), \\ x \odot y &= \alpha\big(\alpha^{-1}(x) \cdot \alpha^{-1}(y)\big), \\ x \oslash y &= \alpha\big(\alpha^{-1}(x) / \alpha^{-1}(y)\big). \end{aligned} \tag{1}$$

If α is chosen to be the identity function and $X = \mathbb{R}$, then (1) reduces to the four operators studied in school; i.e., one gets the standard arithmetic, from which the traditional (Newton–Leibniz) calculus is developed. For other choices of α and X, we can get an infinitude of other arithmetics from which Grossman and Katz produced a series of non-Newton calculi, compiled in the seminal book of 1972 [1]. Among all such non-Newton calculi, recently great interest has been focused on the Grossman–Katz calculus obtained when we fix $\alpha(x) = e^x$, $\alpha^{-1}(x) = \ln(x)$ and $X = \mathbb{R}^+$ for the set of real numbers strictly greater than zero [7,12,14,15]. We shall concentrate here on one option, originally called by Grossman and Katz the geometric/exponential/bigeometric calculus [1,2,13,34–37], but from which other different terminology and small variations of the original calculus have grown up in the literature, in particular, the multiplicative calculus [3–5,8,12,15,36,38–40], and more recently, the proportional calculus [7,11,14,41], which is essentially the bigeometric calculus of [35]. Here we follow closely this last approach—in particular, the exposition of the non-Newton calculus as found in [7,14,35], because it is appealing to scientists who seek ways to express laws in a scale-free form.

Throughout the text, we fix $\alpha(x) = e^x$, $\alpha^{-1}(x) = \ln(x)$, and $X = \mathbb{R}^+$. Then we get from (1) the following operations:

$$\begin{aligned} x \oplus y &= x \cdot y, \\ x \ominus y &= \frac{x}{y}, \\ x \odot y &= x^{\ln(y)}, \\ x \oslash y &= x^{1/\ln(y)}, \quad y \neq 1. \end{aligned} \tag{2}$$

Let $a, b, c \in \mathbb{R}^+$. In the non-Newtonian arithmetic given by (2), the following properties of the $\odot$ operation hold (cf. Proposition 2.1 of [14]):

(i) $a \odot b = b \odot a$ (commutativity);
(ii) $a \odot (b \odot c) = (a \odot b) \odot c$ (associativity);
(iii) $a \odot e = a$ (Euler's/Napier's transcendent number e is the neutral element for $\odot$);
(iv) if $a \neq 1$ and we define $a^{\{-1\}} = e \oslash a$, then $a \odot a^{\{-1\}} = e$ (inverse element).

We see that in this non-Newtonian algebra, $a = 1$ is the traditional "zero" (in the current arithmetic, 0 represents $-\infty$). In fact, see Proposition 2.2 of [14], one has

(v) $b \odot a^{\{-1\}} = b \oslash a$;
(vi) $\left(a^{\{-1\}}\right)^{\{-1\}} = a$;
(vii) $\ln(a \odot b) = \ln(a) \oplus \ln(b)$;
(viii) $(a \odot b)^{\{-1\}} = a^{\{-1\}} \odot b^{\{-1\}}$.

Based on the mentioned properties, one easily proves that $(\mathbb{R}^+, \oplus, \odot)$ is a field (see Theorem 2.3 of [14]). In this field, the following calculus has been developed [7,14].

Definition 1 (absolute value). *The absolute value of $x \in \mathbb{R}^+$, denoted by $[[x]]$, is given by*

$$[[x]] = \begin{cases} x & \text{if } x \geq 1, \\ 1 \ominus x & \text{if } x \in (0,1). \end{cases}$$

Let x, y, z be positive real numbers and define $d : \mathbb{R}^+ \times \mathbb{R}^+ \to \mathbb{R}^+$ as

$$d(x,y) = [[x \ominus y]].$$

The following properties are simple to prove:

- $d(x,y) \geq 1$;
- $d(x,y) = 1$ if, and only if, $x = y$;
- $d(x,y) = d(y,x)$;
- $d(x,z) \leq d(x,y) + d(y,z)$.

We can now introduce the notion of limit.

Definition 2 (limit). *We write $\lim_{x \to x_0} f(x) = L$, $L \in \mathbb{R}^+$, as: if for all $\epsilon > 1$ there exists $\delta > 1$ such that if $1 < d(x, x_0) < \delta$, then $d(f(x), L) < \epsilon$.*

According with Definition 2, it is possible to specify the meaning of equality $\lim_{x\to x_0} f(x) = f(x_0)$, and therefore, the notion of continuity in the non-Newtonian calculus.

Definition 3 (continuity). *We say that f is continuous at x_0 or*

$$\lim_{x\to x_0} f(x) = f(x_0),$$

if $\forall\ \epsilon > 1,\ \exists\ \delta > 1$ *such that* $d(x, x_0) < \delta \implies d(f(x), f(x_0)) < \epsilon$.

We proceed by reviewing the essentials on non-Newtonian differentiation and integration.

2.1. Derivatives

The derivative of a function is introduced in the following terms.

Definition 4 (derivative [7,14,35,41]). *A positive function f is differentiable at x_0 if*

$$\lim_{x\to x_0}[(f(x) \ominus f(x_0)) \oslash (x \ominus x_0)] = \lim_{h\to 1}[(f(x_0 \oplus h) \ominus f(x_0)) \oslash h]$$

exists. In this case, the limit is denoted by $\widetilde{f}(x_0)$ and receives the name of derivative of f at x_0. Moreover, we say that f is differentiable if f is differentiable at x_0 for all x_0 in the domain of f.

It is not difficult to prove that if f is differentiable at x_0, then f is continuous at x_0. Define

$$x^{\{0\}} = e,$$
$$x^{\{n\}} = x \odot \cdots \odot x \ n \text{ times}, \quad n \in \mathbb{N}.$$

We have

$$\widetilde{x^{\{n\}}} = \left(x^{\{n-1\}}\right)^n = e^n \odot x^{\{n-1\}}, \quad n \in \mathbb{N}. \tag{3}$$

In particular, if $n = 1$ in (3), we get:

- If $f(x) = x$, then $\widetilde{f}(x) = e$.

More examples of derivatives of a function in the sense of Definition 4 follow:

- If $f(x) = e^x$, then $\widetilde{f}(x) = e^x$;
- If $f(x) = \ln(x)$, then $\widetilde{f}(x) = e \oslash x$;
- If $\cos_e(x) := e^{\cos(\ln(x))}$ and $\sin_e(x) := e^{\sin(\ln(x))}$, then

$$\widetilde{\cos_e(x)} = 1 \ominus \sin_e(x) \quad \text{and} \quad \widetilde{\sin_e(x)} = \cos_e(x).$$

The basic rules of differentiation (keep recalling that 1 is the "zero" of the non-Newtonian calculus) follow:

(a) If $f(x) = c$, and c is a positive constant, then $\widetilde{f}(x) = 1$ (derivative of a constant);
(b) $\widetilde{f \oplus g} = \widetilde{f} \oplus \widetilde{g}$ (derivative of a sum);
(c) $\widetilde{f \ominus g} = \widetilde{f} \ominus \widetilde{g}$ (derivative of a difference);
(d) $\widetilde{f \odot g} = \left(\widetilde{f} \odot g\right) \oplus (f \odot \widetilde{g})$ (derivative of a product);
(e) $\widetilde{f \circ g} = \left(\widetilde{f} \circ g\right) \odot \widetilde{g}$ (chain rule).

If $\widetilde{f}(x) = 1$ for all $x \in (a, b)$, then $f(x) = c$ for all $x \in (a, b)$, where c is a constant. Moreover, if $\widetilde{f}(x) = \widetilde{g}(x)$ for all $x \in (a, b)$, then there exists a constant c such that $f(x) = cg(x)$ for all $x \in (a, b)$; that is, $f(x) = g(x) \oplus c$.

Higher-order derivatives are defined as usual:

$$\widetilde{f}^{(0)}(x) = f(x),$$
$$\widetilde{f}^{(n)}(x) = \frac{\tilde{d}}{\tilde{d}x}\left[\widetilde{f}^{(n-1)}(x)\right], \quad n \in \mathbb{N}.$$

In the sequel, we use the following notation:

$$\widetilde{\sum_{i=0}^{n}} a_i = a_0 \oplus \cdots \oplus a_n.$$

Theorem 1 (Taylor's theorem)**.** *Let f be a function such that $\widetilde{f}^{(n+1)}(x)$ exists for all x in a range that contains the number a. Then,*

$$f(x) = P_n(x) \oplus R_n(x)$$

for all x, where

$$P_n(x) = \widetilde{\sum_{k=0}^{n}} e^{\frac{1}{k!}} \odot \widetilde{f}^{(k)}(a) \odot (x \ominus a)^{\{k\}}$$

is the Taylor polynomial of degree n and

$$R_n(x) = e^{\frac{1}{(n+1)!}} \odot \widetilde{f}^{(n+1)}(c) \odot (x \ominus a)^{\{n+1\}}$$

is the remainder term in Lagrange form, for some number c between a and x.

Suppose f is a function that has derivatives of all orders over an interval centered on a. If $\lim_{n\to+\infty} R_n(x) = 1$ for all x in the interval, then the Taylor series is convergent and converges to $f(x)$:

$$f(x) = \widetilde{\sum_{k=0}^{+\infty}} e^{\frac{1}{k!}} \odot \widetilde{f}^{(k)}(a) \odot (x \ominus a)^{\{k\}}.$$

As examples of convergent series, one has:

$$e^x = \widetilde{\sum_{k=0}^{+\infty}} e^{\frac{1}{k!}} \odot x^{\{k\}},$$
$$\cos_e(x) = \widetilde{\sum_{k=0}^{+\infty}} e^{\frac{(-1)^k}{2k!}} \odot x^{\{2k\}}.$$

The Taylor's theorem given by Theorem 1 has a natural extension for functions of several variables [6,42]. Here we proceed by briefly reviewing integration. For more on the alpha-arithmetic, its topology and analysis, we refer the reader to the literature. For example, mean value theorems can be found in the original book of Grossman and Katz of 1972 [1]; for a recent reference with detailed proofs, see [43].

2.2. Integrals

The notion of integral for the non-Newtonian calculus under consideration is a type of product integration [10]. As expected, the function $F(x)$ is an antiderivative of function $f(x)$ on the interval I if $\widetilde{F}(x) = f(x)$ for all $x \in I$. The indefinite integral of $f(x)$ is denoted by

$$\widetilde{\int} f(x)\tilde{d}x = F(x) \oplus c,$$

where c is a constant. Examples are (see [7,14,35]):

- If k is a positive constant, then $\fint k\tilde{d}x = (k \odot x) \oplus c$—in particular, if $k = 1$, then $\fint 1\tilde{d}x = c$;
- $\fint e^n \tilde{d}x = x^n \oplus c$;
- $\fint x^{\{n\}} \tilde{d}x = e^{\frac{1}{n+1}} \odot x^{\{n+1\}} \oplus c$;
- $\fint e \oslash x \, \tilde{d}x = \ln(x) \oplus c$;
- $\fint e^{a \odot x} \tilde{d}x = \left(a^{\{-1\}} \odot e^{a \odot x}\right) \oplus c$;
- $\fint e^{rx^2} \tilde{d}x = e^{\frac{rx^2}{2}} \oplus c$.

The definite integral of f on $[a,b]$ is denoted by

$$\fint_a^b f(x)\tilde{d}x.$$

If f is positive and continuous on $[a,b]$, then f is integrable in $[a,b]$. The following properties hold:

(i) $\fint_a^b f(x)\tilde{d}x = \fint_a^c f(x)\tilde{d}x \oplus \fint_c^b f(x)\tilde{d}x$;
(ii) $\fint_a^b (f(x) \oplus g(x))\tilde{d}x = \fint_a^b f(x)\tilde{d}x \oplus \fint_a^b g(x)\tilde{d}x$;
(iii) $\fint_a^b (f(x) \ominus g(x))\tilde{d}x = \fint_a^b f(x)\tilde{d}x \ominus \fint_a^b g(x)\tilde{d}x$;
(iv) $\fint_a^b c \odot f(x)\tilde{d}x = c \odot \fint_a^b f(x)\tilde{d}x$;
(v) If $m \le f(x) \le M$ for all $x \in [a,b]$, then $m \odot (b \ominus a) \le \fint_a^b f(x)\tilde{d}x \le M \odot (b \ominus a)$;
(vi) If $f(x) \le g(x)$ for all $x \in [a,b]$, then $\fint_a^b f(x)\tilde{d}x \le \fint_a^b g(x)\tilde{d}x$.

If f is positive and integrable on $[a,b]$, then F defined on $[a,b]$ by

$$F(x) = \fint_a^x f(t)\tilde{d}t$$

is continuous over $[a,b]$. Moreover, the fundamental theorems of integral calculus hold: if f is continuous in $x \in [a,b]$, then F is differentiable at x with

$$\widetilde{F(x)} = f(x);$$

if $f = \widetilde{h}$ for some function h, then

$$\fint_a^b f(x)\tilde{d}x = h(b) \ominus h(a).$$

For more on the α-arithmetic, its generalized real analysis, its fundamental topological properties related to non-Newtonian metric spaces and its calculus, including non-Newtonian differential equations and its applications, see [7,33,44–48]. For gentle, thorough and modern introduction to the subject of non-Newtonian calculi, we also refer the reader to the recent book [49]. Now we proceed with our original results.

3. Results

In order to develop a non-Newtonian calculus of variations (dynamic optimization), we begin by first proving some necessary results of static optimization.

3.1. Static Optimization

Given $\epsilon > 1$, let

$$\mathcal{B}(\bar{x},\epsilon) := \{x \in \mathbb{R}^+ : d(x,\bar{x}) \leq \epsilon\} = \{x \in \mathbb{R}^+ : [[x \ominus \bar{x}]] \leq \epsilon\}.$$

Note that for $a, b \in \mathbb{R}^+$, one has

$$a = b \Leftrightarrow \frac{a}{b} = 1 \left(\text{or } \frac{b}{a} = 1\right) \Leftrightarrow a \ominus b = 1 \text{ (or } b \ominus a = 1).$$

Similarly for inequalities, for example,

$$a < b \Leftrightarrow \frac{a}{b} < 1 \Leftrightarrow a \ominus b < 1.$$

This means that

$$\mathcal{B}(\bar{x},\epsilon) = \{x \in \mathbb{R}^+ : d(x,\bar{x}) \ominus \epsilon \leq 1\}.$$

Definition 5 (local minimizer). *Let $f : (a,b) \to \mathbb{R}^+$ and consider the problem of minimizing $f(x)$, $x \in (a,b)$. We say that $x \in (a,b)$ is a (local) minimizer of f in (a,b) if there exists $\epsilon > 1$ such that $f(x) \leq f(y)$ (i.e., $f(x) \ominus f(y) \leq 1$) for all $y \in \mathcal{B}(x,\epsilon) \cap (a,b)$. In this case, we say that $f(x)$ is a (local) minimum.*

Another important concept in optimization is that of descent direction.

Definition 6 (descent direction). *A $d \in \mathbb{R}^+$ is said to be a descent direction of f at x if $f(x \oplus \epsilon \odot d) < f(x)$ $\forall$ $\epsilon > 1$ sufficiently close to 1, or equivalently, if $f(x \oplus \epsilon \odot d) \ominus f(x) < 1$ for all $\epsilon > 1$ sufficiently close to 1.*

Remark 1. *From the chain rule and other properties of Section 2, it follows that*

$$\frac{\tilde{d}}{\tilde{d}x}[f(x \oplus \epsilon \odot d)] = \tilde{f}(x \oplus \epsilon \odot d)$$

and

$$\frac{\tilde{d}}{\tilde{d}\epsilon}[f(x \oplus \epsilon \odot d)] = \tilde{f}(x \oplus \epsilon \odot d) \odot d. \tag{4}$$

In particular, we get from (4) that

$$\frac{\tilde{d}}{\tilde{d}\epsilon}[f(x \oplus \epsilon \odot d)]\bigg|_{\epsilon=1} = \tilde{f}(x) \odot d.$$

Our first result allow us to identify a descent direction of f at x based on the derivative of f at x.

Theorem 2. *Let f be differentiable. If there exists $d \in \mathbb{R}^+$ such that $\tilde{f}(x) \odot d < 1$, then d is a descent direction of f at x.*

Proof. We know from Taylor's theorem (Theorem 1) that

$$f(x \oplus \epsilon \odot d) = f(x) \oplus \epsilon \odot \tilde{f}(x) \odot d \oplus R_1(x \oplus \epsilon \odot d), \tag{5}$$

where

$$\begin{aligned} R_1(x \oplus \epsilon \odot d) &= \left(\tilde{f}^{(2)}(c)\right)^{\frac{1}{2}} \odot (\epsilon \odot d)^{\{2\}} \\ &= \epsilon^{\{2\}} \odot \left(\tilde{f}^{(2)}(c)\right)^{\frac{1}{2}} \odot d^{\{2\}} \end{aligned}$$

with c being in the interval between x and $x \oplus \epsilon \odot d$. The equality (5) can be written in the following equivalent form:

$$f(x \oplus \epsilon \odot d) \ominus f(x) = \epsilon \odot \widetilde{f}(x) \odot d \oplus R_1(x \oplus \epsilon \odot d). \tag{6}$$

Recalling that $a \odot b^{\{-1\}} = a \oslash b$, $a \odot a^{\{-1\}} = e$ and $\odot$ is distributive over $\oplus$, we get from (6) that

$$\begin{aligned}(f(x \oplus \epsilon \odot d) \ominus f(x)) \oslash \epsilon &= \widetilde{f}(x) \odot d \oplus \epsilon^{\{-1\}} \odot R_1(x \oplus \epsilon \odot d) \\ &= \widetilde{f}(x) \odot d \oplus \epsilon \odot \left(\widetilde{f}^{(2)}(c)\right)^{\frac{1}{2}} \odot d^{\{2\}}.\end{aligned} \tag{7}$$

Now we note that as $\epsilon \to 1$, one has $\epsilon \odot \left(\widetilde{f}^{(2)}(c)\right)^{\frac{1}{2}} \odot d^{\{2\}} \to 1$ so that the right-hand side of (7) converges to $\widetilde{f}(x) \odot d$. From the hypothesis $\widetilde{f}(x) \odot d < 1$ of our theorem, this means that, for $\epsilon > 1$ sufficiently close to 1, the right-hand side of (7) is strictly less than one. Thus, for $\epsilon > 1$ sufficiently close to 1,

$$(f(x \oplus \epsilon \odot d) \ominus f(x)) \oslash \epsilon < 1. \tag{8}$$

Recalling that $a \oslash \epsilon < 1 \Leftrightarrow a^{1/\ln(\epsilon)} < 1$, we conclude from (8) that for ϵ sufficiently close to 1 we have $f(x \oplus \epsilon \odot d) \ominus f(x) < 1$; that is, d is a descent direction of f at x. □

As a corollary of Theorem 2, we obtain Fermat's necessary optimality condition, which gives us a method to find local minimizers (or maximizers) of differentiable functions on open sets, by showing that every local extremizer of the function is a stationary point (the non-Newtonian function's derivative is one at that point).

Theorem 3 (Fermat's theorem–stationary points). *Let $f : (a, b) \to \mathbb{R}^+$ be differentiable. If $x \in (a, b)$ is a minimizer of f, then $\widetilde{f}(x) = 1$.*

Proof. We want to prove that for a minimizer x we must have $\widetilde{f}(x) = 1 \Leftrightarrow 1 \ominus \widetilde{f}(x) = 1$. We do the proof by contradiction. Assume that $\widetilde{f}(x) \neq 1$; that is, $1 \ominus \widetilde{f}(x) \neq 1$. Let $d = 1 \ominus \widetilde{f}(x)$. Then,

$$\begin{aligned}\widetilde{f}(x) \odot d &= \widetilde{f}(x) \odot \left(1 \ominus \widetilde{f}(x)\right) = 1 \ominus \widetilde{f}(x)^{\{2\}} \\ &= \frac{1}{\widetilde{f}(x) \odot \widetilde{f}(x)} = \left(\frac{1}{\widetilde{f}(x)}\right)^{\ln\left(\widetilde{f}(x)\right)}\end{aligned}$$

and since $g(y) = \left(\frac{1}{y}\right)^{\ln(y)}$ is a function with $0 < g(y) < 1$ for all $y \neq 1$, we conclude that $\widetilde{f}(x) \odot d < 1$. It follows from Theorem 2 that d is a descent direction of f at x, and therefore, from the definition of descent direction, x is not a local minimizer. □

In the next section, we make use of Theorem 3 to prove the non-Newtonian Euler–Lagrange equation.

3.2. Dynamic Optimization

A central tool in dynamic optimization, both in the calculus of variations and optimal control [16], is integration by parts. In what follows, we use the following notation:

$$\psi(x)\big|_a^b = \psi(b) \ominus \psi(a).$$

Theorem 4 (integration by parts). *Let $f : [a,b] \to \mathbb{R}^+$ and $g : [a,b] \to \mathbb{R}^+$ be differentiable. The following formula of integration by parts holds:*

$$\oint_a^b \tilde{f}(x) \odot g(x) \tilde{d}x = f(x) \odot g(x)\Big|_a^b \ominus \oint_a^b f(x) \odot \tilde{g}(x) \tilde{d}x. \tag{9}$$

Proof. From the derivative of a product, we know that

$$\frac{\tilde{d}}{\tilde{d}x}[f(x) \odot g(x)] = \tilde{f}(x) \odot g(x) \oplus f(x) \odot \tilde{g}(x). \tag{10}$$

On the other hand, the fundamental theorem of integral calculus tell us that

$$\oint_a^b \frac{\tilde{d}}{\tilde{d}x}[f(x) \odot g(x)] \tilde{d}x = f(x) \odot g(x)\Big|_a^b.$$

Therefore, by integrating (10) from a to b, we conclude that

$$\oint_a^b \left[\tilde{f}(x) \odot g(x)\right] \tilde{d}x \oplus \oint_a^b [f(x) \odot \tilde{g}(x)] \tilde{d}x = f(x) \odot g(x)\Big|_a^b,$$

which is equivalent to (9). □

We are now in a condition to formulate the fundamental problem of the calculus of variations: to minimize the integral functional

$$\mathcal{F}[y] = \oint_a^b L(x, y(x), \tilde{y}(x)) \tilde{d}x$$

over all smooth functions y on $[a,b]$ with fixed end points $y(a) = y_a$ and $y(b) = y_b$. The central result of the calculus of variations and classical mechanics is the celebrated Euler–Lagrange equation, whose solutions are stationary points of the given action functional. We restrict ourselves here to the classical framework of the calculus of variations, where both the Lagrangian L and admissible functions y are smooth enough: typically, one considers $L \in C^2$ and $y \in C^2$, so that one can look to the Euler–Lagrange equation as a second-order ordinary differential equation [17]. We adopt such assumptions here. We denote the problem by (P).

Before proving the Euler–Lagrange equation (the necessary optimality condition for problem (P)), we first need to prove a non-Newtonian analogue of the fundamental lemma of the calculus of variations.

Lemma 1 (fundamental lemma of the calculus of variations). *If $f : [a,b] \to \mathbb{R}^+$ is a positive continuous function such that*

$$\oint_a^b f(x) \odot h(x) \tilde{d}x = 1 \tag{11}$$

for all functions $h(x)$ that are continuous for $a \le x \le b$ with $h(a) = h(b) = 1$, then $f(x) = 1$ for all $x \in [a,b]$.

Proof. We do the proof by contradiction. Suppose the function f is not one—$f(x) > 1$—at some point $x \in [a,b]$. Then, by continuity, $f(x) > 1$ for all x in some interval $[x_1, x_2] \subset [a,b]$. If we set

$$h(x) = \begin{cases} (x \ominus x_1) \odot (x_2 \ominus x) & \text{if } x \in [x_1, x_2], \\ 1 & \text{if } x \in [a,b] \setminus [x_1, x_2], \end{cases}$$

then $h(x)$ satisfies the assumptions of the lemma; i.e., $h(x)$ is continuous for $x \in [a,b]$ with $h(a) = 1$ and $h(b) = 1$. We have

$$\begin{aligned} \oint_a^b f(x) \odot h(x) \tilde{d}x &= \oint_a^{x_1} 1 \tilde{d}x \oplus \oint_{x_1}^{x_2} f(x) \odot (x \ominus x_1) \odot (x_2 \ominus x) \tilde{d}x \oplus \oint_{x_2}^b 1 \tilde{d}x \\ &= \oint_{x_1}^{x_2} f(x) \odot (x \ominus x_1) \odot (x_2 \ominus x) \tilde{d}x. \end{aligned} \tag{12}$$

Let us analyze the integrand $\beta(x)$ of (12):

$$\begin{aligned} \beta(x) &= f(x) \odot [(x \ominus x_1) \odot (x_2 \ominus x)] = f(x) \odot \left[\frac{x}{x_1} \odot \frac{x_2}{x}\right] \\ &= f(x) \odot \left[\left(\frac{x}{x_1}\right)^{\ln\left(\frac{x_2}{x}\right)}\right]. \end{aligned}$$

Since $f(x) > 1$ and $\left(\frac{x}{x_1}\right)^{\ln\left(\frac{x_2}{x}\right)} > 1$ for any $x \in (x_1, x_2)$, we also have that $\beta(x) > 1$ for any $x \in (x_1, x_2)$, with $\beta(x) = 1$ at $x = x_1$ and $x = x_2$. It follows that

$$\oint_a^b f(x) \odot h(x) \tilde{d}x = \oint_{x_1}^{x_2} \beta(x) \tilde{d}x > 1 \odot (x_2 \ominus x_1) = 1.$$

This contradicts (11) and proves the lemma. □

Now we formulate and prove the analog of the Euler–Lagrange differential equation for our problem (P).

Theorem 5 (Euler–Lagrange equation). *If $y(x)$, $x \in [a,b]$, is a solution to the problem (P)*

$$\mathcal{F}[y] = \oint_a^b L(x, y(x), \widetilde{y}(x)) \tilde{d}x \longrightarrow \min_{y \in \mathcal{Y}(y_a; y_b)}$$

with

$$\mathcal{Y}(y_a; y_b) := \left\{y \in C^2\left([a,b]; \mathbb{R}^+\right) : y(a) = y_a,\ y(b) = y_b,\ y(x) > 0\ \forall\ x \in [a,b]\right\},$$

then $y(x)$ satisfies the Euler–Lagrange equation

$$\widetilde{L}_y(x, y(x), \widetilde{y}(x)) = \frac{\tilde{d}}{\tilde{d}x} \widetilde{L}_{\widetilde{y}}(x, y(x), \widetilde{y}(x)) \tag{13}$$

for all $x \in [a,b]$.

Proof. Let $y(x)$, $x \in [a,b]$, be a minimizer of (5). Then, function $(y \oplus \epsilon \odot h)(x)$, $x \in [a,b]$, belongs to $\mathcal{Y}(y_a; y_b)$ for any function $h \in \mathcal{Y}(1;1)$ and for any ϵ in an open neighborhood of 1. Note that $(y \oplus \epsilon \odot h)(x) = y(x)$ for $\epsilon = 1$. This means that for any smooth function $h(x)$, $x \in [a,b]$, satisfying $h(a) = h(b) = 1$, the function $\varphi(\epsilon)$ defined by

$$\begin{aligned} \varphi(\epsilon) = \mathcal{F}[y \oplus \epsilon \odot h] &= \oint_a^b L\Big(x, (y \oplus \epsilon \odot h)(x), (\widetilde{y \oplus \epsilon \odot h})(x)\Big) \tilde{d}x \\ &= \oint_a^b L\Big(x, y(x) \oplus \epsilon \odot h(x), \widetilde{y}(x) \oplus \epsilon \odot \widetilde{h}(x)\Big) \tilde{d}x \end{aligned} \tag{14}$$

has a minimizer for $\epsilon = 1$. It follows from Fermat's theorem (Theorem 3) that

$$\widetilde{\varphi}(1) = \frac{\tilde{d}}{\tilde{d}\epsilon} \varphi(\epsilon)\bigg|_{\epsilon=1} = 1.$$

By differentiating (14) with respect to ϵ, and then putting $\epsilon = 1$, we get from the chain rule and relations

$$\frac{\tilde{d}}{\tilde{d\epsilon}}(y(x) \oplus \epsilon \odot h(x)) = h(x), \quad \frac{\tilde{d}}{\tilde{d\epsilon}}\left(\tilde{y}(x) \oplus \epsilon \odot \tilde{h}(x)\right) = \tilde{h}(x),$$

that

$$1 = \int_a^b \left[\tilde{L}_y(x, y(x), \tilde{y}(x)) \odot h(x) \oplus \tilde{L}_{\tilde{y}}(x, y(x), \tilde{y}(x)) \odot \tilde{h}(x)\right] \tilde{d}x. \tag{15}$$

From integration by parts (Theorem 4), and the fact that $h(a) = 1$ and $h(b) = 1$, one has

$$\begin{aligned}\int_a^b \tilde{L}_{\tilde{y}}(x, y(x), \tilde{y}(x)) \odot \tilde{h}(x) \tilde{d}x = h(x) \odot \tilde{L}_{\tilde{y}}(x, y(x), \tilde{y}(x))\Big|_a^b \\ \ominus \int_a^b \frac{\tilde{d}}{\tilde{d}x}[\tilde{L}_{\tilde{y}}(x, y(x), \tilde{y}(x))] \odot h(x) \tilde{d}x,\end{aligned}$$

that is,

$$\int_a^b \tilde{L}_{\tilde{y}}(x, y(x), \tilde{y}(x)) \odot \tilde{h}(x) \tilde{d}x = 1 \ominus \int_a^b \frac{\tilde{d}}{\tilde{d}x}[\tilde{L}_{\tilde{y}}(x, y(x), \tilde{y}(x))] \odot h(x) \tilde{d}x. \tag{16}$$

Using equality (16) in the necessary condition (15), we get that

$$\begin{aligned}1 = \int_a^b \tilde{L}_y(x, y(x), \tilde{y}(x)) \odot h(x) \tilde{d}x \oplus 1 \ominus \int_a^b \frac{\tilde{d}}{\tilde{d}x}[\tilde{L}_{\tilde{y}}(x, y(x), \tilde{y}(x))] \odot h(x) \tilde{d}x \\ \Leftrightarrow \int_a^b \left[\tilde{L}_y(x, y(x), \tilde{y}(x)) \ominus \frac{\tilde{d}}{\tilde{d}x}\left(\tilde{L}_{\tilde{y}}(x, y(x), \tilde{y}(x))\right)\right] \odot h(x) \tilde{d}x = 1.\end{aligned} \tag{17}$$

The result follows from the fundamental lemma of the calculus of variations (Lemma 1) applied to (17): $\tilde{L}_y(x, y(x), \tilde{y}(x)) \ominus \frac{\tilde{d}}{\tilde{d}x}\left(\tilde{L}_{\tilde{y}}(x, y(x), \tilde{y}(x))\right) = 1$ for all $x \in [a, b]$. □

To illustrate our main result, let us see an example. Consider the following problem of the calculus of variations:

$$\begin{gathered}\mathcal{F}[y] = \sqrt{e} \odot \int_1^{e^{2\pi}} \left[\tilde{y}^{\{2\}}(x) \ominus y^{\{2\}}(x)\right] \tilde{d}x \longrightarrow \min, \\ y(1) = e, \quad y(e^{2\pi}) = e^{-1}.\end{gathered} \tag{18}$$

Theorem 5 tell us that the solution of (18) must satisfy the Euler–Lagrange Equation (13). In this example, the Lagrangian L is given by $L(x, y, \tilde{y}) = \sqrt{e} \odot \left(\tilde{y}^{\{2\}} \ominus y^{\{2\}}\right)$, so that

$$\begin{aligned}\tilde{L}_y(x, y, \tilde{y}) &= \sqrt{e} \odot \left(1 \ominus e^2 \odot y\right), \\ \tilde{L}_{\tilde{y}}(x, y, \tilde{y}) &= \sqrt{e} \odot \left(e^2 \odot \tilde{y} \ominus 1\right) = \sqrt{e} \odot \left(e^2 \odot \tilde{y}\right).\end{aligned} \tag{19}$$

Noting that $\sqrt{e} = e \oslash e^2 = \left(e^2\right)^{\{-1\}}$, the equalities (19) simplify to

$$\tilde{L}_y(x, y, \tilde{y}) = 1 \ominus y, \quad \tilde{L}_{\tilde{y}}(x, y, \tilde{y}) = \tilde{y}$$

and the Euler–Lagrange Equation (13) takes the form

$$1 \ominus y(x) = \tilde{y}^{(2)}(x) \Leftrightarrow \tilde{y}^{(2)}(x) \oplus y(x) = 1. \tag{20}$$

The second-order differential Equation (20) has solutions of the form

$$y(x) = c_1 \odot \cos_e(x) \oplus c_2 \odot \sin_e(x),$$

where c_1 and c_2 are constants. Given the boundary conditions $y(1) = e$ and $y(e^{2\pi}) = e^{-1}$, we conclude that the Euler–Lagrange extremal for problem (18) is given by

$$y(x) = e \odot \cos_e(x) \oplus e^2 \odot \sin_e(x) \Leftrightarrow y(x) = e^{2\sin(\ln(x))+\cos(\ln(x))}.$$

4. Discussion

One can say that the calculus of variations began in 1687 with Newton's minimal resistance problem [25,50,51]. It immediately occupied the attention of Bernoulli (1655–1705), but it was Euler (1707–1783) who first elaborated mathematically on the subject, beginning in 1733. Lagrange (1736–1813) was influenced by Euler's work and contributed significantly to the theory, introducing a purely analytic approach to the subject based on additive variations $y + \epsilon h$, whose essence we still follow today [17]. The calculus of variations is concerned with the minimization of integral functionals:

$$\int_a^b L(x, y(x), y'(x))dx \longrightarrow \min. \tag{21}$$

However, as observed in [52], there are other interesting problems that arise in applications in which the functionals to be minimized are not of the form (21). For example, the planning of a firm trying to program its production and investment policies to reach a given production rate and to maximize its future market competitiveness at a given time horizon, can be mathematically stated in the form

$$\left(\int_a^b L_1(x, y(x), y'(x))dx\right) \cdot \left(\int_a^b L_2(x, y(x), y'(x))dx\right) \longrightarrow \min \tag{22}$$

(see [52]). Another example, also given in [52], appears when dealing with the so called "slope stability problem", which is described mathematically as minimizing a quotient functional:

$$\frac{\int_a^b L_1(x, y(x), y'(x))dx}{\int_a^b L_2(x, y(x), y'(x))dx} \longrightarrow \min. \tag{23}$$

Such multiplicative integral minimization problems that arise in different applications are nonstandard problems of the calculus of variations, but they can be naturally modeled in the non-Newtonian calculus of variations, and then solved in a rather standard way, using non-Newtonian Lagrange variations of the form $y \oplus \epsilon \odot h$, as we have proposed here. Therefore, we claim that the non-Newtonian calculus of variations just introduced may be useful for dealing with multiplicative functionals that arise in economics, physics and biology.

In this paper we have restricted ourselves to the ideas and to the central results of any calculus of variations: the celebrated Euler–Lagrange equation, which is a first-order necessary optimality condition. Of course our results can be extended, for example, by relaxing the considered hypotheses and enlarging the space of admissible functions, which we have considered here to be C^2, or considering vector functions instead of scalar ones. We leave such generalizations to the interested and curious reader. In fact, much remains now to be done. As possible future research directions we can mention: obtaining natural boundary conditions (sometimes also called transversality conditions) to be satisfied at a boundary point a and/or b, when $y(a)$ and/or $y(b)$ are free or restricted to take values on a given curve; obtaining second-order necessary conditions; obtaining sufficient conditions; to investigate nonadditive isoperimetric problems; etc.

5. Conclusions

In this work, a new calculus of variations was proposed, based on the non-Newtonian approach introduced by Grossman and Katz, thereby avoiding problems about nonnegativity. A new relation was proved, the multiplicative Euler–Lagrange differential Equation (13), which each solution of a non-Newtonian variational problem, with admissible functions taking positive values only, must satisfy. An example was provided for illustration purposes.

Grossman and Katz have shown that infinitely many calculi can be constructed independently [1]. Each of these calculi provide different perspectives for approaching many problems in science and engineering [53]. Additionally, a mathematical problem, which is difficult or impossible to solve in one calculus, can be easily revealed through another calculus [14,39].

Since the pioneering work of Grossman and Katz, non-Newtonian calculi have been a topic for new study areas in mathematics and its applications [15,54]. Particularly, Stanley [55], Córdova-Lepe [11], Slavík [10], Pap [44] and Bashirov et al. [12], have called the attention of mathematicians to the topic. More recently, non-Newtonian calculi have become a hot topic in economic and finance [56], quantum calculus [54], complex analysis [57–59], numerical analysis [2,37,42], inequalities [40,60], biomathematics [7,61] and mathematical education [14].

Here we adopted the non-Newtonian calculus as originally introduced by Grossman and Katz [1,34,35] and recently developed by Córdova-Lepe [11,41] and collaborators [4]: see the recent reviews in [7,14]. Roughly speaking, the key to understanding such calculus, valid for positive functions, is a formal substitution, where one replaces addition and subtraction with multiplication and division, respectively; multiplication in standard calculus is replaced by exponentiation in the non-Newtonian case, and thus, division by exponentiation with the reciprocal exponent. Our main contribution here was to develop, for the first time in the literature, a suitable non-Newtonian calculus of variations that minimizes a non-Newtonian integral functional with a Lagrangian that depends on the non-Newtonian derivative. The main result is a first-order necessary optimality condition of Euler–Lagrange type.

We trust that the present paper marks the beginning of a fruitful road for Non-Newtonian (NN) mechanics, NN calculus of variations and NN optimal control, thereby calling attention to and serving as inspiration for a new generation of researchers. Currently, we are investigating the validity of Emmy Noether's principle in the NN/multiplicative calculus of variations here introduced.

Funding: This research was funded by The Center for Research and Development in Mathematics and Applications (CIDMA) through the Portuguese Foundation for Science and Technology (FCT–Fundação para a Ciência e a Tecnologia), grant number UIDB/04106/2020.

Institutional Review Board Statement: Not applicable.

Informed Consent Statement: Not applicable.

Data Availability Statement: Not applicable.

Acknowledgments: The author is grateful to Luna Shen, Managing Editor of Axioms, for proposing a volume/special issue dedicated to his 50th anniversary, and to Natália Martins, Ricardo Almeida, Cristiana J. Silva and M. Rchid Sidi Ammi, who kindly accepted the invitation of Axioms to lead the project.

Conflicts of Interest: The author declares no conflict of interest. The funders had no role in the design of the study; in the collection, analyses, or interpretation of data; in the writing of the manuscript or in the decision to publish the results.

References

1. Grossman, M.; Katz, R. *Non-Newtonian Calculus*; Lee Press: Pigeon Cove, MA, USA, 1972.
2. Boruah, K.; Hazarika, B.; Bashirov, A.E. Solvability of bigeometric differential equations by numerical methods. *Bol. Soc. Parana. Mat.* **2021**, *39*, 203–222. [CrossRef]
3. Bashirov, A.E.; Mı sırlı, E.; Tandoğdu, Y.; Özyapıcı, A. On modeling with multiplicative differential equations. *Appl. Math. J. Chin. Univ. Ser. B* **2011**, *26*, 425–438. [CrossRef]
4. Mora, M.; Córdova-Lepe, F.; Del-Valle, R. A non-Newtonian gradient for contour detection in images with multiplicative noise. *Pattern Recognit. Lett.* **2012**, *33*, 1245–1256. [CrossRef]

5. Ozyapici, A.; Bilgehan, B. Finite product representation via multiplicative calculus and its applications to exponential signal processing. *Numer. Algorithms* **2016**, *71*, 475–489. [CrossRef]
6. Czachor, M. Unifying Aspects of Generalized Calculus. *Entropy* **2020**, *22*, 1180. [CrossRef] [PubMed]
7. Pinto, M.; Torres, R.; Campillay-Llanos, W.; Guevara-Morales, F. Applications of proportional calculus and a non-Newtonian logistic growth model. *Proyecciones* **2020**, *39*, 1471–1513. [CrossRef]
8. Özyapıcı, A.; Riza, M.; Bilgehan, B.; Bashirov, A.E. On multiplicative and Volterra minimization methods. *Numer. Algorithms* **2014**, *67*, 623–636. [CrossRef]
9. Jaëck, F. Calcul différentiel et intégral adapté aux substitutions par Volterra. *Hist. Math.* **2019**, *48*, 29–68. [CrossRef]
10. Slavík, A. *Product Integration, Its History and Applications;* Matfyzpress: Prague, Czech Republic, 2007; p. iv+147.
11. Córdova-Lepe, F. The multiplicative derivative as a measure of elasticity in economics. *TMAT Rev. Latinoam. Cienc. E Ing.* **2006**, *2*, 1–8.
12. Bashirov, A.E.; Kurpı nar, E.M.; Özyapıcı, A. Multiplicative calculus and its applications. *J. Math. Anal. Appl.* **2008**, *337*, 36–48. [CrossRef]
13. Boruah, K.; Hazarika, B. On some generalized geometric difference sequence spaces. *Proyecciones* **2017**, *36*, 373–395. [CrossRef]
14. Campillay-Llanos, W.; Guevara, F.; Pinto, M.; Torres, R. Differential and integral proportional calculus: How to find a primitive for $f(x) = 1/sqrt2\pi e^{-(1/2)x^2}$. *Int. J. Math. Ed. Sci. Technol.* **2021**, *52*, 463–476. [CrossRef]
15. Gurefe, Y.; Kadak, U.; Misirli, E.; Kurdi, A. A new look at the classical sequence spaces by using multiplicative calculus. *Politehn. Univ. Buchar. Sci. Bull. Ser. A Appl. Math. Phys.* **2016**, *78*, 9–20.
16. Leitmann, G. *The Calculus of Variations and Optimal Control; Mathematical Concepts and Methods in Science and Engineering;* Plenum Press: New York, NY, USA; London, UK, 1981; p. xvi+311.
17. van Brunt, B. *The Calculus of Variations;* Universitext, Springer: New York, NY, USA, 2004; p. xiv+290. [CrossRef]
18. Sidi Ammi, M.R.; Torres, D.F.M. Regularity of solutions to higher-order integrals of the calculus of variations. *Int. J. Syst. Sci.* **2008**, *39*, 889–895. [CrossRef]
19. Almeida, R.; Tavares, D.; Torres, D.F.M. *The Variable-Order Fractional Calculus of Variations;* Springer: Cham, Switzerland, 2019; p. xiv+124. [CrossRef]
20. Almeida, R.; Martins, N. Fractional variational principle of Herglotz for a new class of problems with dependence on the boundaries and a real parameter. *J. Math. Phys.* **2020**, *61*, 102701. [CrossRef]
21. Malinowska, A.B.; Torres, D.F.M. *Quantum Variational Calculus;* Springer: Cham, Switzerland, 2014. [CrossRef]
22. Brito da Cruz, A.M.C.; Martins, N. General quantum variational calculus. *Stat. Optim. Inf. Comput.* **2018**, *6*, 22–41. [CrossRef]
23. Martins, N.; Torres, D.F.M. Calculus of variations on time scales with nabla derivatives. *Nonlinear Anal.* **2009**, *71*, e763–e773. [CrossRef]
24. Dryl, M.; Torres, D.F.M. Direct and inverse variational problems on time scales: A survey. In *Modeling, Dynamics, Optimization and Bioeconomics, II, Proceedings of the International Conference on Dynamics, Games and Science, Porto, Portugal, 17–21 February 2014;* Springer Proceedings in Mathematics & Statistics; Springer: Cham, Switzerland, 2017; Volume 195, pp. 223–265. [CrossRef]
25. Silva, C.J.; Torres, D.F.M. Two-dimensional Newton's problem of minimal resistance. *Control Cybernet.* **2006**, *35*, 965–975.
26. Frederico, G.S.F.; Odzijewicz, T.; Torres, D.F.M. Noether's theorem for non-smooth extremals of variational problems with time delay. *Appl. Anal.* **2014**, *93*, 153–170. [CrossRef]
27. Ferreira, R.A.C.; Torres, D.F.M. Remarks on the calculus of variations on time scales. *Int. J. Ecol. Econ. Stat.* **2007**, *9*, 65–73.
28. Dryl, M.; Torres, D.F.M. A general delta-nabla calculus of variations on time scales with application to economics. *Int. J. Dyn. Syst. Differ. Equ.* **2014**, *5*, 42–71. [CrossRef]
29. Jost, J. *Mathematical Methods in Biology and Neurobiology;* Universitext, Springer: London, UK, 2014; p. x+226. [CrossRef]
30. Lemos-Paião, A.P.; Silva, C.J.; Torres, D.F.M.; Venturino, E. Optimal control of aquatic diseases: A case study of Yemen's cholera outbreak. *J. Optim. Theory Appl.* **2020**, *185*, 1008–1030. [CrossRef]
31. Grbić, T.; Medić, S.; Perović, A.; Mihailović, B.; Novković, N.; Duraković, N. A premium principle based on the g-integral. *Stoch. Anal. Appl.* **2017**, *35*, 465–477. [CrossRef]
32. Ünlüyol, E.; Salaş, S.; İşcan, I. A new view of some operators and their properties in terms of the non-Newtonian calculus. *Topol. Algebra Appl.* **2017**, *5*, 49–54. [CrossRef]
33. Tekin, S.; Başar, F. Certain sequence spaces over the non-Newtonian complex field. *Abstr. Appl. Anal.* **2013**, *2013*, 739319. [CrossRef]
34. Grossman, M. *The First Nonlinear System of Differential and Integral Calculus;* MATHCO: Rockport, MA, USA, 1979; p. xi+85.
35. Grossman, M. *Bigeometric Calculus;* Archimedes Foundation: Rockport, MA, USA, 1983; p. vii+100.
36. Riza, M.; Aktöre, H. The Runge-Kutta method in geometric multiplicative calculus. *LMS J. Comput. Math.* **2015**, *18*, 539–554. [CrossRef]
37. Boruah, K.; Hazarika, B. Some basic properties of bigeometric calculus and its applications in numerical analysis. *Afr. Mat.* **2021**, *32*, 211–227. [CrossRef]
38. Bashirov, A.E.; Norozpour, S. On complex multiplicative integration. *TWMS J. Appl. Eng. Math.* **2017**, *7*, 82–93.
39. Waseem, M.; Aslam Noor, M.; Ahmed Shah, F.; Inayat Noor, K. An efficient technique to solve nonlinear equations using multiplicative calculus. *Turk. J. Math.* **2018**, *42*, 679–691. [CrossRef]

40. Ali, M.A.; Abbas, M.; Zafar, A.A. On some Hermite-Hadamard integral inequalities in multiplicative calculus. *J. Inequal. Spec. Funct.* **2019**, *10*, 111–122.
41. Córdova-Lepe, F. From quotient operation toward a proportional calculus. *Int. J. Math. Game Theory Algebra* **2009**, *18*, 527–536.
42. Yazıcı, M.; Selvitopi, H. Numerical methods for the multiplicative partial differential equations. *Open Math.* **2017**, *15*, 1344–1350. [CrossRef]
43. Boruah, K.; Hazarika, B. *G*-calculus. *TWMS J. Appl. Eng. Math.* **2018**, *8*, 94–105.
44. Pap, E. Generalized real analysis and its applications. *Int. J. Approx. Reason.* **2008**, *47*, 368–386. [CrossRef]
45. Duyar, C.; Sağlr, B.; Oğur, O. Some basic topological properties on Non-Newtonian real line. *Br. J. Math. Comput. Sci.* **2015**, *9*, 300–307. [CrossRef]
46. Binbaşıoğlu, D.; Demiriz, S.; Türkoğlu, D. Fixed points of non-Newtonian contraction mappings on non-Newtonian metric spaces. *J. Fixed Point Theory Appl.* **2016**, *18*, 213–224. [CrossRef]
47. Güngör, N. Some geometric properties of the non-Newtonian sequence spaces $l_p(N)$. *Math. Slovaca* **2020**, *70*, 689–696. [CrossRef]
48. Sağır, B.; Erdoğan, F. On non-Newtonian power series and its applications. *Konuralp J. Math.* **2020**, *8*, 294–303.
49. Burgin, M.; Czachor, M. *Non-Diophantine Arithmetics in Mathematics, Physics and Psychology*; World Scientific: Singapore, 2021.
50. Plakhov, A.Y.; Torres, D.F.M. Newton's aerodynamic problem in media of chaotically moving particles. *Mat. Sb.* **2005**, *196*, 111–160. [CrossRef]
51. Torres, D.F.M.; Plakhov, A.Y. Optimal control of Newton-type problems of minimal resistance. *Rend. Semin. Mat. Univ. Politec. Torino* **2006**, *64*, 79–95.
52. Castillo, E.; Luceño, A.; Pedregal, P. Composition functionals in calculus of variations. Application to products and quotients. *Math. Model. Methods Appl. Sci.* **2008**, *18*, 47–75. [CrossRef]
53. Riza, M.; Özyapici, A.; Misirli, E. Multiplicative finite difference methods. *Quart. Appl. Math.* **2009**, *67*, 745–754. [CrossRef]
54. Yener, G.; Emiroglu, I. A q-analogue of the multiplicative calculus: q-multiplicative calculus. *Discret. Contin. Dyn. Syst. Ser. S* **2015**, *8*, 1435–1450. [CrossRef]
55. Stanley, D. A multiplicative calculus. *Primus* **1999**, *9*, 310–326. [CrossRef]
56. Filip, D.A.; Piatecki, C. A non-newtonian examination of the theory of exogenous economic growth. *Math. Aeterna* **2014**, *4*, 101–117.
57. Uzer, A. Multiplicative type complex calculus as an alternative to the classical calculus. *Comput. Math. Appl.* **2010**, *60*, 2725–2737. [CrossRef]
58. Çakmak, A.F.; Başar, F. Certain spaces of functions over the field of non-Newtonian complex numbers. *Abstr. Appl. Anal.* **2014**, *2014*, 236124. [CrossRef]
59. Bashirov, A.E.; Norozpour, S. On an alternative view to complex calculus. *Math. Methods Appl. Sci.* **2018**, *41*, 7313–7324. [CrossRef]
60. Çakmak, A.F.; Başar, F. Some new results on sequence spaces with respect to non-Newtonian calculus. *J. Inequal. Appl.* **2012**, *2012*, 228. [CrossRef]
61. Florack, L.; van Assen, H. Multiplicative calculus in biomedical image analysis. *J. Math. Imaging Vis.* **2012**, *42*, 64–75. [CrossRef]

Short Biography of Author

Delfim Fernando Marado Torres is a Portuguese Mathematician born 16 August 1971 in Nampula, Portuguese Mozambique. He obtained a PhD in Mathematics from the University of Aveiro (UA) in 2002, and habilitation in Mathematics, UA, in 2011. He is a full professor of mathematics since 9 March 2015. He has been the Director of the R&D Unit CIDMA, the largest Portuguese research center for mathematics, and Coordinator of its Systems and Control Group. His main research areas are calculus of variations and optimal control; optimization; fractional derivatives and integrals; dynamic equations on time scales; and mathematical biology. Torres has written outstanding scientific and pedagogical publications. In particular, he has co-authored two books with Imperial College Press and three books with Springer. He has strong experience in graduate and post-graduate student supervision and teaching in mathematics. Twenty PhD students in mathematics have successfully finished under his supervision. Moreover, he has been the leading member in several national and international R&D projects, including EU projects and networks. Professor Torres has been, since 2013, the Director of the Doctoral Programme Consortium in Mathematics and Applications (MAP-PDMA) of Universities of Minho, Aveiro, and Porto. Delfim married 2003 and has one daughter and two sons.

MDPI
St. Alban-Anlage 66
4052 Basel
Switzerland
Tel. +41 61 683 77 34
Fax +41 61 302 89 18
www.mdpi.com

Axioms Editorial Office
E-mail: axioms@mdpi.com
www.mdpi.com/journal/axioms

www.ingramcontent.com/pod-product-compliance
Lightning Source LLC
LaVergne TN
LVHW071526170726
843515LV00008B/2070

* 9 7 8 3 0 3 6 5 6 8 5 6 0 *